Pollination and Floral Ecology

Pollination and Floral Ecology

Editor: Blake Hughes

MURPHY & MOORE
www.murphy-moorepublishing.com

Murphy & Moore Publishing,
1 Rockefeller Plaza,
New York City, NY 10020, USA

Visit us on the World Wide Web at:
www.murphy-moorepublishing.com

ISBN: 978-1-63987-439-2 (Hardback)

Cataloging-in-Publication Data

Pollination and floral ecology / edited by Blake Hughes.
p. cm.
Includes bibliographical references and index.
ISBN 978-1-63987-439-2
1. Pollination. 2. Plant ecology. 3. Fertilization of plants. 4. Ecology. 5. Botany. I. Hughes, Blake.
QK926 .P65 2022
571.864 2--dc23

Table of Contents

Preface

This book aims to highlight the current researches and provides a platform to further the scope of innovations in this area. This book is a product of the combined efforts of many researchers and scientists, after going through thorough studies and analysis from different parts of the world. The objective of this book is to provide the readers with the latest information of the field.

Flowers are structures produced by angiosperms or flowering plants. Their evolutionary history has been shaped by their association with pollinators, primarily insects. The transfer of pollen grains from the stamens to the ovules, or pollination, is an essential aspect of plant reproduction. In flowering plants, the ovules are contained within the pistil, and the pollen is generally deposited on the stigma, which is the receptive surface of the pistil. In conifers and cycads, the ovules are exposed and the pollen is caught in the fluid secreted by the ovule. This process is mediated by insects, vertebrates (bats and birds), certain mammals and some lizards. Such interaction between flowers and their pollinators are mutualistic. These influence floral ecology and are believed to have led to the wide diversity of pollinators today. This book contains some path-breaking studies in pollination and floral ecology. It will also provide interesting topics for research which interested readers can take up. A number of latest researches have been included to keep the readers up-to-date with the global concepts in this area of study.

I would like to express my sincere thanks to the authors for their dedicated efforts in the completion of this book. I acknowledge the efforts of the publisher for providing constant support. Lastly, I would like to thank my family for their support in all academic endeavors.

Editor

1

POLLINATION ECOLOGY IN THE 21ST CENTURY

Carolin Mayer*[1], Lynn Adler[2], W. Scott Armbruster[3], Amots Dafni[4], Connal Eardley[5], Shuang-Quan Huang[6], Peter G. Kevan[7], Jeff Ollerton[8], Laurence Packer[9], Axel Ssymank[10], Jane C. Stout[11], Simon G. Potts[12]

[1]Earth and Life Institute, Université Catholique de Louvain, Croix du Sud 2 box 14, 1348 Louvain-la-Neuve, Belgium

[2]Department of Plant, Soil and Insect Science, 270 Stockbridge Rd., University of Massachusetts, Amherst, MA 01002, USA

[3]School of Biological Sciences, University of Portsmouth, Portsmouth PO1 2DY, UK / Institute of Arctic Biology, University of Alaska, Fairbanks, AK 99775, USA

[4]Department of Evolutionary and Environmental Biology, Institute of Evolution, Faculty of Sciences, Haifa University, Haifa 31905, Israel

[5]Agricultural Research Council, Private Bag X134, Queenswood 0121, South Africa / School of Biological and Conservation Sciences, University of KwaZulu-Natal, Private Bag X01, Scottsville, Pietermaritzburg, 3209, South Africa

[6]College of Life Sciences, Wuhan University, Wuhan 430072, China

[7]School of Environmental Sciences, University of Guelph, 50 Stone Road East, Guelph, Ontario, N1G 2W1 Canada

[8]School of Science and Technology, University of Northampton, Newton Building, Avenue Campus, Northampton, NN2 6JD, UK

[9]Department of Biology, York University, 4700 Keele St, Toronto, Ontario, M3J 1P3 Canada

[10]Federal Agency for Nature Conservation (BfN), Konstantinstrasse 110, 53179 Bonn, Germany

[11]School of Natural Sciences, Trinity College Dublin, Dublin 2, Republic of Ireland

[12]School of Agriculture, Policy and Development, University of Reading, RG6 6AR, UK

Abstract—To inspire new ideas in research on pollination ecology, we list the most important unanswered questions in the field. This list was drawn up by contacting 170 scientists from different areas of pollination ecology and asking them to contribute their opinion on the greatest knowledge gaps that need to be addressed. Almost 40% of them took part in our email poll and we received more than 650 questions and comments, which we classified into different categories representing various aspects of pollination research. The original questions were merged and synthesised, and a final vote and ranking led to the resultant list. The categories cover plant sexual reproduction, pollen and stigma biology, abiotic pollination, evolution of animal-mediated pollination, interactions of pollinators and floral antagonists, pollinator behaviour, taxonomy, plant-pollinator assemblages, geographical trends in diversity, drivers of pollinator loss, ecosystem services, management of pollination, and conservation issues such as the implementation of pollinator conservation. We focused on questions that were of a broad scope rather than case-specific; thus, addressing some questions may not be feasible within single research projects but constitute a general guide for future directions. With this compilation we hope to raise awareness of pollination-related topics not only among researchers but also among non-specialists including policy makers, funding agencies and the public at large.

Keywords: *angiosperms, conservation, evolution, ecosystem services, diversity, mutualism, species interactions*

INTRODUCTION

Since Sprengel (1793) published his famous findings on the interactions between flowers and insects that transport pollen grains to the stigmas of plants, demonstrating that they are essential for the production of seeds and fruit, our knowledge of pollination as an essential ecosystem process has vastly increased. The large majority of flowering plant species in almost all angiosperm-dominated communities studied so far interact with pollinators, ranging on average from 78% in temperate communities to 94% in tropical communities, equalling over 308,000 species or more than 87% of all angiosperms (Ollerton et al. 2011). Moreover, it has been suggested that over 300,000 animal species are directly dependent upon floral resources (Buchmann & Nabhan 1996).

Plant-pollinator interactions have always provided excellent model systems to test and develop new theories in ecology and evolution (Mitchell et al. 2009). Nevertheless, because of the huge breadth (for a few examples see Fig. 1), depth and scope of this discipline, there is no doubt that many issues remain unresolved or have not been fully explored. All of the questions that we present in this paper

*Corresponding author; email: carolin.mayer@uclouvain.be

have previously been addressed by researchers to some degree and we do not wish to give the impression that they have never been investigated. However, these are questions that, in the minds of a significant sample of researchers in the field, are not yet fully resolved. By defining the most urgent questions, we seek to identify knowledge gaps in pollination ecology to inspire future research. Today, we are facing a global decline of biodiversity, fuelled by the immense destruction and degradation of habitats and climate change through human activities, which also heavily affects and interrupts plant – pollinator relationships (Kevan & Baker 1983; Aizen & Feinsinger 1994; Kearns et al. 1998; Kevan 1999; Kremen et al. 2002; Biesmeijer et al. 2006; Winfree et al. 2007; Memmott et al. 2007). This spurs the need to define priorities for plant and pollinator conservation and to invoke responses from policy makers and stakeholders to seize the initiative for action.

Previous 'brainstorms' have been highly successful in informing researchers and decision makers (National Research Council 2006; Sutherland et al. 2006, 2009). Such overarching efforts can inspire lively discussions not only from scientists but also from conservationists, politicians and funders. Our aim is to organize our efforts into thematic approaches, to raise consciousness among non-specialists and to draw attention to research topics that are still largely outside public awareness. In this paper we are using the term "pollination ecology" in its broadest sense to cover a range of topics associated with plant reproduction and the role of pollinators (and other interacting species) in influencing that reproduction. Scientists must help to define priorities for plant and pollinator conservation in order to engage policy makers, conservation organisations and other stakeholders.

Figure 1. A few examples of the amazing breadth of pollination ecology: The dove tree (*Davidia involucrata*) has "naked" flowers that lack a perianth but are surrounded by two large bracts. These turn from green (a) to white (b) when flowers open which increases pollinator attraction (see Sun et al. 2008). *Calypso bulbosa* (c), a rewardless orchid, has evolved to deceive naïve bumblebees in the early spring to effect pollination without payment in nectar (see Ackerman 1981). The perennial vine, *Gelsemium sempervirens* (d), contains alkaloids in all plant parts including corollas and nectar which deter floral herbivores and nectar robbers but - in high concentrations - also pollinators (Adler & Irwin 2005). Hoverflies (e) visit flowers in a large range of habitats, they even pollinate cultivated plants. Though they are numerous, species-rich and wide-spread, their role as pollinators is often still neglected. (f) In the Succulent Karoo of South Africa, monkey beetles (Hopliini) are important pollinators of many plant species (Mayer et al. 2006). This one is diving into a flower of *Ruschia goodiae* (Aizoaceae) to reach nectar. (g) Flowers of *Ceropegia arabica* (Apocynaceae) are pollinated by small Diptera in common with all other members of the genus studied to date (Ollerton et al. 2009). Photograph by Sage Reynolds. *Rhododendron ponticum* (h) is a highly invasive plant in Ireland. Though it provides food resources for generalist bumblebees, effects on native plant pollinator interactions vary with intensity of invasion.

METHODS

The authors of this paper compiled a list of 170 researchers in pollination ecology around the world with whose work they were familiar and who had made (in their opinions) significant contributions to the field of pollination ecology over the last two decades. This was not meant to be an exhaustive polling but rather a representative sample, and the exclusion of any researchers from the process in no way implied that the authors did not value their work. In March 2010, these researchers were consulted by email and asked to formulate what they thought were the crucial questions in pollination ecology that should be explored within the coming decades. Most (>80%) of those polled are based in Europe or Northern America (Appendix I).

We asked for questions that should (a) address important knowledge gaps, (b) not be too case-specific or detailed, (c) be of interest to non-specialists, and (d) answerable in time and space under reasonable budgetary constraints. We proposed 17 preliminary broad thematic categories that were meant to assist in structuring one's opinion, but there was no demand for respondents to contribute to all of them. However, when compiling the final list we set value on the significance and scope of the questions rather than feasibility within the timeframes of single research projects.

From the 170 scientists contacted, 66 were able to respond (39 % of those asked). They provided a total of 663 queries and comments that were assigned to the categories (if not done so by the consultees themselves). Of the 17 preliminary categories, four were discarded since there were hardly any replies which could not be fitted into other topics (the discarded topics were: "Status and trends of global pollinators", "Pollinators and technological development", "Methodology in pollination" and "Environmental effects on pollination"). On the other hand, one new category was created ("Evolution of animal-mediated pollination") corresponding to the demand from several contributors. The questions were then merged and rephrased within categories, which were handled by different members of our team. This procedure narrowed the initial list to about 170 questions (Appendix II). The authors of this paper (who comprise most of the Editorial Board of the Journal of Pollination Ecology) then indicated their preferences by vote, either giving one credit to each of 100 questions or allocating more credits to fewer questions (to a total of 100 credits per voter). After voting, the questions were ranked according to preference (i.e. number of credits) and the first one hundred selected. Based upon feedback from the respondents and discussions among the authors and because of persisting overlap between the general topics of some questions after voting, we narrowed down the number of questions to a final list of 86. In October 2010 these questions were presented to delegates at the 24th Scandinavian Association for Pollination Ecology (SCAPE) meeting at Tovetorp, Sweden. The long and detailed discussion that followed was used as the basis for the final refinement of the list of questions. In this final list, we have not ranked nor ascribed an order of importance to the questions. The contributors are acknowledged at the end of this paper.

RESULTS

For each of the 14 categories of questions given below, we present a short piece of text to give the questions context and then list the questions themselves. The categories are sorted so as to begin with more basic aspects of plant sexual reproduction followed by topics on the interaction of plants and animals. Risks for plant-pollinator interactions are then addressed, leading to more applied fields of study such as management and conservation.

1. Plant sexual reproduction (Box 1)

Unlike most animals, many plants are capable of reproduction via sexual and asexual means. Since plants are sessile, in order to successfully reproduce sexually they have evolved specialised floral structures, many of which promote out-crossing, driven by the benefits associated with population level genetic diversity (such as the ability of a population to adapt in response to environmental change, improved resistance to disease, etc.; Barrett & Harder 1996). Despite this, self-fertilisation is possible, frequent and even promoted in some flowers (within and among individuals, populations and species; Uyenoyama et al. 1993; Redbotorstensson & Berg 1995; Culley 2002). However, it is not clear what drives plants to adopt autonomous pollination strategies, although possibilities include reproductive assurance in the face of pollen limitation, maintenance of locally-adapted genotypes and lack of requirement for a mate during colonisation of new habitats (Lloyd 1979; Kalisz & Vogler 2003).

Recent work suggests that floral traits are not just driven by the breeding strategy of the plant (whether its strategy is out-crossing or selfing), but are the result of selection by abiotic and biotic factors, which can in turn affect the breeding strategy of the plant. For example, water availability can inflict constraints on the evolution of floral size, which then influences the likelihood of selfing (Elle 2004); other abiotic factors, such as soil nutrients, light, and temperature, could play similar roles. Flower shapes may affect intrafloral temperature, or even function as pollen shelter to reduce rain damage (Mao & Huang 2009). The role of abiotic factors in floral evolution is largely underappreciated.

1. Why are some plant species capable of autonomous pollination – and what are the consequences regarding inbreeding depression and population viability?
2. Does pollen limitation affect breeding systems, genetic diversity and speciation rates and, if so, how?
3. What is the relative contribution of abiotic versus biotic factors in shaping the evolution of flowers?
4. Is male reproductive success more variable than female reproductive success?
5. Is there sexual selection in plants?
6. What is the role of hybridization in plant speciation?

Box 1: Questions on plant sexual reproduction

Since the majority of flowers are hermaphrodite, fitness can be determined as the result of female (seed set) and male (pollen dispersal) reproductive success. Male success can be more variable since, theoretically, a male (or male phase) flower can fertilize numerous female (or female phase) flowers. Although the importance of male function was emphasized by Bell (1985), tracking pollen movement, male success and fitness is technically difficult. Furthermore, variation in female fitness according to structural properties of flowers remains to be investigated. For example, we do not know whether apocarpous species with distinct, single-carpelled pistils are more likely to experience pollen limitation than syncarpous species with united carpels (Endress 1982; Armbruster et al. 2002). Or, whether there is a functional difference between polypetalous (separate petals) and sympetalous (united petals) flowers.

2. Pollen and stigma biology (Box 2)

Assessments of the duration of pollen viability and stigma receptivity are critical for many aspects of pollination biology (Dafni et al. 2005 and references therein), including: monitoring pollen and stigma life spans, the success of pollination at different stages in the flower life-cycle, the relative importance of various pollinators, the interference between male and female functions, the rate of competition through improper pollen transfer, the chances of gametophytic selection, and the assessment of pollination efficiency. Evaluation of pollen viability and stigma receptivity is the first step towards an evaluation of the chances of a given pollen grain to germinate. A pollen grain that fails to germinate on a conspecific and genetically appropriate receptive stigma, and later to fertilize an ovule, results in an unsuccessful pollination event.

7. How can we assess pollen viability and stigma receptivity under field conditions?
8. How often is "stigma clogging" (incompatible pollen physically blocking the stigma) an important effect in nature?
9. What influences the pollen availability of a population?
10. What proportion of pollen grains from a plant are viable?
11. How much viable pollen is transferred to flowers?
12. What is the lifespan of pollen grains?

Box 2: Questions on pollen and stigma biology

The pollen-stigma relationship depends on pollen viability, stigmatic receptivity, and genetic interaction of both partners as dictated by the incompatibility system and may vary according to environmental conditions such as relative humidity or temperature (Franchi et al. 2007; Douglas & Freyre 2010). Even a successful pollen germination event does not ensure later success. For example, pollen grains compete with each other or interact with style tissues which may interfere and block fertilization, as may late-acting incompatibility processes, post-zygotic abortion and pollen allelopathy. Any study on the quality and quantity of pollen grains, their fate and chances of pollination success in the context of environmental variation and pollinator performance should try to understand these components as a basic background (Ne'eman et al. 2010).

3. Abiotic pollination (Box 3)

About 20% of all angiosperm families contain plant species that do not rely on animal vectors for pollen transport, but use physical agents such as wind and water for pollen transfer (Ackerman 2000). Anemophily (wind pollination) has evolved many times and is quite common (estimated at about 10 to 13 % of all angiosperm species; Friedman & Barrett 2009; Ollerton et al. 2011). Hydrophily (water pollination) is mainly confined to monocotyledons and is found in only 2.7% of all angiosperm species (Ackerman 2000). Phylogenetic analyses propose that wind pollination evolves from insect pollination, but our knowledge about the underlying mechanisms or evolutionary pathways is still rudimentary (Culley et al. 2002; Friedman & Barrett 2009). However, a few cases exist suggesting some wind pollinated plants reverted to become insect pollinated (Norman et al. 1997; Peeters & Totland 1999).

Increasing records report anemophily in otherwise entomophilous families and species (Weller et al. 2006). Ambophily (both biotic and abiotic pollination occurring in the same flower) may occur under certain circumstances that favour either wind or insect pollination, providing an alternative to self-pollination to ensure successful reproduction (Culley et al. 2002). Ambophily may also be favoured where pollinator abundance varies and wind can transport supplementary pollen to pollinate otherwise unvisited flowers (Duan et al. 2009). One might hypothesise that pollen limitation is scarce among wind-pollinated plants if pollen transport by wind for some species serves as a reproductive assurance in the absence of pollinators (Friedman & Barrett 2009). However, apart from a few studies on species in fragmented habitats (e.g. Knapp et al. 2001; Eppley & Pannell 2009), investigations of pollen limitation or population densities and plant reproductive success in wind pollinated species are still lacking (see Friedman & Barrett 2009 and references therein).

13. How many animal-pollinated plant species have cryptic or partial wind pollination?
14. When, where and how did evolutionary shifts from biotic to abiotic pollination systems, and vice versa, occur?
15. How are biophysical mechanisms involved in abiotic pollination?
16. What are the relative proportions of biotic vs. abiotic pollination services in crop and wild plants?
17. How frequent is pollen limitation in wind pollinated plants?

Box 3: Questions on abiotic pollination

Early work concerning pollen release and dispersal described models of boundary-layer air flow around pollen grains and morphological structures of flowers adapted to direct microcurrents to deposit pollen on stigmas (Niklas 1985). Since high proportions of conspecific pollen can be found on stigmas (>40%, Linder & Midgley 1996), biophysical proportions such as settling speed may enhance filtering of pollen during pollen capture (Aylor et al. 2005; Friedman & Barrett 2009).

4. Evolution of animal-mediated pollination (Box 4)

Pollinators exert selective pressures on plants and their floral traits, and, similarly, plants may influence the evolution of pollinating animals. The evolution of floral traits has been proposed to be moulded by the most frequent and effective pollinators (Stebbins 1970; although see Aigner 2001). This can lead to co-evolution and co-adaptation resulting in varying degrees of specialisation, which can increase pollination accuracy and/or reduce heterospecific pollen interference. For example, mechanical fit between pollinators and flowers can partly explain the evolution of floral traits (such as spur length) and complementary animal traits (such as tongue length; Nilsson 1988; Whittall & Hodges 2007). However, it is increasingly recognised that the evolution of flowers is probably not so straightforward because many plants have more than one type of pollinator (Waser et al. 1996; Waser & Ollerton 2006) and floral evolution can be driven by conflicting selection by these pollinators, as well as by herbivores and other antagonists. The relative contributions of mutualists, antagonists (see topic 5) and abiotic factors in shaping the evolution of flowers is only beginning to be studied for most species.

18. What roles do pollinators play in the processes that lead to plant speciation?
19. What role does coevolution play in plant diversification?
20. How, and how frequently, do positive correlations emerge between floral specialisation and species diversity within a clade?
21. How and why does specialisation in pollination systems evolve and what are the driving factors?
22. Why does floral deception evolve?
23. How do plants avoid heterospecific pollen interference given that generalist pollinators visit other sympatric flowering plants?
24. How frequently do floral traits converge on traditional pollination syndromes?
25. What are the ultimate factors determining taxonomic diversity of visitors to a plant species?
26. How is floral diversity generated, or why are there so many kinds of flowers?

Box 4: Questions on evolution of animal-mediated pollination

Three methodological approaches and lines of enquiry have developed in the study of the evolution of plants, pollinators, and pollination systems: i) genetic and/or experimental research focusing on speciation and reproductive isolation, ii) genetic and/or experimental research focusing on microevolution, and iii) comparative research inferring evolutionary process from macroevolutionary patterns.

Traditionally, shifts between pollination systems or pollinators have been viewed in light of the establishment and maintenance of reproductive isolation and the speciation process (Grant 1949; Stebbins 1970). This perspective has enjoyed a renaissance with the advent of new molecular and analytical approaches (Bradshaw et al. 1995; Schemske & Bradshaw 1999; Ramsey et al. 2003; Kay & Schemske 2008). As the process of speciation is still poorly understood, it is critical that we understand the contributions of pollination ecology to the establishment and reinforcement of reproductive isolation and other speciation processes. At the same time, associations between specialised pollination and clade species richness (biodiversity) suggest that specialised pollination may increase speciation rates or reduce extinction rates (Sargent 2004; but see also Armbruster & Muchhala 2009).

Various aspects of specialisation in pollination ecology have been addressed in the past using ecological, microevolutionary, and comparative approaches, but there is still much debate and uncertainty about the broad applicability of the traditional pollination syndrome concept (in which the main pollinators can be predicted from flower colour, shape, scent and so forth) and the relative frequency and importance of specialised and generalised plant pollination ecologies (see Waser et al. 1996; Ollerton & Watts 2000; Fenster et al. 2004; Ollerton et al. 2009).

A final question concerns the evolutionary and ecological processes that have generated and maintained the diversity of flower types across species of angiosperms; this issue inspired a significant part of Darwin's research and contributed to his insights into natural selection and the origin of species. Comparative analysis in the context of phylogeny has been the main approach used to tackle this question, and the evolution of floral diversity remains a major theme in evolutionary research.

5. Interactions of plants, pollinators and floral antagonists (Box 5)

The evolution of floral traits is not only shaped by interactions with pollinators, but also with antagonists (Strauss & Whittall 2006; Adler 2007), and floral traits probably represent a compromise between selection by mutualists and antagonists (e.g. Gómez 2003). The outcome of interactions with pollinators and antagonists may depend on community context and vary geographically (e.g. Thompson & Cunningham 2002), but the extent to which this occurs is unknown for most systems. Furthermore, not much is known about how visits by one pollinator species affect floral interactions with other pollinators (Morris et al. 2007) or how floral antagonists affect interactions with other floral antagonists. Finally, compared to herbivores, the role of community context such as bottom-up (resource availability) and top down factors (predators or parasites Dukas 2001; Gegear et al. 2006; Otti & Schmid-Hempel 2008) on pollinator behaviour and population dynamics is relatively understudied.

> 27. What is the relative importance of selection on floral traits and mating systems by mutualists (pollinators) vs. antagonists (florivores, nectar robbers, pathogens, seed predators, pollinating herbivores)?
> 28. How does community context alter the outcome of floral interactions with pollinators and antagonists?
> 29. How do floral traits mediate interactions with pollinators and floral antagonists, and how do pollinators and floral antagonists alter these traits?
> 30. How do pathogens transmitted at flowers affect population dynamics and evolution of plants and pollinators?

Box 5: Questions on interactions of plants, pollinators and floral antagonists

Apart from floral cues and nectar rewards, floral volatiles and secondary compounds may also play important roles in pollinator behaviour (Adler 2000; Kessler & Baldwin 2007; Theis et al. 2007). However, the examination of how traits like scent, colour, morphology, floral longevity, and defence and reward traits (including chemical defences in both nectar and pollen) mediate interactions with floral mutualists and antagonists is in its infancy. Such traits can have synergistic effects on visitor behaviour (Raguso & Willis 2005), and can be induced or altered in response to leaf and floral damage (e.g. Adler et al. 2006; McCall & Karban 2006), but this has been examined in only a handful of systems.

Flowers have the potential to host and transmit a wide range of micro-organisms (Herrera et al. 2009) with consequent effects on both insect and plant hosts. For example, some of the most serious pathogens of bumble bees, such as *Crithidia*, are transmitted at flowers (Durrer & Schmid-Hempel 1994), as are major plant agricultural diseases such as fire blight in apples (Johnson et al. 2006). Bacterial wilt (*Erwinia tracheiphila*), a lethal disease of cucurbit crops, was thought to be transmitted only through leaf damage from a specialist beetle herbivore. However, beetles also consume cucurbit pollen, and recent work found that defecation in flowers can cause infection through nectaries (Sasu et al. 2010). More work is needed to examine how plant pathogens transmitted via flowers affect the evolution of mating systems and floral traits (Shykoff et al. 1997; Liu & Carlsson 2002; Hood et al. 2010).

6. *Pollinator behaviour (Box 6)*

To receive the service of pollen transfer, plants often offer rewards to flower visiting animals, such as nectar, oil, resin, pollen, breeding sites, etc. Flowers attract pollinators via various stimuli, whether they are olfactory or visual cues acting from a distance or tactile cues to guide pollinators to rewarding resources at close vicinity. Floral traits, resource distribution, and cognitive and learning abilities of pollinators influence their behaviour, which in turn is strongly linked to plant mating patterns and gene flow within and among plant populations (Goulson 1999; Karron et al. 2009). Interdisciplinary studies that combine all the different aspects of stimuli, pollination biology and genetics are still lacking (Whitehead & Peakall 2009).

> 31. What influences the pattern of movement of pollinators across landscapes?
> 32. How does pollinator behaviour affect pollen dispersal, gene flow, pollination, and plant reproductive success?
> 33. Are pollinator preferences mainly driven by learned or innate behaviour?
> 34. How and to what extent do pollinator species differ from one another in their cognitive abilities?
> 35. What factors determine the foraging choices of pollinators, and to what extent do these approximate optimal choices?
> 36. How and at which distances do floral traits influence foraging choices of pollinators?
> 37. How does competition for pollinator services influence patterns of gene dispersal?

Box 6: Questions on pollinator behaviour

Apart from social bees and birds (including work by Heinrich 1979; Pyke 1981; Chittka & Raine 2006), our knowledge about the different aspects of pollinator behaviour, such as foraging strategies or distances, cognitive or learning abilities, and flower constancy remain rudimentary (Weiss 2001; Riffell et al. 2008). We know that sensory perception of floral cues varies among taxa (Weiss 2001), but we have little knowledge about colour vision, smell or taste, their interaction, and or learning speed in individuals in space and time and how these affect the evolution of floral traits, mediated through behavioural differences. Innate colour preferences, for example, can be found over a broad taxonomic range of insect families but may differ between genera, species or even sexes (Weiss 2001; Kandori et al. 2009; Alarcón et al. 2010). Spatial processes, especially at large scales, and their consequences for pollen distribution in and between plant populations are even less well understood (Westrich 1996; Schulke & Waser 2001; Greenleaf et al. 2007; Beil et al. 2008).

Inter- and intraspecific competition among flower visiting animals can change the availability of resources and lead to changes in behaviour through enlargement of pollinator diets, though this can vary according to species, individuals and body size of pollinators (Inouye 1978; Walther-Hellwig et al. 2006; Fontaine et al. 2008). The consequences for plant reproductive success are still to be investigated (Dohzono & Yokoyama 2010).

7. *Taxonomy (Box 7)*

Pollinators do not need taxonomy but communicating scientific results on pollination research does. Therefore, taxonomy is essential for people to understand the intricate interactions in nature and to appreciate the world around them. Taxonomy is also needed for more practical endeavours, like the study of biodiversity, conservation and improving agricultural production. This is because taxonomy does not only provide names; it provides a systematic framework to our understanding of the natural world. In addition, taxonomic revisions usually give a lot more information; they document colour and morphological variability, enable predictability (closely related species often have similar behaviour and ecology), give distribution patterns and phenological information, and document associated organisms, such as parasites and food plants.

38. How do we solve the taxonomic impediment?
39. Which molecular methods are useful to evaluate inter- and intraspecific diversity of pollinators?
40. What are the ecological, social and economic impacts resulting from an inability to identify and manage pollinators?
41. How many species of pollinators have been described and how many undiscovered species remain to be described?

Box 7: Questions on taxonomy

However, enormous knowledge gaps still exist in our taxonomic system, and taxonomic expertise is in decline. Through the Convention on Biological Diversity (CBD), governments have acknowledged the existence of a "taxonomic impediment" in 1998 on the fourth Conference of the Parties and launched the Global Taxonomic Initiative (COP 4 Decision IV/1). Thereby it is hoped to help alleviate the problem of shortage of taxonomic workforce which hinders sound work and management on biodiversity. Identification guides that can be easily used by non-taxonomists are still rare, and available for relatively few taxonomic groups and geographic areas (e.g. Eardley et al. 2010). For bees, there exist fully-illustrated keys to family level (Packer & Ratti 2009, Fig. 2) and the monograph of Michener (2007) with keys to genera and subgenera, but for many important groups of pollinators like Diptera (see Ssymank et al. 2008), there remains a need for keys and catalogues. DNA barcoding has shown promise for the identification of bees and other pollinators (Packer et al. 2009). However, additional genetic markers may be required to obtain clearer patterns of species differentiation.

Figure 2. Detailed images taken from a user-friendly key to identify the bee families of the world (Packer and Ratti, 2009). Distinguishing characters (here: scopae on female hind legs) are described and highlighted. Further images show different forms of the character making correct identification much easier.

8. The breadth and depth of our current understanding of plant-pollinator assemblages (Box 8)

Plants and their pollinators do not exist in isolation, but rather in highly connected assemblages of interacting species (Memmott 1999; Jordano et al. 2003). These assemblages vary temporally within flowering seasons (requiring some matching between plant and pollinator activities) and between seasons, as pollinator populations fluctuate. There is also spatial variation, with similar plant communities having rather different groups of pollinators which may vary in their relative effectiveness. Spatial and temporal variation in a plant species' pollinators means that for many plants the terms "specialised" or "generalised" are relative ones, highly dependent upon assemblage/community context (Ollerton et al. 2007).

42. To what extent are local plant communities dependent upon pollinators?
43. To what extent is functional replacement of different species of pollinators possible?
44. What is the relationship between pollinator diversity and plant diversity?
45. How do pollinator population fluctuations affect pollination?
46. What are the proximate, ecological determinants of pollination system specialisation or generalisation?
47. What ecological and evolutionary processes determine the structure of interactions in a network?
48. What proportion of pollination is undertaken by the different functional groups of pollinators in a community?
49. To what extent are pollinator life cycles synchronised to the phenologies of their forage plants?
50. What factors influence the composition of pollinator species at the community level?
51. How commonly do pollinators compete for floral resources?
52. How do plants of different species interact through competition or facilitation via common pollinators?

Box 8: Questions on plant-pollinator assemblages

How this variation in pollinators subsequently feeds into immediate effects on plant reproduction and longer term effects on plant diversity is only now being addressed experimentally, and even then, experiments are necessarily constrained and may not represent the true ecological situation (e.g. Fontaine et al. 2006).

Although the advent of sophisticated network analytical techniques has allowed us to probe these assemblages at levels not anticipated even 20 years ago, there are significant gaps in our knowledge of how such assemblages come about over ecological and evolutionary time scales. For instance, we know little about how increases in plant diversity result in greater species richness of pollinators, although recent research shows that they may be correlated (e.g. Ebeling et al. 2008; Fründ et al. 2010).

Finally, the interactions of plants and pollinators within their respective groups, via competition for resources (nectar, pollen or pollinator services) or facilitation is an area currently under-studied in community pollination ecology but where novel approaches are revealing some interesting results regarding the frequency of positive versus negative effects, both within and between species (Hegland et al. 2009).

9. *Geographical trends in pollinator diversity (Box 9)*

A glance at the history of pollination ecology shows us that research within Europe and North America dominated the field from the 18th to the first half of the 20th centuries (work by Sprengel, Delpino, Darwin, Müller, Robertson, etc., with notable exceptions such as Burkill's work in India and some of the early research in Southern Africa and South America, see Faegri & Pijl 1979; Proctor et al. 1996). This has resulted in a rather Euro-American view of the diversity of plant-pollinator interactions that has only begun to be challenged in the last two decades (Johnson and Steiner 2000). Even now, our understanding about geographical patterns of the diversity of pollinators and pollination systems in large parts of the world is virtually non-existent or at most patchy.

53. How does the diversity of pollinators vary geographically at the level of species and functional groups?
54. How and why do plant and pollinator specialisation, ecological redundancy, and other network characteristics vary geographically?
55. What are the geographic units of functional relevance to pollinator diversity, e.g. local, landscape, regional, or continental?
56. How common are geographical mosaics of plant-pollinator interactions?
57. What is the scale of the biotic homogenisation of pollinators, as observed in Europe?

Box 9: Questions on geographical trends of pollinator diversity

Research in such areas each year bring accounts of pollination by taxa previously considered unlikely to be pollinators of specialised plants, for example crickets (Micheneau et al. 2010), spider hunting wasps, fruit chafers (Ollerton et al. 2003) and cockroaches (Nagamitsu & Inoue 1997). In some instances, these taxa are dominant pollinators within communities, not simply marginal curiosities of natural history, and there is the need to more fully understand the roles played by the full complement of pollinators.

Questions also remain on processes and their ultimate consequences. For example, we strongly suspect that the ecological isolation of oceanic islands reduces the diversity of pollinators that can reach such islands (Woodell 1979), resulting in the evolution of "unusual" pollination systems such as lizard pollination (e.g. Hansen & Müller 2009). We do not know what the consequences are of habitat loss and species extinctions for such isolated biota. Such interaction networks could either be more fragile than those on the continents, because there are fewer species which can act as substitute pollinators for the plants, or they might be relatively robust because the more brittle interactions are already lost. The latter idea is backed up by work showing that "super-generalist" plants and pollinators may be common on islands, increasing the robustness of island interaction networks (Olesen et al. 2002).

Geographic patterns, however, occur at all spatial scales and adjacent regions may harbour distinctly different groups of pollinators, though this is little studied (but see Galen et al. 1987; Armbruster & Guinn 1989). Consequently, a significant concern of pollination ecologists is that in highly agro-industrialised parts of the world, the removal of natural habitats, the extinction of rare species and the introduction of non-native pollinators is resulting in the biological homogenisation of regional pollinator faunas (see also topic 10).

10. Drivers of pollinator loss (Box 10)

Pollinator decline and loss of pollination services have become issues of political, media and scientific focus worldwide. Many studies have identified relationships between various human activities and changes in pollinator behaviour, population structure, overall abundance, richness and diversity (Biesmeijer et al. 2006; Winfree et al. 2007), their interactions with plants and pollination services (Ricketts et al. 2008; Carvalheiro et al. 2010). However, although some social bees have been widely studied (e.g. Williams & Osborne 2009), the extent to which other taxa are declining, and geographical variations in the decline, are not as widely understood (Kevan & Fonseca 2002).

As with biodiversity losses in other taxa, key drivers of decline are identified as: (1) habitat destruction, degradation and fragmentation (resulting in a loss of foraging, mating and nesting sites - particularly driven by changes in agricultural management practices; Kearns et al. 1998; Taki et al. 2008; Brown & Paxton 2009), (2) pollution (in particular by agri-chemicals including neonicotinoids; Kevan 1999; Brittain et al. 2010), (3) invasive alien species (including introduced plants, pollinators, pests and diseases; Stout & Morales 2009; Dafni et al. 2010), and (4) climate change (which affects the spatio-temporal dynamics of plant-pollinator interactions; Memmott et al. 2007; Hegland et al. 2009). In the future, new drivers of decline may emerge, either as a result of the introduction of novel species into the environment (such as bioenergy crops and genetically modified organisms) or other technological and social changes (associated with renewable energy generation, electromagnetic fields, urbanisation or other factors).

However, many questions remain, and it is likely that drivers are spatio-temporally variable, differ among species and/or functional groups, and interact synergistically (Schweiger et al., 2010). One feature that is of particular relevance to bee declines is that the genetic load caused by their unusual sex determining mechanism makes them particularly susceptible to environmental impacts that reduce population sizes (Zayed & Packer 2005; Packer 2010).

Furthermore, many studies are limited to field-scale, pair-wise interactions, and it may be difficult to assess impacts on entire ecological systems at the landscape scale (Taki et al. 2007) or to determine how resilient pollinators (from individuals, populations and species to communities) are to change (Kevan & LaBerge 1979). For a meta-analysis of the ecological traits of bees that make different groups susceptible to different environmental impacts, see Williams et al. (2010).

58. Besides habitat destruction, pollution, invasive species, and climate change – how can we best identify future risks to plant-pollinator interactions?
59. Which pollinator taxa and functional groups are in decline?
60. What are the geographical patterns of decline?
61. What are the rates of decline?
62. What is the relative importance of the various drivers of pollinator decline?
63. How do drivers of loss interact, and how do they vary in space and time?
64. Do drivers of pollinator decline also drive loss of pollination services and, if so, what is the rate and shape of change?
65. How do we quantify drivers of decline experimentally and at the landscape scale?
66. What features of the life histories of pollinators (e.g. haplodiploidy in bees, or specialist larval food plants in Lepidoptera) make them more susceptible to extirpation?

Box 10: Questions on drivers of pollinator loss

11. Pollination as an ecosystem service (Box 11)

The Millennium Ecosystem Assessment (MEA 2005) characterises pollination as a regulating service which underpins the reproduction of the most flowering plant species and the productivity of many of the world's crops. Despite many decades of study, there is still a fundamental gap in our understanding of which animals are the primary pollinators of wild flowers and crops. Although some landmark agriculturally-focussed publications (e.g. Free 1993; Klein et al. 2007) explore these relationships, there remains an outstanding challenge to compile a systematic catalogue including variation with location, plant variety, season, and time of day.

Reviews estimate that 62-73% of wild plant species are pollen limited (e.g. Burd 1994; Ashman et al. 2004), however, for most wild plants and for many crops this basic information is still missing.

Globally it is estimated that the value of pollinators for crop production was €153 billion per year in 2005 (Gallai et al. 2009). Although this figure provides a useful initial overview, the calculation does not include all of the world's pollinator-dependent crops and also excludes subsistence farming, the value of non-cultivated plant products (e.g. wild harvested fruits and berries), and indirect values of livestock reliant on pollination products such as clover or alfalfa. The aesthetic and cultural values of pollination, e.g. by providing florally rich environments for pleasure and recreation, are also of high scientific and policy relevance (Eardley et al. 2006).

Most of the world's staple crops (e.g. rice, wheat, maize and sorghum) are abiotically pollinated. However, most fruits and vegetables benefit from biotic pollination (Roubik 1995; Klein et al. 2007). They provide many of the micronutrients (e.g. vitamins and minerals) essential for healthy diets (Johns 2007). There is, therefore, an increasing demand for pollination services for world agriculture, as the area planted with these crops expands (Aizen & Harder 2009). This suggests an increasing divergence in the demand and supply of pollination services. Future challenges will centre on ways of ensuring sufficient and reliable pollination services for food production and security.

Wild plant communities themselves also contribute to a suite of other essential ecosystem services such as water filtration, maintenance of soil quality, fertility and structure, carbon balance and oxygen production. Plants also provide habitats and resources for much of terrestrial biodiversity. The role of pollination in supporting other ecosystem services has already begun to be valued by initiatives such as TEEB (The Economics of Ecosystems and Biodiversity, TEEB 2010). Still, how these ecosystem services may be threatened by pollinator loss is not well understood. This knowledge gap must be bridged if we are to manage ecosystems for human livelihoods and biodiversity conservation (Kevan & Phillips 2001; Potts et al. 2010).

67. What are the most important pollinators of crops and wild plants?
68. How widespread are pollination deficits in crops and wild plants?
69. What is the economic value of pollination services?
70. How will we supply the growing demand from agriculture for pollination services?
71. What is the role of pollination in global food security?
72. What other ecosystem services would be affected by the loss of pollinators?

Box 11: Questions on pollination as an ecosystem service

12. Managing pollination services (Box 12)

Whenever 'free' naturally provided pollination services are insufficient in magnitude and reliability, pollination needs to be managed through: (i) managing habitats, including agro-ecosystems, to support wild pollinators; and/or (ii) introducing domesticated pollinators. In order to best match pollinators and crops, detailed studies of pollinator behaviour, morphology and physiology and the availability of crop forage rewards, breeding system and pollination requirements are needed (Delaplane & Mayer 2000). For example, Sheffield et al. (2008) have shown that alternative food sources can help maintain pollinator populations outside of crop flowering time, but this needs further investigation.

Honey bees (*Apis* spp.) are the most widely used managed pollinators across the globe. However, other bee species are also being managed for crop pollination (Delaplane & Mayer 2000; Mader et al. 2010), including some stingless bees (Meliponini), bumblebees (*Bombus* spp.), mason bees (*Osmia* spp.), leafcutter bees (e.g. *Megachile rotundata*) and ground nesting solitary bees (e.g. *Nomia melanderi*). Many of these are more efficient pollinators than honey bees on specific crops but are generally not managed on the same scale (Free 1993). Undoubtedly other wild species could be developed for crop pollination, but the species requirements and techniques for managing them need to be underpinned by fundamental research into their ecology and behaviour.

Large-scale commercial beekeeping is well established in the USA (NRC 2006), and some other developed countries, whereby growers pay directly for hive rental and benefit from the enhanced service. However, throughout much of the rest of the world beekeepers often place hives in cultivated areas and only derive indirect benefits through honey production, unconscious of the value of the service they provide to agriculture. The challenge is to quantify the value of pollination services and develop fiscal mechanisms whereby service providers can be compensated for the real value of the service provided. The development of Payment for Ecosystem Services (PES, Jack et al. 2008) for pollination is a key future research and policy need that could be extended to other parts of the world, not just the USA.

73. How do we match the right pollinators to individual crops?
74. When do we need to use managed pollinators?
75. What alternative species of pollinators can we manage for pollination services?
76. How can beekeepers, and providers of other pollinators, be paid for delivering pollination services?
77. How can the potentially conflicting demands for pollinator diversity conservation and crop pollination be reconciled?

Box 12: Questions on managing pollination services

There are demands on our landscapes to provide multiple functions simultaneously and some of these demands may be in conflict. For food production, the objective may be to maximise productivity and profit, while conservation of pollinators may require more extensive farming practices (Tscharntke et al. 2005). There is a need for research to identify and test other management practices which can produce win-win options where both productivity and pollinators co-benefit.

13. Conservation (Box 13)

Plants and pollinators may have very close links and examples show that the loss of pollinators lead to the decline of their associated plants, as recently reported by Anderson et al. (2011) for a New Zealand shrub species (see also Biesmeijer et al. 2006). Thus, also the conservation of plants and pollinators is strongly associated, and so strategies to promote pollinator conservation will have knock on benefits for plant conservation and *vice versa*. Classic conservation measures involve site protection in national parks and nature conservation areas, but these alone cannot stop plant or pollinator decline over broader landscapes. Strict species protection regimes have been set up in several countries for some plants, butterflies and bees. These may be useful for reducing direct threats such as collection and trade but are of minor help for most biodiversity since they do not prevent general major threats such as habitat destruction, degradation and fragmentation. Plants often suffer from inbreeding depression when population size becomes too small (Honnay et al. 2005; Aguilar et al. 2006). Most pollinating insects are dependent on a habitat mosaic to fulfil different needs (for example, nest sites as well as floral resources for bees, and larval food plants as well as the flowers required for adult feeding for butterflies and moths). Therefore, a "Green Infrastructure" with habitat connectivity is needed for conservation of plants and their pollinators (Benedict & McMahon 2006). Stepping stones, corridors, buffers or nodes at different spatial scales offer possibilities to connect fragmented habitats (Dover & Settele 2008). Agri-environment schemes are good examples where potential conflicts between economic profit and conservation can be resolved; non-cultivated areas, such as flower-rich field margins provide pollinator habitat, which in turn may enhance local pollination services for plants, and farmers receive financial support for establishing the margins (Carvell et al. 2007; Potts et al. 2009). Migratory species, such as some butterflies, bats, hoverflies, and hummingbirds, especially need nectar corridors for conservation (see the Migratory Pollinators Program of the Arizona-Sonora Desert Museum, http://www.desertmuseum.org/pollination).

Comprehensive monitoring of the pollinator status and the threats and pressures on plants and pollinators is indispensible to develop conservation measures. Red Data books are a possibility, but a shortage of resources and expertise may result in smaller or taxonomically difficult groups being ignored. Targeted species action plans for plants or pollinators are still rare, and general monitoring schemes are missing (except for honey bees and other managed bee species in some countries).

78. What conservation laws, policy requirements and adaptation of existing agri-environmental funding schemes are needed to sustainably conserve pollination services, as well as plant and pollinator diversity?
79. How can we best monitor and document the status, threats and pressures on pollinators including effects on plants and biodiversity as a whole?
80. What essential modifications in land use management and practices are needed to halt and reverse plant and pollinator declines?
81. How can we ensure adequate prioritisation, sufficient action and implementation?

Box 13: Questions on conservation

The 2010 target of reducing the loss of biodiversity could not be reached worldwide under the Convention on Biological Diversity (CDB) nor within the EU, despite the latter's common nature conservation law such as the Habitats Directive (i.e. Council Directive 92/43/EEC of 21 May 1992 on the conservation of natural habitats and of wild fauna and flora). To achieve the goal of stemming the loss of plant and pollinator biodiversity, major questions for research must focus on monitoring their status, and designing effective laws, policies and financial incentives as well as the necessary adaptations in land use.

14. Implementation of plant-pollinator interaction conservation (Box 14)

The first political steps to acknowledge the importance of pollinators and their interactions with plants and to raise awareness were undertaken within the CBD on the 5th Conference of Parties (in 2000) with the "Sao Paulo Declaration on Pollinators" (International Pollinator Initiative 1999). An action plan (decision VI/5) was developed, and an International Pollinator Initiative was formed under the leadership of the Food and Agriculture Organisation (FAO). Several regional initiatives in Europe, North America, Canada, Africa, Brazil and Oceania have been established since then (Ssymank et al. 2009). FAO's rapid assessment of pollinator status (FAO 2008) gave an important input for politics. Recently the International Risk Governance Council in Geneva addressed the policy needs in pollination services (IRGC 2009), pointing out risks related to the loss of pollination services, regulatory and governance shortcomings and knowledge deficits. The ALARM project of the EU (Assessing LArge-scale environmental Risks for biodiversity with tested Methods) included projects on diversity of plants and pollinators and follow-up projects under the EU research programme will be started. Further, the United Nations Environment Programme (UNEP), Global Environment Facility (GEF) and FAO have launched a Global Pollination Project on "Conservation and Management of Pollinators for Sustainable Agriculture, through an Ecosystem Approach" for almost $27 million.

82. How can we promote strategic networking on pollination issues in a broad, integrated, and interdisciplinary approach?
83. How can we effectively raise awareness among society about plants, pollinators and pollination services?
84. What kinds of training, education and capacities are needed to protect the diversity of pollinators and their food plants?
85. How can we better employ plants and their pollinators as educational tools for public awareness?
86. How can pollination ecologists learn from other fields to communicate effectively about pollinators with policy makers and the general public?

Box 14: Questions on implementation of plant-pollinator interaction conservation

Despite these efforts, more education, capacity building and networking for the implementation of plant and pollinator management are desperately needed. Taxonomists, plant and pollination ecologists, training for students, materials, and identification keys are missing for many pollinator groups (Packer et al. 2009). Museum and research capacities as well as capacities for practical implementation and management are still low and neglected. The main questions are therefore expressing the concern on raising awareness, capacity building and closing the enormous knowledge gaps on pollination services in a systematic way including all major pollinator groups over all major terrestrial ecosystems worldwide. This will be beneficial not only for plant diversity and agriculture, but also for forestry, gardening, medicine and human health.

Discussion

We conclude that sufficient unanswered questions remain to feed research for several generations to come. This exercise has some shortcomings as does every scientific study: only a limited number of consultees were contacted and only a fraction of them (<40%) responded. However, there was substantial overlap in the more than 650 contributions we received, and the themes of the most urgent subjects were easily identified. We found it much more difficult to select from the mass of good questions which might have been too case-specific to be included in a publication also intended for non-specialists. We decided to focus on a broader level, which makes it difficult to answer single questions with single projects. Nevertheless, many of the questions can be split or broken down into digestible parts and offer objectives for studies within a defined time frame. On the other hand, it may be argued that 86 questions are too many to represent the "key" issues, but the final number of questions reflects the breadth of the topic and the current debates within the field. Pollination ecology is a complex area and despite more than 200 years of study, many topics are not yet fully understood, whilst all the time new questions arise as our environment changes and novel mechanisms for investigation and interpretation appear.

We do not claim our list to be comprehensive nor to have dealt with the various thematic areas with exhaustive profundity. There may be other good ways to categorize the questions; we chose the thematic groupings that made the most sense to us. Reviews or special issues can provide much more detailed and specific information about research needs in a particular topic (e.g. Raguso 2009; Friedman & Barrett 2009). We are also aware that we have focussed primarily on the angiosperms, though many of these questions would also apply to wind and biotic pollination in the gymnosperms. Still, we hope that we can inspire new tracks in research on pollination, thereby offering politicians, stakeholders, and donors insight into research and conservation priorities with the goal of conserving plant and pollinator biodiversity and preserving an ecosystem service essential for our well-being. An understanding of patterns and processes is necessary if we are to fully comprehend the rich ecological and evolutionary interplay between pollinators and the plants they service, and how these interactions can be conserved in an anthropogenically changing world.

Acknowledgements

We would like to express our sincere thanks to all of the pollination scientists who freely and generously shared their ideas with us by taking part in our poll:

Josef Ackerman, Canada; Klaus Amman, Switzerland; Bruce Anderson, RSA; Jordi Bascompte, Spain; Peter Bernhardt, USA; Andrew Bourke, UK; Mark Brown, UK; Stefanie Christmann, Germany; Richard Corlett, Singapore; James Cresswell, UK; Saul Cunningham, Australia; Ben Darvill, UK; Robin Dean, UK; Antonio Felicioli, Italy; Charles Fenster, USA; Una Fitzpatrick, Ireland; Ted Fleming, USA; Vera Fonseca, Brazil; Francis Gilbert, UK; José María Gómez , Spain; Dave Goulson, UK; Melanie Hagen, Germany; Andre Hamm, Germany; Carlos Herrera, Spain; David Inouye, USA; Rebecca Irwin, USA; Anne-Laure Jaquemart, Belgium; Steve Johnson, RSA; Jeffrey Karron, USA; Makoto Kato, Japan; Danny Kessler, Germany; Vicky Kindemba, UK; Michael Kuhlmann, UK; Peter Kwapong, Ghana; John Manning, RSA; Maria Angeles Marcos García, Spain; Dino Martins, USA/Kenya; Andrew McCall, USA; Jane Memmott, UK; Denis Michez, Belgium; Pierre Mineau, Canada; Randall Mitchell, USA; Carolina Morales, Argentina; Linda Newstrom-Lloyd, NZ; Kazuharu Ohashi, Japan; Anton Pauw, RSA; Robert Paxton, UK; Theodora Petanidou, Greece; Marino Quaranta, Italy; Robert Raguso, USA; Nelson Ramirez, Venezuela; T'ai H. Roulston, USA; Lauren Ruane, USA; Chris Saure, Germany; Florian Schiestl, Switzerland; Josef Settele, Germany; Matt Shardlow, UK; Andrew Stephenson, USA; Juergen Tautz, Germany; John Thompson, USA; James D. Thomson, Canada; Teja Tscharntke, Germany; Diego Vázquez, Argentina; Nick Waser, USA; Dieter Wittmann, Germany; Lorne Wolfe, USA.

We further thank the participants of the SCAPE meeting at Tovetorp (Sweden), 2010, for their highly appreciated comments on the questions. Thanks go also to Koos Biesmeijer and two anonymous reviewers for helping us to improve earlier versions of the manuscript.

Appendices

Additional supporting information may be found in the online version of this article:

Appendix I. Number of researchers who were contacted and who replied in the different regions of the world.

Appendix II. List of questions after first compression.

Literature Cited

Ackerman JD (1981) Pollination biology of *Calypso bulbosa* var. *occidentalis* (Orchidaceae): a food-deception system. Madroño 3:101-110.

Ackerman JD (2000) Abiotic pollen and pollination: Ecological, functional, and evolutionary perspectives. Plant Systematics and Evolution 222:167-185.

Adler LS (2000) The ecological significance of toxic nectar. Oikos 91:409-420.

Adler LS (2007) Selection by pollinators and herbivores on attraction and defense. In: Tilmon K (ed) Specialization, speciation and radiation: The evolutionary biology of herbivorous insects. University of California Press, Berkeley, pp 162-173.

Adler LS, Irwin RE (2005) Ecological costs and benefits of defenses in nectar. Ecology 86:2968-2978.

Adler LS, Wink M, Distl M, Lentz AJ (2006) Leaf herbivory and nutrients increase nectar alkaloids. Ecology Letters 9:960-967.

Aguilar R, Ashworth L, Galetto L, Aizen M (2006) Plant reproductive susceptibility to habitat fragmentation: review and synthesis through a meta-analysis. Ecology Letters 9:968-980.

Aizen M, Feinsinger P (1994) Forest fragmentation, pollination, and plant reproduction in a Chaco dry forest, Argentina. Ecology 75:330-351.

Aizen MA, Harder LD (2009) The global stock of domesticated honey bees is growing slower than agricultural demand for pollination. Current Biology 19:915-918.

Alarcón R, Riffell J, Davidowitz G, Hildebrand J, Bronstein J (2010) Sex-dependent variation in the floral preferences of the hawkmoth *Manduca sexta*. Animal Behaviour 80:289-296.

Anderson SH, Kelly D, Ladley JJ, Molloy S, Terry J (2011) Cascading effects of bird functional extinction reduce pollination and plant density. Science:DOI: 10.1126/science.1199092.

Armbruster WS, Debevec EM, Willson MF (2002) Evolution of syncarpy in angiosperms: theoretical and phylogenetic analyses of the effects of carpel fusion on offspring quantity and quality. Journal of Evolutionary Biology 15:657-672.

Armbruster WS, Guinn DA (1989) The solitary bee fauna (Hymenoptera: Apoidea) of interior and arctic Alaska: flower associations, habitat use, and phenology. Journal of the Kansas Entomological Society 62:468-483.

Armbruster WS, Muchhala N (2009) Associations between floral specialization and species diversity: cause, effect, or correlation? Evolutionary Ecology 23:159-179.

Ashman T, Knight T, Steets J, Amarasekare P, Burd M, Campbell D, Dudash M, Johnston M, Mazer S, Mitchell R, Morgan M, Wilson W (2004) Pollen limitation of plant reproduction: Ecological and evolutionary causes and consequences. Ecology 85:2408-2421.

Aylor DE, Baltazar BM, Schoper JB (2005) Some physical properties of teosinte (*Zea mays* subsp. *parviglumis*) pollen. Journal of Experimental Botany 56:2401 -2407.

Barrett SCH, Harder LD (1996) Ecology and evolution of plant mating. Trends in Ecology and Evolution 11:73-79.

Beil M, Horn H, Schwabe A (2008) Analysis of pollen loads in a wild bee community (Hymenoptera: Apidae) – a method for elucidating habitat use and foraging distances. Apidologie 39:456-467.

Bell G (1985) On the function of flowers. Proceedings of the Royal Society of London. Series B: Biological Sciences 224:223-266.

Benedict MA, McMahon E (2006) Green infrastructure: linking landscapes and communities. Island Press, Washington D.C.

Biesmeijer J, Roberts S, Reemer M, Ohlemuller R, Edwards M, Peeters T, Schaffers A, Potts S, Kleukers R, Thomas C, Settele J, Kunin W (2006) Parallel declines in pollinators and insect-pollinated plants in Britain and the Netherlands. Science 313:351-354.

Bradshaw HD, Wilbert SM, Otto KG, Schemske DW (1995) Genetic mapping of floral traits associated with reproductive isolation in monkeyflowers (*Mimulus*). Nature 376:762-765.

Brittain C, Vighi M, Bommarco R, Settele J, Potts S (2010) Impacts of a pesticide on pollinator species richness at different spatial scales. Basic and Applied Ecology 11:106-115.

Brown MJ, Paxton RJ (2009) The conservation of bees: A global perspective. Apidologie 40:7.

Buchmann S, Nabhan G (1996) The forgotten pollinators. Island Press, Washington D.C.

Burd M (1994) Bateman's principle and plant reproduction: The role of pollen limitation in fruit and seed set. Botanical Review 60:83-139.

Carvalheiro LG, Seymour CL, Veldtman R, Nicolson SW (2010) Pollination services decline with distance from natural habitat even in biodiversity-rich areas. Journal of Applied Ecology 47:810-820.

Carvell C, Meek W, Pywell R, Goulson D, Nowakowski M (2007) Comparing the efficacy of agri-environment schemes to enhance bumble bee abundance and diversity on arable field margins. Journal of Applied Ecology 44:29-40.

Chittka L, Raine NE (2006) Recognition of flowers by pollinators. Current Opinion in Plant Biology 9:428-435.

Culley T, Weller S, Sakai A (2002) The evolution of wind pollination in angiosperms. Trends in Ecology and Evolution 17:361-369.

Culley TM (2002) Reproductive biology and delayed selfing in *Viola pubescens* (Violaceae), an understory herb with chasmogamous and cleistogamous flowers. International Journal of Plant Sciences 163:113-122.

Dafni A, Kevan P, Gross CL, Goka K (2010) *Bombus terrestris*, pollinator, invasive and pest: An assessment of problems associated with its widespread introductions for commercial purposes. Applied Entomology and Zoology 45:101-113.

Dafni A, Nepi M, Pacini E (2005) Pollen and stigma biology. In: Practical Pollination Biology. Enviroquest Ltd., Cambridge, Ontario, Canada, pp 83-146.

Delaplane KS, Mayer DF (2000) Crop pollination by bees. CABI, New York.

Dohzono I, Yokoyama J (2010) Impacts of alien bees on native plant-pollinator relationships: A review with special emphasis on plant reproduction. Applied Entomology and Zoology 45:37-47.

Douglas AC, Freyre R (2010) Floral development, stigma receptivity and pollen viability in eight *Nolana* (Solanaceae) species. Euphytica 174:105-117.

Dover J, Settele J (2008) The influences of landscape structure on butterfly distribution and movement: A review. Journal of Insect Conservation 13:3-27.

Duan YW, Zhang TF, He YP, Liu JQ (2009) Insect and wind pollination of an alpine biennial *Aconitum gymnandrum* (Ranunculaceae). Plant Biology 11:796-802.

Dukas R (2001) Effects of perceived danger on flower choice by bees. Ecology Letters 4:327-333.

Durrer S, Schmid-Hempel P (1994) Shared use of flowers leads to horizontal pathogen transmission. Proceedings of the Royal Society of London. Series B: Biological Sciences258:299-302.

Eardley C, Kuhlmann M, Pauly A (2010) The bee genera and subgenera of sub-Saharan Africa. ABC Taxa, Brussels.

Eardley C, Roth D, Clarke J, Buchmann S, Gemmill B (eds) (2006) Pollinators and pollination: A resource book for policy and practice. African Pollinator Initiative, Pretoria.

Ebeling A, Klein A, Schumacher J, Weisser WW, Tscharntke T (2008) How does plant richness affect pollinator richness and temporal stability of flower visits? Oikos 117:1808-1815.

Elle E (2004) Floral adaptations and biotic and abiotic selection pressures. In: Cronk Q, Ree R, Taylor I, Whitton J (eds) Plant adaptation: Molecular genetics and ecology. National Research Council of Canada, Ottawa, Ontario, pp 111-118.

Endress PK (1982) Syncarpy and alternative modes of escaping disadvantages of apocarpy in primitive angiosperms. Taxon 31:48-52.

Eppley SM, Pannell JR (2009) Inbreeding depression in dioecious populations of the plant *Mercurialis annua*: comparisons between outcrossed progeny and the progeny of self-fertilized feminized males. Heredity 102:600-608.

Faegri K, Pijl L (1979) The principles of pollination ecology. Pergamon Press, Oxford.

FAO (2008) A contribution to the International Initiative for the conservation and sustainable use of pollinators – Rapid assessment of pollinators' status. FAO, Rome.

Fenster C, Armbruster S, Wilson P, Dudash M, Thomson J (2004) Pollination syndromes and floral specialization. Annual Review of Ecology and Systematics 35:375-403.

Fontaine C, Dajoz I, Meriguet J, Loreau M (2006) Functional diversity of plant-pollinator interaction webs enhances the persistence of plant communities. PLoS Biology 4:e1.

Fontaine C, Collin CL, Dajoz I (2008) Generalist foraging of pollinators: diet expansion at high density. Journal of Ecology 96:1002-1010.

Franchi GG, Nepi M, Matthews ML, Pacini E (2007) Anther opening, pollen biology and stigma receptivity in the long blooming species, *Parietaria judaica* L. (Urticaceae). Flora 202:118-127.

Free JB (1993) Insect pollination of crops. Academic Press, London.

Friedman J, Barrett SCH (2009) Wind of change: New insights on the ecology and evolution of pollination and mating in wind-pollinated plants. Annals of Botany 103:1515-1527.

Fründ J, Linsenmair KE, Blüthgen N (2010) Pollinator diversity and specialization in relation to flower diversity. Oikos 119:1581-1590.

Galen C, Zimmer KA, Newport ME (1987) Pollination in floral scent morphs of *Polemonium viscosum*: A mechanism for disruptive selection on flower size. Evolution 41:599-606.

Gallai N, Salles JM, Settele J, Vaissière BE (2009) Economic valuation of the vulnerability of world agriculture confronted with pollinator decline. Ecological Economics 68:810-821.

Gegear RJ, Otterstatter MC, Thomson JD (2006) Bumble-bee foragers infected by a gut parasite have an impaired ability to utilize floral information. Proceedings of the Royal Society of London. Series B: Biological Sciences 273:1073-1078.

Gómez JM (2003) Herbivory reduces the strength of pollinator-mediated selection in the Mediterranean herb *Erysimum mediohispanicum*: Consequences for plant specialization. The American Naturalist 162:242-256.

Goulson D (1999) Foraging strategies of insects for gathering nectar and pollen, and implications for plant ecology and evolution. Perspectives in Plant Ecology, Evolution and Systematics 2:185-209.

Grant V (1949) Pollination systems as isolating mechanisms in angiosperms. Evolution 3:82-97.

Greenleaf S, Williams N, Winfree R, Kremen C (2007) Bee foraging ranges and their relationship to body size. Oecologia 153:589-596.

Hansen DM, Müller CB (2009) Reproductive ecology of the endangered enigmatic Mauritian endemic *Roussea simplex* (Rousseaceae). International Journal of Plant Sciences 170:42-52.

Hegland SJ, Grytnes J, Totland Ø (2009) The relative importance of positive and negative interactions for pollinator attraction in a plant community. Ecological Research 24:929-936.

Heinrich B (1979) Bumblebee economics. Harvard University Press, Cambridge.

Herrera CM, de Vega C, Canto A, Pozo MI (2009) Yeasts in floral nectar: A quantitative survey. Annals of Botany 103:1415-1423.

Honnay O, Jaquemyn H, Bossuyt B, Hermy M (2005) Forest fragmentation effects on patch occupancy and population viability of herbaceous plant species. New Phytologist 166:723-736.

Hood ME, Mena-Alí JI, Gibson AK, Oxelman B, Giraud T, Yockteng R, Arroyo MTK, Conti F, Pedersen AB, Gladieux P, Antonovics J (2010) Distribution of the anther-smut pathogen *Microbotryum* on species of the Caryophyllaceae. New Phytologist 187:217-229.

Inouye DW (1978) Resource partitioning in bumblebees: Experimental studies of foraging behavior. Ecology 59:672-678.

International Pollinator Initiative (1999) The São Paulo declaration on pollinators. Report on the recommendations of the workshop on the conservation and sustainable use of pollinators in agriculture with emphasis on bees. Brazilian Ministry of the Environment (MMA), Brasília.

IRGC (International Risk Governance Council) (2009) Concept note: Risk governance of pollination services. IRGC, Geneva.

Jack BK, Kousky C, Sims KRE (2008) Designing payments for ecosystem services: Lessons from previous experience with incentive-based mechanisms. Proceedings of the National Academy of Sciences USA 105:9465 -9470.

Johns T (2007) Agrobiodiversity, diet, and human health. In: Jarvis DI, Padoch C, Cooper HD (eds) Managing biodiversity in agricultural ecosystems. Columbia University Press, New York, pp 382-406.

Johnson KB, Sawyer TL, Temple TN (2006) Rates of epiphytic growth of *Erwinia amylovora* on flowers common in the landscape. Plant Disease 90:1331-1336.

Jordano P, Bascompte J, Olesen JM (2003) Invariant properties in coevolutionary networks of plant-animal interactions. Ecology Letters 6:69-81.

Kalisz S, Vogler D (2003) Benefits of autonomous selfing under unpredictable pollinator environments. Ecology 84:2928-2942.

Kandori I, Yamaki T, Okuyama S, Sakamoto N, Yokoi T (2009) Interspecific and intersexual learning rate differences in four butterfly species. Journal of Experimental Biology 212:3810-3816.

Karron JD, Holmquist KG, Flanagan RJ, Mitchell RJ (2009) Pollinator visitation patterns strongly influence among-flower variation in selfing rate. Annals of Botany 103:1379-1383.

Kay KM, Schemske DW (2008) Natural selection reinforces speciation in a radiation of neotropical rainforest plants. Evolution 62:2628-2642.

Kearns C, Inouye D, Waser N (1998) Endangered mutualisms: The conservation of plant-pollinator interactions. Annual Review of Ecology and Systematics 29:83-112.

Kessler D, Baldwin IT (2007) Making sense of nectar scents: the effects of nectar secondary metabolites on floral visitors of *Nicotiana attenuata*. The Plant Journal 49:840-854.

Kevan PG, Baker HG (1983) Insects as flower visitors and pollinators. Annual Review of Entomology 28:407-453.

Kevan P (1999) Pollinators as bioindicators of the state of the environment: species, activity and diversity. Agriculture, Ecosystems & Environment 74:373-395.

Kevan P, Phillips T (2001) The economic impacts of pollinator declines: An approach to assessing the consequences. Conservation Ecology 5:8. [online] URL: http://www.consecol.org/vol5/iss1/art8/.

Kevan P, Fonseca LI (Eds) (2002) Pollinating bees - The conservation link between agriculture and nature. Ministry of Environment, Brasília.

Kevan P, LaBerge W (1979) Demise and recovery of native pollinator populations through pesticide use and some economic implications. In: Caron D (ed) Proceedings IV International Symposium on Pollination (Maryland Agricultural Experimental Station Special Miscellaneous Publication). pp 489-508.

Klein A, Vaissière BE, Cane J, Steffan-Dewenter I, Cunningham S, Kremen C, Tscharntke T (2007) Importance of pollinators in changing landscapes for world crops. Proceedings of the Royal Society of London. Series B: Biological Sciences 274:303-313.

Knapp EE, Goedde MA, Rice KJ (2001) Pollen-limited reproduction in blue oak: implications for wind pollination in fragmented populations. Oecologia 128:48-55.

Kremen C, Williams N, Thorp R (2002) Crop pollination from native bees at risk from agricultural intensification. Proceedings of the National Academy of Sciences USA 99:16812-16816.

Linder HP, Midgley J (1996) Anemophilous plants select pollen from their own species from the air. Oecologia 108:85-87.

Liu DY, Carlsson U (2002) Infection of the anther-smut *Microbotryum violaceum* in populations of *Silene dioica*: Variation in floral morphology, patterns of spore deposition and pathogen-pollinator mediated selection. Acta Botanica Sinica 44:88-96.

Lloyd DG (1979) Some reproductive factors affecting the selection of self-fertilization in plants. The American Naturalist 113:67-79.

Mader E, Spivak M, Evans E (2010) Managing alternative pollinators: A handbook for beekeepers, growers, and conservationists. Copublished by SARE and NRAES.

Mao Y, Huang S (2009) Pollen resistance to water in 80 angiosperm species: Flower structures protect rain-susceptible pollen. New Phytologist 183:892-899.

Mayer C, Soka G, Picker M (2006) The importance of monkey beetle (Scarabaeidae: Hopliini) pollination for Aizoaceae and Asteraceae in grazed and ungrazed areas at Paulshoek, Succulent Karoo, South Africa. Journal of Insect Conservation 10:323-333.

McCall AC, Karban R (2006) Induced defense in *Nicotiana attenuata* (Solanaceae) fruit and flowers. Oecologia 146:566-571.

MEA - Millennium Ecosystem Assessment (2005) Ecosystems and human well-being, synthesis. Island Press, Washington D.C.

Memmott J (1999) The structure of a plant-pollinator food web. Ecology Letters 2:276-280.

Memmott J, Craze PG, Waser NM, Price MV (2007) Global warming and the disruption of plant-pollinator interactions. Ecology Letters 10:710-717.

Micheneau C, Fournel J, Warren BH, Hugel S, Gauvin-Bialecki A, Pailler T, Strasberg D, Chase MW (2010) Orthoptera, a new order of pollinator. Annals of Botany 105:355-364.

Michener CD (2007) The bees of the world. The Johns Hopkins University Press, Baltimore and London.

Mitchell RJ, Irwin RE, Flanagan RJ, Karron JD (2009) Ecology and evolution of plant-pollinator interactions. Annals of Botany 103:1355-1363.

Morris WF, Hufbauer RA, Agrawal AA, Bever JD, Borowicz VA, Gilbert GS, Maron JL, Mitchell CE, Parker IM, Alison G. Power, Torchin ME, Vázquez DP (2007) Direct and interactive effects of enemies and mutualists on plant performance: A meta-analysis. Ecology 88:1021-1029.

Nagamitsu T, Inoue T (1997) Cockroach pollination and breeding system of *Uvaria elmeri* (Annonaceae) in a lowland mixed-dipterocarp forest in Sarawak. American Journal of Botany 84:208-213.

National Research Council (2006) Status of pollinators in North America. National Academic Press, Washington D.C.

Ne'eman G, Jürgens A, Newstrom-Lloyd L, Potts SG, Dafni A (2010) A framework for comparing pollinator performance: Effectiveness and efficiency. Biological Reviews:435-451.

Niklas KJ (1985) The aerodynamics of wind pollination. The Botanical Review 51:328-386.

Nilsson LA (1988) The evolution of flowers with deep corolla tubes. Nature 334:147-149.

Norman J, Weller S, Sakai A (1997) Pollination biology and outcrossing rates in hermaphroditic *Schiedea lydgatei* (Caryophyllaceae). American Journal of Botany 84:641.

Olesen JM, Eskildsen LI, Venkatasamy S (2002) Invasion of pollination networks on oceanic islands: importance of invader complexes and endemic super generalists. Diversity and Distributions 8:181-192.

Ollerton J, Johnson S, Cranmer L, Kellie S (2003) The pollination ecology of an assemblage of grassland asclepiads in South Africa. Annals of Botany 92:807-834.

Ollerton J, Alarcon R, Waser NM, Price MV, Watts S, Cranmer L, Hingston A, Peter CI, Rotenberry J (2009) A global test of the pollination syndrome hypothesis. Annals of Botany 103:1471-1480.

Ollerton J, Killick A, Lamborn E, Watts S, Whiston M (2007) Multiple meanings and modes: On the many ways to be a generalist flower. Taxon 56:717-728.

Ollerton J, Masinde S, Meve U, Picker M, Whittington A (2009) Fly pollination in *Ceropegia* (Apocynaceae: Asclepiadoideae): biogeographic and phylogenetic perspectives. Annals of Botany 103:1501 -1514.

Ollerton J, Watts S (2000) Phenotypic space and floral typology: Towards an objective assessment of pollination syndromes. Norske Vidensakps-Akademi Mat-Naturv Klasse Skrifter Nye:149-159.

Ollerton J, Winfree R, Tarrant S (2011) How many flowering plants are pollinated by animals? Oikos 120:321-326.

Otti O, Schmid-Hempel P (2008) A field experiment on the effect of *Nosema bombi* in colonies of the bumblebee *Bombus terrestris*. Ecological Entomology 33:577-582.

Packer L, Grixti J, Roughley R, Hanner R (2009) The status of taxonomy in Canada and the impact of DNA barcoding. Canadian Journal of Zoology 87:1097-1110.

Packer L, Ratti C (2009) Key to the bee families of the world. http://www.yorku.ca/bugsrus/BFoW/Images/Introduction/Introduction.html.

Packer L (2010) Keeping the bees. HarperCollins Publishers Ltd, Toronto, Ontario, Canada.

Packer L, Gibbs J, Sheffield C, Hanner R (2009) DNA barcoding and the mediocrity of morphology. Molecular Ecology Resources 9:42-50.

Peeters L, Totland Ø (1999) Wind to insect pollination ratios and floral traits in five alpine *Salix* species. Canadian Journal of Botany 77:556-563.

Potts SG, Woodcock BA, Roberts SPM, Tscheulin T, Pilgrim ES, Brown VK, Tallowin JR (2009) Enhancing pollinator biodiversity in intensive grasslands. Journal of Applied Ecology 46:369-379.

Potts SG, Biesmeijer JC, Kremen C, Neumann P, Schweiger O, Kunin WE (2010) Global pollinator declines: Trends, impacts and drivers. Trend of Ecology and Evolution 25:345-353.

Proctor M, Yeo P, A. L (1996) The natural history of pollination. Harper Collins Publishers (The New Naturalist), London.

Pyke GH (1981) Optimal foraging in hummingbirds: Rule of movement between inflorescences. Animal Behaviour 29:889-896.

Raguso RA (2009) Floral scent in a whole-plant context: Moving beyond pollinator attraction. Functional Ecology 23:837-840.

Raguso RA, Willis MA (2005) Synergy between visual and olfactory cues in nectar feeding by wild hawkmoths, *Manduca sexta*. Animal Behaviour 69:407-418.

Ramsey J, Bradshaw HD, Schemske DW (2003) Components of Reproductive Isolation between the monkeyflowers *Mimulus lewisii* and *M. cardinalis* (Phrymaceae). Evolution 57:1520-1534.

Redbotorstensson P, Berg H (1995) Seasonal cleistogamy - a conditional strategy to provide reproductive assurance. Acta Botanica Neerlandica 44:247-256.

Ricketts TH, Regetz J, Steffan-Dewenter I, Cunningham SA, Kremen C, Bogdanski A, Gemmill-Herren B, Greenleaf SS, Klein AM, Mayfield MM, Morandin LA, Ochieng' A, Potts SG, Viana BF (2008) Landscape effects on crop pollination services: Are there general patterns? Ecology Letters 11:499-515.

Riffell J, Alarcon R, Abrell L, Davidowitz G, Bronstein J, Hildebrand J (2008) Behavioral consequences of innate preferences and olfactory learning in hawkmoth-flower interactions. Proceedings of the National Academy of Sciences USA 105:3404-3409.

Roubik D (1995) Pollination of cultivated plants in the tropics. Food and Agriculture Organization, Rome, Italy.

Sargent RD (2004) Floral symmetry affects speciation rates in angiosperms. Proceedings of the Royal Society of London. Series B: Biological Sciences 271:603-608.

Sasu MA, Seidl-Adams I, Wall K, Winsor JA, Stephenson AG (2010) Floral transmission of *Erwinia tracheiphila* by cucumber beetles in a wild *Cucurbita pepo*. Environmental Entomology 39:140-148.

Schemske DW, Bradshaw HD (1999) Pollinator preference and the evolution of floral traits in monkeyflowers (*Mimulus*). Proceedings of the National Academy of Sciences USA 96:11910 -11915.

Schulke B, Waser N (2001) Long-distance pollinator flights and pollen dispersal between populations of *Delphinium nuttallianum*. Oecologia 127:239-245.

Sheffield C, Westby S, Smith R, Kevan P (2008) Potential of bigleaf lupine for building and sustaining *Osmia lignaria* populations for pollination of apple. The Canadian Entomologist 140:589-599.

Shykoff JA, Bucheli E, Kaltz O (1997) Anther smut disease in *Dianthus silvester* (Caryophyllaceae): Natural selection on floral traits. Evolution 51:383-392.

Sprengel C (1793) Das entdeckte Geheimnis der Natur im Bau und in der Befruchtung der

Blumen. F. Vieweg der Ältere, Berlin.

Ssymank A, Hamm A, Vischer-Leopold M (2009) Caring for pollinators. Safeguarding agro-biodiversity and wild plant diversity. - Results of a workshop and research project commissioned by the German Federal Agency for Nature Conservation. Bundesamt für Naturschutz, Bonn.

Ssymank A, Kearns C, Pape T, Thompson CF (2008) Pollinating flies (Diptera): A major contribution to plant diversity and agricultural production. Biodiversity 9:86-89.

Stebbins L (1970) Adaptive radiation or reproductive characteristics in angiosperms, I: Pollination mechanisms. Annual Review of Ecology and Systematics 1:307-326.

Stout JC, Morales CL (2009) Ecological impacts of invasive alien species on bees. Apidologie 40:388-409.

Strauss S, Whittall J (2006) Non-pollinator agents of selection on floral traits. In: Harder L, Barrett S (eds) Ecology and evolution of flowers. Oxford University Press, Oxford, pp 120-138.

Sun J, Gong Y, Renner SS, Huang S (2008) Multifunctional bracts in the dove tree *Davidia involucrata* (Nyssaceae: Cornales): rain protection and pollinator attraction. The American Naturalist 171:119-124.

Sutherland WJ, Adams WM, Aronson RB, Aveling R, Blackburn TM, Broad S, Ceballos G, CôTé IM, Cowling RM, Da Fonseca GAB, Dinerstein E, Ferraro PJ, Fleishman E, Gascon C, Hunter M, Hutton J, Kareiva P, Kuria A, Macdonald DW, Mackinnon K, Madgwick FJ, Mascia MB, Mcneely J, Milner-Gulland EJ, Moon S, Morley CG, Nelson S, Osborn D, Pai M, Parsons ECM, Peck LS, Possingham H, Prior SV, Pullin AS, Rands MRW, Ranganathan J, Redford KH, Rodriguez JP, Seymour F, Sobel J, Sodhi NS, Stott A, Vance-Borland K, Watkinson AR (2009) One hundred questions of importance to the conservation of global biological diversity. Conservation Biology 23:557-567.

Sutherland W, Armstrong-Brown S, Armsworth P, Tom B, Brickland J, Campbell C, Chamberlain D, Cooke A, Dulvy N, Dusic N, Fitton M, Freckleton R, Godfray H, Grout N, Harvey H, Hedley C, Hopkins J, Kift N, Kirby J, Kunin W, Macdonald D, Marker B, Naura M, Neale A, Oliver T, Osborn D, Pullin A, Shardlow M, Showler D, Smith P, Smithers R, Solandt J, Spencer J, Spray C, Thomas C, Thompson J (2006) The identification of 100 ecological questions of high policy relevance in the UK. Journal of Applied Ecology 43:617-627.

Taki H, Kevan P, Ascher J (2007) Landscape effects of forest loss in a pollination system. Landscape Ecology 22:1575-1587.

Taki H, Viana B, Kevan P, Silva F, Buck M (2008) Does forest loss affect the communities of trap-nesting wasps (Hymenoptera: Aculeata) in forests? Landscape vs. local habitat conditions. Journal of Insect Conservation 12:15-21.

TEEB (2010) The economics of ecosystems and biodiversity: mainstreaming the economics of nature: A synthesis of the approach, conclusions and recommendations of TEEB.:36pp; available at: http://www.teebweb.org/.

Theis N, Lerdau M, Raguso RA (2007) The challenge of attracting pollinators while evading floral herbivores: Patterns of fragrance emission in *Cirsium arvense* and *Cirsium repandum* (Asteraceae). International Journal of Plant Sciences 168:587-601.

Thompson JN, Cunningham BM (2002) Geographic structure and dynamics of coevolutionary selection. Nature 417:735-738.

Tscharntke T, Klein A, Kruess A, Steffan-Dewenter I, Thies C (2005) Landscape perspectives on agricultural intensification and biodiversity - ecosystem service management. Ecology Letters 8:857-874.

Uyenoyama M, Holsinger K, Waller D (1993) Ecological and genetic factors directing the evolution of self-fertilization. Oxford Surveys in Evolutionary Biology 9:327-381.

Walther-Hellwig K, Fokhul G, Frankl R, Büchler R, Ekschmitt K, Wolters V (2006) Increased density of honeybee colonies affects foraging bumblebees. Apidologie 37:517-532.

Waser NM, Ollerton J (2006) Plant-pollinator interactions: from specialization to generalization. University of Chicago Press, Chicago.

Waser NM, Chittka L, Price MV, Williams NM, Ollerton J (1996) Generalization in pollination systems, and why it matters. Ecology 77:1043-1060.

Weiss MR (2001) Vision and learning in some neglected pollinators: Beetles, flies, moths, and butterflies. In: Chittka L, Thomson J (eds) Cognitive ecology of pollination. Cambridge Univ Press, Cambridge, UK, pp 179-190.

Weller SG, Sakai AK, Culley TM, Campbell DR, Dunbar-Wallis AK (2006) Predicting the pathway to wind pollination: Heritabilities and genetic correlations of inflorescence traits associated with wind pollination in *Schiedea salicaria* (Caryophyllaceae). Journal of Evolutionary Biology 19:331-342.

Westrich P (1996) Habitat requirements of central European bees and the problems of partial habitats. In: Matheson A, Buchmann S, O'Toole C, Westrich P, Williams I (eds) The conservation of bees. Academic Press, New York, pp 1-16.

Whitehead MR, Peakall R (2009) Integrating floral scent, pollination ecology and population genetics. Functional Ecology 23:863-874.

Whittall JB, Hodges SA (2007) Pollinator shifts drive increasingly long nectar spurs in columbine flowers. Nature 447:706-709.

Williams NM, Crone EE, Roulston TH, Minckley RL, Packer L, Potts SG (2010) Ecological and life-history traits predict bee species responses to environmental disturbances. Biological Conservation 143:2280-2291.

Williams PH, Osborne JL (2009) Bumblebee vulnerability and conservation world-wide. Apidologie 40:367-387.

Winfree R, Griswold T, Kremen C (2007) Effect of human disturbance on bee communities in a forested ecosystem. Conservation Biology 21:213-223.

Woodell SRJ (1979) The role of unspecialized pollinators in the reproductive success of Aldabran plants. Philosophical Transactions of the Royal Society of London. Series B, Biological Sciences 286:99-108.

Zayed A, Packer L (2005) Complementary sex determination substantially increases extinction proneness of haplodiploid populations. Proceedings of the National Academy of Sciences USA 102:10742-10746.

2

A TEST OF DENSITY-DEPENDENT POLLINATION WITHIN THREE POPULATIONS OF ENDANGERED *PENTACHAETA LYONII*

Jocelyn R. Holt[1], Paul Wilson[1]*, & Christy A. Brigham[2]

[1]*Department of Biology, California State University, 18111 Nordhoff Street, Northridge, CA 91330-8303;*
[2]*Santa Monica Mountains National Recreation Area, 401 West Hillcrest Drive, Thousand Oaks, CA 91360*

Abstract—A major concern with endangered plants is that they might attract insufficient numbers of pollinators, produce low numbers of seeds, and decline towards extinction. We examined effects of density as it varied within populations on the pollination of *Pentachaeta lyonii*, an endangered species that requires pollinators for seed set. Generalist bee-flies and bees were abundant pollinators at three sites for two years. Per-capita visitation rates did not decline at sparse points or for plants placed on the order of 10 m away from other flowering individuals. Seed production was not pollinator-limited within patches, but seed set was low beyond 10 m from neighbours. Considering prior findings, factors such as habitat loss, competition with alien plants, and poor establishment of new populations likely contribute to the rarity of *P. lyonii* more than pollination failure.

Keywords: Allee effect; California endemic; endangered species; Pentachaeta lyonii; *phytometer; pollination*

INTRODUCTION

A major conservation concern surrounds the possibility that organisms at unnaturally low densities might reproduce poorly causing their populations to decline towards extinction (Courchamp et al. 1999, 2008). The kind of density-dependence in which sparse populations fail is called "inverse density dependence"; it is a positive relationship between density and success. In contrast, "negative density dependence" is when vital rates dampen at high densities. Considering both kinds of density dependence, patches of flowers at both low density and at high density ought to, in theory, receive insufficient numbers of pollen grains to fully set seed, whereas patches of flowers at medium density ought to receive the most visits per capita (Rathcke 1983). The question then becomes, for any particular endangered species, are densities too low, overwhelming, or at adequate medium levels?

Many researchers have studied pollination success among populations of different sizes (Groom 1998, 2001; Steven et al. 2003; Field et al. 2005; Feldman 2006). In contrast, we are concerned with density as it varies within a population and within the field of view of a pollinator (Kunin 1997a,b; Dauber et al. 2010). As pollinators decide where they will forage, they assess not just individual blossoms but aggregates of flowers (Sih & Baltus 1987). Flowers at low-density points may be unable to attract visitors and fail to set seed. Plants growing some distance from the main population may likewise fail to reproduce. The population may be unable to grow outwardly at its margins for lack of pollination (or for other reasons).

*Corresponding author; email: paul.wilson@csun.edu

Pentachaeta lyonii A. Gray (Asteraceae) is an endangered species. The U.S. Fish and Wildlife Service (1999) specifically suggested that reduced pollinator visitation might contribute to its decline. The species is self-incompatible and dependent on generalist pollinators (Fotheringham & Keeley 1998). One visit by either a bee or a bee-fly is about equally effective at causing a flowering head to set seed (Braker & Verhoeven 2000, 2002). The concern is that *P. lyonii* populations, after having declined for other reasons, have entered into a downward cycle in which sparse and/or small populations attract few pollinators and set few seeds (Moroney et al. 2011). Here we tested for declines in visitation rate and in seed production at decreasing point densities as surveyed throughout three natural populations of *P. lyonii*. In 2008 wild plants were studied as they occurred at varying densities throughout the populations. In 2009 we placed plants in pots throughout the populations to measure pollination success.

MATERIALS AND METHODS

Pentachaeta lyonii (Fig. 1) is an annual sunflower that occurs in openings in coastal sage scrub, a vegetation type that is shrinking due to suburban development (Davis et al. 1994; Minnich & Dezzani 1998; Rundel 2007) and due to type-conversion into alien annual prairie (D'Antonio & Vitousek 1992; Keeley 2004). We studied three populations, all near Thousand Oaks, California: EO24 (0325061, 3787047 UTM), EO11 (0329184, 3778060 UTM), and EO27 (0332072, 3778067 UTM). Each population was composed of patches that varied in area and density, ranging from plants isolated from others by distances on the order of 10 m up to a few hundred flowering heads per m^2.

FIGURE 1. *Pentachaeta lyonii.*

Visitation in 2008

In 2008 from March to June, we quantified insect visitation as it related to flowering density. On any given day, we positioned a number of 1 × 1 m quadrats so as to study a range of densities, and we scored density as the number of open flower heads of *P. lyonii* in a quadrat. Quadrats in the three natural populations contained from 2 to 270 heads. The number of visits was counted as the number of times an insect landed on a head and contacted disk florets during a 15-minute census. Censuses were done between 10:00 a.m. and 4:30 p.m. We accumulated n = 135 censuses in 135 quadrats.

Seed Set in 2008

In each quadrat, we tested for pollen limitation of seed set. After the visitor census, two flowering heads were chosen to be similar in size and blooming stage. Each was marked with a numbered tag. One was randomly assigned to be hand-supplemented; the other was open-pollinated. We used a paintbrush to pick up pollen from flowers in the surrounding area and to transfer it to the supplemented head. Approximately two days later, when the ray petals of a tagged head began to wither, we covered the plant with a mesh bag supported by a wooden stake. After the pappus began to protrude from the bracts, about a week later, we collected the fruiting head. We counted the number of developed seeds and the number of undeveloped ovules (either not fertilized or aborted). We were able to evaluate seed set for n = 52 pairs of hand-supplemented and open-pollinated heads. For 24 additional cases one head of a pair could be scored.

Visitation in 2009

In 2009, we performed a phytometer study. Phytometer studies use plants to measure the quality of the environment at each of many points (Clements & Goldsmith 1924; Wilson 1995; Kniskern & Rausher 2006). In our case, each phytometer was a plant in a pot trimmed to present one flower head. When a plant had a head that was about to open, it was trimmed of any other open heads. Phytometers were placed so that the surrounding densities of natural *P. lyonii* varied. Phytometers were left in place to develop until the head was fully open, which took 1 to 3 days. Then, density was measured in one of two ways. (1) For phytometers with no other *P. lyonii* within a surrounding 1 × 1 m quadrat, the average distance to the three nearest blooming *P. lyonii* plants was measured. (2) For phytometers placed such that other *P. lyonii* shared a 1 × 1 m quadrat, density was measured as the number of open heads in the quadrat surrounding the phytometer and including it. The latter measure gives a more discerning index of density when it is seen at the scale of meters. The former measure prevents ties when no other flower heads are within the quadrat. We combined these two measures into a variable that we call "rank density." The phytometer with the lowest rank density was the one with the greatest average distance to its neighbours. The phytometer with the highest rank density was the one that had the most neighbours in its quadrat. This rank ordering of density nearly eliminates ties. We had 129 phytometers positioned outside of quadrats (n = 36 from 18.2 m down to 10 m distant from naturally occurring *P. lyonii*; n = 40 from 10 m down to 5 m distant from other *P. lyonii*; n = 53 from 5 m down to 0.56 m distant from other *P. lyonii*). We had 119 phytometers located inside of quadrats (which varied from having 2 to 455 open heads). Visitor censuses were done from April to July in the three study populations. We censused visitors for 30 minutes per phytometer (i.e. two 15-minute visitor censuses, usually back to back).

Seed Production in 2009

Phytometers were left in place until the corollas began to wither. Then the head was bagged until seeds had time to set, which took 1 to 2 weeks. Mature seeds were counted as in 2008, but undeveloped ovaries were not counted because maturing heads were collected at a less uniform stage than in 2008. Without the number of undeveloped ovules, seed production in 2009 was merely the number of mature seeds without a denominator.

Data Analysis

Three analyses were performed on the 2008 visitation data. First, a Spearman's rank correlation related visitation to number of flower heads in a quadrat, i.e., to density. Second, a logistic regression used log(density) as the predictor variable and whether or not a quadrat was visited as the response variable. Associated with logistic regression, McFadden's ρ^2 quantified the strength of the relationship. Third, for those quadrats that received at least one visit, a model II regression on log-log transformed data was used to test whether the effect of density was allometrically above or below 1, i.e., accelerating or decelerating.

The 2008 seed-set data were squared to improve normality of residuals, so the dependent variable was [seeds/florets]^2. Squaring is a simple instance of a power transformation (Sokal & Rohlf 1995). It resulted in residuals that were more normally distributed than the more commonly used arcsine-square root transformation. For numbers greater than zero, squaring is a monotonic transformation. Some analyses were based on differences, i.e., [seed set in the hand-supplemented head]^2 – [seed set in the open-pollinated head]^2. We transformed before subtraction because subtracting is equivalent to including a random blocking variable in the analysis, and the variates before subtraction are therefore more fundamental (Sokal & Rohlf 1995). The differences were also approximately normally distributed. We started with an analysis of covariance in which the dependent variable was the difference between the squared seed set values and the independent variables were log(density) and population. Next, we used a paired *t*-test comparing hand-supplemented flower heads to open-pollination heads. Pairing focused the analysis on testing for an effect of supplemental hand pollination, blocking out variation among quadrats and populations. Average squared seed set was also analysed. We used a general linear model with three predictor variables: log(visits + 1), density, and population. This model was simplified after log(visits + 1) yielded $P > 0.25$ (the threshold for dropping terms recommended by Quinn & Keough 2002).

For our 2009 visitation analyses, we report only on visits to phytometers. A generalized linear model with a Poisson error structure was used to relate the number of visits during 30 minutes to rank density, simultaneously accounting for which of the three populations the phytometer was placed in. The use of ranks is very similar to the use of non-parametric statistics (Conover & Iman 1981). Ranks are robust protection against uneven distributions, and when ties are few, the use of ranks is nearly as powerful as a parametric test (Quinn & Keough 2002). We transformed only the independent variable and only in analyses for which we are primarily testing the null hypothesis of no relationship (not trying to parameterize a function for data that tightly follow a curve).

For 2009 seed production, 70 fruiting heads from phytometers were scorable. Number of seeds produced was plotted against rank density. A generalized linear model with a Poisson error structure was used, also accounting for the three populations.

Significance testing was performed in SYSTAT 11, except for the generalized linear models involving Poisson distributions, which were performed in SPSS 20.

Results

Over the course of two years, we observed 5,720 insect visits while watching 24,012 flower heads during 15-minute visitor censuses. In other words, the number of visits per head was 0.238, which extrapolates to 7.6 visits per 8 mid-day hours during which any given stigma would have been receptive. The common visitors, in order of abundance, belonged to the families Bombyliidae, Megachilidae, Apidae, Syrphidae, and Melyridae. Particularly common were bee-flies (in the genera *Lepidanthrax, Paravilla, Exoprosopa*) and bees (in the genera *Ashmeadiella, Ceratina, Exomalopsis*). All these animals carried pollen and contacted stigmas. They are generalist foragers that visit many other kinds of flowers.

Visitation in 2008

Number of flower heads in a quadrat significantly and positively predicted number of visits to those flower heads; this correlation included zero-visit censuses ($r_s = 0.259$, $n = 135$, $P < 0.005$). The logistic regression showed that less dense patches were less likely to be visited than more dense patches (deviance $G^2 = 10.59$, $df = 1$, $P = 0.001$), but this effect was very weak (McFadden's $\rho^2 = 0.057$). The model II regression yielded a reduced major axis slope of 1.072, a number that did not significantly differ from 1 ($t = 0.624$, $df = 72$, $P = 0.535$). In other words, as density increased, number of visits increased only as a proportion of number of heads, with no per-capita increase in visitation rate associated with increasing densities.

Seed Set in 2008

The average number of florets in a head was 50, and the average seed set was 72%. In the analysis of covariance of the differences between supplemented and open-pollinated heads, log(density) had no significant effect ($P = 0.348$; Fig. 2), nor did population ($P = 0.401$). In the paired *t*-test, seed set did not differ significantly between hand-supplemented and open-pollinated heads (respectively 70% versus 74% back-transformed from mean of squared proportions 0.493, SE = 0.0259, $n = 65$ versus 0.544, SE = 0.0291, $n = 64$; paired $t = 0.603$, $n = 52$ pairs, $P = 0.549$). In the analysis of average squared seed set, log(visits +1) was not a significant predictor and was dropped ($P = 0.682$). In the simplified model, density did not significantly contribute to predicting seed set ($P = 0.100$); that model accounted for differences among the three populations ($P < 0.001$).

Visitation in 2009

Phytometers measured whether there was a change in pollination success from points 18.2 m from neighbours to quadrats with hundreds of other flowering heads (Fig. 3). We found negative density dependence, i.e., the denser the

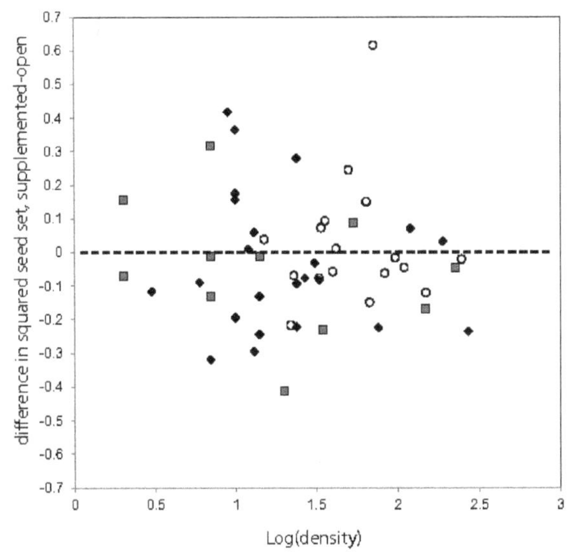

FIGURE 2. Pollen-limitation not found and not dependent on density. The three symbols denote the three study populations (diamonds=EO24; squares=EO11; circles=EO27). The dashed line indicates the null hypothesis for the paired *t*-test.

point, the fewer the visits to the phytometer (Wald χ^2 = 8.68, df = 1, n = 256, P = 0.003); the three populations also varied significantly in visitation rate (Wald χ^2 = 58.89, df = 2, P < 0.001).

Seed Production in 2009

Only 27% of the phytometers used to measure visitation were undamaged and available to be scored for seed production, and only 6 seeds per head were produced on average. Contrary to the visitation results, no seeds were produced on the phytometers that were placed farthest from neighbours, whereas seeds were more often produced on phytometers that had been placed amid other flowering *P. lyonii* (Fig. 4). Inverse density dependence was detected (Wald χ^2 = 125.16, df = 1, n = 70, P < 0.001), with the three populations also varying significantly (Wald χ^2 = 116.35, df = 2, P < 0.001).

DISCUSSION

The populations of *P. lyonii* that we studied received plenty of pollinator visits. Flowers were generally visited repeatedly over the course of a day. The visitors belonged to various species of insects that seem to be effective pollinators and that are generalist floral foragers (Holt 2011). In 2008 visitation increased in proportion to density. In 2009 isolated phytometers received more visits than phytometers placed within a patch.

Consistent with the visitation data, seed set in 2008 was not pollen limited and was not related to density. However, in 2009 the ten most distant phytometers failed to set seed, a finding that is at odds with our other results. One possible explanation is that visitors to the most isolated phytometers may not have been carrying pollen. Unfortunately, our data on seed production in 2009 are weak. The phytometers were water stressed after pollination, and they had to be handled several times. Many phytometers had to be eliminated from the analysis because their fruiting heads were damaged. Thus, the apparent inverse density dependence in seed production in 2009 could be an artifact, or it could be due to the distance the farthest phytometers were from other flowering *P. lyonii*. In theory, there is some distance that is too far for pollinators to be both attracted and carrying conspecific pollen.

Our study relates to topics that are often discussed as the Allee effect, which is when small or sparse populations collapse. A distinction is sometimes made between a "demographic Allee effect"—a decrease in average fitness in small populations—versus a "component Allee effect"—a decrease in individual fitness with decreasing density (Stephens et al. 1999). Not all failures to mate at the individual level translate into decreases in population growth (Ashman et al. 2004; Price et al. 2008), and not all demographic Allee effects among populations can be detected by merely studying variation within populations (Gascoigne et al. 2009). Our results indicate that poor visitation was not causing declines within the sparse areas of the three large populations that we studied. Poor pollination is unlikely to be the reason *P. lyonii* does not fill in the

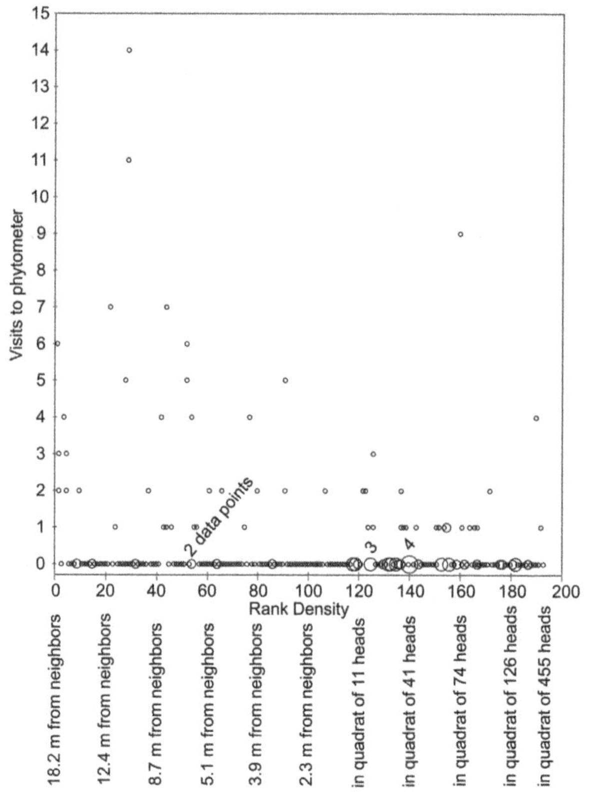

FIGURE 3. Phytometer visits in 2009 scattered against rank density. Bubble size indicates the number of overlapping data points. Density had a negative effect on visitation.

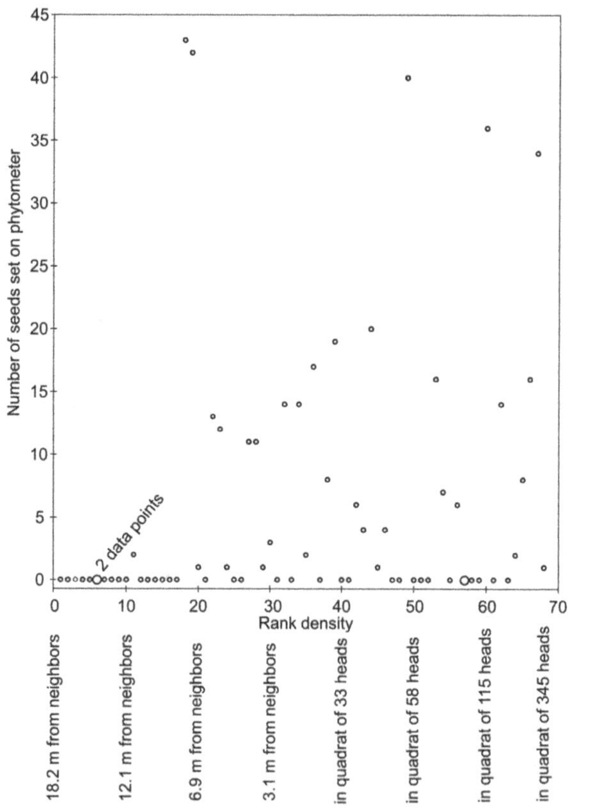

FIGURE 4. Seed production in 2009 scattered against rank density. Bubble size indicates number of overlapping data points. Phytometers placed the farthest from other flowering *P. lyonii* produced zero seeds.

sparse areas between patches and expand outwardly from the margin. Nevertheless, we did not study truly isolated, small populations, which might suffer from a low rate of pollinator visitation.

Experiments with other plants have found considerable pollination of plants isolated by greater distances than we studied. For *Dianella revoluta* R. Br., conspecific pollen deposition declined from 0 to 50 m from a pollen source; however, at greater spatial scales pollen deposition, if anything, increased as arrays were placed 1 to 5 km away from a nature reserve (Duncan et al. 2004). For *Delphinium nuttallianum* Pritz., arrays of 16 emasculated plants isolated by 50 to 400 m received less pollen than control plants in natural patches, but pollen receipt was still far from zero (Schulke & Waser 2001). Given such findings, it is not surprising that we found that pollinators visit flowers only meters from dense patches.

The potential for pollination to depend on density has been studied in other systems, often with results more alarming than ours. In *Senecio integrifolius* (L.) Clairv., a species that has declined due to changes in land use, decreased density was found to reduce reproduction (Widén 1993). In the narrow endemic *Banksia goodii* R. Br., plants in small populations were found to reproduce less than plants in larger populations, and the smallest populations produced no fertile cones at all (Lamont et al. 1993). In *Argyroxiphium sandwicense* DC. subsp. *macrocephalum* Meyrat, a taxon in which most plants flower together in high-bloom years, plants that flowered during low-bloom years had decreases reproductive success (Forsyth 2003).

Low density does not necessarily lead to poor pollinator visitation. Isolated plants of *Echinacea angustifolia* DC. have been shown to be visited more than plants in patches, although that increase did not result in increased seed production (Wagenius & Lyon 2010). Such results mirror our phytometer results with *P. lyonii.* A similar dynamic was found for *Hymenoxys herbacea* (Greene) Cronquist—small patches received increased visitation per flowering head, but because of a decrease in mate availability, pollen limitation was only negligibly density dependent (Campbell & Husband 2007). In our study, the window of intermediate densities within which flowers were well pollinated ranged from plants isolated by about 10 m up to densities of much more than 100 plants per m^2.

To comprehend the limits on population growth of any endangered species, one should consider each segment of the lifecycle (Caughley 1994; Holt 2011). (1) *Pentachaeta lyonii* might suffer from poor seed dispersal with new patches rarely becoming established even in suitable habitat. Dispersal might amount to little more than seeds dropping on the ground next to the parent plant. (2) Seed germination might be problematic. Seeds remain viable for years in the lab, but they might germinate too readily in nature, not waiting until a year when the weather promises to allow *P. lyonii* to do well (Keeley 1995). Seed banks might end up being depleted by years in which many seeds germinate but few plants mature. (3) The growth of plants and the number of flowers produced is negatively impacted by competition from alien plants (Moroney et al. 2011). Extant populations of *P. lyonii* live in compact inorganic soils, often in places where competitors are sparse, yet when raised in pots and given ample water, plants can grow many times larger than individuals found in the field. These various clues suggest that competition is restricting the populations to isolated areas where the soil is too harsh for alien dominance. (4) Since the pollination portion of the lifecycle is working well, efforts to protect *P. lyonii* should address other stages.

An obvious first-priority for conservation is preventing destruction of any more populations of *P. lyonii.* Adding new populations should also be considered. The meta-population dynamics of *P. lyonii* have been more toward the loss of populations than the establishment of new populations. A method should be developed for multiplying seeds in a nursery and then starting new populations in nature from seed. Pucci (2006) attempted to plant one new population from seed. After the first season's bloom, the number of plants at that site declined. Future researchers should use a multiple-regression approach to find factors that will predict successful establishment of new populations.

ACKNOWLEDGEMENTS

B. Kanno and P. Lee-Gabriel helped grow phytometers. J. Hogue helped with insect identifications. Field assistants included P. Miranda, M. Colussi, V. Rifkin, C. Law, and D. Miranda, C. Bowman-Prideaux and L. Zung. T. Karels, and P. Schiffman provided invaluable suggestions. Reviewers, including N. Waser, went beyond the call of duty to iteratively improve the manuscript. Funding was contributed by the Western National Parks Association, a Norris Grant, a Newhall Land and Farming Company Ecology Research Grant, and the California State University Northridge Associated Students.

REFERENCES

Ashman T-L, Knight TM, Steers JA, Amarasekare P, Burd M, Campbell DR, Dudash MR, Johnston MO, Mazer SJ, Mitchell RJ, Morgan MT, Wilson WG (2004) Pollen limitation of plant reproduction: ecological and evolutionary causes and consequences. Ecology 85:2408-2421.

Braker E, Verhoeven L (2000) Relative success of flower visitors in pollination of the endangered Lyon's Pentachaeta (*Pentachaeta lyonii*). Lay report. National Park Service, Thousand Oaks, CA.

Braker E, Verhoeven L (2002) Influence of plant neighborhood on pollinator behavior for the endangered Lyon's Pentachaeta (*Pentachaeta lyonii*). Lay report. National Park Service, Thousand Oaks, CA.

Campbell LG, Husband BC (2007) Small populations are mate-poor but pollinator-rich in a rare, self-incompatible plant, *Hymenoxys herbacea* (Asteraceae). New Phytologist 174:915-925.

Caughley G (1994) Directions in conservation biology. Journal of Animal Ecology 63:215-244.

Clements FE, Goldsmith GW (1924) The phytometer method in ecology: the plant and community as instruments. Carnegie Institution of Washington, publication no. 356, Washington, DC.

Conover WJ, Iman RL. 1981. Rank transformation as a bridge between parametric and nonparametric statistics. American Statistician 35:123-129.

Courchamp F, Berec L, Gascoigne J (2008) Allee effects in ecology and conservation. Oxford University Press, Oxford, UK.

Courchamp F, Clutton-Brock T, Grenfell B (1999) Inverse density dependence and the Allee effect. Trends in Ecology and Evolution 14:405-410.

D'Antonio CM, Vitousek PM (1992) Biological invasions by exotic grasses, the grass/fire cycle, and global change. Annual Review of Ecology and Systematics 23:63-87.

Dauber J, Biesmeijer JC, Gabriel D, Kunin WE, Lamborn E, Meyer B, Nielsen A, Potts SG, Roberts SPM, Sõber V, Settele J, Steffan-Dewenter I, Stout JC, Teder T, Tscheulin T, Vavarelle D, Petanidou T (2010) Effects of patch size and density on flower visitation and seed set of wild plants: a pan-European approach. Journal of Ecology 98:188-19.

Davis FW, Stine PA, Stoms DM (1994) Distribution and conservation status of coastal sage scrub in southwestern California. Journal of Vegetation Science 5:743-756.

Duncan DH, Nicotra AB, Wood JT, Cunningham SA. 2004. Plant isolation reduces outcross pollen receipt in a partially self-compatible herb. Journal of Ecology 92:977-985.

Feldman TS (2006) Pollinator aggregative and functional responses to flower density: does pollinator response to patches of plants accelerate at low-densities? Oikos 115:128-140.

Field DL, Ayre DJ, Whelan RJ (2005) The effect of local plant density on pollinator behavior and the breeding system of *Persoonia bargoensis* (Proteaceae). International Journal of Plant Sciences 166:969-977.

Forsyth SA (2003) Density-dependent seed set in the Haleakala silversword: evidence for an Allee effect. Oecologia 136:551-557.

Fotheringham CJ, Keeley JE (1998) Ecology and distribution of Braunton's milkvetch (*Astragalus brauntonii*) and Lyon's pentachaeta (*Pentachaeta lyonii*). California Department of Fish and Game. Pp. 1-58, Sacramento, CA.

Gascoigne J, Berec L, Gregory S, Courchamp F (2009) Dangerously few liaisons: a review of mate-finding Allee effects. Population Ecology 51:355-372.

Groom MJ (1998) Allee effects limit population viability of an annual plant. American Naturalist 151:487-496.

Groom MJ (2001) Consequences of subpopulation isolation for pollination, herbivory, and population growth in *Clarkia concinna concinna* (Onagraceae). Biological Conservation 100:55-63.

Holt JR (2011) No Allee effects in Lyon's pentachaeta, a federally listed endangered sunflower. M.S. thesis, California State University, Northridge. Northridge, CA. http://hdl.handle.net/10211.2/684

Keeley JE (1995) Seed germination and dormancy of *Pentachaeta lyonii*. Final report to California Department of Fish and Game, Sacramento, CA.

Keeley JE (2004) Invasive plants and fire management in California mediterranean-climate ecosystems. In: Arianoutsou M, Papanastasis VP (ed) *Proceedings 10th MEDECOS Conference*. Mill Press, Rotterdam, Netherlands, pp. 1-10.

Kniskern JM, Rausher MD (2006) Environmental variation mediates the deleterious effects of *Coleosporium ipomoeae* on *Ipomoea purpurea*. Ecology 87:675-685.

Kunin WE (1997a) Population biology and rarity: on the complexity of density dependence in insect-plant interactions. In: Kunin WE, Gaston KJ (eds) The biology of rarity. Chapman & Hall, London, pp. 150-173.

Kunin WE (1997b) Population size and density effects in pollination: pollinator foraging and plant reproductive success in experimental arrays of *Brasica kaber*. Journal of Ecology 85:225-234.

Lamont BB, Klinkhamer PGL, Witkowski ETF (1993) Population fragmentation may reduce fertility to zero in *Banksia goodii*—a demonstration of the Allee effect. Oecologia 94:446-450.

Minnich RA, Dezzani RJ (1998) Historical decline of coastal sage scrub in the Riverside-Perris Plain, California. Western Birds 29:366-391.

Moroney JR, Schiffman PM, Brigham CA (2011) Invasive European annual plants impact a rare endemic sunflower. Madroño 58:69-77.

Price MV, Campbell DR, Waser NM, Brody AK. 2008. Bridging the generation gap in plants: pollination, parental fecundity, and offspring demography. Ecology 89:1596-1604.

Pucci JR (2006) The effects of invasive plants on the endangered sunflower, *Pentachaeta lyonii* Gray. M.S. thesis. California State University, Northridge. Northridge, CA. http://hdl.handle.net/10211.2/844

Quinn G, Keough M (2002) Experimental design and data analysis for biologists. Cambridge University Press, Cambridge.

Rathcke B (1983) Competition and facilitation among plants for pollination. In: Real L (ed) Pollination biology. Academic Press, Orlando, Florida, pp. 305-329.

Rundel PW (2007) Sage scrub. In: Barbour MG, Keeler-Wolf T., Schoenherr AA (eds) *Terrestrial vegetation of California*. University of California Press Berkeley, CA, pp. 208-228.

Schulke B, Waser NM (2001) Long-distance pollinator flights and pollen dispersal between populations of *Delphinium nuttallianum*. Oecologia 127:239-245.

Sih A, Baltus M-S (1987) Patch size, pollinator behavior, and pollinator limitation in catnip. Ecology 68:1679-1690.

Sokal RR, Rohlf FJ (1995) Biometry, 3rd ed. W. H. Freeman and Company, New York.

Steven JC, Rooney TP, Boyle OD, Waller DM (2003) Density-dependent pollinator visitation and self-incompatibility in upper Great Lakes populations of *Trillium grandiflorum*. Journal of the Torrey Botanical Society 130:23-29.

Stephens PA, Sutherland WJ, Freckleton RP (1999) What is the Allee effect? Oikos 87:185-190.

U.S. Fish and Wildlife Service (1999) Recovery plan for six plants from the mountains surrounding the Los Angeles basin. Portland, Oregon.

Wagenius S, Lyon SP (2010) Reproduction of *Echinacea angustifolia* in fragmented prairie is pollen-limited but not pollinator-limited. Ecology 91:733-742.

Widén B (1993) Demographic and genetic effects on reproduction as related to population size in a rare, perennial herb, *Senecio integrifolius* (Asteraceae). Biological Journal of the Linnean Society 50:179-195.

Wilson P (1995) Variation in the intensity of pollination in *Drosera tracyi*: selection is strongest when resources are intermediate. Evolutionary Ecology 9:382-396.

3

Do Native Bees Have the Potential to Promote Interspecific Pollination in Introduced *Impatiens* Species?

Paola Ugoletti*[1, 2], Darren Reidy[2, 3], Michael B. Jones[1, 2], Jane C. Stout[1, 2]

[1]*Trinity Centre for Biodiversity Research, Trinity College Dublin, Dublin 2, Ireland*
[2]*School of Natural Sciences, Botany Building, Trinity College Dublin, Dublin 2, Ireland*
[3]*Killarney National Park Education Centre, Knockreer House, Killarney, Ireland*

Abstract— Artificial introductions of non-native species by humans can remove geographical barriers between species. In the absence of reproductive barriers, closely related introduced taxa may be able to hybridize, resulting in the formation of novel genotypes. These may be more suited to the new environment than either of their parent taxa, and have the potential to become more invasive. We investigated potential reproductive barriers between the non-native invasive *Impatiens glandulifera* and its less aggressive non-native congener *I. balfourii.* We examined behaviour of pollinators, including their foraging preferences and whether they switched from one species to the other. Moreover, conspecific and interspecific artificial crosses were performed between these species and seed production and the germination success of hybrid seeds were assessed.

Both *I. balfourii* and *I. glandulifera* had relatively long flowering periods which mostly overlapped. Insect visitors were observed to switch from flowers of one species to the other during a single foraging bout, confirming that natural pollen transfer between species is possible. Artificial interspecific pollination resulted in the production of seeds, suggesting the presence of incomplete reproductive barriers between the two species. However, hybrid seeds mostly failed to germinate making the production of invasive hybrids unlikely. Nevertheless, because of the prolific reproductive output of these species we would not exclude the possibility that some viable hybrid seeds could be formed and become established. Further investigation of the fitness of hybrid seeds is required as hybridisation may allow *Impatiens* to exploit different ecological niches.

Keywords: Bombus, *heterospecific pollen transfer, invasive alien species, non-native species, reproductive barriers*

Introduction

One important indirect consequence of the increasing number of introduced plant species is the possible hybridisation between closely related taxa (Abbott 1992; Bleeker 2007). While natural hybridisation can be responsible for a greater genetic diversity, hybridisation between anthropogenically introduced species may lead to increased invasiveness for some introduced taxa, since it may result in genotypes more adapted to or fitter in novel habitats (Anderson and Stebbins 1954; Ellstrand and Schierenbeck 2000). Although many interspecific hybrids are sterile, more than half of all hybrid plants between non-native species in the UK were found to be fertile (Abbott 1992). Fertile hybrids may become invasive because they are more versatile than parents and/or able to tolerate a wider range of environments, including habitats which may not have been colonised by the species originally introduced. Alternatively, fertile hybrids may backcross with parental taxa with subsequent introgression, which could result in increased invasiveness in the parental species (e.g. Milne and Abbott 2000). For example, fertile allotetraploid hybrids resulting from interbreeding between non-native *Tragopogon* species have become invasive in the USA (Novak et al. 1991). Also Tiébré et al. (2007) concluded that sexual reproduction among non-native *Fallopia* species can increase the invasive capacity of the genus. Furthermore, Ellenstrad and Schierenbeck (2000) reported 28 examples across 12 families of hybrid plants that had become invasive, mostly in Europe and North America.

In order for hybridisation to occur between non-native plant species, they must overlap in their flowering phenology, and their pollination mechanisms and breeding systems must allow interspecific pollen transfer. In addition, animal-pollinated species require appropriate flower visitors who include both taxa in their diet and who switch between taxa during a single foraging bout, thus mediating interspecific pollen transfer. Most non-native plants are well served by generalist native pollinators (Valentine 1978). Some generalists such as bumblebees (*Bombus* spp.) visit a range of taxa during a single foraging bout (i.e. are not flower constant; Heinrich 1976); particularly if those taxa are morphologically similar (Goulson 2000) as this increases their foraging efficiency. Hence, co-flowering, closely-related, non-native plant species may well have pollen transferred between them as a result of pollinator foraging strategies. However, the transfer of pollen is not sufficient for hybridisation in itself. Interspecific pollen must be able to germinate on the stigma and grow a pollen tube through the style, fertilize the ovule and form a functional seed, and the

*Corresponding author: ugolettp@tcd.ie

seed must be able to germinate and grow into a new plant (Wolf et al. 2001).

The genus *Impatiens* contains several biotically pollinated species which vary in their invasive capacity throughout their introduced range (Perrins et al. 1993). A great number of hybrids have been bred for ornamental purposes. Besides the many ornamental *Impatiens,* natural hybridisations are reported to occur in wild populations of *Impatiens* species, both in tropical (Grey-Wilson 1980; Tsukaya 2004) and temperate regions (Zika 2006). Although frequently co-occurring native species in north-east USA do not appear to hybridize (*I. capensis* and *I. pallida*; Randall and Hilu 1990; Tabak and von Wettberg 2008), hybridisation between *I. capensis* and the closely related introduced species, *I. noli-tangere* (native to Europe) is possible (Tabak and von Wettberg 2008). Ornduff (1967) reported that *I. capensis* can hybridize with native *I. ecalcarata* in the north-west USA. However, little information is available on boundaries between other temperate *Impatiens* species (Tabak and von Wettberg 2008), particularly regarding those which are problematic invaders.

The most common *Impatiens* species in Ireland, Britain and continental Europe is the aggressive invader *I. glandulifera* (DAISIE 2009; NOBANIS 2010), which was first introduced from the Himalayas to Europe (to Kew Gardens) in 1839 (Britten 1900, Beerling and Perrins 1993). Originating from the same native range, *I. balfourii* was introduced to the south of France about 50 years later (Beerling and Perrins 1993; Adamowski 2009) and since then has spread to southern and central Europe (Schmitz and Dericks 2010). *Impatiens glandulifera* is currently much more invasive than *I. balfourii.* Even though it appears that *I. glandulifera* requires higher soil moisture (Beerling and Perrins 1993; Schmitz and Dericks 2010), both species have similar ecological requirements and have been observed to grow together in the same habitat (Fig. 1). It is possible that *I. balfourii* is only in an early stage of invasion and may become more established and widespread in the future (Adamowski 2009; Schmitz and Dericks 2010). This might result in the frequent occurrence of this species in habitats already invaded by *I. glandulifera.* In its introduced range, *I. glandulifera* has been reported to flower from July to October (Beerling and Perrins 1993) and although there is little information available on the phenology of *I. balfourii* in its introduced range, this species flowers from the end of July to August in its native range (Adamowski 2009).

Unlike the perennial *Impatiens* characteristic of the tropical regions which reproduce vegetatively, the annual *Impatiens* species, including *I. glandulifera* and *I. balfourii,* completely rely on sexual reproduction (Grey-Wilson 1980; Beerling and Perrins 1993). The understanding of the hybridisation potential of these species is therefore important to evaluate their possible invasion dynamics (Ellstrand and Schierenbeck, 2000, Richardson and Pysek, 2006).

Impatiens glandulifera produces both flowers and inflorescences that are slightly larger in dimension than *I. balfourii* (Beerling and Perrins 1993; Adamowski 2009) (Fig. 1). However, the two species present a similar flower morphology and reproductive system. The male phase precedes the female phase with the stigma becoming visible when the androecium has completely dehisced and fallen off (Wilson and Thomson 1991; 1996). *Impatiens glandulifera*'s flowers have traditionally been described as strongly protandrous (Valentine 1978; Bell et al. 1984; Titze 2000). However, Vervoort et al. (2011) reported that *I. glandulifera*'s stigma is already receptive at bud stage but flower morphology prevents self pollination and pollinator activity is needed for this species to maximise fruit set. *Impatiens glandulifera* has been reported to attract great numbers of insect visitors in its introduced range, especially generalist *Bombus* species, attracted by its plentiful nectar production (Lopezaraiza-Mikel et al. 2007; Nienhuis and Stout 2009). As far as we are aware, little is known about the insect visitors or the floral rewards of *I. balfourii* (but see Elias and Gelband 1977; and Vervoort et al. 2011).

FIG. 1. *Impatiens glandulifera* (the larger species) and *I. balfourii* growing and flowering together in Saint Christophe en Oisans (Rhone-Alpes, Isere, France), August 2010. Photograph by P. Ugoletti.

Impatiens glandulifera is not reported to hybridize with other species (Beerling and Perrins 1993). In "Flora of Pakistan" the possibility of hybridisation between *I. balfourii* and *I. bicolour* Royle is hypothesized, but the potential for crosses between *I. balfourii* and *I. glandulifera* is not considered (Nasir 1980).

By focussing on these two species in Ireland, we examined the potential for hybridisation and for production of hybrid offspring. We hypothesised that: 1) insect visitor guilds overlap between *I. glandulifera* and *I. balfourii* allowing natural pollen transfer between the two species to occur; 2) pollen from *I. balfourii* and *I. glandulifera* are capable of germinating on and penetrating through stigmas of the other species, and producing hybrid seeds; and 3) that hybrid seeds are viable and can germinate.

MATERIALS AND METHODS

Insect behaviour

Seeds of *I. glandulifera* and *I. balfourii* were obtained from several Botanic Gardens across Europe (Appendix 1). Approximately equal numbers of seeds from each source were mixed within species and stratified at 4°C for about 50 days to break dormancy (Mumford 1988). Seeds were then germinated on moist filter paper in 90 mm diameter Petri dishes in a growth chamber (Ugoletti et al. 2011). Seedlings were kept inside in a greenhouse at Trinity College Botanic Garden, Dublin, for an initial period to avoid any frost damage. Sixty plants each of *I. glandulifera* and *I. balfourii* were randomly chosen and potted in 20-litre pots (one seedling per pot). In early June 2008, the plants from the two species were randomly interspersed and positioned approximately 1 m apart from each other in an open, unshaded area in Trinity College Botanic Garden. During July, 56 individual bees of three species (*Apis mellifera, Bombus hortorum* and *Bombus pascuorum*) were observed for approximately 7 minutes each. The number of visits made by each individual bee to flowers of each *Impatiens* species and the number of switches between *Impatiens* species were recorded during the 7-minute period. Observations were made opportunistically, and so the number of bees of each species observed depended on the abundance of those bees at the site on days with appropriate weather conditions for insect foraging (i.e. no or light wind, temperatures >15°C).

Experimental crosses

In 2009, 15 plants of *I. glandulifera* and 15 plants of *I. balfourii* were grown as described above and again placed outside in the Trinity College Botanic Gardens. On each plant, 30 flower buds were bagged with bridal veil material (with a mesh < 1 mm) to exclude pollinators. If more than 30 flowers were produced per plant, excess flowers were removed. Each plant was randomly allocated to one of three treatments (5 plants per treatment): interspecific cross pollination (treatments B×G or G×B); conspecific cross pollination (treatments B×B or G×G); and control flowers receiving no outcross pollen (treatments B0 and G0; included to confirm absence of autogamous selfing and the effectiveness of the pollen exclusion bags). When, at the end of the male phase of flowering, the androecium fell off exposing the gynoecium (Wilson and Thomson 1991), flowers were hand pollinated by removing the whole anthers from flowers of pollen donors and applying the pollen directly to receptive stigmas of test flowers (Nienhuis and Stout 2009). Hand pollinated flowers were marked by applying a strip of black adhesive tape around the flower stem and re-bagged. Once mature, fruits were collected and the number of seeds per fruit was counted. Seeds were stored dry in paper bags at 20°C for a period of 8 months, after which they received a stratification of 20 days at 4°C to break dormancy and they were subsequently placed to germinate at 20°C (Ugoletti et al. 2011). Germination was compared among four treatments (B×G, G×B, B×B and G×G) since the unpollinated flowers (B0 and G0) produced only very low numbers of seeds. For each treatment, different numbers of seeds were germinated depending on seed availability. Seed production was higher when flowers had been pollinated with conspecific pollen. For the conspecific treatments B×B and G×G, 300 and 273 seeds were germinated, while for the heterospecific G×B and B×G treatments, only 91 and 84 seeds were available for germination tests. Seed germination was followed for eight weeks. However, no seeds germinated after the fifth week.

Data analysis

For all bee species, the numbers of visits by each individual bee to flowers of *I. glandulifera* or *I. balfourii* were compared between plants using a non-parametric Wilcoxon test (Z) for two related samples (since the same individual bees were visiting both plants). For the most abundant bee species (*B. pascuorum*), numbers of switches of each bee from *I. glandulifera* to *I. balfourii* and from *I. balfourii* to *I. glandulifera*, were compared with Wilcoxon test (Z) for two related samples. For *B. pascuorum*, also the number of consecutive visits to *I. glandulifera* and the number of consecutive visits to *I. balfourii* were compared with Mann-Whitney test (U) for two unrelated samples.

Data on the number of seeds per fruit consisted of integer counts; therefore differences between treatments were tested with a generalized linear model (GLM) with Poisson distribution and log-link function (Crawley 1993). Since data showed overdispersion an overdispersed Poisson model was fitted, including the inverse of Pearson chi-square/df as scale weight, which increased the standard error and made the test more conservative. Treatment (B×B, B×G, B0, G×G, G×B, and G0) was included in the model as categorical predictor (factor). To investigate differences in the fruit set (proportion of fruits containing seeds) according to the different types of cross, a GLM with binomial distribution and logit-link function was used. The type of cross (B×B, B×G, B0, G×G, G×B, and G0) was included in the model as categorical predictor (factor).

Germination rates after 8 weeks were calculated for each treatment and compared with a binomial logistic GLM with logit-link function with the type of cross (B×B, B×G, G×G and G×B) as categorical predictor.

Pair-wise comparisons of number of seeds per fruit, fruit set and germination percentage between each pair of crosses were performed using sequential Bonferroni *post-hoc* test. All the data analyses were run using SPSS 16 (SPSS, Inc. Chicago IL).

RESULTS

Bee behaviour

Visitation patterns were recorded for eight *Apis mellifera*, six *Bombus hortorum* and 42 *Bombus pascuorum* workers, the number of bees of each species reflecting its

	Visits to flowers				Switches between plants				Consecutive visits		
	Total	To IG	To IB	Wilcoxon test	Total	From IG to IB	From IB to IG	Wilcoxon test	To IG	To IB	MWU test
A. mellifera	11.88±1.90	11.25±2.04	0.63±0.63	Z=2.37, P=0.018	0.13±0.13	0.13±0.13	0		11.25±2.04	2	
B. hortorum	22.00±7.03	17.50±4.91	4.50±4.31	Z=1.78, P =0.075	0.83±0.54	0.50±0.34	0.33±0.21		13.13±4.25	9.00±7.50	
B. pascuorum	15.83±1.31	10.02±1.03	5.81±1.14	Z=2.31, P =0.021	1.76±0.31	0.79±0.16	0.98±0.16	Z=34, P =0.0455	6.34±0.66	4.75±0.50	U=1861.5, P =0.194

TABLE I. Number of visits to flowers of *I. glandulifera* (IG) and I. *balfourii*, (IB) average number of switches between different plant species, and average number of consecutive visits to *I. glandulifera* and to *I. balfourii* recorded for the three bee species during single foraging bouts. Data are presented as mean ± SE.

relative abundance at the site. During a 7-minute foraging bout, all three bee species visited more *I. glandulifera* flowers than *I. balfourii*, but this difference was only significant for *A. mellifera* and *B. pascuorum* (Tab. I)

Individuals of all bee species switched between the two *Impatiens* species. The honeybees (*A. mellifera*) were observed foraging mainly on *I. glandulifera* flowers and were seen switching only from *I. glandulifera* to *I. balfourii*. The other two bumblebee species switched either from *I. glandulifera* to *I. balfourii* or from *I. balfourii* to *I. glandulifera*. The most abundant bee species, *B. pascuorum*, showed a preference regarding the direction of the switch; it was observed to switch more often from *I. balfourii* to *I. glandulifera* than vice versa (Tab. I). This bumble bee species further showed the lowest (though not significantly) average number of consecutive visits to the same plant species than the other two bee species (Tab. I).

Fruit set and seed production

Most fruits from flowers which did not receive any pollen did not develop (Fig. 2A), and those that did reach maturity contained very low numbers of seeds per fruit for both species (Fig. 2B). This confirms a low autogamous self-pollination for both species and the effectiveness of bags at excluding pollen from other sources. No differences between *I. glandulifera* and *I. balfourii* were found in fruit set in

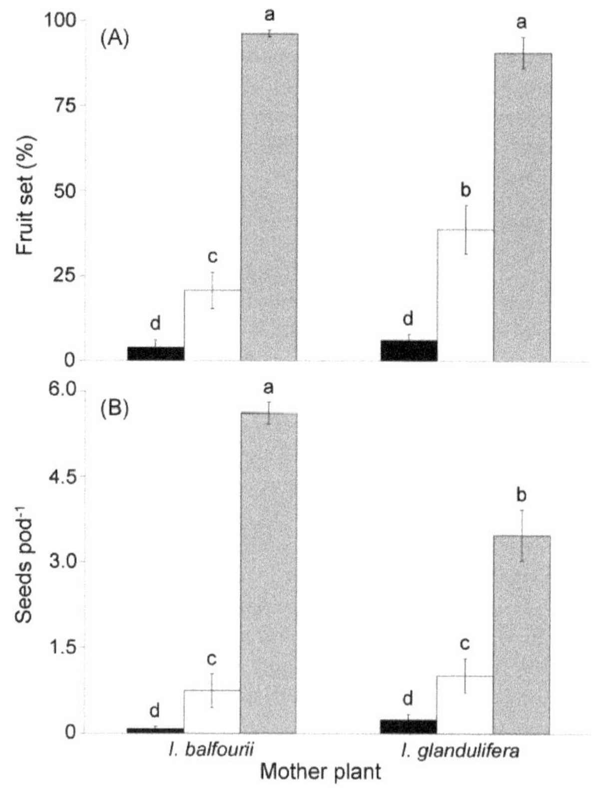

FIG. 2. Mean (± SE) fruit (A) and seed (B) production following pollen exclusion (black bars, treatments B0 and G0), interspecific crosses (white bars, treatments B×G and G×B) and conspecific crosses (grey bars, treatments B×B and G×G). Letters above columns in the same graph indicate significant differences among types of cross (Sequential Bonferroni test, $P < 0.05$).

flowers pollinated with conspecific pollen (mean difference = -0.06 ± 0.03, P = 0.063), and conspecific crosses produced higher fruit set than interspecific crosses (-0.58 ± 0.04 to 0.76 ± 0.04, P < 0.001, Fig. 2A). Differences in fruit set were significant between the two interspecific cross types: *Impatiens glandulifera* flowers pollinated with *I. balfourii* pollen had higher fruit set than *I. balfourii* flowers pollinated with *I. glandulifera* pollen (0.18 ± 0.05, P = 0.001, Fig. 2A).

Fruits resulting from conspecific crosses contained significantly more seeds than those from interspecific crosses (2.46 ± 0.30 to 4.87 ± 0.36, P < 0.001). Conspecific crosses resulted in higher seed numbers for *I. balfourii* than *I. glandulifera* (2.15 ± 0.43, P < 0.001). There was no difference in seed numbers between the two interspecific cross types (B×G and G×B) (-0.27 ± 0.19, P= 0.155, Fig. 2B).

Seed germination

None of the seeds produced from *I. balfourii* flowers pollinated with *I. glandulifera* pollen (B×G) germinated. Hybrid seeds with *I. glandulifera* as mother and *I. balfourii* as the pollen donor showed negligible germination (1.1% ± 1.1, N = 91, Fig. 3). In the group of control seeds, produced by pollinating flowers with conspecific pollen, germination reached 28.3% ± 7.5 (N = 273) for BxB and 14.1% ± 6.2 (N = 300) for G×G (Fig. 3).

DISCUSSION

The reproductive biology of *I. glandulifera* has been extensively studied because of its widespread invasive capacity (Stary and Tkalcu 1998; Titze 2000; Nienhuis and Stout 2009). Some attention has been given to *I. balfourii*, which has recently been suggested as a potential invader (Schmitz and Dericks 2010). However, so far no study has investigated the potential for sexual reproduction between these two species in their introduced range.

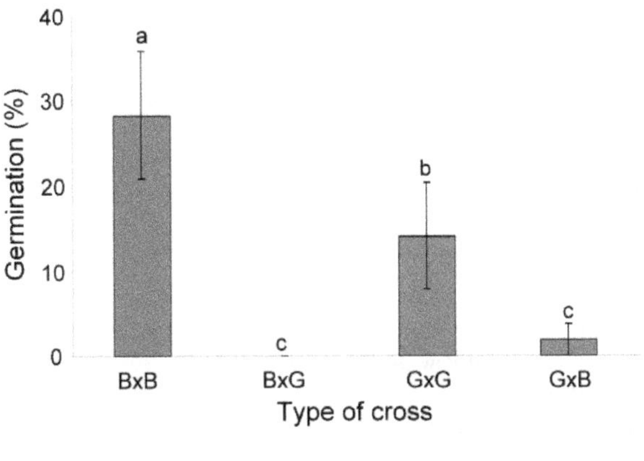

FIG. 3. Mean percentage of germination (± SE) for the conspecific (B×B and G×G) and interspecific (B×G and G×B) crosses. Different letters above columns indicate significant differences among species (Sequential Bonferroni test, P< 0.05).

Impatiens balfourii is reported as established in a few countries in the south of Europe but not in Ireland (DAISIE 2009). However, during our experiments this species was able to grow, flower and produce seeds under local climate, suggesting that it might also become established in Ireland. We observed a longer flowering period for *I. balfourii* (from mid June to October) than has been reported for this species in its native range (end of July to August, Adamowski 2009). Its flowering period overlapped with that of *I. glandulifera*, which began flowering only slightly later in our common garden experiment. Perrins et al. (1993) also reported *I. glandulifera* flowering from the end of June to October. For *I. balfourii* they found that, in England, plants of this species started flowering more than a month later and continued flowering until they were killed by frost (Perrins et al. 1993). Our experiment took place in 2008, when mean air temperature and sunshine totals for May were above normal. As a consequence, *I. balfourii* might have flowered earlier than in years with a cooler spring. There is no information available on *I. balfourii*'s flowering phenology in its introduced range in the south of Europe.

Bee visitation

Native honey- and bumblebees visited both species in our study. *Impatiens glandulifera* seemed to outcompete *I. balfourii* for pollinators, which could promote its reproductive success and invasion capacity. Two of the three bee species visited significantly more *I. glandulifera* flowers, perhaps because of the large nectar and pollen reward of this species (Nienhuis et al., 2009) and/or because of other flower characteristics (e.g. larger flower size, brighter colour and possibly stronger odour). No study has been done to assess *I. balfourii*'s floral rewards. *Apis mellifera* seemed to express fidelity to *I. glandulifera*, but due to the weather conditions during observations, and possibly because of the suburban location of the study site, not many honeybees were observed. However, especially the two bumblebee species were frequently observed to switch between plant species during a single foraging bout. Since pollen placement on the bees is identical for both *Impatiens*, there is a high probability that a switch between species by a foraging bee results in interspecific pollen transfer. *Bombus pascuorum*'s observed preference to switch from *I. balfourii* to *I. glandulifera* makes it more likely that hybrid seeds will be formed when *I. glandulifera* is the mother plant.

Fruit set and number of seeds per fruit

A previous comparison of reproductive output showed no significant difference in natural levels of fruit set and seed production per plant between *I. glandulifera* and *I. balfourii* (Ugoletti et al., 2011). Perrins et al. (1993) found that seed set was approximately double in *I. glandulifera* compared with *I. balfourii* (6 seeds pod^{-1} in *I. glandulifera* and less than 3 seeds pod^{-1} in *I. balfourii*). Conversely, in our current study, we found that conspecific crosses resulted in more seeds per pod in *I. balfourii* than in *I. glandulifera*. Knowing *I.balfourii*'s ovule number would give a better insight into this species' reproductive traits. Vervoort et al. (2011) reported an ovule number of 6.8 for *I. glandulifera*, but the authors did not include *I. balfourii* in their study. However, besides ovule number, other factors (e.g. environmental stress) could affect the fertilization or post-fertilization abortion of an ovule. *Impatiens glandulifera* is a larger, more moisture-loving plant, which might have suffered more than *I. balfourii* from experimental conditions consequently reducing seed set.

Following heterospecific crosses, we found that fruit formation was higher when *I. glandulifera* was the maternal plant. This could be due to mechanical or chemical characteristics of the stigma affecting the degree of adhesion and germination of interspecific pollen. Seed production following interspecific pollen deposition was lower than after conspecific crosses, suggesting that interspecific pollen deposition would reduce the reproductive success/output of plants. This has been shown for other plant species (e.g. Levin et al. 1996). Nagamitsu et al (2006) hypothesised that hybrid seeds could be produced at the expense of conspecific ones. Consistently, Randall and Hilu (1990) found that when interspecific crosses were made between *I. capensis* and *I. pallida,* pollen mixtures of the two species reduced fruit set, which was greater in *I. capensis* than in *I. pallida.* Improper pollen transfer has been indicated as a primary competition mechanism also for other species that rely on generalist pollinators; for instance, *Claytonia virginica* and *Stellaria pubera* (Campbell and Motten 1985). The negative effects of improper pollen transfer can even lead to the exclusion of one of the two species, representing an important selective force (Waser 1978). Nevertheless, though large quantities of *I. glandulifera* pollen were detected on two co-flowering native species, their seed set was not reduced (Cawoy et al. 2012). We observed that conspecific flowers were frequently visited consecutively, resulting in conspecific pollination. So far, there is no evidence that the co-occurrence of *I. glandulifera* and *I. balfourii* would reduce (or even exclude) one or both species through improper pollen transfer. However, our hand pollination experiments showed that interspecific pollen can adhere and germinate on the other species' stigmas and progress through the style to reach the ovary to produce hybrid seeds.

Seeds produced from interspecific crosses mostly failed to germinate, suggesting the presence of post-fertilization reproductive barriers. Hybrid seeds were only able to germinate when *I. glandulifera* was the mother plant but the germination rate was very low (1.1%). Germination rates of seeds produced by conspecific crosses were 3-4 times lower than previously observed for these two species (Ugoletti et al. 2009; 2011). This could be due to collection of seeds before fruits were fully ripe, which was done to avoid seed loss when fruits explode. Consequently, we may have underestimated the potential for hybrid seed germination. Considering the total number of *I. glandulifera* flowers for each treatment (150) and the average number of ovules in this species (6.8, Vervoort et al., 2011), we can estimate that per treatment approximately 1020 ovules could have been fertilized. According to our germination rate, the probability of a seedling originating from a conspecific cross is approximately 0.035 while the probability of a new individual originating from the interspecific crosses is only approximately 0.001. Considering this ratio of conspecific to hybrid seedlings of 35:1, but given the vast number of *I. glandulifera* plants in its introduced range, we cannot exclude the possibility that hybrids could establish.

Conclusion

Natural pollen transfer between *I. glandulifera* and *I. balfourii* is possible due to overlapping flowering periods and to the presence of generalist pollinators that include both species in their diet and switch between species during single foraging bouts.

Seeds developed from interspecific crosses. Thus, the formation and establishment of hybrids with *I. glandulifera* as mother plant is possible. However, hybrid seeds mostly failed to germinate making it unlikely that hybridisation between these two taxa will result in a more aggressive invasive hybrid taxon. The probability that seedlings originate from hybrid crosses is much lower than the chance that seedlings originate from conspecific crosses. It is more likely that interspecific pollination could reduce conspecific seed production in either species through improper pollen transfer. From a conservation point of view, it would be worth to further investigate the possibility that the coexistence of these two introduced species could limit each other's spread.

Acknowledgements

We are grateful to the Natural History Museum of Denmark, Botanical Garden and Museum, University of Copenhagen; the Botanical Garden University of Helsinki, Finland; the Jardin Botanique National de Belgique, Meise, Belgium; MNHS Department des Jardins Botaniques et Voologiques, Jardin des Plantes, Arboretum National de Chevreloup, Paris; the Hortus Botanicus Ljubljana, Slovenia; Cambridge University Botanic Garden and Dublin Botanic Garden for providing the seeds. Thanks also to Elizabeth Bird, Mike McCann and the late Cathy Macken for plant maintenance at the Trinity College Botanic Gardens and to Caroline Nienhuis for the helpful suggestions and Muriel Thomasset for help with potting. We thank three anonymous referees for comments on the manuscript. We are grateful to the Irish Research Council for Science, Engineering & Technology (IRCSET) for supporting this study via an Embark Postgraduate Studentship held by P.U. [grant number: 05_EDIV_FP095_PARALISE].

Appendix

Additional supporting information may be found in the online version of this article:

Appendix I. Seeds providers of the *Impatiens* plants used for the experiment.

References

Abbott RJ (1992) Plant invasions, interspecific hybridisation and the evolution of new plant taxa. Trends in Ecology & Evolution 7:401-405.

Adamowski W (2009) *Impatiens balfourii* as an emerging invader in Europe. In: Pyšek P, Pergl J (eds) Biological Invasions: Towards a Synthesis, Proceedings of the 5th Neobiota conference, 23rd-26th September 2008, Prague, Czech Republic, pp 183-194.

Anderson E, Stebbins GL (1954) Hybridization as an evolutionary stimulus. Evolution 8:378-388.

Beerling DJ, Perrins JM (1993) *Impatiens glandulifera* Royle (*Impatiens roylei* Walp). Journal of Ecology 81:367-382.

Bell G, Lefebvre L, Giraldeau L-A, Weary D (1984) Partial preference of insects for male flowers of an annual herb. Oecologia 64:287-294.

Bleeker W (2007) Interspecific hybridization in *Rorippa* (Brassicaceae): patterns and processes. Systematics and Biodiversity 5:311-319.

Britten J (1900) *Impatien roylei* in England. Journal of Ecology, 38:50-51.

Campbell DR, Motten AF (1985) The mechanism of competition for pollination between 2 forest herbs. Ecology 66:554-563.

Cawoy V, Jonard M, Mayer C, Jacquemart A-L (2012) Do abundance and proximity of the alien *Impatiens glandulifera* affect pollination and reproductive success of two sympatric co-flowering native species? Journal of Pollination Ecology 10:130–139.

Crawley MJ (1993) GLIM for Ecologists. Blackwell, Oxford.

DAISIE (2009) Delivering Alien Invasive Species Inventions for Europe - European invasive alien species gateway. URL: http://www.europe-aliens.org (last accessed on 20 January 2009).

Elias TS, Gelband H (1977) Morphology, anatomy, and relationship of extra-floral nectaries and hydathodes in 2 species of *Impatiens* (Balsaminaceae). Botanical Gazette 138: 206-212.

Ellstrand NC, Schierenbeck KA (2000) Hybridization as a stimulus for the evolution of invasiveness in plants? Proceedings of the National Academy of Sciences of the United States of America 97:7043-7050.

Goulson D (2000) Are insects flower constant because they use search images to find flowers? Oikos 88:547-552.

Grey-Wilson C (1980) Hybridization in African *Impatiens*: Studies in Balsaminaceae: II. Kew Bulletin 34:689-722.

Heinrich B (1976) The foraging specializations of individual bumblebees. Ecological Monographs 46:105.

Levin DA, Ortega FJ, Jansen RK (1996) Hybridization and the extinction of rare plant species. Conservation Biology 10:10-16.

Lopezaraiza-Mikel ME, Hayes RB, Whalley MR, Memmott J (2007) The impact of an alien plant on a native plant-pollinator network: an experimental approach. Ecology Letters 10:539-550.

Milne RI, Abbott RJ (2000) Origin and evolution of invasive naturalized material of *Rhododendron ponticum* L. in the British Isles. Molecular Ecology 9:541-556.

Mumford PM (1988) Alleviation and induction of dormancy by temperature in *Impatiens glandulifera* Royle. New Phytologist 109:107-110.

Nagamitsu T, Kawahara T, Kanazashi A (2006) Pollen-limited production of viable seeds in an endemic dwarf birch, *Betula apoiensis*, and incomplete reproductive barriers to a sympatric congener, *B. ermanii*. Biological Conservation 129:91-99.

Nasir YJ (1980) Balsaminaceae. In: Nasir & Ali SI (eds) Flora of Pakistan. Agricultural Research Counsils, Islamabad, pp 1-17.

Nienhuis CM, Dietzsch AC, Stout JC (2009) The impacts of an invasive alien plant and its removal on native bees. Apidologie 40: 450-463.

Nienhuis CM, Stout JC (2009) Effectiveness of native bumblebees as pollinators of the alien invasive plant *Impatiens glandulifera* (Balsaminaceae) in Ireland. Journal of Pollination Ecology 1:1-11.

NOBANIS (2010) European network on invasive alien species - gateway to information on invasive alien species in North and Central Europe. URL: http://www.nobanis.org/default.asp.

Novak SJ, Soltis DE, Soltis PS (1991) Ownbey's Tragopogons: 40 Years Later. American Journal of Botany 78:1586.

Ornduff R (1967) Hybridization and regional variation in Pacific North-western *Impatiens* (Balsaminaceae). Brittonia 19:122-128.

Perrins J, Fitter A, Williamson M (1993) Population biology and rates of invasion of three introduced *Impatiens* species in the British Isles. Journal of Biogeography 20:33-44.

Randall JL, Hilu KW (1990) Interference through improper pollen transfer in mixed stands of *Impatiens capensis* and *I. pallida* (Balsaminaceae). American Journal of Botany 77:939-944.

Richardson DM, Pysek P (2006) Plant invasions: merging the concepts of species invasiveness and community invasibility. Progress in Physical Geography 30: 409-431.

Schmitz U, Dericks G (2010) Spread of alien invasive *Impatiens balfourii* in Europe and its temperature, light and soil moisture demands. Flora - Morphology, Distribution, Functional Ecology of Plants 205:772-776.

Stary P, Tkalcu B (1998) Bumble-bees (Hym., Bombidae) associated with the expansive touch-me-not, *Impatiens glandulifera* in wetland biocorridors. Anzeiger für Schädlingskunde Pflanzenschutz Umweltschutz 71:85-87.

Tabak NM, von Wettberg E (2008) Native and introduced jewelweeds of the Northeast. Northeastern Naturalist 15:159-176.

Tiébré MS, Vanderhoeven S, Saad L, Mahy G (2007) Hybridization and sexual reproduction in the invasive alien *Fallopia* (Polygonaceae) complex in Belgium. Annals of Botany 99:193-203.

Titze A (2000) The efficiency of insect pollination of the neophyte *Impatiens glandulifera* (Balsaminaceae). Nordic Journal of Botany 20:33-42.

Tsukaya H (2004) Gene flow between *Impatiens radicans* and *I. javensis* (Balsaminaceae) in Gunung Pangrango, Central Java, Indonesia. American Journal of Botany 91:2119-2123.

Ugoletti P, Jones MB, Stout JC (2009) A comparison among *Impatiens* species with different invasive capacities. Comparative Biochemistry and Physiology A) Molecular & Integrative Physiology 153A:S216-S216.

Ugoletti P (2011) *Ecophysiology of introduced* Impatiens *species*. PhD Thesis, Trinity College, Dublin.

Ugoletti P, Stout JC, Jones MB (2011) Ecophysiological traits of invasive and non-invasive introduced *Impatiens* species. *Biology and Environment-Proceedings of the Royal Irish Academy,* 111B: 143-156.

Valentine DH (1978) The pollination of introduced species, with special reference to the British Isles and the genus *Impatiens*. In: Richards AJ (eds) The pollination of flowers by insects. Academic Press, London, pp 117-123.

Vervoort A, Cawoy V, Jacquemart AL (2011) Comparative reproductive biology in co-occurring invasive and native *Impatiens* species. International Journal of Plant Sciences 172: 366-377

Vilà M, Weber E, Antonio CMD (2000) Conservation implications of invasion by plant hybridization. Biological Invasions 2:207-217.

Waser NM (1978) Interspecific pollen transfer and competition between co-occurring plant species. Oecologia 36: 223-236.

Wilson P, Thomson JD (1991) Heterogeneity among floral visitors leads to discordance between removal and deposition of pollen. Ecology 72:1503-1507.

Wilson P, Thomson JD (1996) How do flowers diverge? In: Lloyd DG, Barrett SCH (eds) Floral biology: studies on floral evolution in animal-pollinated systems. Chapman and Hall, New York, pp 88-111.

Wolf PG, Campbell DR, Waser NM, Sipes SD, Toler TR, Archibald JK (2001) Tests of pre- and postpollination barriers to hybridization between sympatric species of *Ipomopsis* (Polemoniaceae). American Journal of Botany 88:213-219.

Zika PF (2006) The status of *Impatiens capensis* (Balsaminaceae) on the Pacific Northwest coast. Journal of the Torrey Botanical Society 133:593-600.

4

POLLINATION ECOLOGY AND FLORAL FUNCTION OF BROWN'S PEONY (*PAEONIA BROWNII*) IN THE BLUE MOUNTAINS OF NORTHEASTERN OREGON

Peter Bernhardt*[1], Retha Meier[2], Nan Vance[3]

[1]*Department of Biology, Saint Louis University, St. Louis, Missouri 63103, USA*
[2]*Department of Educational Studies, Saint Louis University, St. Louis, Missouri 63103*
[3]*USDA Forest Service, Pacific Northwest Research Station, 3200 SW Jefferson Way, Corvallis, Oregon 97331, USA*

Abstract— Brown's peony, *Paeonia brownii* (Paeoniaceae), is one of only two peony species native to the Western Hemisphere, yet its pollination ecology and breeding system have never been documented. Using flowering individuals of an endemic colony in the Blue Mountains of Oregon, U.S., we investigated the peony's pollination system and floral function. We also examined pollen/carpel interactions through experimental pollinations aided by fluorescence microscopy. *Paeonia brownii* appears to be self compatible and mostly protogynous with floral traits of a generalist pollination system. The flowers appear to attract insects by producing abundant floral nectar secreted from lobes of a perigynous disc throughout their 9-15-days of anthesis. The most common pollen vectors were wasp queens (Vespidae), the large flower fly *Criorhina caudata* (Syrphidae), and females of *Lasioglossum* spp. (Halictidae), all of which foraged exclusively for nectar. Whether collected from foraging wasps and flies, anthers, or stigmas, about half the pollen grains appeared fertile. The number of ovules per carpel was about 19. Seed set (seeds/ovule) of naturally pollinated flowers was about 20% with about 4 viable seeds per follicle. The number of fertile pollen grains transferred to the stigma under natural conditions was highly variable but generally low, which may have contributed in part to the low rate of seed set. This study raises further questions about the role of pollen sterility, floral nectar and vespid wasps in shaping a pollinator system that is unusual in *Paeonia*.

Keywords: Breeding system, Floral nectar, Paeonia brownii, Pollen, Pollination

INTRODUCTION

Despite a long history of breeding and propagation that continues today, most wild populations of Eurasian species of the genus *Paeonia* are listed as threatened or under protection. Because of their enduring popularity as ornamental and medicinal plants, continual wild collection and habitat loss threaten the existence of endemic populations (Page 2005). Concerns for the conservation of these Paeonia species' remnant populations are prompting research that leads to a greater understanding of their reproductive biology including studies on pollination and breeding systems (Luo et al. 1998; Jing & Zheng 1999; Zhou et al. 1999; Sánchez-Lafuente et al. 1999; Sánchez-Lafuente 2002; Andrieu et al. 2007).

Paeonia brownii Douglas ex Hook. and *P. californica* Nutt., the two species that comprise section Onaepia, are endemic to Pacific western North America. Unlike the Eurasian peonies, they have not been cultivated and are almost always found in the wild. Schlising (1976) determined that protogyny, floral presentation and abundant pollen in flowers of *P. californica* were indicative of allogamy and described a generalist pollination system that incorporated the naturalised honeybee, *Apis mellifera*.

*Corresponding author; email: bernhap2@slu.edu

Paeonia brownii shares similar morphology and floral traits with *P. californica* (Stern 1946); however, *P. brownii* is endemic to locales at higher latitudes and elevations and the hardier of the two species (Stebbins 1938). Stebbins and Ellerton (1939) found and described cytological differences between *P. brownii* and *P. californica.* These differences between the two species may have evolved through allopatry and endemism to different habitats (Stebbins & Major 1965) while remaining under "phylogenetic constraint" (sensu Baker & Baker 1983).

Cytological studies of *P. california* and *P. brownii* revealed chromosomal irregularities during meiosis, including translocation heterozygosity, which can result in gametophytic sterility (Stebbins & Ellerton 1939; Walters 1962; Zhang and Sang 1998). Pollen viability of *P. brownii* and *P. californica* had been ascertained to be as low as 20% and 10%, respectively (Stebbins & Ellerton 1939; Walters 1956; Zhang & Sang 1998). However, Walters (1962) suggested that the high number of anthers more than compensated for pollen sterility and that ovule fertility is not the limiting factor in seed set as evidenced by high seed production.

Grant (1975) proposed that *P. californica* was probably a self-pollinating species; however, Schlising (1976) noted a higher seed set in experimental cross- than self-pollinated flowers and a high frequency of anthophilous insects. If *P.*

brownii is primarily cross-pollinated, pollen sterility may well be a factor in the quantity and quality of the vector pollen load and the amount of fertile pollen deposited on the receptive stigma (Inouye et al. 1994). However, a first investigation should be to identify the potential pollinator system that is associated with this species' endemism and consider its effectiveness.

The purpose of this study is to investigate floral functions, pollen viability (fertility), and breeding and pollination of *P. brownii* flowers in relation to their potential to influence fecundity. A major objective is to ascertain if this species is self-pollinating or displays active self-incompatibility as has been demonstrated in other species (Zhou et al. 1999). A specific objective is to test the null hypothesis that there are no differences in the pollination system between *P. brownii* and other *Paeonia* species. Our intent is that by characterising floral traits and components of the pollination system unique to *P. brownii*, we will contribute to a broader conceptual model of the pollination systems in the genus *Paeonia*.

MATERIALS AND METHODS

Study species

Paeonia brownii Dougl. ex Hook. (Paeoniaceae) is a long-lived perennial herb (Saunders & Stebbins 1938; Page 2005) that occurs primarily in the semi-arid regions of far-western North America. A relict of mesic forests once widespread in the northern hemisphere during the Paleocene-Pliocene period, it is patchily distributed from southern British Columbia to the Sierra Nevada in California and east to Wyoming and Utah where it is rare (Hitchcock et al. 1964; Stebbins & Major 1965). Similar to other *Paeonia* endemics, *P. brownii* is adapted to habitat conditions that consist primarily of cold winters and warm-hot summers (USDA 1988; Page 2005). It commonly occurs in pine forests and sagebrush steppe, but is also found in cooler fir and aspen communities at higher elevations up to about 2500 m. Individuals are widely scattered and rarely form dense colonies but over time can become large plants. Shoots first appear in early spring often through snow. It flowers from April-June depending upon elevation.

At maturity *P. brownii* is multi-stemmed, growing from a large, fleshy root. Each stem is semi-decumbent and has 5-7 biternate, bluish green, glabrous and somewhat thick leaves with deeply dissected leaflets having purple-tinged edges in cold temperatures (Hitchcock et al. 1964). Flowers are large, nodding, solitary and terminal on a branching stem. The 5-6 sepals are a purplish green, ovate to sub-orbicular, overlapping and cupped; the 5-10 orbicular petals grade in colour from brownish-maroon to yellow on the margins. The polyandrous flower has yellow stamens and a perigynous, fleshy nectariferous disc with greenish-yellow lobes 2.5-3.0 mm high. The large gynoecium consists of 2-6 glabrous yellow-green to yellow-red carpels, each having a short style surmounted by a curved stigma appearing as a narrow ridge at the apex of the style. Fertilised carpels mature into follicles that become leathery at time of dehiscence (Fig. 1). Seeds are yellowish-tan to black, round-ovoid and 6-11 mm in diameter. *Paeonia brownii* is a diploid (2N = 10) species (Stern 1946; Hitchcock et al. 1964).

Study sites

The study was conducted from April-May, 2003-2005 on flowering plants selected from a colony of *P. brownii* composed of 100-200 individuals distributed over 4 ha (45° 32.622' N, 117° 58.680' W) on the property of the Grande Ronde Overlook Wildflower Institute Serving Ecological Restoration. This study area is located within the 2.4 million ha Blue Mountains Ecological Province in northeastern Oregon, U.S. A different colony was located 320 km west of the Blue Mountains study area on the eastern flank of the Oregon Cascades Range at about 1000 m elevation. It was visited to sample nectar from flowering individuals for chromatographic analysis and for additional observation of floral foragers.

The Blue Mountains site consists of an upland prairie adjacent to a coniferous forest located on basalt-substrate, scabland of a shield volcano 1036-1076 m in elevation. The plant community is composed of a mixture of grasses, broadleaf herbs, and shrubs endemic to bunch-grass prairies and open pine forests of the Blue Mountains. Other putatively entomophilous species associated with *P. brownii* include *Aconitum columbianum, Balsamorrhiza hookeri, B. sagitatta, Ceanothus sanguineus, Clarkia pulchella, Clematis hirsutissima, Delphinium* spp., *Eriogonum heracleoides, Lomatium dissectum, L. triternatum, Lupinus sulphureus, Penstemon* spp., *Phacelia heterophylla, Potentilla glandulosa, P. gracilis, Ranunculus glaberrimus, Rosa gymnocarpus, R. nootkana, Symphoricarpos albus, Vicia americana* and *Wyethia amplexicaulus.*

Floral presentation and reproductive traits

For two weeks beginning 09 May 2004, we randomly selected a single bud from each of 16 randomly selected plants and observed daily developmental changes in floral structure using 10X optical glass magnifiers (Donegan Optical Co.). We recorded dates when the following developments appeared in each flower: corolla opening, nectar secretion onset, stigmatic receptivity onset (the fold of the stigmatic crest are open, pale, moist, and papillose) and ending (stigmatic crest folds together, darkens, becomes crustose with senescent papillae), anther dehiscence, and petal abscission. We collected an open flower (male phase) from each of 33 randomly selected plants in late April and May 2004. Fixation and storage of the flowers followed Goldblatt et al. (2004). Under a dissecting microscope we counted the number of lobes of each flower's floral disc. Anthers were counted using enlarged digital images of a subset of 8 undamaged flowers in mid-late anthesis when all anthers were exposed.

In early May 2004 we collected a single flower that completed flowering (petals and anthers senesced and dropped) from each of 42 randomly selected plants. Under the microscope we counted the number of ovules per ovary of a carpel removed from each flower's gynoecium. In mid-May we also counted the number of carpels, and developing follicles taken from each of 18 randomly selected flowers that finished flowering. In mid June, 24 follicles were collected by random selection and seeds were counted and evaluated. Empty seeds appeared shrivelled or poorly developed; when cut, endosperm was lacking or much reduced and necrotic compared to that of developed seeds. Seeds that appeared viable were smooth, yellow to tan-brown, medium-to-large and filled with creamy white endosperm with a rudimentary embryo visible at the basal end.

Observation and collection of floral foragers

Observation of insects visiting the flowers of *P. brownii* in May 2003 and in April-May, 2004-2005 constituted about 52 hours. We noted which insects visited flowers beginning on the first day of anthesis, which was defined as beginning when petals parted enough to expose the tips of the carpels. We recorded whether insect visitors foraged for nectar and/or pollen and whether they could be found on the flower throughout the floral lifespan. We used binocular magnifiers to determine whether insects contacted the stigmas while they visited the flower.

We took pollen load data on 100 insects over three seasons using the following procedures: Individual insects were caught with a butterfly net after they were observed contacting floral organs. After they were killed in a jar containing ethyl acetate, pollen grains were washed from the insect specimens with ethanol, pinned and labelled following the protocol of Goldblatt & Bernhardt (1990). Pollen grains from the insects were mounted on a glass slide, stained with Calberla's fluid (Ogden et al. 1974), identified and counted under a compound microscope. As different insects were killed in the same jar, even though few grains were dislodged from the insects, pollen of a plant species was listed as present only after we counted $\geq$ 20 grains of the same pollen morphotype on the slide. Vouchers of insect specimens were identified and deposited in collections by entomologists at the American Museum of Natural History (New York; J. Carpenter, wasps and J. Rozen, bees), Smithsonian Institution (Washington, DC; C. Thomson, flies), Snow Entomological Museum (Kansas State University; C.D. Michener, bees) and the Museum of Southwestern Biology (University of New Mexico; L. Larkin, bees in the genus *Andrena*).

We also collected caterpillars that were found inside flowers. Some specimens were preserved in 70% ethanol. Other live specimens were brought back to Saint Louis University and reared in insect cages in the laboratory of J.C. Fortier. Moths emerging from pupation were killed and pinned. Pinned moths, along with caterpillars and pupal cases, were sent for identification to M. Pogue, Smithsonian Institution, Washington DC.

Pollen analysis

We removed pollen grains of *P. brownii* from sampled insect visitors, anthers or stigmas and stained them with Calberla's solution, a fuchsin stain that allows the pollen structures to be assessed under the microscope. We classified the grains as appearing either potentially fertile or sterile using the following criteria: pollen walls of fertile grains stained pink as they hydrated or began to form pollen tubes (for grains on stigma); all three exine apertures including gray, granular cytoplasm were visible under light microscopy. In contrast, the walls of sterile grains stained a deep reddish-purple; never hydrated or germinated; one or all exine apertures were not discernible because of the collapsed wall folding up on itself with no visible cytoplasm inside the grain.

To determine whether sterile pollen grains were being transported by insect visitors, we counted the first 50 visible grains on a pollen wash slide of an insect and classified them as either sterile or fertile using the above protocol. Pollen samples were taken from 5 vespid wasps and 9 syrphid flies (*Criorhina caudata*) only because these insects were observed to contact receptive stigmas while foraging for nectar but were never observed ingesting pollen or actively foraging on dehiscent anthers for pollen (sensu Bernhardt 1996). To compare fertile versus sterile grains of insect-born pollen with non-transported grains remaining inside anthers, we took a pollen smear from a dehiscent anther of a flower collected one per day from each of 15 randomly selected plants over a two-week period in May 2004. Pollen grains of each anther were stained and counted as described above. To determine the proportion of sterile pollen that was deposited onto receptive stigmas under natural conditions, from the same flowers we selected 10 visibly receptive stigmas that had a minimum of 10 visible pollen grains. We counted and classified all the grains of *P. brownii* pollen detected on each stigma using the protocol for classification described above.

Pollen-carpel interactions

To determine natural rates of pollination over the flowering period, we collected a flower with a visibly receptive stigma (see above) from each of 5-6 plants on 7 separate dates after 1200 h, from 14 April through 03 May 2004. Flowers with petals expanded sufficiently to expose one or more stigmas were tagged and harvested 48 h later, when they were in the male phase of anthesis. Once picked, each whole flower was fixed and stored following Goldblatt et al. (2004). To view and count the number of pollen grains on the stigmas and the number of pollen tubes penetrating carpel tissue, one carpel from each flower was selected at random, excised and soaked in a 5% sodium sulfite solution in de-ionised water at 60 °C for one hour and 40 min. The carpel was removed from the softening solution and washed in room temperature with de-ionised water. Each carpel was placed on a separate glass slide. A razor blade was used to longitudinally bisect the organ, which caused the thick cuticle to rupture, and the cuticle was stripped off the epidermis with forceps prior to squashing. Final preparation of the carpel halves for spreading tissues, staining and viewing under fluorescence microscopy followed Goldblatt & Bernhardt (1990). We combined counts of pollen tube germination and penetration for each pair of carpel halves. Counts included the number of germinating grains on the stigma, the number of pollen tubes visible in the style, and the number of tubes in the ovary loculus (including those adjacent to and/or penetrating ovules).

To determine whether carpels showed any recognizable early- or late-acting, self-incompatibility responses, in 2004 we selected a single, mature flower bud on each of 33 plants during the flowering period in early May. Each labelled bud was randomly selected for one of four pollen enhancement treatments (Vance et al. 2004): 1) Autogamous pollination (AP). The bud was covered securely in a tulle bag 12-48 h before anthesis and remained isolated for the duration of the experiment. 2) Hand applied self-pollination (HSP). The bud was isolated as for AP until the day of anthesis when the bag was removed briefly and all stigmas in the flower were smeared with a coat of pollen (visible under a binocular magnifier) taken from dehiscent anthers of an open flower on the same plant (geitonogamous pollination). The bag was replaced for the duration of the experiment. 3) Hand applied cross-pollination (HCP). The procedure was the same as for (HSP) except the pollen was transferred from a different flower on a plant separated by a distance > 1m. 4) Natural insect- and/or self-pollination (IP). The mature bud was not bagged and the flowers were collected as described in text above.

All flowers hand-pollinated in this study were harvested and fixed three days after the petals opened when the stigmatic surface appeared necrotic or crustose but before stigmas shrivelled, making it impossible to count both the full number of pollen grains adhering to and germinating on the stigmatic surface and pollen tubes penetrating the transmission tissue (Bernhardt & Dafni 2000). Fixation, storage and preparation for fluorescence microscopy followed the protocol described above.

Nectar analysis

In 2005, a flower bud on each of 40 plants was bagged for floral nectar collection to determine volume and concentration un-confounded by nectar feeders and uncontaminated with insect saliva and debris. After petals opened, we collected nectar from 39 flowers. We withdrew nectar in the concavity at the base of each lobe of the floral nectary (disc) using a 10 μL capillary tube. The dissolved solute content volume and concentration of the extracted nectar were recorded with a Bellingham and Stanley refractometer (Eclipse model 45-03; 0-50%). Volume and concentration of dissolved solutes were co-referenced with flower phase. The following year we sampled nectar from flowering individuals at the eastside Cascades location for analysis of nectar sugars and amino acids. For quantitative analysis of nectar constituents, we drew nectar into 100 μL capillary tubes from the floral nectary in each of 11 flowers that were pre-bagged as buds. The nectar was spotted on filter paper, labelled and express mailed to the Biology and Agriculture Chromatography Laboratory at Brigham Young University, Provo, UT, U.S. Sugars were separated using gas chromatography (Hewlett Packard 5890 GC, Palo Alto, CA, U.S.) equipped with a flame ionization detector and identified by comparing retention times with standards. Amino acids were separated and analysed by HPLC (Agilent 1100 Series, Santa Clara, CA, U.S.) using fluorescence detection following the method of Cohen & Michaud (1993).

Statistical analysis

Data were statistically analysed using Statgraphics Plus statistical software version 5.1. (Manugistics, 2000, Rockville, MD, U.S.). We used one way ANOVA for analysis of pollen tube germination among treatments applied to characterise breeding system and the t-test for differences in pollen viability among different sources of pollen. Significant differences in means were tested at the 95% confidence level. Confidence intervals among the means were determined using Bonferroni intervals. Three statistics (Cochran's C, Hartley's and Levene's tests) were used to test for equal variance at $P \leq 0.05$. Where assumptions of equal variance were not met, values were log transformed to remove dependence of the standard deviation on the mean. Where assumptions were not met for normal distribution, the Kruskal-Wallis or Mann-Whitney nonparametric test was used to compare medians instead of means. Multiple regression analysis was used with the R^2-statistic to develop best-fit linear models that explained variation in factors associated with nectar and seeds. ANOVA was used to test for a statistically significant relationship between the variables at the 95% confidence level.

RESULTS

Floral traits and life span

In early April 2004 at the Blue Mountains site, *P. brownii* had multiple spreading stems (Fig. 1a). Each of the flowering stems (2-10 depending on the size of the plant), bore a single, large, globose terminal bud (Fig. 1a). Anthesis of flowers began in mid April 2004 and 2005, and continued through mid May. The mean floral lifespan of a flower was 11.4 d (n = 16) and ranged from 9-15 d. During the first several days of anthesis, the corolla expanded enough to expose each stigma on the carpel's axis but partially occlude the indehiscent anthers (Fig. 1b). The moist stigmatic ridge of each carpel opened and expanded until the receptive stigma exposed the turgid, unicellular papillae. The mean number of days of stigma receptivity was 2 (range 1-4) or 18.7 % of the floral lifespan. Over the floral lifespan, as carpels increased in size they became more prominent and their colour changed from yellowish green to yellowish red (Fig. 1c,d,e).

The mean number of anthers in the androecium was 84.4 ranging from 63-92 (n = 10). Each anther was born on a thick filament that was elongated at time of dehiscence. The mean number of days in the male or staminate phase was 9.8 or 85.9% of the floral lifespan. In succession centrifugally, 6-12 anthers dehisced each day (Fig. 1d). In 6 of 16 flowers observed daily over the floral lifespan, a few anthers closest to the carpels began to dehisce while surfaces of the stigmatic crests were still fully expanded. The length of overlapping male and female phases of those flowers was about 1 day or 5% of the floral lifespan. Over the course of the male phase while the petals remained attached to the flower, the stigma and short style shrivelled, and any remaining exposed papillae on the closed stigmatic crest turned brown and crusty (Fig. 1d). The fleshy, yellowish, lobed disc that encircled the carpels at the base of the

FIG. 1. (a) *Paeonia brownii* bearing terminal globose buds. (b) Opening bud exposing receptive stigma. (c) Early anthesis with visible carpels, anthers. (d) Flower mid-anthesis, anthers visibly dehiscent, nectariferous disc at base of carpels exposed. (e) Flower with petals and stamens removed to show prominent lobes of nectary (arrow). (f) Opened flower revealing moth larva and damage from herbivory. (g) Solitary bee attracted to nectar. (h) *Criorhina caudata* (Syrphidae), *Bombus* mimic and a primary pollinator of *P. brownii*. (i) *Dolichovespula arenaria* (Vespidae) queen on flower in staminate phase.

perianth formed an incomplete barrier between the androecium and gynoecium (Fig. 1d,e). When the floral phase was over and the lobes were no longer secreting, they also shrivelled and appeared faded (Fig. 1f).

Every disc sampled from flowers in the pistillate or staminate phase secreted nectar. The nectar was sweet with a bitter aftertaste, reminiscent of the floral odour which was weakly bitter and unpleasantly scented. The same odour became far more pronounced in leaf and stem vegetation when it was broken, bruised or damaged by predation (Fig. 1f).

Flower visitors

We identified insects representing 16 different taxa in the Orders of Diptera and Hymenoptera that were collected on open flowers of *P. brownii* from 2003-2005 (Tab. 1). Of 100 visitors recorded 80% were represented by members of 4 taxa: *Lasioglossum* spp. (Halictidae) (27%) (Fig. 1g), *Criorhina caudata* (Syrphidae) (17%) (Fig. 1h), *Dolichovespula arenaria* (Vespinae) (19%) (Fig. 1i) and *Polistes aurifer* (Polistinae) (17%). Six wasp species represented 45% of the visitors recorded.

Insects of other taxa were observed visiting *P. brownii* flowers at the eastside Cascades location (Vance pers. com., data not shown). Among them were large flies in families Tachinidae and Sarcophagidae, and a single *D. Adulterina* queen, an obligatory social parasite of *D. arenaria* (Greene et al. 1978; Buck et al. 2008). Unidentified small to medium bees visited the flowers at all phases of its life span. Ants (*Formica* spp.) were commonly observed crawling on flowers and buds (Vance 2012).

We observed vespid wasps from the last two weeks of April until the end of the flowering period in May. Although wasps were observed on flowers as early as 0930, their peak periods of activity occurred from 1430-1630 h during the warmest part of the day. All of the wasp species visited flowers at all phases of the floral life span. All were observed foraging exclusively for nectar. *Dolichovespula arenaria* tended to visit all open flowers on the same plant and/or flowers on individual plants with overlapping stems. Commonly, we observed as many as 3 wasps concurrently visiting the same flower. All wasp visitors were native except for *P. dominula,* which we observed building a nest near *P. brownii.* All wasps collected on *P. brownii* were fertile, egg-laying females ranging 13-20 mm in length (Tab. I)

Female, solitary bees in the genus *Andrena* about 11 mm in length and *Lasioglossum* (*Evylaeus*) ranging from 6-7 mm were most often observed foraging for nectar in staminate-phase flowers. Those recorded were seldom seen contacting the elevated, receptive stigmas in early female phase flowers. Bee visits to female-phase flowers were observed a total of four times over two seasons at the Blue Mountains site. Other medium-large bees including *Apis mellifera* (12 mm), *Eucera* spp. (6-14 mm), and *Bombus mixtus* (16 mm) were infrequent visitors to *P. brownii* over the two seasons.

Of the dipterans collected, 19 were syrphid flies, 18 of which were *Criorhina caudata.* These *Bombus* mimics are mostly black, thickly pilose and large, ranging from 13 to 18 mm in length (Fig. 1 h). *Criorhina caudata* was active on *P. brownii* from April until the beginning of the first week in May in both years. These flies remained in May but appeared to forage preferentially on *Balsamorhiza* spp. and *Delphinium nuttallianum.* Throughout early April, on overcast days with temperatures ranging from 0 to 16 °C, male and female *C. caudata* were observed moving from flower to flower (< 1 min/flower) taking up nectar through their siphon-like probosces. Peak activity was approximately from 0900–1200 h. We observed flies contacting receptive stigmas as they moved from lobe to lobe of the nectary. They also attempted to probe mature buds in which the petals had not yet separated. All *C. caudata* that were analysed for pollen had > 200 pollen grains of *P. brownii* typically on their heads and thoraces.

Of 85 insect specimens that carried pollen from *P. brownii,* 41% also carried pollen from other plant species (Tab. 1). The pollen grains of co-flowering species most frequently found on insects were from *B. hookeri, B. sagittata,* and *D. nuttallianum.* Species in which the majority of the specimens carried only *P. brownii* pollen were the fly *C. caudata* and the wasps *D. arenaria* and *P. aurifer,* none of which foraged for pollen. Of the 37 wasps that carried pollen, 6 carried > 100 grains on their heads, legs and thoraces; 24% had grains of species other than *P. brownii.*

Insect taxon	*P. brownii* only	*P. brownii* + other	Other only	No pollen
Diptera (Syrphidae)				
Criorhina caudate	12	5	0	0
Eristalis hirta	1	1	0	0
Subtotals	13	6	0	0
Hymenoptera (bees)				
Andrena astragali	1	2	0	0
Andrena lawrenci	1	0	0	0
Andrena nigrocaerulea	0	1	0	0
Apis mellifera	1	0	0	0
Bombus mixtus	0	1	0	0
Eucera actuosa	0	1	0	0
Eucera sp.	0	0	1	0
Lasioglossum spp.	6	16	2	3
Subtotals	9	21	3	3
Hymenoptera (wasps)				
Dolichovespula arenaria	10	5	0	4
Polistes aurifer	10	2	1	4
Polistes dominula	1	0	0	0
Vespula atropilosa	3	0	0	0
Vespula consobrina	2	1	0	0
Vespula pensylvanica	2	0	0	0
Subtotals	28	8	1	8
Grand totals	50	35	4	11

TABLE 1. Pollen load analyses of insects collected on flowers of *Paeonia brownii* in the Blue Mountains of northeastern Oregon, April-May 2003-2005.

Of 19 flies that carried pollen, *C. caudata* generally carried more pollen than the wasps; the number of grains ranged from > 100 to > 200. About 32% of the *C. caudata* also carried pollen from other co-blooming species (Tab. 1). The number of grains that bees carried varied widely among bee species; grains carried by *Andrena* and *Lasioglossum* species ranged from 10 to >100; about 72 % of the bees carried pollen grains of species other than *P. brownii.*

Pollen analysis

Light and fluorescence microscopy showed that pollen grains from each of the sources (flies, wasps, anthers, stigmas) consisted of a mixture of sterile and fertile pollen. Fertile pollen from all sources combined was 51.8 ± 5% and highly variable ranging from 20 to 92% (Tab. 2). The difference between wasps and flies in the mean number of potentially fertile pollen grains they carried was significant ($t = 3.6$, $P < 0.004$) (Tab. 2). Otherwise, no significant differences were detected in the mean percentage of potentially viable pollen between each source carried by the insects or found on stigmas or anthers (Tab. 2).

Pollen-carpel interactions

The mean number of fertile pollen grains on the stigmatic surface of *P. brownii* flowers differed significantly with sampling date ($F = 3.58$, d.f.= 32, $P = 0.01$) (Fig. 2), and was highest on April 22, ranging from 21-157 ($n = 5$). The pollen tubes in the transmission tissue of the style were fewer in number than on the stigmas and their means did not differ significantly among the sampling dates ($F = 0.22$, $P = 0.96$). In the ovary, pollen tubes were only detected 48 h after anthesis on three dates in late April and one in early-May (Fig. 2). Zhou et al. (1999) reported that at least 48 hours were required for pollen tubes to penetrate the ovary of *P. jishanensis.* Pollen tube growth may have been slowed

TABLE 2. Mean percentage of potentially fertile pollen deposited on stigmas, and of 50 pollen grains collected from flies (Syrphidae), wasps (Vespidae) and anthers of *Paeonia brownii* and analysed under a compound microscope. Results of t-tests for significant differences between the means at $P < 0.05$ are shown by different letters; c.i. = 95% confidence intervals for means.

Pollen source	Fertile pollen (%)	(c.i.)	*n*
Fly	63.3[a]	(57.7, 68.9)	9
Wasp	43.2[b]	(25.9, 60.5)	5
Anther	50.3[ab]	(41.4, 59.2)	15
Stigma	58.5[ab]	(43.4, 73.6)	10

by the low night-time temperatures at the site (mean $t_L = 0.5$ °C). Minimum daily temperatures were never higher than 4 °C over the entire flowering period from mid April to mid May (Oregon Climate Service 2004 data, Elgin, Oregon Climate Station, Oregon State University).

Significant differences were detected among the four pollen treatments used to evaluate pollen-pistil interactions ($F = 7.13$, $P = 0.002$). The hand pollinated stigmas (HCP and HSP) contained significantly more fertile pollen grains than did stigmas exposed to insect pollinators (IP) or bagged flowers that only allowed autogamy (AP) (Tab. 3). Four of the AP-treated stigmas completely lacked pollen grains. Also the number of pollen tubes counted in the styles differed significantly among the treatments (Tab. 3). In general, fewer pollen tubes were counted growing through the transmission tissue of the style than on the stigma, but there were significantly more pollen tubes in the styles and ovaries of the HSP- and HCP- treated carpels than the IP or AP treated carpels (Tab. 3, Fig. 3 c,d).

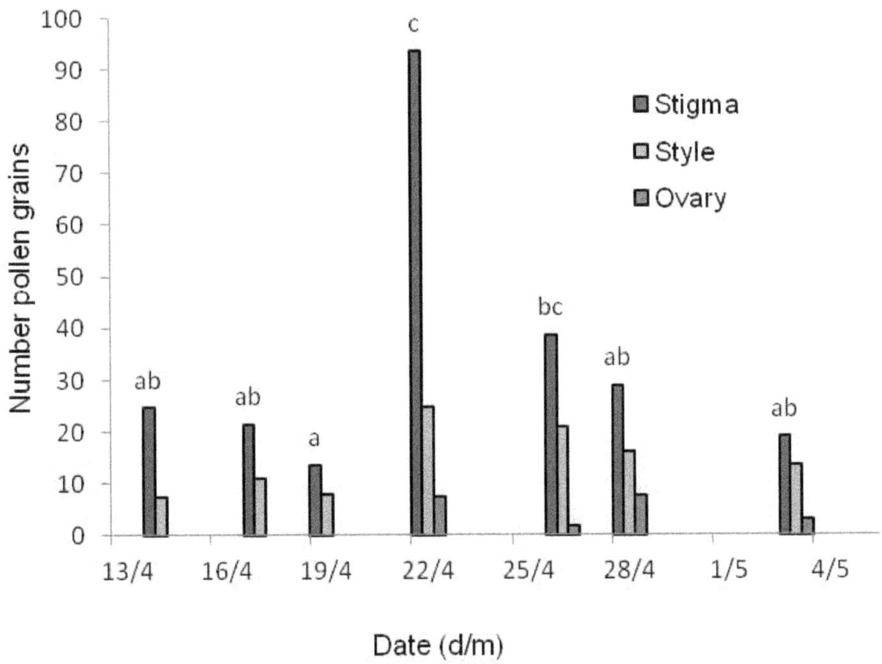

FIGURE 2. Mean number of visibly fertile and/or germinated pollen grains on stigmas, and pollen tubes in styles and ovaries of *Paeonia brownii* on various dates from 14 April – 3 May 2004. Different letters denote significant difference in mean number of pollen grains on stigmas among different dates determined by multiple range tests using LSD at 95% confidence on log-transformed data.

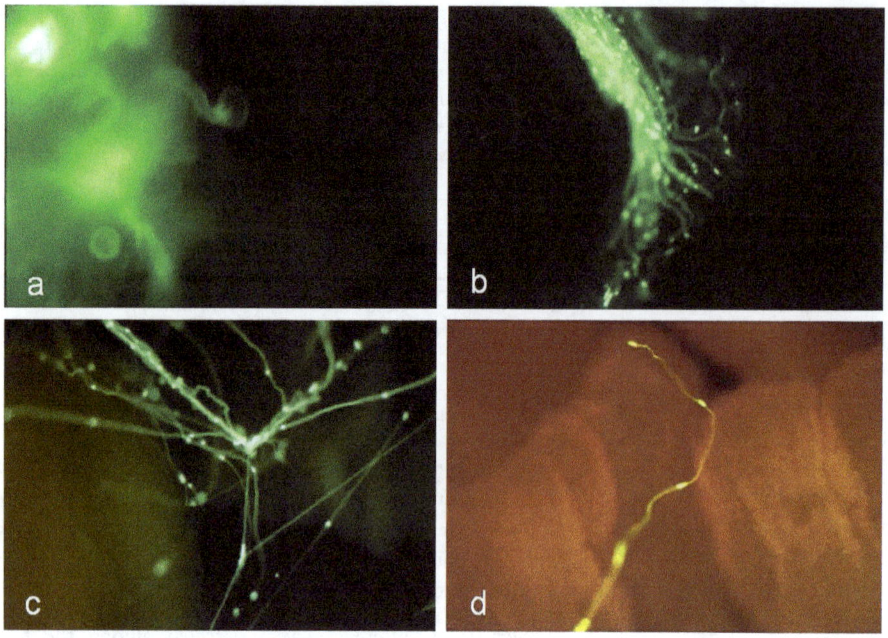

FIGURE 3. Microphotographs of pollen tubes of *Paeonia brownii* hand self-pollinated (a-b) and autogamously pollinated (c-d). (a) Pollen grains germinating between stigmatic papillae (x 40, 1/8 s). (b) Pollen tubes penetrating the transmission tissue of the style (x 20, 1/8 s). (c) Pollen tubes visible among ovules in the ovary (x10, 1/8 s). (d) Ovary of bagged flower with a single, visible pollen tube between the ovules (x10, ¼ s).

Nectar secretion and composition

Over the entire floral lifespan of *P. brownii,* nectar was secreted only from secretory tissue in the individual lobes of the perigynous disc and not from ruptured phloem in the staminal filaments of *P. brownii* as reported by Hiepko (1964) in Eurasian peonies. Neither nectar concentration nor volume nor the number of secreting lobes on a flower differed significantly between 2004 and 2005, thus, the data were combined for further analysis. On average, out of 12 lobes per disc, 3 visibly secreted nectar which collected in the concavity at the base of each lobe (Tab. 4). The mean volume of fluid collected from an individual lobe was 7.1 µL ranging from 1.0 to 32.8 µL. The number of lobes secreting nectar partially explained the total volume of floral nectar described in a linear model (r^2 = 0.33, d.f. = 38, $P <$ 0.001).

TABLE 3. Mean number of germinated pollen grains on the stigma and pollen tubes in the style and ovary of *Paeonia brownii* flowers that received hand cross- (HCP), self- (HSP), natural insect (IP), or autogamous (AP) pollination. Different letters denote significant differences among treatments at $P \leq 0.05$ by analysis of variance on log-transformed data (untransformed means and standard deviations are shown).

	Treat-ment	*n*	Pollen no. s.d.	Range
Stigma				
	HCP	8	80.9 ± 52.6 [a]	21 – 156
	HSP	8	80.0 ± 56.4 [a]	37 – 21
	IP	8	16.6 ± 8.4 [b]	6 – 26
	AP	9	7.7 ± 10.9 [b]	0 – 28
Style				
	HCP	8	36.8 ± 18.6 [a]	17 – 61
	HSP	8	38.1 ± 24.8 [a]	8 – 80
	IP	8	5.8 ± 6.5 [b]	0 – 17
	AP	9	2.9 ± 5.7 [b]	0 – 17
Ovary				
	HCP	8	10.3 ± 11.9 [a]	0 – 35
	HSP	8	13.0 ± 9.9 [a]	0 – 28
	IP	8	2.6 ± 5.9 [b]	0 – 17
	AP	9	0.6 ± 1.3 [b]	0 – 4

No significant difference was detected in total volume of floral disc or concentration between the female (first) and male (second) phase of the unbagged or bagged flowers (Tab. 5). Although the nectar volume of unbagged flowers was only about 25% of the volume of bagged flowers in male or female phase, the concentration changed little (Tab. 5). A linear model that shows concentration weakly related to volume (r^2 = 0.14, d.f. = 38, P < 0.05) indicates that concentration was minimally affected by volume differences (Nicolson & Thornburg 2007).

Of the two hexose sugars identified, the most abundant in the nectar was glucose. The mean concentration of glucose was 37.35 µg/µL which was 3-times the mean concentration of fructose (11.33 µg/µL). Two disaccharides were found: maltose with a low mean concentration of 2.15 µg/µL and an unidentified sugar that we presume to be sucrose. The mean concentration of this sugar was 13.8µg/µL. Because sucrose can elute virtually at the same time as lactose, a lactose standard may have interfered with its positive identification (Halket & Zaikin 2003; Madeiros & Simoneit 2007). These sugars were not qualitatively analysed by mass spectrometry (MS). However, the fructose and glucose concentrations were fitted in a multiple linear regression model in which the r^2 adjusted for 10 d.f. explained 92.4% of the variation in the concentrations of the unconfirmed disaccharide (F = 62.2, $P <$ 0.0001).

The concentrations of sixteen amino acids detected by chromatography varied widely among the 11 samples (Tab.

TABLE 4. Floral and fruit traits of *Paeonia brownii* of sampled individuals at the Blue Mountains location in northeastern Oregon, May-June 2004. Data are presented as means (± 1 s.d).

Floral trait	Mean (s.d.)	*n*
Lobes/floral nectary	12.0 (1.0)	33
Lobes with nectar/disc	3.2 (1.4)	39
Carpels/gynoecium	4.9 (0.8)	52
Ovules/carpel	18.9 (3.1)	42
Developed follicles/gynoecium	4.1 (1.3)	24
Follicle length (cm)	4.3 (6.2)	24
Total seeds/follicle	4.0 (1.3)	24
Filled seeds/follicle	3.7 (1.6)	24

TABLE 5. Mean volume and concentration percentage of secreted nectar in *Paeonia brownii* flowers sampled at the Blue Mountains location in northeastern Oregon (2004-2005 data combined). Nectary of female- and male-phase flowers visited by insects (Open) and with insects excluded (Closed).

Nectary	Phase	Volume µL ± s.d.	Concentration % ± s.d.	*n*
Closed	Female	26.45 ± 21.99	18.35 ± 6.54	20
	Male	23.63 ± 30.24	23.66 ± 10.54	19
Open*	Female	6.75 ± 6.59	23.83 ± 11.28	12
	Male	6.23 ± 2.13	32.23 ± 11.20	13

TABLE 6. Mean concentration of 16 amino acids in nectar analysed by HPLC from flowers of *Paeonia brownii* collected May 2006 at the eastside Cascades Range in central Oregon. Shown are means and standard deviations, *n* = 11 flowers.

Amino acid	Concentration µg/µL ± s.d.
Alanine	0.0040 ± 0.0031
Arginine	0.0223 ± 0.0226
Aspartic acid	0.0038 ± 0.0028
Glutamic acid	0.0026 ± 0.0014
Glycine	0.0038 ± 0.0052
Histidine	0.0050 ± 0.0073
Isoleucine	0.0006 ± 0.0011
Leucine	0.0043 ± 0.0021
Lysine	0.0012 ± 0.0021
Methionine	0.0013 ± 0.0013
Pheylalanine	0.0008 ± 0.0006
Proline	0.0109 ± 0.0052
Serine	0.0021 ± 0.0015
Threonine	0.0092 ± 0.0043
Tyrosine	0.0005 ± 0.0008
Valine	0.0018 ± 0.0013

6). In some samples the amino acid concentration fell below detectable levels; however, 10 of the 16 amino acids were detected in all samples. The mean concentration of 16 amino acids combined was 0.074 µg/µL. The mean of the individual amino acids was 0.0046 µg/µL and ranged from 0.0005-0.0223 µg/µL. The amino acids with highest mean concentration (n = 11) were arginine, proline, and threonine (Tab. 6).

Seed production and herbivory

The number of expanded carpels with green, swollen ovaries collected from flowers in an early stage of maturity averaged about 5 and ranged from 3-7 (n = 52) with 5 also the most common number of carpels (35/52). The mean number of ovules in a carpel was 18.9 and ranged from 14-28 (n = 42) (Tab. 4)

Of follicles collected from naturally pollinated flowers, the mean number of fully developed seeds per follicle was 4 (Tab. 4). The ratio of the mean number of filled seed per follicle (3.7 ± 1.6) to the mean number of ovules per carpel (18.9 ± 3.1) was about 0.20 (Tab. 4). Although the ratio of seed/ovule was low, the great majority (92.0% ± 16.7) of the seeds were viable. Filled seeds were irregularly ovoid and varied in size ranging from 5-12 mm at their widest diameter (n = 24). The mean seed weight was 0.19 ± 0.05 g at 10% moisture content (n = 20). The number of seeds in the follicle partially explained the variation in follicle length ($r^2 = 0.43$, $P < 0.001$).

Buds and opened flowers showed evidence of herbivory by at least three species of moth caterpillars in the family Noctuidae s.l. (Hardwick 1996). Flowers that had been predated by moth larvae bore tiny (1-2 mm) holes in the calyx and corolla. Caterpillars of all three species were often found under sepals or in damaged carpels (Fig. 1 f). Winged adults reared from captured larvae were identified as *Euxoa ustula* La Fontaine, *Fishia yosemitae* (Grote) and *Heliothodes diminutives* (Grote). For all species, it is the first record of *Paeonia brownii* as host plant (M. Pogue, Smithsonian, pers. com.). Crumb (1956) records *Solidago* and *Verbascum* as the most common hosts of *F. yosemitae.* Hardwick (1996) identifies *Layia platyglossa, Lasthenia californica* and *Madia* spp. as food plants of larvae of *H. diminutivus.*

DISCUSSION

Flower visitors and relation to floral nectar

Paeonia brownii bears hermaphroditic, mostly protogynous flowers that exhibit floral traits consistent with allogamy and a generalist pollination system. That autogamy occurs infrequently is evident in the low incidence of pollen tubes in the style and ovary of flowers denied insect contact (Tab. 3).

Floral nectar is secreted throughout the individual flower's long (9-15 days) lifespan attracting a unique variety of hymenopteran and dipteran insects that including large wasps and flies and a variety of small-medium sized bees. The floral nectary found in *P. brownii* is uncommon in Paeoniaceae, and outside of section Onaepia, is secretory only in the *P. delavayi* complex (sensu Hong et al. 1998). Although during the male phase the flowers produce abundant pollen, unlike the Eurasian peonies they do not attract the assemblage of pollen-foraging bee and beetle visitors, as do, for example, the flowers of *P. broteroi, P. jishanensis, P. suffruticosis* and *P. officinalis* (Luo et al.1998; Sánchez-Lafuente et al. 1999; Zhou et al. 1999; Andrieu et al. 2007) or even of its closest congener *P. californica* (Schlising 1976).

Polistine and vespine wasp queens constituted more than half of the hymenopteran visitors that carried pollen of *P. brownii*. The post-hibernating wasps in early spring actively seek nectar that is easily accessed and in large volumes (Matsuura & Yamane 1990; Kearns & Inouye 1994; Mackensie et al. 2006). The high volume and concentration of carbohydrates and amino acids in the floral nectar of *P. brownii* are nutritionally important to the post-hibernating wasp queens, which can lose up to about 40% of their body weight and up to 75% of their stored fat (Spradbery 1973; Matsuura & Yamane 1984; Baker & Baker 1986; Terrab et al. 2007). They repeatedly foraged on the flowers of *P. brownii* exclusively for nectar even as the flower's presentation changed from a semi-upright bud to a nodding, open flower over the course of anthesis (Fig. 1 b-d), thus behaving like facultative specialists (Richter 2000; Mackensie et al. 2006).

Wasp dominance at the Blue Mountains site may be due in part to other floral resources' availability to the endemic solitary bees, but also because the wasps themselves present a threat. Vespid wasps have a large range of arthropod prey that includes Hymenoptera and Diptera (Akre & Myhre 1992; Richter 2000). They also fiercely guard their food sources and have been documented attacking other hymenopterans including ants and smaller wasps (Kurczewski 1968; Matsuura & Yamane 1990). One example is a pinned wasp (*Polistes*) in the data collection grasping an ant's head in its tarsal claws (not shown). Even though wasps typically prey on larvae following nectar consumption (Kurczewski 1968; Domínguez et al. 1989), we never observed them preying on the identified herbivores. Nevertheless, since seed production is impacted through damage by moth larvae, wasps may also provide a defensive function for *P. brownii*.

Despite their potential to predate larval herbivores (Stamp & Bowers 1988), it is doubtful that *P. brownii* flowers co-evolved with vespid wasps because their body morphology, foraging and food defending traits do not support effective cross-pollination (Domínguez et al. 1989; Galindo-González & Ornelas 2002). Vespid wasps do not necessarily emerge and forage synchronously with anthesis of *P. brownii* flowers (Akre & Myhre 1992). When vespid females do emerge in the spring, they tend to find the closest and most convenient source of nectar (Matsuura & Yamane 1990; Liebert et al. 2006). The wasps' relation to *P. brownii* more likely may be one of opportunistic mutualism. The protection that the wasp may confer to the flower through its attraction to floral nectar would be facultative (Domínguez et al. 1989).

Criorhina caudata may forage on *P. brownii* flowers guided by olfactory cues as reported elsewhere (Proctor et al. 1996; Sutherland et al. 1999). The large *Bombus* mimic visited *P. brownii* in early April before the appearance of wasp queens which may have not yet emerged in this particular region of eastern Oregon (Akre & Myhre 1992). Later in April and up to the first week in May, *C. caudata* was a frequent and constant visitor as has been reported elsewhere (Proctor et al. 1996; Goulson & Wright 1997) primarily in the morning hours when wasps were not as active. The pollen grains collected from the pilose coat of *C. caudata* were almost all from *P. brownii*, generally >200 and had a higher rate of viability than those carried by wasps (Tab. 2). *Criorhina* species may not only fill a niche in the pollination ecology of *P. brownii* at the Blue Mountains site but appear to be the more potentially effective pollen vector.

Flowers of *P. brownii* were visited by female native bees, especially small-to-medium *Lasioglossum* (Halictidae), foraging for nectar. They almost always bore pollen grains of *P. brownii* but were observed visiting the flower often after the stigma was no longer receptive (data not shown). The heterogeneous mix of pollen grains found on these bees is evidence that they visited a variety of co-flowering bee-pollinated shrubs and herbs. Such flower inconstant behaviour could increase the likelihood of pollen contamination (Rhoades & Bergdahl 1981; Shuttleworth & Johnson 2006; Sahli & Conner 2007). However, Zhou et al. (1999) found that contamination with alien pollen had no effect on seed set of *P. jishanensis*. Further investigation may show that the bees as alternative pollinators help to "buffer" the pollination system against fluctuations in their environment (sensu Sahli & Conner 2007).

Pollen carpel interactions and fecundity

The mechanical cross- and self-pollination experiments resulted in no significant difference in pollen tubes germinating on stigmas and growing in the stylar transmission tissue indicating no apparent expression of early self-incompatability (Fig. 3, Tab. 3).

Although highly variable over several weeks of sampling, the rate of fertile pollen transferred to the carpels naturally through insect (and autonomous) pollination was generally low (Fig. 2) and may help to explain the low seed set. The number of pollen tubes reaching the style under the hand pollinations was significantly higher than by natural pollination which is in contrast to other *Paeonia* species, where supplementary pollination did not necessarily increase seed set (Luo et al. 1998; Sánchez-Lafuente et al. 1999; Zhou et al. 1999; Andrieu et al. 2007). Meiotic dysfunction as well as late-acting incompatibility may be expressed when the gametes form a zygote which would account for aborted ovules (Camp & Hubbard 1963). In fact, immature follicles taken from the Blue Mountains population showed evidence of early aborted seeds in addition to underdeveloped seeds.

Similar to other *Paeonia* congeners (Schlising 1976; Sánchez-Lafuente et al. 1999; Zhou et al. 1999; Andrieu et al. 2007), the seeds of *P. brownii* are large. A low rate of conversion from ovules into viable seed is common in large-seeded plants (Guo et al. 2000). Seed number and development may be limited by costs of prior allocation to reproductive resources; however, no such causal evidence explained sub-maximal seed production in *P. officinalis* (Andrieu et al. 2007). Perhaps female function provides fitness for this long-lived perennial herb with less than maximal seed/ovule ratios as was presented for other *Paeonia* species (Schlising 1976; Andrieu et al. 2007). Based on the relictual vasculature of *P. brownii*, Camp & Hubbard (1963) proposed that *P. brownii* may have evolved from an ancestor that produced a large and more complex carpel and presumably more seeds. Thus evolutionary selection may have been toward fewer, larger seeds. Additional studies that examine resource and genetic influences on seed set may provide a clearer explanation.

Conclusion

Flowers of the allogamous *P. brownii*, despite about half its pollen showing sterility, retained different pollination vectors over its long flowering season through the provision of floral nectar. *Paeonia brownii* blooms in a colder environment and at higher elevations and latitudes than its California congener. Vespid wasps and large, hairy syrphid flies that tolerate cool temperatures (Edwards 1980; Matsuura & Yamane 1984; Kearns 1992) appear to dominate the pollinator function. However, other flies and female bees such as *Lasioglossum* species that carry *P. brownii* pollen could potentially function as secondary pollinators, as described by Kearns & Inouye (1994) and Kearns (2001). This study in the Blue Mountains of Oregon has shed light on what appears to be a unique pollination system in *Paeonia* that warrants further investigation at other locations within its range. How the aromatic and nutrient composition of the floral nectar attracts specific insects (Nicolson & Thornburg 2007), the influence of predatory wasps in shaping the pollinator system, and how exogenous factors influence this species' low fecundity are salient questions arising from this study that warrant investigation.

Acknowledgements

The authors thank Andy Huber, Oregon State University for providing the Blue Mountains site, GROWISER, its facilities and its population of *Paeonia brownii* for three years as well as for his personal assistance and advice; Gary Booth and Jiping Zou, Brigham Young University, for the analysis of nectar sugars and amino acids; Kyra Krakos at Maryville College and Peter Goldblatt, senior curator, Missouri Botanical Garden, St. Louis, MO for taking additional field data and providing valuable input on the study. We thank Joseph Fortier SJ, Department of Biology Saint Louis University for generously overseeing the live collections; and Michael Pogue, USDA-ARS, Museum of Natural History, Smithsonian Institution, Washington, DC, for his time and expertise in insect identification. We thank Robert Schlising, professor emeritus, California State University Chico for his careful review and cogent advice and Rebekah Beyer for her skilful editing. This research was supported in part by funding from the USDA Forest Service, Pacific Northwest Research Station.

References

Akre RD, Myhre EA (1992) Nesting biology and behavior of the baldfaced hornet, *Dolichovespula maculata* (L.) Hymenoptera: Vespidae in the Pacific Northwest. Melanderia 47:1-30.

Andrieu E, Debussche M, Galloni M, Thompson JD (2007) The interplay of pollination, costs of reproduction and plant size in maternal fertility limitation in perennial *Paeonia officinalis*. Oecologia 152:515-524.

Baker HG, Baker I (1983) A brief historical review of the chemistry of floral nectar. In: Bentley B, Elias T (eds) The Biology of Nectaries. Columbia University Press, New York, 126-152.

Baker HG, Baker I (1986) The occurrence and significance of amino acids in floral nectars. Plant Systematics and Evolution 151:175-186.

Bernhardt P. 1996. Anther adaptations in animal pollination. In: D'Arcy WG, Keating RC (eds) The Anther: Form, Function and Phylogeny. Cambridge University Press, Cambridge, 192–220.

Bernhardt P, Dafni A (2000) Breeding system and pollination biology of *Mandragora officinarum* L. (Solanaceae) in northern Israel. Det Norske Videnskaps-Akadami. I. Mat.-Naturv, Klasse Skrifter 39:215-234.

Buck M, Marshall SA, Cheung DKB (2008) Identification Atlas of the Vespidae (Hymenoptera, Aculeata) of the northeastern Nearctic region. Canadian Journal of Arthropod Identification No 5 (PDF version).

Camp WH, Hubbard MM (1963) Vascular supply and structure of the ovule and aril in peony and the aril in nutmeg. American Journal of Botany 50:174-178.

Cohen SA, Michaud DP (1993) Synthesis of a fluorescent derivatizing reagent 6-aminoquinolyl-N-hydroxysuccinimidyl carbamate, and its application for analysis of hydrolysate amino acids via high performance liquid chromatography. Analytical Biochemistry 211:279-287.

Crumb SE (1956) The larvae of the Phalaenidae. United States Department of Agriculture Technical Bulletin No1135, Washington DC.

Domínguez CA, Dirzo R, Bullock SH (1989) On the function of floral nectar in *Croton suberosus* (Euphorbiaceae). Oikos 56:109–114.

Edwards, R (1980) Social wasps their biology and control. Rentokil Library, East Grinstead UK.

Galindo- González J, Ornelas JF (2002) Interference between wasps and hummingbirds at a feeder. Foresta Veracruzana 4:39-44.

Goldblatt P, Bernhardt P (1990) Pollination biology of *Nivenia* (Iridaceae) and the presence of heterostylous self-compatibility. Israel Journal of Botany 39:93-111.

Goldblatt P, Bernhardt P, Vogan P, Manning JC (2004) Pollination by fungus gnats (Diptera: Mycetophilidae) and self-recognition sites in *Tolmiea menziesii* (Saxifragaceae). Plant Systematics and Evolution 244:55-¬67.

Goulson D, Wright NP (1997) Flower constancy in the hoverflies *Episyrphus balteatus* (Degeer) and *Syrphus ribesii* (L.) (Syrphidae). Behavioral Ecology 9:213-219.

Grant, V (1975) Genetics of flowering plants. Columbia University Press, New York.

Greene A, Akre RD, Landolt PJ (1978) Behavior of the yellowjacket social parasite, *Dolichovespula arctica* (Rohwer) (Hymenoptera: Vespidae). Melanderia 29:1-28.

Guo Q, Brown JH, Valone TJ, Kachman SD (2000) Constraints of seed size on plant distribution and abundance. Ecology 81:2149-2155.

Halket JM, Zaikin VG (2003) Derivatization in mass spectrometry -1. Silylation. European Journal of Mass Spectrometry 9:1-21.

Hardwick DF (1996) A monograph of the North American Heliothentinae (Lepidoptera; Noctuidae). Hardwick DF, Ottowa.

Hiepko P (1964) Das zentrifugale androecium der Paeoniaceae. Deutsche Botanische Gesellschaft 77:427-435.

Hitchcock CL, Cronquist A, Ownbey M, Thompson JW (1964) Vascular plants of the Pacific Northwest Part 2. Salicaceae to Saxifragaceae. University of Washington Press, Seattle.

Hong D, Pan K, Hong Y (1998) Taxonomy of the *Paeonia delavayi* complex (*Paeoniaceae*). Annals of Missouri Botanical Garden 85:554-564.

Inouye DW, Gill DE, Dudash MR, Fenster, CB (1994) A model and lexicon for pollen fate. American Journal of Botany 81:1517-1530.

Jing XM, Zheng GH (1999) Four species of wild peony seed dormancy and germination characteristics, and its relationship to the brink. Acta Phytophysiologica Sinica 25. (Translated from Chinese, no pagination.).

Kearns CA (1992) Anthophilous fly distribution across an elevation gradient. American Midland Naturalist 127:172-182.

Kearns CA (2001) North American dipteran pollinators: assessing their value and conservation status. Conservation Ecology 5: [online] URL: http://www.consecol.org/vol5/iss1/art5/.

Kearns CA, Inouye DW (1994) Fly pollination of *Linum lewisii* (Linaceae). American Journal of Botany 8:1091-1095.

Kurczewski FE (1968) *Vespula maculifrons* (Hymenoptera: Vespidae) preying on the European earwig *Forficula auicularia.* Journal of the New York Entomological Society 76:84-86.

Liebert AE, Gamboa GJ, Stamp NE, Curtis TR, Monnet KM, Turillazzi S, Starks PT (2006) Genetics, behavior, and ecology of a wasp invasion: *Polistes dominulus* in North America. Annales Zoologici Fennici 43:595-624.

Luo YB, Pei YL, Pan KY, Hong DY (1998) A study on pollination biology of *Paeonia suffruticosa* subsp. *spontanea* (*Paeoniaceae*). Acta Phytotaxonomica Sinica 36:134-144.

MacKensie JK, Landolt PJ, Zack RS (2006) Attraction to ornamental peony (*Paeonia lactiflora, Paeoniaceae*) by *Polistes dominulus* Christ (Hymenoptera: Vespidae) demonstrated using olfactometers. Journal of the Kansas Entomological Society 79:231-238.

Madeiros PM, Simoneit RT (2007) Analysis of sugars in environmental samples by gas chromatography-mass spectrometry. Journal of Chromatography 1141:271-278.

Manugistics (2000) STATGRAPHICS 5 Plus. Manugistics, Inc, Rockville, MD.

Matsuura M, Yamane S (1990) Biology of the vespine wasps. Springer-Verlag, Berlin.

Nicolson, SW, Thornburg, RW (2007) Nectar chemistry In: Nicolson SW *et al.* (eds) Nectaries and Nectar. Springer, New York, 215-263.

Ogden, EC, Raynor GS, JHayers JV, Lewis DM (1974) Manual of Sampling Airborne Pollen. Hafner Press, London.

Page M (2005) The Gardener's Peony: Herbaceous and Tree Peonies. Timber Press, Portland, OR.

Proctor M, Yeo P, Lack A (19960 The Natural History of Pollination. Timber Press, Portland, OR.

Rhoades DF, Bergdahl JC (1981) Adaptive significance of toxic nectar. The American Naturalist. 117:798-803.

Richter MR (2000) Social wasp (Hymenoptera: Vespidae) foraging behavior. Annual Review of Entomology 45:121-150.

Sahli HF, Conner, JK (2007) Visitation, effectiveness, and efficiency of 15 genera of visitors to wild radish, *Raphanus raphanistrum* (Brassicaceae). American Journal of Botany 94:203-209.

Sánchez-Lafuente AM (2002) Floral variation in the generalist perennial herb *Paeonia broteroi* (Paeoniaceae): Differences between regions with different pollinators and herbivores. American Journal of Botany 89:1260-1269.

Sánchez-Lafuente AM, Rey PJ, Alcántara JM, Valera F (1999) Breeding system and the role of floral visitors in seed production of a "few-flowered" perennial herb., *Paeonia broteroi* Boiss and Reut. (Paeoniaceae). Ecoscience 6:163-172.

Saunders, AP, Stebbins, LP (1938) Cytogenetic studies in *Paeonia* I. the compatibility of the species and the appearance of the hybrids. Genetics 23:65-82.

Schlising RA (1976) Reproductive proficiency in *Paeonia californica* (*Paeoniaceae*). *American Journal of Botany* 63:1095-1103.

Shuttleworth A, Johnson SD (2006) Specialised pollination by large spider-hunting wasps and self-incompatibility in the African milkweed *Pachycarpus asperifolius.* International Journal of Plant Science 167:177-1186.

Spradbery JP (1973) Wasps. University of Washington Press, Seattle.

Stamp NE, Bowers MD (1988) Direct and indirect effects of predatory wasps (*Polistes* sp.: Vespidae) on gregarious caterpillars (*Hemileuca lucina*: Saturniidae). Oecologia 75:619-624.

Stebbins GL (1938) The Western American species of *Paeonia.* Madrono 4:252-260.

Stebbins GL, Ellerton S (1939) Structural hybridity in *Paeonia californica* and *P. brownii*. Journal of Genetics 38:1-36.

Stebbins GL, Major J (1965) Endemism and speciation in the California flora. Ecological Monographs 35:1-35.

Stern WT (1946) A study of the genus *Paeonia.* The Royal Horticultural Society, London.

Sutherland JP, Sullivan MS, Poppy GM (1999) The influence of floral character on the foraging behaviour of the hoverfly, *Episyrphus balteatus.* Entomologia Experimentalis et Applicata 93:157-164.

Terrab A, García-Castaño JL, Romero JM, Berjano R, DeVega C, Talavera S (2007) Analysis of amino acids in nectar from *Silene colorata* Poiret (Caryophyllaceae). Botanical Journal of the Linnean Society 155:49-56.

USDA Forest Service (1988) Range plant handbook. (Reprint of original 1937 publication). Dover Press, New York.

Vance NC (2012) Finding Brown's peony a sweet attraction. Kalmiopsis 19:1-6.

Vance NC, Bernhardt P, Edens R (2004) Pollination and seed production in *Xerophyllum tenax* (Melanthiaceae) in the Cascade Range of central Oregon. American Journal of Botany 91:2060-2068.

Walters JL (1956) Spontaneous meiotic chromosome breakage in natural populations of *Paeonia californica.* American journal of Botany 43:342-354.

Walters JL (1962) Megasporogenesis and gametophyte selection in *Paeonia californica.* American Journal of Botany 49:789-794.

Zhang D, Sang T (1998) Chromosomal rearrangement of *P. brownii* and *P. californica* revealed by fluorescence *in situ* hybridization. Genome 41: 848-853.

Zhou S, Hong D, Pan K (1999) Pollination biology of *Paeonia jishanensis* T. Hong and W. Z. Zhao (*Paeoniaceae*) with special emphasis on pollen and stigma biology. Botanical Journal of the Linnean Society 130:43-52

5

DO ABUNDANCE AND PROXIMITY OF THE ALIEN *IMPATIENS GLANDULIFERA* AFFECT POLLINATION AND REPRODUCTIVE SUCCESS OF TWO SYMPATRIC CO-FLOWERING NATIVE SPECIES?

Valérie Cawoy[1], Mathieu Jonard[2], Carolin Mayer[1] and Anne-Laure Jacquemart[1,*]

[1]*Research group « Genetics, reproduction, populations » - Earth and Life Institute - Université catholique de Louvain, 2 (box L7.05.14) Croix du Sud, 1348 Louvain-la-Neuve, Belgium*

[2]*Research group « Forest sciences » - Earth and Life Institute - Université catholique de Louvain, 2 (box L7.05.09) Croix du Sud, 1348 Louvain-la-Neuve, Belgium*

Abstract— In invasion ecology, potential impacts of aliens on native flora are still under debate. Our aim was to determine the pollinator mediated effects of both proximity and abundance of an alien species on the reproductive success of natives. We chose the highly invasive *Impatiens glandulifera* and two native species: *Epilobium angustifolium* and *Aconitum napellus* ssp. *lusitanicum*. These species share characteristics allowing for pollination interactions: similar biotopes, overlapping flowering periods and same main pollinators. The effects of abundance (5, 25 and 100 individuals) and proximity (0 and 15 m) of the alien on visitation rate, insect behaviour, pollen deposition and reproductive success of both natives were investigated during 2 flowering seasons. We used centred visitation rates as they can be directly interpreted as a positive or negative effect of the invasive.

Both abundance and proximity of the alien increased bumblebee visitation rates to both natives. On the other hand, abundance of the exotic species had a slight negative effect on honeybee visits to natives while its proximity had no effect. The behaviour of bumblebees changed as visitors left significantly more often the native plants for *I. glandulifera* when its abundance increased. As a consequence of this "inconstancy", bees deposited considerable quantities of alien pollen on native stigmas. Nevertheless, this interspecific pollen transfer did not decrease seed set in natives. Self-compatibility and high attractiveness of both native species probably alleviate the risk of altered pollinator services and reproductive success due to the invader in natural populations.

Keywords: flower constancy, facilitation, pollen deposition, seed set, visitation rate

INTRODUCTION

Entomophilous invasive plant species usually present floral traits that are highly attractive to pollinators like showy and numerous flowers, an extended flowering period, a high nectar production and large amounts of pollen (Ghazoul 2002; Traveset & Richardson 2006). Examples include species introduced for ornamental purposes such as *Heracleum mantegazzianum* (Nielsen et al. 2008), *Impatiens glandulifera* (Chittka & Schürkens 2001; Lopezaraiza-Mikel et al. 2007) or *Lythrum salicaria* (Brown et al. 2002). These 'mass flowering' species are supergeneralists as they attract polylectic generalist insects present in their neighbourhood. Their potential impacts on pollen transfer and subsequent seed set of sympatric co-flowering natives they share pollinators with have been increasingly studied. Insect visitors of native plants are often generalists and readily include aliens in their diets (Memmott & Waser 2002; Lopezaraiza-Mikel et al. 2007; Aizen et al. 2008; Vilà et al. 2009). Therefore, extended resources offered by mass-flowering invaders and the prevalence of generalist pollinators in natural communities may evoke changes in pollinator services to native plants.

The floral composition of the local neighbourhood influences plant-pollinator interactions (Rathcke 1983). An alien plant species might alter these interactions (Bjerknes et al. 2007) but it remains difficult to draw general conclusions. So far, three types of interactions between alien and native plant species have been described: first, a competitive interaction, where an alien reduces pollinator visits to native plants (e.g. Brown et al. 2002; Totland et al. 2006; Kandori et al. 2009); second, a facilitative interaction where a 'magnet-invader' attracts pollinators which would visit natives less in the absence of the invader (e.g. Lopezaraiza-Mikel et al. 2007; Nielsen et al. 2008; Bartomeus, Vilà, et al. 2008); and third, a neutral interaction where an alien does not influence pollinator visits to natives (e.g. Bjerknes et al. 2007; Kaiser-Bunbury & Müller 2009; Vanparys et al. 2011).

The presence of an alien may affect both quantity and compatibility of pollen grains deposited on native stigmas. Less conspecific pollen is deposited in case of competition (Larson et al. 2006; Matsumoto et al. 2010) or due to pollen loss during their flights between species (Flanagan et al. 2009). On the other hand, in case of facilitation, more

*Corresponding author: Anne-Laure.Jacquemart@uclouvain.be

conspecific pollen is deposited. Still, if pollinators also forage on the invader and show inconstant foraging behaviour, loads of alien pollen have been observed on native stigmas (Moragues & Traveset 2005; Cariveau & Norton 2009).

A recent quantitative synthesis compiling data from 40 studies revealed an overall negative impact of aliens on visitation rates and reproductive success of native species (Morales & Traveset 2009). Thus, identifying invasion-sensitive native plant species is crucial to improve conservation strategies.

With numerous showy flowers producing large amounts of nectar and pollen, *Impatiens glandulifera* has been reported to effectively outcompete native species for pollination services in riparian sites and to represent a significant threat on the reproductive success of natives (Chittka & Schürkens 2001; DAISIE 2009; Nienhuis et al. 2009; Vervoort et al. 2011). However, also facilitation effects of *I. glandulifera* on an entire community of co-flowering natives have been observed in field experiments by Lopezaraiza-Mikel et al. (2007), while Bartomeus et al. (2010) did not detect any effect.

Experimental studies usually investigated the effects on visitation rates on a selection of native species or on the entire community of co-flowering natives but have rarely examined seed set (e.g. Chittka & Schürkens 2001; Lopezaraiza-Mikel et al. 2007; Vilà et al. 2009; but see Bartomeus et al. 2010). Diverging results between visitation rates and reproductive success highlight the importance to study both components simultaneously, and to consider changes in conspecific and heterospecific pollen deposits (Moragues & Traveset 2005; Morales & Traveset 2009). Moreover, the effect of the presence of the alien may differ according to its abundance (number of individuals or flowers) and to proximity of alien individuals to natives (Muñoz & Cavieres 2008; Cariveau & Norton 2009; Morales & Traveset 2009). It is therefore necessary to disentangle these two aspects when looking at pollination and reproductive success of natives. To our knowledge, no study has assessed the effect of an alien species at all reproduction levels, from visitation rates to seed set focusing on native species sharing biotopes and presenting similar flowering traits.

We investigated the impact of *Impatiens glandulifera* on pollinator visits, pollen transfer and reproductive success of two sympatric co-flowering native species, the rare *Aconitum napellus* ssp. *lusitanicum* and the common *Epilobium angustifolium*. We chose these species because (1) they share biotopes with *I. glandulifera* like wet meadows, edges of marshes and river banks (Lambinon et al. 2004), (2) their flowering periods overlap (July-August) and (3) their main pollinators are bumblebees and honeybees (Myerscough 1980; Le Cadre 2005; Vervoort et al. 2011). We might thus expect competition for pollinator services and that the alien might affect the reproductive success of the natives, especially threatening the rare native *A. napellus*. All three species are considered valuable sources of nectar and, despite being protandrous, all are self-compatible (Myerscough 1980; Le Cadre 2005; Vervoort et al. 2011).

Using an experimental design, we examined whether the effects of interaction may increase with the abundance and the proximity of the alien. Four main questions were addressed: Does the abundance and/or proximity of the alien *I. glandulifera* (1) modify visitation rates of the main pollinators to the native individuals? (2) change the behaviour of their main pollinators? (3) lead to deposition of alien pollen on native stigmas? and (4) affect the seed set of the natives?

MATERIALS AND METHODS

Studied species

Impatiens glandulifera Royle (Balsaminaceae) is an herbaceous annual introduced from the Himalayas in Europe for horticulture in the 1830s (Beerling & Perrins 1993; Titze 2000). Due to its numerous large flowers with copious nectar (up to 8µl per flower.day) attracting large numbers of insects, the species has been favoured by beekeepers and nursery gardeners (Beerling & Perrins 1993; Nienhuis et al. 2009). *Impatiens glandulifera* has spread in the majority of temperate communities in Europe, growing in riparian biotopes and in other disturbed sites with good water and nutrient supply. It is now considered as one of the 100 worst invasive species in Europe (DAISIE 2009). Its inflorescences consist of 3-12 white to pink or purple flowers. Flowers are 3-4 cm long, zygomorphic, helmet-like and spurred (Beerling & Perrins 1993; Titze 2000; Fig. 1A). Anthesis lasts two days.

Aconitum napellus L. ssp. *lusitanicum* Rouy (Ranunculaceae) is a herbaceous perennial native to Europe. This rare taxon grows in wet biotopes, primarily riparian zones, along rivers and marshes, or in wet forest edges (Lambinon et al. 2004; Le Cadre et al. 2008). Its recent decline has been attributed to destruction and fragmentation of suitable biotopes (Le Cadre 2005; Le Cadre et al. 2008). The main raceme bears up to 40 zygomorphic dark blue or violet helmet shaped flowers (Lambinon et al. 2004; Le Cadre 2005; Fig. 1B). Anthesis lasts 1-2 weeks and flowers produce large amounts of nectar (up to 8µl per flower.day; Heinrich 1979; Marden 1984).

Epilobium angustifolium L. (Onagraceae) is a common herbaceous perennial native to Europe, growing in open and disturbed biotopes like clear-cuttings, wood edges and wet meadows (Myerscough 1980; Lambinon et al. 2004). The main raceme bears up to 400 actinomorphic pink flowers (Sargent & Roitberg 2000; Fig. 1C). Anthesis lasts 2-7 days. *Epilobium angustifolium* is also considered an attractive source of nectar for bees (up to 12µl per flower.day; Schmid-Hempel & Speiser 1988; Sargent & Roitberg 2000).

Cultivation

Native species were cultivated from pieces of rhizomes collected in April 2008 in three large populations from South Belgium (50 fragments per population). The annual *I. glandulifera* was cultivated from seeds collected in 2007 in two river bank populations in Central Belgium. Seeds were sown each year and germinated in a climate chamber as described by Vervoort et al. (2011).

FIG. 1. Morphology of the three plant species and visiting behaviour of bumblebees (*Bombus pascuorum*). (A) When a bumblebee saps nectar from the spur of *I. glandulifera*, it touches anthers with dorsal parts (head, thorax and abdomen). Prior to insect visits, blue fluorescent dye had been deposited on open anthers of *I. glandulifera* and a dye load is visible on the back of the insect. (B) Bumblebee legitimately visiting *A. napellus* in search for nectar. When it enters the flower, dorsal parts touch anthers and stigmas. (C) Bumblebee drinking nectar on *E. angustifolium*. The ventral parts of the insect body are in contact with stigmatic lobes and anthers.

All plants were grown outdoors in 5 L pots filled with peat compost at the Université catholique de Louvain (UCL, Louvain-la-Neuve). During summer 2008 and 2009, flowering individuals were placed in two experimental sites (see below). The same native individuals were used during the two years.

Site description and experimental design

Independent experiments with both alien - native pairs (*I. glandulifera – E. angustifolium* and *I. glandulifera – A. napellus*) were carried out in two experimental sites separated by 5 km, in Central Belgium, Louvain-la-Neuve (50°40' N, 4°37' E). Both sites were mown grasslands of approximately 170 x 55 m. They were homogeneous and poor in entomophilous flowering plants. The vegetation was dominated by grasses, 10-25 cm high. *Trifolium repens* was the only insect pollinated species flowering during experiments, at maximum covering 10% of the surface.

The experimental design combined different degrees of *I. glandulifera* abundance and proximity. In 2008, we tested two degrees of *I. glandulifera* abundance (5 and 25 individuals) and two degrees of proximity (0 and 15 m). In 2009, we tested three degrees of *I. glandulifera* abundance (5, 25 and 100 individuals) at close proximity (0 m). Both years, we added plots of natives without *I. glandulifera* (see below centred visitation rate calculation). Native species were grouped in 1 m^2 clumps of seven individuals (Fig. 2). In both years, experiments started with the pair *I. glandulifera – E. angustifolium* in mid July. Observations on the second pair *I. glandulifera – A. napellus* began immediately afterwards until mid August.

Insect observations

Insect visits and foraging behaviour were recorded during 10-min periods on 23 days for the pair *I. glandulifera – E. angustifolium* and on 28 days for *I. glandulifera – A. napellus*. Observations were carried out at both sites simultaneously on sunny days between 10h00 and 17h00. Between five and 42 observations (24 on average) were conducted per site per day, resulting in total a observation time of 181 hrs. Open flowers per individual were counted daily. Number of visitors and visitor category were

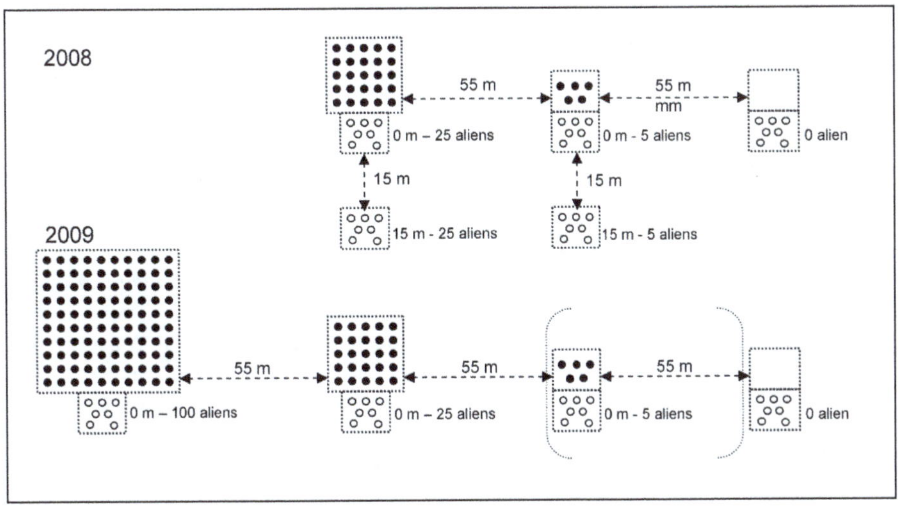

FIG. 2. Per year (2008 and 2009), experimental design for both alien-native pairs (*I. glandulifera - A. napellus*; *I. glandulifera - E. angustifolium*). Each dot represents one individual of an alien (black dots) or a native plant (white dots). In 2009, the treatment "0 m -5 aliens" was only present in the site A.

noted. Insect categories were limited to the four main previously determined visitor groups: honeybees, bumblebees, small-sized syrphids and other insects. During observations, a total of 14 349 insects were recorded. Insect category proportions varied among sites and years but social bees and syrphids were the most frequent for all three studied species. Observations confirmed that *A. napellus, E. angustifolium* and *I. glandulifera* shared the main visitors: bumblebees (65.5% of visits), small-sized syrphids (17.2%), and honeybees (15.3%). An insect was considered a pollinator when it touched the reproductive organs. Social bees were effective pollinators due to high abundances (74.4% on *A. napellus,* 55.7% on *E. angustifolium* and 93.5% on *I. glandulifera*) and frequent contacts with anthers and stigmas (Fig. 1B-C). Honeybees (*Apis mellifera*) and two bumblebee species were identified in the field: *Bombus pascuorum* and *B. terrestris* s.l. On the contrary, small-sized syrphids, mainly *Episyrphus balteatus,* despite their large numbers (21.3% on *A. napellus,* 38.5% on *E. angustifolium* and 5.8% on *I. glandulifera,*) did not act as pollinators and were not included in analyses.

Total visitation rate was calculated as the total number of visits per 10-min period divided by the number of open flowers observed. We also calculated visitation rates for honeybees and for bumblebees separately. For both natives, all flowers of the seven individuals from each patch were observed simultaneously. For *I. glandulifera,* due to the numerous flowers and the high visitation rates, observations were made on a group of 20-30 flowers.

To detect a possible change in insect behaviour in 2009, the direction of insect transitions was recorded when they visited native plants (i.e. if they came from or left a native for an alien).

Pollen load on native stigmas

In order to test if alien pollen contaminated native stigmas as a result of interspecific insect movements, we collected native flowers in female phase (12 per treatment and per site) in 2008. Flowers were fixed in FAA before staining stigmas in acid fuchsin solution (Kearns & Inouye 1993). The red-coloured pollen grains were immediately observed under a light microscope. Pollen grains of the three species were easily distinguished according to size and shape, and counted.

Reproductive success: seed set of natives

On each inflorescence of the native individual plants, all flowers open during observations were marked. Of these, five ripe fruits per plant were collected in August for *E. angustifolium,* and eight ripe fruits per plant were collected in September for *A. napellus.* Seeds of *A. napellus* were counted by naked eye. Wind-dispersed plumed seeds of *E. angustifolium* were scattered on a sticky transparent sheet and scanned at 800 dpi. Seeds were automatically counted using the MacBiophotonics ImageJ software (Plugins: Particle Analysis, Cell counter; Abramoff et al. 2004).

For both species, ripe viable seeds, aborted seeds (flattened and smaller size) and unfertilized ovules were distinguished. Viability was confirmed via germination tests after stratification during three months at 4°C. The reproductive success (seed set) was estimated as number of viable seeds divided per number of total ovules per fruit.

Statistical analyses

Visitation rates were transformed into centred visitation rates (VR_C) by subtracting each visitation rate (VR) by the visitation rate of the plot without *I. glandulifera* within a radius of 55m (VR_0) on the same date and time of day.

$$VR_C = VR - VR_0 \qquad \text{[eq. 1]}$$

The advantage of the centred visitation rate is that it can be directly interpreted as a positive or negative effect of the alien and that its absolute value provides the magnitude of the effect. This transformation removes part of the site effects independent from *I. glandulifera* abundance (a comparable method was used by Price et al. 2005). Centred visitation rates were analyzed using linear mixed models with the MIXED procedure in SAS. Mixed models allow for correct analysis of such intercorrelated data sets (here: repeated measurements in time) and effectively deal with pseudoreplication (van de Pol & Wright 2009), which has been defined as the misanalyses or misinterpretation of replicates that are not statistically independent (Hurlbert 1984).

The effect of *I. glandulifera* abundance on the centred visitation rate was analyzed using only the data from 2009 (three levels of abundance). A linear mixed model describing the variation of the centred visitation rate with the natural logarithm of *I. glandulifera* abundance was adjusted for each native plant species and pollinator category. We log-transformed abundance to take into account the fact that an increase of 10 individuals has a much stronger effect when it occurs between 0 and 10 than between 100 and 110 or 1000 and 1010. The formula of the model is:

$$VR_C = a \times \ln(n\_imp) + \delta(0, \sigma^2_\delta) + \varepsilon(0, \sigma^2) \qquad \text{[eq. 2]}$$

where:

- VR_C is the centered visitation rate,
- *n_imp* is the abundance of *I. glandulifera* (number of plants in the patch),
- *a* is a model parameter,
- δ is a random effect accounting for the census date effect,
- ε is the residual term whose normality was checked.

The relevance of introducing a random effect in the model was evaluated on the basis of the AIC (Brown & Prescott 1999).

The impact of the proximity of *I. glandulifera* on the effect of its abundance on the centred visitation rate was analyzed based on the data collected in 2008. A linear model of the same form was used except that the parameter *a* varied according to the proximity level (0 or 15 m). Contrasts were used to test the *I. glandulifera* proximity effect on the parameter *a* of the model (Brown & Prescott 1999).

To test whether changes of visitation rates to the native species would simply mirror an increase in patch size, we compared visitation rates of bumble and honeybees to *I. glandulifera* for the different abundances with a generalized linear mixed model with the different abundance classes and years as fixed effects and included date and site as random factors (Procedure GLIMMIX). To test the effects of *I. glandulifera* abundance on the proportions of bumblebee transitions from and towards *I. glandulifera* we used a Chi-square test using data collected in 2009 (not enough data for honeybee movements). Effects of *I. glandulifera* abundance on the number of pollen grains deposited on native stigmas in 2008 were tested for pollen origins, alien (*I. glandulifera*), conspecific (*A. napellus* or *E. angustifolium*), or other (other species) with one-way ANOVA. Finally, to test the effects of *I. glandulifera* abundance and proximity on seed set we used one-way ANOVA. Data for seed sets were arcsin (*A. napellus*) or log (*E. angustifolium*) transformed to achieve normality. All statistical analyses were performed using SAS (version 9.1; SAS Institute Inc., Cary, N.C.). Means are given with their standard errors.

RESULTS

Effects of I. glandulifera *on the visitation rates to natives*

The total visitation rate to the alien (0.34 ± 0.1 visits 10 min^{-1} flower^{-1}; years and sites pooled) was 10.4 times higher than that to *A. napellus* and 6.1 times higher than that to *E. angustifolium*. Visitation rates to *I. glandulifera* did not change with increasing abundance ($F_{1,741}$ = 0.60, P = 0.549). However, the abundance of the alien modified visitation rates to natives.

Bumblebee and honeybee visitation rates responded differently to the abundance and proximity of the alien (Fig. 3). Higher abundance of *I. glandulifera* increased the centered visitation rates of bumblebees to native plants, especially for *E. angustifolium* (P = 0.137 for *A. napellus* and P < 0.001 for *E. angustifolium*). On the contrary, higher abundance of *I. glandulifera* decreased the visitation rate of honeybees to both natives (Fig. 3; P = 0.022 for *A. napellus* and P = 0.003 for *E. angustifolium*).

Proximity of *I. glandulifera* significantly increased the effect of its abundance on bumblebee visitation rate mainly for *E. angustifolium* (Fig. 4; P < 0.001) but also for *A. napellus* (P = 0.007). Not surprisingly, the abundance effect was much larger when *I. glandulifera* plants were adjacent to the natives. For *A. napellus*, the abundance effect disappeared at 15 m distance between the alien and the native but for *E. angustifolium*, a small facilitation effect could still be shown at 15 m (Fig. 4; P = 0.035). Regarding honeybees, proximity had no significant impact on the effect of alien abundance on the visitation rate (P = 0.102 for *A. napellus* and P = 0.980 for *E. angustifolium*).

Behavioural changes for pollinators

Bumblebees shifted frequently between *I. glandulifera* and the native species as indicated by the proportions of individuals coming from or leaving for an alien flower (Table 1). When alien abundance increased, a significantly higher proportion of bumblebees left both natives for the alien but proportions of bumblebees coming from the alien did not change significantly (Table 1). In all cases, the majority of the observed movements were interspecific.

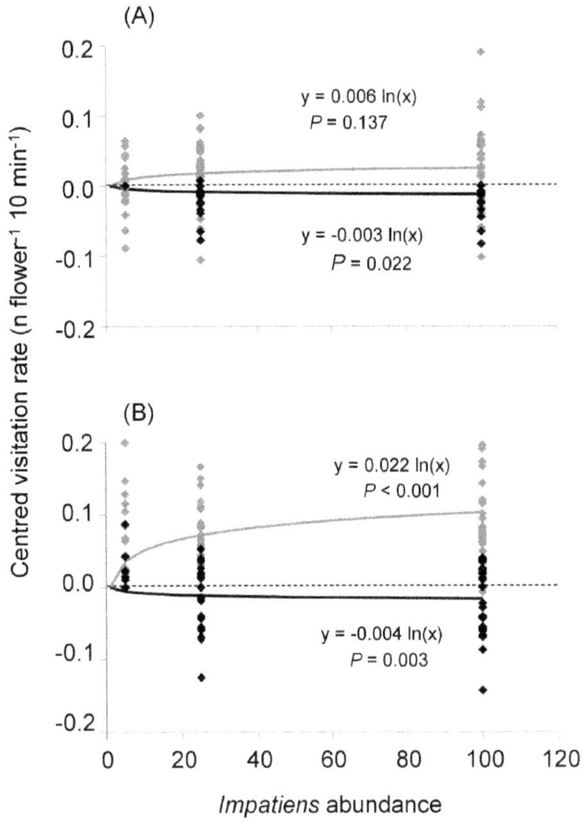

FIG. 3. Centred visitation rate in 2009 as a function of *I. glandulifera* abundance (5, 25 and 100 individuals) for both native plant species, (A) *A. napellus* and (B) *E. angustifolium* and for the two main pollinator categories (grey diamonds: bumblebees; black diamonds: honeybees). For both pollinator categories, model predictions are represented by a solid line and described by the equation with the corresponding *P*-value indicating significance.

Pollen loads on native stigmas

Insects deposited considerable quantities of pollen from *I. glandulifera* on native stigmas (Fig. 5). The abundance of the alien affected heterospecific pollen deposition since the quantities of alien pollen on stigmas of both natives increased with higher *I. glandulifera* numbers of individuals (Fig. 5; $F_{1,84}$ = 15.42; P < 0.001 for *A. napellus* and $F_{1,85}$ = 48.67; P < 0.001 for *E. angustifolium*). This increase of heterospecific pollen was much lower for *A. napellus* than for *E. angustifolium*. At the same time, quantities of conspecific pollen on stigmas of both natives remained more or less equal and unaffected by alien abundance ($F_{1,84}$ = 0.86; P = 0.427 for *A. napellus* and $F_{1,85}$ = 0.25; P = 0.783 for *E. angustifolium* respectively). As a consequence, total pollen loads on stigmas of *A. napellus* remained mainly conspecific, even at high abundances of *I. glandulifera* (Fig. 5A). On the

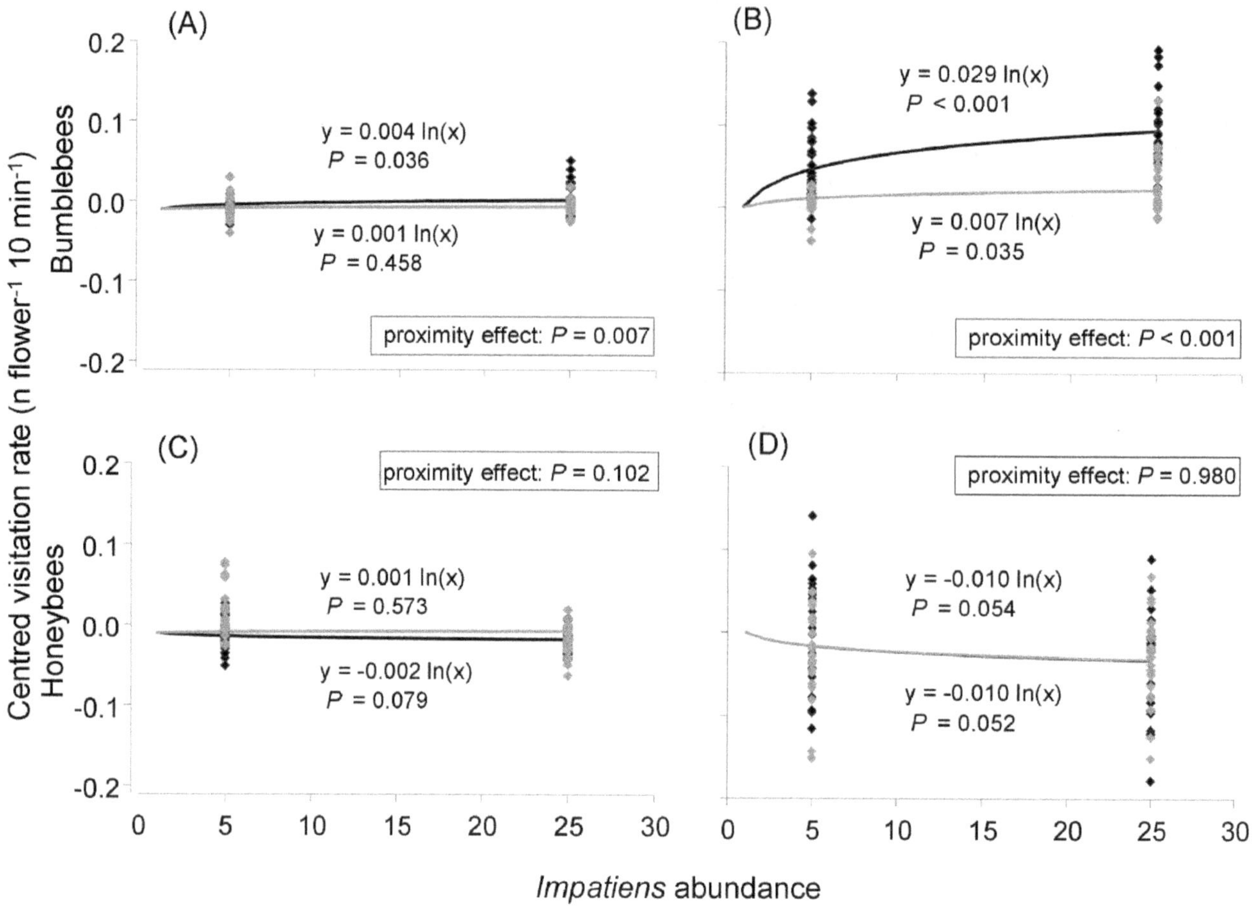

FIG. 4. Impact of the proximity (black diamonds: 0 m distance, grey diamonds: 15 m distance) of *I. glandulifera* on the effect of its abundance on the centred visitation rate in 2008 for both native plant species, *A. napellus* (A) & (C) and *E. angustifolium* (B) & (D), and for the two main pollinator categories (bumblebees and honeybees). For each combination of plant species, pollinator category and proximity level, model predictions are represented by a solid line and described by the equation with the corresponding *P*-value indicating significance. Another *P*-value indicates the significance of the proximity effect.

contrary, total pollen loads on stigmas of *E. angustifolium* increased with growing alien abundance and were finally dominated by alien pollen (Fig. 5B).

Impacts on seed set of natives

Seed set ranged between 71 and 85% for *A. napellus* and between 53 and 64% for *E. angustifolium* (Fig. 6). The alien abundance did not affect the seed set of *A. napellus* ($F_{1,191}$ = 0.46; P = 0.781) and *E. angustifolium* ($F_{1,119}$ = 0.42; P = 0.762). Neither proximity nor abundance of the alien influenced the seed sets of both natives in both sites (Fig. 6).

DISCUSSION

To determine a possible influence of the alien species on the reproductive success of natives, all steps during the pollination process, from insect visitation rates to pollen deposition and seed set, will be discussed sequentially.

Do abundance and/or proximity of the alien modify the visitation rates to natives?

Also other studies observed high visitation rates to *I. glandulifera,* probably due to its valuable sources of both

	I. glandulifera abundance			χ^2	*P*
	5	25	100		
A. napellus					
Coming from alien	93.6	92.5	90.9	0.79	NS
Leaving for alien	69.2	83.3	87.8	15.41	<0.0001***
E. angustifolium					
Coming from alien	86.2	89.8	87.8	0.72	NS
Leaving for alien	73.1	90.4	90.6	22.00	<0.0001***

TABLE 1. Total percentage of bumblebee transitions from ("coming from") or towards ("leaving for") *I. glandulifera* observed on *A. napellus* (total number of transitions observed: n = 1 045) and *E. angustifolium* (n = 898) for the three abundance levels of the alien (5, 25 and 100 individuals) in 2009.

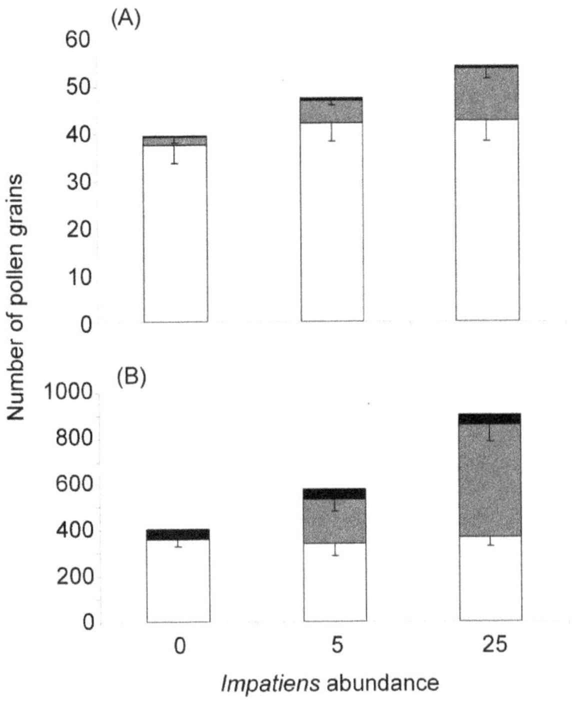

FIG. 5. Origin (white: conspecific, grey: alien, black: other) and quantity of pollen grains deposited on stigmas of native species according to *I. glandulifera* abundance (0, 5 and 25 individuals). (A) *A. napellus* (one stigma per flower). (B) *E. angustifolium* (one stigmatic lobe per flower). Data were collected in 2008 in plots adjacent to *I. glandulifera* and are presented as mean ± s.e. (n = 24-30 flowers).

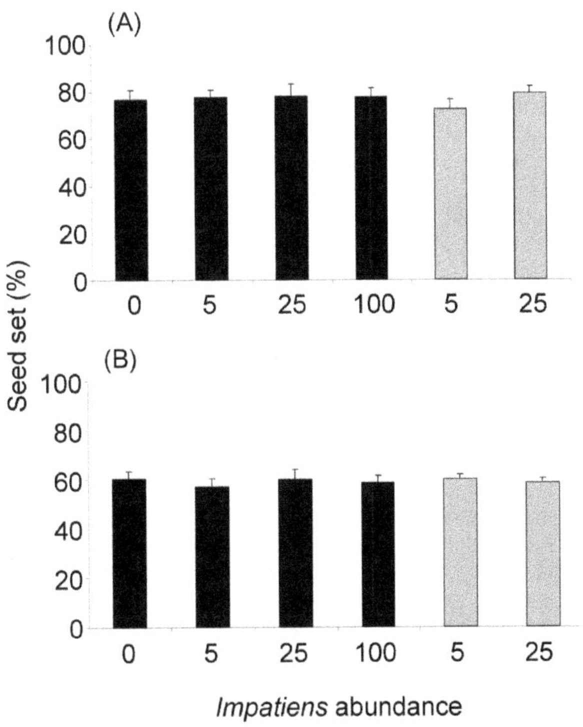

FIG. 6. Seed set of (A) *A. napellus* and (B) *E. angustifolium* for both years and sites pooled, according to the abundance (0, 5, 25 and 100 individuals) and proximity (black bars: 0 m distance, grey bars: 15 m distance) of the alien.

nectar and pollen (Lopezaraiza-Mikel et al. 2007; Nienhuis et al. 2009; Bartomeus et al. 2010; Vervoort et al. 2011). As already reported elsewhere, *Bombus pascuorum* was the main pollinator we observed (Chittka & Schürkens 2001; Nienhuis et al. 2009; Vervoort et al. 2011). Our results confirmed that the alien *I. glandulifera* received higher total visitation rates than both natives studied, comparable to other comparisons, e.g. with the native *I. noli-tangere* (Vervoort et al. 2011). Therefore, *I. glandulifera* can be considered as a potential competitor for pollinators of natives (Chittka & Schürkens 2001). In case of competition, the reproductive success of natives could be reduced. This would imply a greater threat for *A. napellus* ssp. *lusitanicum*, a rare species of patrimonial interest mainly pollinated by long-tongued bumblebees (Le Cadre 2005).

However, our results demonstrated a facilitative effect for *A. napellus* with an increase of visitation rates at least for bumblebee visitors (mainly *B. pascuorum*, a long-tongued bumblebee). On the contrary, for honeybees, a small competitive effect with lower visitation rates was detected. Similarly, for *E. angustifolium*, an increase of visitation rates was observed for bumblebees (*Bombus pascuorum* and *B. terrestris* s.l.) whereas lower visitation rates were registered for honeybees. The alien was more attractive than both natives, even for honeybees. We can posit that the alien provided such large quantities of nectar and pollen that hive recruitment increased honeybee worker visits. Honeybees might further have difficulties to exploit the native *A. napellus*, and they were often observed to rob nectar. For bumblebees, the facilitation effect obviously outweighed possible competition which might be due to the otherwise poor flower resources at the experimental sites and could be different in natural populations.

Five *I. glandulifera* individuals were sufficient to facilitate visitation of bumblebees to the investigated native species, an effect that strongly increased with higher alien abundance. This facilitation effect was not simply due to a rise in patch size, since at the same time, the visitation rate to *I. glandulifera* did not increase with higher abundances. We did, however, not control for increased abundance of native floral resources. Our results show effects at the local scale but data are still needed for non invaded large natural populations. Even though our patches were spatially not independent (bees can forage over larger ranges than 50 m), the potential of *I. glandulifera* to influence pollination of natives increased at short distances. Similar trends have been reported in previous studies using experimental design with potted plants (Nielsen et al. 2008; Cariveau & Norton 2009). Also the meta-analytical approach of Morales and Traveset (2009) confirmed the importance of alien abundance for interactions for insect visits. Of course, facilitation and competition for pollinator services between invasive and native species are extremes on a continuum that is a function of population density or relative abundance of floral resources (Bernhardt et al. 2008). And they definitely depend upon the insect species that visit. It is therefore not surprising that we obtained different results for honeybees and bumblebees.

Does pollinator behaviour change according to abundance and/or proximity of the alien?

The majority of bumblebees showed inconstant foraging behaviour and switched between the alien and the adjacent native. Moreover, interspecific movements increased with the abundance of the alien, and higher proportions of bumblebees left natives for the alien. Bumblebee constancy is often higher when pollinators forage among flowers that differ in size and odours than when flowers differ in colour only (Goulson 2003; Gegear & Thomson 2004; Gegear 2005; Raine & Chittka 2007). Flower constancy in pollinators is related to their inability to perceive process or recall multicomponent floral signals (Gegear 2005). In our case, bumblebee constancy was low at close vicinity of the alien (contrary to honeybees). This infidelity could be due to flower similarities among the three species. However, colours are more similar between the alien and *E. angustifolium* than with the violet *A. napellus*. This fact is compensated by flower morphology which is more similar between the alien and *A. napellus* (helmet-shaped and zygomorphic). More detailed analyses about flower odour or size could disentangle attractiveness components among our studied species.

Do visitors deposit alien pollen on native stigmas?

The amount of alien pollen deposited by pollinators on native stigmas is usually considered rather low, suggesting that differences in pollen placement on insect bodies might avoid interspecific pollen deposits (Lopezaraiza-Mikel et al. 2007; Jakobsson et al. 2008). In our study, we observed that honeybees crawled into the flower of *I. glandulifera* in order to reach the nectar without touching the anthers (also reported by Sowig 1989) and they did not visit this species for pollen. It was therefore unlikely that they contributed significantly to alien pollen transfer to natives. On the contrary, dorsal parts (head, thorax and abdomen) of visiting bumblebees were covered with alien pollen (see also Nienhuis & Stout 2009). When they visited natives, they turned inside the flowers thereby touching stigmas with their dorsal parts. As they moved from invasive to native flowers, they deposited pollen of *I. glandulifera* on native stigmas. As a consequence, when both natives were placed close to *I. glandulifera*, alien pollen was deposited on their stigmas. Therefore, conspecific native pollen could experience competition with alien pollen that can impede its placement, its germination or the growth of pollen tubes by stylar inhibition or stigma clogging (Brown & Mitchell 2001; Morales & Traveset 2008; Nielsen et al. 2008).

Variations in the visitation rates of bumblebees and honeybees can induce fluctuations in the amount of conspecific pollen deposited on native stigmas (Ghazoul 2002; Larson et al. 2006; Bartomeus, Bosch, et al. 2008; Flanagan et al. 2009). However, as reported for other alien-native pairs, even if the amount of alien pollen increased with alien abundance, the amount of conspecific pollen on *A. napellus* and *E. angustifolium* stigmas was globally not affected (Grabas & Laverty 1999; Larson et al. 2006).

What are the consequences on the seed set of natives?

Conspecific pollen deposition allowed for sufficient native fertilization as seed set of both natives was not affected by the abundance and/or the proximity of the alien. Thus, even if the quantity of heterospecific pollen increased due to the increase of alien pollen at high alien abundances, this did not reduce subsequent seed set. It has already been reported that in some cases, even when a large amount of heterospecific pollen was deposited, no germination interference and no reduction of reproductive success were observed (Morales & Traveset 2008). Closer investigation of pollen germination and pollen tube growth could give further insights into possible lack of competition.

Conclusions

Abundance and proximity of the alien increased bumblebee visitation rates to natives. Even if bumblebees moved frequently between species allowing for alien pollen deposits on stigmas of the natives *A. napellus* and *E. angustifolium*, no negative impacts of *I. glandulifera* on native reproductive success were found. Probably owed to self-compatibility and high attractiveness of both native species, the risk that the invader impedes their reproductive success by altering pollinator services may assumed to be low in natural populations. Still, effects of *I. glandulifera* on bumblebee population dynamics need further evaluation. Social bees, like bumblebees and honeybees, have broad foraging ranges and benefit from mass-flowering resources at large spatial scales (Westphal et al. 2003). Therefore, the alien *I. glandulifera* could even have a positive effect on colony fitness of *Bombus* species (Westphal et al. 2009; Bartomeus et al. 2010).

ACKNOWLEDGEMENTS

The authors thank P. Lhoir and C. Decamps for providing access to the experimental sites and G. Collignon for water supplying. We thank Y. Storder for *A. napellus* location information and to D. Jacquemart and L. Saad for *E. angustifolium* populations. We received authorisation (ref. DNF/DN/PB/SLI/MLB/802.7) to collect *A. napellus* in extended and no endangered populations. We gratefully thank people who helped us to transport the plants, especially the team of GPEX (UCLouvain), and all the students and members of different labs who participated to the insect observations: E. Affrye, A. Alonso, M. André, P. Baret, A. Bauwens, A. Burton, N. Chevalier, A. Defgnée, M. Descamps, M. De Toffoli, L. Dhont, J. F. Dumasy, B. Franck, C. Gibon, B. Host, A. Jones, G. Kelly, M. Lanotte, J. Lebrun, O. Lehanse, L. C. Kouonon, Y. Martin, S. Meurée, L. A. Minsart, C. Noel, B. Pernechele, J. Pirotte, M. and E. Senterre, J. Tigel-Pourtois, V. Vanparys, F. Vanwindekens, A. Verhegghen and X. Vroman. This study was financed by the Belgian Science Policy (Alien Impact project) [grant number SD/BD/01A-B] and the Université catholique de Louvain (FSR project). Three anonymous reviewers gave helpful comments on earlier versions of this manuscript.

REFERENCES

Abramoff MD, Magalhães PJ, Ram SJ (2004) Image processing with ImageJ. Biophotonics international 11:36–42.

Aizen MA, Morales CL, Morales JM (2008) Invasive Mutualists Erode Native Pollination Webs. PLoS Biology 6:e31.

Bartomeus I, Bosch J, Vilà M (2008) High invasive pollen transfer, yet low deposition on native stigmas in a *Carpobrotus*-invaded community. Annals of Botany 102:417–24.

Bartomeus I, Vilà M, Santamaría L (2008) Contrasting effects of invasive plants in plant-pollinator networks. Oecologia 155:761–770.

Bartomeus I, Vilà M, Steffan-Dewenter I (2010) Combined effects of *Impatiens glandulifera* invasion and landscape structure on native plant pollination. Journal of Ecology 98:440–450.

Beerling DJ, Perrins JM (1993) *Impatiens glandulifera* Royle (*Impatiens roylei* Walp.). Journal of Ecology 81:367–382.

Bernhardt CE, Mitchell RJ, Michaels HJ (2008) Effects of population size and density on pollinator visitation, pollinator behavior, and pollen tube abundance in *Lupinus perennis*. International Journal of Plant Sciences 169:944–953.

Bjerknes A-L, Totland Ø, Hegland SJ, Nielsen A (2007) Do alien plant invasions really affect pollination success in native plant species? Biological Conservation 138:1–12.

Brown BJ, Mitchell RJ (2001) Competition for pollination: effects of pollen of an invasive plant on seed set of a native congener. Oecologia 129:43–49.

Brown BJ, Mitchell RJ, Graham SA (2002) Competition for pollination between an invasive species (purple loosestrife) and a native congener. Ecology 83:2328–2336.

Brown H, Prescott R (1999) Applied Mixed Models in Medicine. John Wiley and Sons Ltd., Chichester, England.

Le Cadre S (2005) Allee effects on plants. The case of *Aconitum napellus* subsp. *lusitanicum*. Thesis. Université Pierre et Marie Curie (Paris VI). Paris, France.

Le Cadre S, Tully T, Mazer SJ, Ferdy J-B, Moret J, Machon N (2008) Allee effects within small populations of *Aconitum napellus* ssp. *lusitanicum*, a protected subspecies in northern France. The New Phytologist 179:1171–1182.

Cariveau DP, Norton AP (2009) Spatially contingent interactions between an exotic and native plant mediated through flower visitors. Oikos 118:107–114.

Chittka L, Schürkens S (2001) Successful invasion of a floral market. Nature 411:653.

DAISIE (2009) Handbook of alien species in Europe. Springer, Dordrecht.

Flanagan RJ, Mitchell RJ, Knutowski D, Karron JD (2009) Interspecific pollinator movements reduce pollen deposition and seed production in *Mimulus ringens* (Phrymaceae). American Journal of Botany 96:809–815.

Gegear RJ (2005) Multicomponent floral signals elicit selective foraging in bumblebees. Naturwissenschaften 92:269–271.

Gegear RJ, Thomson JD (2004) Does the flower constancy of bumble bees reflect foraging economics? Ethology 110:793–805.

Ghazoul J (2002) Flowers at the front line of invasion? Ecological Entomology 27:638–640.

Goulson D (2003) Bumblebees: Their behaviour and ecology. Oxford University Press, Oxford.

Grabas GP, Laverty TMDA (1999) The effect of purple loosestrife (*Lythrum salicaria* L,; Lythraceae) on the pollination and reproductive success of sympatric co-flowering wetland plants. Ecoscience 6:230–242.

Heinrich B (1979) Resource heterogeneity and patterns of movement in foraging bumblebees. Oecologia 40:235–245.

Hurlbert SH (1984) Pseudoreplication and the design of ecological field experiments. Ecological Monographs 54:187–211.

Jakobsson A, Padrón B, Traveset A (2008) Pollen transfer from invasive *Carpobrotus* spp. to natives – A study of pollinator behaviour and reproduction success. Biological Conservation 141:136–145.

Kaiser-Bunbury CN, Müller CB (2009) Indirect interactions between invasive and native plants via pollinators. Naturwissenschaften 96:339–346.

Kandori I, Hirao T, Matsunaga S, Kurosaki T (2009) An invasive dandelion unilaterally reduces the reproduction of a native congener through competition for pollination. Oecologia 159:559–569.

Kearns CA, Inouye DW (1993) Techniques for pollination biologists. University Press of Colorado, Niwot, Colorado.

Lambinon J, Delvosalle L, Duvigneaud J (2004) Nouvelle flore de la Belgique, du Grand-Duche de Luxembourg, du nord de la France et des regions voisines: Pteridophytes et spermatophytes. Jardin botanique de Belgique.

Larson DL, Royer RA, Royer MRDAJDOJB (2006) Insect visitation and pollen deposition in an invaded prairie plant community. Biological Conservation 130:148–159.

Lopezaraiza-Mikel ME, Hayes RB, Whalley MR, Memmott J (2007) The impact of an alien plant on a native plant-pollinator network: an experimental approach. Ecology Letters 10:539–550.

Marden JH (1984) Remote perception of floral nectar by bumblebees. Oecologia 64:232–240.

Matsumoto T, Takakura K-I, Nishida T (2010) Alien pollen grains interfere with the reproductive success of native congener. Biological Invasions 12:1617–1626.

Memmott J, Waser NM (2002) Integration of alien plants into a native flower-pollinator visitation web. Proceedings of the Royal Society B: Biological Sciences 269:2395–2399.

Moragues E, Traveset A (2005) Effect of *Carpobrotus* spp. on the pollination success of native plant species of the Balearic Islands. Biological Conservation 122:611–619.

Morales CL, Traveset A (2008) Interspecific pollen transfer: magnitude, prevalence and consequences for plant fitness. Critical Reviews in Plant Sciences 27:221–238.

Morales CL, Traveset A (2009) A meta-analysis of impacts of alien vs. native plants on pollinator visitation and reproductive success of co-flowering native plants. Ecology Letters 12:716–728.

Muñoz AA, Cavieres LA (2008) The presence of a showy invasive plant disrupts pollinator service and reproductive output in native alpine species only at high densities. Journal of Ecology 96:459–467.

Myerscough PJ (1980) *Epilobium angustifolium* L. Journal of Ecology 68:1047–1074.

Nielsen C, Heimes C, Kollmann J (2008) Little evidence for negative effects of an invasive alien plant on pollinator services. Biological Invasions 10:1353–1363.

Nienhuis CM, Dietzsch AC, Stout JC (2009) The impacts of an invasive alien plant and its removal on native bees. Apidologie 40:450–463.

Nienhuis C, Stout J (2009) Effectiveness of native bumblebees as pollinators of the alien invasive plant *Impatiens glandulifera* (Balsaminaceae) in Ireland. Journal of Pollination Ecology 1:1–11.

van de Pol M, Wright J (2009) A simple method for distinguishing within- versus between-subject effects using mixed models. Animal Behaviour 77:753–758.

Price MV, Waser NM, Irwin RE, Campbell DR, Brody AK (2005) Temporal and spatial variation in pollination of a montane herb: a seven-year study. Ecology 86:2106–2116.

Raine NE, Chittka L (2007) Flower constancy and memory dynamics in bumblebees (Hymenoptera: Apidae: *Bombus*). Entomologia Generalis 29:179–199.

Rathcke B (1983) Competition and facilitation among plants for pollination. In: Real L (ed) Pollination Biology. Academic Press, London, pp 305–329.

Sargent RD, Roitberg BD (2000) Seasonal decline in male-phase duration in a protandrous plant: a response to increased mating opportunities? Functional Ecology 14:484–489.

Schmid-Hempel P, Speiser B (1988) Effects of inflorescence size on pollination in *Epilobium angustifolium*. Oikos 53:98–104.

Sowig P (1989) Effects of flowering plant's patch size on species composition of pollinator communities, foraging strategies, and resource partitioning in bumblebees (Hymenoptera: Apidae). Oecologia 78:550–558.

Titze A (2000) The efficiency of insect pollination of the neophyte *Impatiens glandulifera* (Balsaminaceae). Nordic Journal of Botany 20:33–42.

Totland Ø, Nielsen A, Bjerknes A-L, Ohlson M (2006) Effects of an exotic plant and habitat disturbance on pollinator visitation and reproduction in a boreal forest herb. American Journal of Botany 93:868–873.

Traveset A, Richardson DM (2006) Biological invasions as disruptors of plant reproductive mutualisms. Trends in Ecology and Evolution 21:208–216.

Vanparys V, Cawoy V, Mahaux O, Jacquemart A-L (2011) Comparative study of the reproductive ecology of two co-occurring related plant species: the invasive *Senecio inaequidens* and the native *Jacobaea vulgaris*. Plant Ecology and Evolution 144:3–11.

Vervoort A, Cawoy V, Jacquemart A-L (2011) Comparative reproductive biology in co-occurring invasive and native *Impatiens* species. International Journal of Plant Sciences 172:366–377.

Vilà M, Bartomeus I, Dietzsch AC, Petanidou T, Steffan-Dewenter I, Stout JC, Tscheulin T (2009) Invasive plant integration into native plant–pollinator networks across Europe. Proceedings of the Royal Society B: Biological Sciences 276:3887 –3893.

Westphal C, Steffan-Dewenter I, Tscharntke T (2003) Mass flowering crops enhance pollinator densities at a landscape scale. Ecology Letters 6:961–965.

Westphal C, Steffan-Dewenter I, Tscharntke T (2009) Mass flowering oilseed rape improves early colony growth but not sexual reproduction of bumblebees. Journal of Applied Ecology 46:187–193.

6

POLLINATOR CONSERVATION AND MANAGEMENT ON ELECTRICAL TRANSMISSION AND ROADSIDE RIGHTS-OF-WAY

Victoria A. Wojcik[1] and Stephen Buchmann[1,2]

[1]*Pollinator Partnership, 423 Washington Street, 5th Floor San Francisco, CA 94111*
[2]*Departments of Entomology, and Ecology and Evolutionary Biology, The University of Arizona, Tucson, AZ 85743*

Abstract—Early successional landscapes created through right-of-way management are increasingly being viewed as potential pollinator conservation zones. Habitat development initiatives in these landscapes are active, but vetted support for particular techniques and strategies is limited and technical information is diffuse. Our review examined 34 published works in order to outline the current depth and breadth of investigation into the conservation and management of pollinators on rights-of-way along roadsides and underneath electrical transmission corridors. Taxonomic surveys dominate the literature and are focused on diversity patterns in butterflies and moths; the importance of host plant presence as a determinant of abundance and diversity patterns is highlighted in select cases. Keystone agricultural pollinators, including managed and wild bees, have only been examined in a handful of studies. Investigations of pollinator service provisioning within or adjacent to rights-of-way are also lacking. There are no studies of vertebrate or migratory pollinating species. Contrasting results are reported for the impact of disturbance regimes associated with management (mowing and herbicide use), and there is only limited consideration of any potential negative impacts. Studies were also focused on Europe and North America, omitting rapidly developing regions that are experiencing the highest rates of landscape conversion, and where dependence on wild pollinators for food production is high. Successful pollinator species management requires more refined information, and significant gaps exist in the understanding of how rights-of-way can benefit all pollinators, we therefore encourage further management-based investigations in order to develop best practices.

Keywords: right-of-way, roadside, electrical transmission corridor, utility landscape, habitat management, pollinator conservation

INTRODUCTION

There is growing interest in developing and preserving pollinator habitats within managed infrastructure landscapes such as roadside verges and electrical utility corridors. Although landscape conversion is a leading cause of pollinator decline (Kremen et al. 2002; Buchmann & Ascher 2005; Murray et al. 2009) correctly managed green spaces within anthropogenic systems can provide a full range of habitat requirements and can act as reserves (Angold et al. 2006; McFrederick & LeBuhn 2006). Rights-of-way alongside or underneath transportation and electrical utility installations are suggested to be particularly influential because of the extent of this landscape type – over 25 000 000 kilometres of road and 300 000 kilometres of electrical utility corridors in the United States alone (see Table I for details and for information on networks in other countries); connectivity and intersection with multiple habitats (Soulé 1991; Smith 1993; Weber & Allen 2010); and the prescribed and often mandated management regimes (e.g., NERC 2009) that maintain an early successional landscape. The realization that slight modification to existing management practices within electrical utility and transportation landscapes could both save resources (financial and physical) and improve environmental quality for pollinators, combined with the growing public understanding of the important roles that these species play in human survival and livelihoods, has also spurred the broad adoption of adaptive management and *pollinator-friendly* techniques. Successful implementation of landscape management within these systems can be invaluable, creating millions of hectares of pollinator landscape spanning continents, but the availability of vetted guidelines and techniques will be a limiting factor.

Pollinating insects (beetles, flies, wasps, ants, bees, butterflies, moths, thrips, and a few others), birds, and mammals have specific food, nesting, and mating requirements that are fulfilled better in some habitats than others (Schmidt et al. 1995; Kearns and Inouye 1997; Kremen et al. 2002; Kremen et al. 2004). A visual inspection of most rights-of-way would suggest them to be ideal or potentially functional pollinator habitats. Naturalists and wildlife ecologists have documented species along these habitat remnants for quite some time (Soulé 1991; Nabhan 2001; Russell et al. 2005; WHC-USDA 2005). Even rare and endangered pollinators such as the Karner Blue Butterfly have made a re-appearance in the electrical utility corridors in Pennsylvania, USA (Smallidge et al. 1996, Lowell & Loundsbury 2000). But has there been significant investigation into habitat development techniques and

*Corresponding author: vw@pollinator.org; victoriawojcik@hotmail.com

management regimes to prescribe treatments? Inspired by our recent activities with resource managers and by scarce published evidence encountered in our own investigations we undertook a review of the body of scientific and technical literature relating to pollinators on infrastructure supporting landscapes.

In this review we aim to amalgamate the diffuse literature on the topic of pollinators and rights-of-way associated with transportation and electrical utility infrastructure. We work to assess the scope and detail of investigation to develop a summary of the discipline. We also aim to outline elements of current practice and management that both are and are not supported by science and highlight areas of roadside and electrical utility pollinator ecology that require further investigation. Before we present our review we provide some background on the general ecology of electrical utility and transportation installations as they may relate to pollinators and other important wildlife species, as well as summary of the survival needs of each pollinator guild.

TABLE 1: Estimates of the total potential pollinator habitat that could be created given the extent of national roadways in the 20 countries with the most extensive roadway networks. Countries marked with an † are those for which peer reviewed works accounting pollinations on rights-of-way exist. (*Estimates of potential habitat were calculated using management areas equal to 3 metres on each side of the roadway.)

Global Rank (extent of road network)	Country	Amount of roadways (kilometres)	Estimated habitat potential (hectares)*
1	United States†	6 506 204	3 903 722
2	China	3 860 800	2 316 480
3	India	3 320 410	1 992 246
4	Brazil	1 751 868	1 051 121
5	Japan	1 203 777	722 266
6	Canada†	1 042 300	625 380
7	France	1 027 183	616 310
8	Russia	982 000	589 200
9	Australia†	812 972	487 783
10	Spain	681 298	408 779
11	Germany†	644 480	386 688
12	Sweden	572 900	343 740
13	Italy	487 700	292 620
14	Indonesia	437 759	262 655
15	Poland	423 997	254 3982
16	United Kingdom†	394 428	236 657
17	Mexico	366 095	219 657
18	South Africa	362 099	217 259
19	Turkey	352 046	211 228
20	Pakistan	260 760	156 456
	Total	25 491 076	15 294 646

The survival needs of pollinators

Among the most ancient pollinators are the beetles (Coleoptera) and flies (Diptera). Species in both orders range from extreme generalist floral visitors to extreme specialist that visit and pollinate but a few or even one kind of flowering plants (Kevan 2002). Generally, beetle pollinators (including scarabs, staphylinids and sap beetles) are somewhat indiscriminate in the flowers they visit, foraging for pollen and sometimes nectar on open bowl-shaped blossoms such as magnolia flowers or the capitula of flowers in the sunflower family (Asteraceae) (Kevan & Baker 1998). Similarly, flies can be generalist visitors and sometimes highly effective pollinators due to their hairy bodies. Other flies are floral specialists on various trap flowers (e.g., *Arisaema*). One of the most important of all fly pollinator groups are the flower flies in the family Syrphidae (Kevan 2002; Shepherd et al. 2003). Flower flies are dominant floral visitors and important pollinators wherever they occur. Females require one or more protein meals (from pollen and nectar) before they can develop eggs, which drives their floral visitation.

Butterflies and moths use separate host plants as adults and larvae. Nectar from a diverse set of flowers provides the energy required for flight. Adult females search for and oviposit on or near preferred host plants and larvae later emerge from eggs and begin immediate feeding, remaining on their food plants until they move to the soil or a limb or crevice in which to pupate. Sunning locations, such as bare ground or rocks, are important to maintain elevated temperatures for active flight; within managed landscapes there are reported correlations with increased sunlight and increased butterfly numbers (Carter & Anderson 1987). In North America, monarch butterflies (*Danaus plexippus*) undertake a continental migration and require appropriate high energy food sources distributed across a vast landscape (Brower et al. 2006). With habitat loss resulting in significant declines in monarch numbers in recent years, the roles that linear corridor habitats can play in migration support are under investigation.

Bees are the dominant pollinators of flowering plants (Buchmann & Nabhan 1996; Kevan 2001; Buchmann & Asher 2005; Ollerton et al. 2011) and are vital to agricultural productivity. All bees require high quality sources of pollen and nectar. This is especially true for long-lived eusocial species in annual and perennial colonies such as bumble bees (*Bombus* spp.) and honey bees (*Apis mellifera*) whose carrying capacity requirements have been studied (Schmidt et al. 1995). The proximity of nesting sites to food resources confines local populations as bees are limited by relative body size (Greenleaf et al. 2007) and the metabolic demands of flight in their dispersal (Heinrich 1979). The vast majority of bee species are ground-nesting requiring patches of bare soil of the right texture and moisture located close to their food plants (Michener 2001), but about 10% of bees nest in broken pithy stems such as brambles (*Rubus* etc.) and other plants. Bumble bees require cavities such as the abandoned underground nests of mice and other rodents. Honey bees (*Apis mellifera*) nest within hollow trees, or in cavities at the ground surface or rock cavities. This is especially true for Africanized honey bees (*Apis mellifera scutellata*) (Winston 1992).

Wasps are less hairy than bees and being carnivorous are less interested in blossoms except as nectar sources. Certain wasps can be excellent pollinators; one family of wasps in particular, the Masaridae, behave like bees in that they collect and provision larval cells with mixtures of pollen and nectar like most bees (Gess & Gess 2004). Among the Hymenoptera, the ants are very minor pollinators and may be more antagonistic in their relationships with angiosperms (Ashman & King 2005).

Hummingbirds and bats are vertebrate pollinator species, many of which are migratory, following seasonal patterns of blooms from overwintering to breeding areas. In these cases, the energy requirements of flight need to be provided by nectar rich plants present throughout their migratory range (Berlanga et al. 2010). Hummingbirds specialize on flowers with tubular shapes which are often more woody and shrub-like species including *Salvia* species (Buchmann & Nabhan 1996). In addition to feeding sites, hummingbirds require a landscape with sufficient vertical structure and cover to support nesting. Preferred nesting sites include areas where hardwood trees with ample leafy cover can be found. Insects provide protein for growing hummingbird chicks; again a diverse local ecosystem is needed to support these species. Bats feed more narrowly on cacti and species within desert landscapes (Buchmann et al. 1999; Molina-Freaner et al. 2004). Roadsides are not generally populated by plant species that might attract bats, but electrical utility corridors passing through the Sonoran Desert and other regions of the American southwest are, and could potentially be important migratory habitats. Bats in particular have suffered declines in the range of their food plants along north-south migration routes within North America (Buchmann et al. 1999).

Roadways and their associated verges

Roadways, whether they are rural dirt roads, paved city streets, or interstate super highways, all produce varying effects, both negative and positive on landscapes, wildlife, and people. The most obvious and far-reaching impact on landscapes and their biota is the simple bisection and fragmentation of habits that comes with road-building and maintenance. Roads of all types and sizes fragment natural, urban, and agricultural landscapes into ever-smaller areas, irregular polygons, often referred to as habitat islands (Forman 1995; Dramstad et al. 1996). Dividing landscapes into smaller parcels has profound consequences on the plants and animals of those regions and neighbouring areas (Fahrig 2003). Fewer species can live and reproduce on smaller habitat islands (MacArthur & Wilson 1967). Thus, roadways divide habitats up into smaller irregular areas that then often become impoverished in terms of the number of plants and animals that can live there, unless re-colonized from adjacent source areas, in the oceanic island analogy, the mainland (Smith 1993). Range requirements for plants and animals, rather than strict abundance, may determine the likelihood of local extinctions (Ney-Nifle & Mangel 2000). It should be remembered, however, that habitat diversity and environmental structure are often more important than size alone when creating a nature preserve or managing existing habitats; the results of our review support this.

Along with the direct effects, roadways often act as barriers or channels to move or direct the daily movements of animals in their paths, or more dramatically, the migration routes of migratory species, especially for mammals. Alternatively, they have been shown to promote species movement, although in some cases this could account for the spread of invasive species (Getz et al. 1978; Garland & Bradley 1984; Von Der Lippe & Kowarik 2007). Anthropogenic inputs are common along roads. Vehicular traffics introduces and spreads gasoline, motor oil, ethylene glycol (anti-freeze), other chemicals and vast amounts of pulverized bits of automobile tires across the landscape everywhere there are roads. Furthermore, roadways are managed with chemical and mechanical disturbances that can cause potential negative impacts on species.

There are also positive impacts that roadways have on their nearby landscapes and on pollinator communities. The greening of roadside plants, both native and exotics, is due to rain water harvesting by the paved surfaces and channeling of runoff waters to the roadside verges. Plant growth is often exuberant, thick, and tall among roadsides during the growing seasons. As will be discussed in more detail, roadside flowers have documented associations with many pollinating insects (Way 1977), including butterflies and moths (Ries et al. 2000; Saarinen et al. 2005; Valtonen et al. 2006; Williams 2008), flies (Free et al. 1975; Raemakers et al. 2001), and especially bees (Way 1969; Hopwood et al. 2010). The utility of roadsides had been recognized by beekeepers that have these landscapes as transient bee pastures to capture honey flows (Harper-Lore & Wilson 2000). Roadways could also possibly create new areas, strips or patches of open compacted ground and may enhance nesting of certain native ground-nesting bees. Some authors have suggested that roads and their associated verges can play significant roles in habitat connectivity, perhaps compensating for the some of the negative impacts (Forman et al. 2002). Thus roads have both beneficial and negative effects for plants, animals, and people. Table 2 lists and contrasts many of these ecological effects.

Electronic power transmission lines and their maintenance corridors

High voltage electric power transmission lines, their towers, and associated access roadways provide most of the same negative and positive effects of regular roadways discussed above. There are differences, however, in their total effects on urban, rural, and wilderness landscapes. Like roads, electrical transmission line corridors impact landscapes and wildlife habitats directly and most importantly by fragmenting them into smaller and smaller pieces as discussed for island habitats. For some species, they may also connect habitats and allow movement between feeding or breeding areas, especially for larger mammals (Yahner 2004; Weber & Allen 2010). Electric and magnetic field (EMF) radiation from high voltage lines may pose a threat to animals living under or near these lines; both anecdotal and published accounts suggest that honey bees exhibit aggression and lower productivity when they are underneath

Table 2. Postulated ecological effects of creating new roadways and electric power transmission line corridors on insects and other wildlife.

Negative Effects	*Positive Effects*
Roadways and Roadsides:	
• Direct elimination of habitat area (new installation). • Bisect and fragment landscapes into habitat islands (new installation). • Conduits for dispersal of weeds and exotic animals. • May alter migratory patterns, especially larger animals. • Allows deep access into wildlands for further exploitation. • Increases frequency of wildfires (tossed cigarettes). • Introduction and spread of gasoline, exhaust fumes, rubber particles from tires. • Mortality due to interactions with vehicles and traffic (road kill).	• Water runoff creates hedgerow effect of new growth. • Increased flowering promotes bees, other pollinators, grazing by herbivores. • Thoughtful management fosters or resets succession promoting colonizing species. • Creates new bare ground along edges, promoting bee nesting. • Road-killed individuals provide important food for avian raptors, other birds, mammals and certain invertebrates.
Transmission Power Line Corridors and Maintenance Roads:	
• Bisect otherwise unbroken habitats into habit islands (new installation). • Raptors may be electrocuted when using transmission towers as hunting perches. • Electric and Magnetic Field (EMF) radiation from high voltages may pose threats to animals living under them. Honey bees have been implicated.[1]	• Chemical or mechanical thinning opens up habitats. • Many wildlife species require and favour these early successional stages. Many animals live and prefer to forage along these new edge habitats. • Bare soil in maintenance roads favours ground-nesting bees and wasps. Increased rodent nesting may favour bumble bees. • Flowering and fruiting may increase food for birds and mammals. • Avian raptors often use transmission towers as perches from which to hunt.

[1](Wellenstein, 1973; Rogers et al. 1982; Lee and Reiner, 1983)

high voltage electrical utility wires (Greenberg et al. 1981; Rogers et al. 1982; Lee & Reiner 1983).

Although both road and electrical utility landscapes are managed with safety in mind, electrical utility rights-of-way, and especially those underneath high voltage wires, are systematically and vigorously cleared of taller vegetation that might interact with power lines. The impacts of a tree-wire interaction can be significant as was seen in the 2003 Northeastern black-out in the United States and Canada where nearly 30 million people lost power – the culprit was a tree growing too close to overhead lines and an eventual blow out when extreme heat caused the wires to sag into the tree. Strict guidelines and policies are now in place within the electrical transmission industry to prevent such an incident from repeating (NERC 2009). To eliminate the risk of trees interacting with overhead lines, transmission line corridors are either mechanically or chemically treated for vegetation management. This adds physical structure and edges to habitats, and allows the growth and reproduction of understory plants that will exclude taller growing species. Many of these plants produce flowers for pollinators along with fruits, seeds and berries for birds, mammals and other wildlife species. Electrical corridors are thus continually reset to earlier seral or successional stages. Many types of wildlife thrive in such altered habitats including favoured game species such as quail, pheasants, and wild turkeys, which in the United States have made a significant comeback due to electrical utility corridor conservation programmes (King & Byers 2002; Willyard et al 2004; Goodrich-Mahoney et al. 2008). Conversely, management that is too aggressive and frequent can cause ecological deserts. Again, Table 2 presents the anticipated positive and negative impacts of electrical utility rights-of-way.

METHODS

A keyword literature search was conducted using a combination of electronic scholarly databases and world-wide-web search engines. Our interest was in peer reviewed scientific literature, technical publications, or any other form of literature that land managers might look to for guidelines. The key words used in this search were any combination of: *pollinator, pollinators, bee, bees, butterfly, butterflies, moth, moths, bat, bats, bird, birds, hummingbird, hummingbirds, beetle, beetles, fly, flies,* and *beneficial insects* in conjunction with either: *right-of-way, ROW, utility corridor, electrical utility corridor, power lines, verge, road side, transportation corridor, transmission corridor,* and *IVM* or *Integrate Vegetation Management.* The reference lists cited in each paper were also searched for further relevant publications that were not acquired in the primary search. Contacts were made with experts in the fields of utility and roadside management to widen the search. Only references that were explicitly focused on pollinators and pollinator habitat management in electrical utility and roadside rights-of-way were retained for analysis.

To better assess the taxonomic diversity of pollinators that have received attention in right-of-way landscapes, the taxonomic resolution of each investigation was noted, including the general pollinator group studied and, if applicable the number of species, guilds, or groups examined. Studies were classified as surveys, comparative studies, or experimental. The selection of works was categorized by geographic region and the type of right-of-way (electrical utility or roadside). The duration of each study was noted in order to gauge how well temporally variability in populations and landscape change over time has been addressed. Finally, published works were sorted as either scientific or technical to determine the proportion of data generating work versus management-oriented guidelines.

RESULTS

Our literature search resulted in only 29 peer reviewed studies spanning a time period from 1969 to 2011 that were retained as either pollinator specific or pollinator relevant, and an additional 5 published technical guides that are aimed at assisting land managers in producing pollinator-friendly landscapes. The majority of works on the subject of pollinators in managed infrastructure landscapes were produced from 2000 onward. Most studies are very limited in temporal scope, focusing on one season or one year, which is not considered sufficient to adequately assess pollinator populations that are known to be highly temporally variable (Williams et al. 2001; Hilton & Miller 2003; Cane et al. 2005). Only two studies have exceeded the short time frame, one is an ongoing study of butterfly habitat development along roadside for threatened butterfly species in the Netherlands (Wynhoff et al. 2011); the other is a 51-year study of a famous Pennsylvania electrical transmission corridor (Yahner 2004). Table 3 presents a summary of the various foci of the recovered research and Appendix I provides a full summary of each study.

The majority of studies (19 of 29, or 66%) examined roadsides. Studies of electrical utility rights-of-way focused only on overhead electrical transmission corridors. Three of the pieces were reviews of landscape management techniques, one of which presented a combined review of both roadside and electrical landscapes (Smallidge & Leopold 1997). The geographic focus of work is heavily biased to North America (62%) and Europe (31%); however in North American there are only studies from Canada and the United States, and none from Mexico. Within the United States works are focused in the east and mid-west, with one study in the Pacific Northwest. Work from Canada comes from Quebec. European studies have been conducted in Finland, the Netherlands, Germany, and the United Kingdom. There is one study from Australia. No published accounts from Central and South America, Africa, or Asia were recovered.

The majority of investigations into pollinators on rights-of-way have been comparative, contrasting the fauna found along rights-of-way with that found in natural habitats (Smallidge et al. 1996; Lanham & Nichols 2000; Hopwood 2008; Larsen 2010), or comparing managed rights-of-ways to unmanaged systems (Ries et al. 2000; Russell et al. 2005; Hopwood et al. 2010; Wynhoff et al. 2011). The earlier studies recorded surveys of plant-insect interactions within right-of-way landscapes (Way 1969; Way 1977; Free et al. 1975). Most studies focus on understanding the positive impacts of electrical utility instillations, but a handful do examined the potential negative impacts associated with pollinator mortality and roadways (Seibert & Convoer 1991; McKenna et al. 2001). Experimental approaches are limited and have focused on three subject areas: understanding the responses of honeybees to high voltage (Wellenstein 1973; Rogers et al. 1982; Lee & Reiner 1983), testing the impacts of varied management regimes on butterflies and moths (Noordijk et al. 2010), and testing if targeted habitat development techniques are successful in increasing butterfly occurrence (Forrester et al. 2005).

Butterflies and moths were the most rigorously studied, including three studies of the endangered Karner Blue Butterfly (*Lycaeides melissa samuelis*) in the United States (Smallidge et al. 1996; Forrester et al. 2005; Lowell & Loundbury 2000) and one study of the success of reintroduced *Phengaris* (*Maculinea*) *teleius* and *P. nausithous* species in the Netherlands (Wynhoff et al. 2011). Studies of bees examined communities of native bee communities on electrical utility corridors (Russell et al. 2005) and along roadsides (Hopwood 2008), the response of honey bees to electric and magnetic waves underneath high voltage power lines (Wellenstein 1973; Greenberg et al. 1981; Rogers et al. 1982) and one study focused on bumble bees along roadsides (Hopwood et al. 2010). Studies of pollinators in general included flies and focused on Syrphidae (Free et al. 1975). No studies examined the interactions that hummingbirds and pollinating bats might have with managed electrical utility and roadside landscapes. Studies on pollinating beetles were also lacking.

The technical guidelines recovered in this review include habitat provisioning for all native pollinators (NAPPC 2010; WHC 2005), as well as some that are specific to butterflies (UK Forest Council 1987) and bees (Xerces Society 2010). Guidelines have been produced by organizations within the United States and in the UK.

DISCUSSION

There are multiple beneficial insect associations on roadsides and underneath electrical corridors that contain native flora. Early and beginning works in the field (Way 1969; Free et al. 1975) indicated that pollinator diversity was much higher than expected in these landscapes. Since then, the pace of work has increased and the presence of pollinators in infrastructure landscapes managed as early successional habitat has been documented in both peer reviewed and grey literature. The three reviews acquired in our overview (Way 1977; Lee & Rener, 1983; Smallidge & Leopold 1997) provide broad summaries of species diversity patterns along roads and underneath power lines. Peer

Table 3: Summary statistics for literature reviewed

Pollinator group studied*							
		Bees	Butterflies/Moths	Beetles	Flies	Bats	Hummingbirds
	no	13	16	0	2	0	0
	%	45%	55%	0%	7%	0%	0%
Geographic area of study							
		North America	South America	Europe	Australia	Africa	Asia
	no	18	0	9	1	0	0
	%	62%	0%	31%	3%	0%	0%
Type of study							
		Compare	Survey	Experiment	Review		
	no	16	4	6	3		
	%	55%	14%	21%	10%		
Landscape type**							
		Road	Electrical				
	no	19	11				
	%	68%	39%				

*some studies included more than one group

**some studies examined both roadsides and electrical landscapes

reviewed accounts of electrical utility rights-of-way and roads provide a verification that landscapes that "appear" to be pollinator positive do in fact contain many species of butterflies, moths, bees, and flies.

All accounts of beneficial habitat associations, especially in managed and anthropogenic landscapes, are important, but what is more valuable to management-based ecosystems is addressing the impact of interventions and the responses of individual species. Successful pollinator species management requires more refined information, and significant gaps exist in the understanding of how roads and rights-of-way can benefit all pollinators. This lack of detailed accounts of autecology indicates that more study is needed in this landscapes type.

Pollinator groups that dominate rights-of-way

Butterflies, moths, and to a lesser extent, bees have been studied in infrastructure landscapes along roads and underneath transmission lines. The majority of works on diurnal moths and butterflies have outlined the importance of host plant presence as a significant factor in determining species occurrence (Smallidge et al. 1996; Valtonen et al. 2006; Larsen 2010). Butterflies in electrical utility rights-of-way responded to similar factors that promoted their occurrence in field boundaries and agricultural landscapes where they were more commonly associated with edge habitats (Sparks & Parish 1995). Management strategies that increased edges (especially ones that created more scalloped edges), favoured trees along sidelines, and increased bare ground provided better butterfly habitat (Carter & Anderson 1987). Sunlight was particularly important in species occurrence patterns, as this is important for basking behaviours that warm the body for flight (Smallidge et al. 1996). Moths and butterflies were often treated together in investigations, and generally responded to similar variables, but butterflies were more influenced by the presence of nectar plants, and diurnal moths responded to vertical structure in the landscape (Saarinen et al. 2005). Natural prairie lands and restored roadsides were found to be equivalent in Lepidopteran richness and diversity (Williams 2008), suggesting again that targeted initiatives produce beneficial results for butterflies. Lahham & Nichols (2000) also compared multiple right-of-way transects for trends in butterfly richness and found no significant difference, although abundance patterns did vary.

Native bee communities have primarily been studied along roadsides that have been planted with wildflowers. Hopwood (2008) provides an account of bees along roadsides that have been managed in various ways, noting that promoting the occurrence of local native plant species creates a landscape in which bumble bees and other native bees actively forage. Bumble bees in particular were positively influenced by roadside mowing (Noordijk et al. 2009). Comparative studies of targeted roadside restoration efforts that recreated prairies were shown to have faunal composition similar to that of nearby native prairies. Although basic wildflower plantings had more diverse bee faunas than mowed roadsides, selective prairie plantings showed the greatest success (Hopwood 2008), supporting the importance of the local floral community in determining pollinator community structure (see Potts et al. 2003).

Only one study of native bees and electrical utility landscapes has been published to date. Russell et al. (2005) examined native bee communities along an electrical right-of-way. This study took place along an electrical corridor that passed through a conservation reserve resulting in a situation where landscape management occurred only in an easement zone underneath the overhead wires. More diverse and more abundant communities of native bees were recorded underneath the overhead wires and this was attributed to a more floristically diverse landscape that was able to develop due to intermediate disturbance from mowing and herbicide treatment (Russell et al. 2005).

The impacts of management on pollinator populations

Electrical utility corridors and roadsides require continuous management for accessibility, visibility, and user safety. As noted above, there are often easements or mandates that allow their management even in environmentally sensitive areas. Most maintenance activities maintain vegetation at an early successionary stage that is presumed to be a positive habitat for pollinators. Noordijk et al. (2009) investigated the impacts of one of the most standard management practices along roadsides – mowing. Overall, mowing was shown to increase bee and butterfly occurrence (Noordijk et al. 2009), but initial decreases were documented after implementation that were attributed to a reduction in food plant availability. Management regimes that limited and timed mowing were the most successful in increasing beneficial species occurrence. For example, increasing mowing to twice a year and removing residues had the greatest positive impact on moth and butterfly diversity in managed systems in the Netherlands (Noordijk et al. 2009). Overall, single-season mowing was better than no mowing, but targeted mowing efforts produced the most pollinator richness (Ries et al. 2000; Wynhott et al. 2011). Bumble bees in particular were positively influenced by roadside mowing in this system (Noordijk et al. 2009). Champagne and Bourassa (2000) conducted an earlier study of mowing regimes along roadside in Quebec, Canada that yielded similar results; the highest biodiversity, including many pollinating species, was seen in areas mowed annually while more intensive mowing schedules (mowing three to five times a season) had lower richness and abundance measures

Weediness correlated with lower beneficial insect diversity (Way 1977; Ries et al. 2000; Valtonen et al. 2006). Herbicide treatment is a common method for weed eradication and the practice is often mandated by local invasive species action plan. The observed direct impact of herbicides on local pollinator communities would come from a direct reduction in their food supply, although the revived literature does not agree on an impact of herbicide application. Yahner (2004) did not find any measurable impacts on butterflies associated with herbicide use. Bramble et al. (1999) found greater butterfly diversity in land management systems that used herbicide treatment, although this treatment was targeted toward the removal of grasses and tree species. Larsen (2010) did, however, find that lower numbers of butterflies were associated with herbicide applications, but that butterfly numbers increased thereafter.

Integrated Vegetation Management, or IVM, programmes are becoming more broadly implemented in right-of-way landscapes as they save resources and have been shown to improve wildlife habitat (Yahner 2004). These programmes include a combination of mowing, burning, hand removal, and localized herbicide use to achieve a desired landscape. Direct studies of pollinator community responses to IVM practices compared against traditional board scale mowing and herbicide management systems were not recovered in our review; however, our field research is currently testing these alternatives to provide quantitative results.

Managing to promote pollinators

A few studies in our review documented targeted management efforts that seek to capture the potential of roadsides and rights-of-way as conservation lands for species of concern. The endangered Karner Blue Butterfly (*Lycaeides melissa samuelis*) has been studied in some detail within the electrical utility corridors of Pennsylvania (Smallidge et al. 1996; Forrester et al. 2005; Lowell & Loundbury 2000). These landscapes were noted as one of the few remaining habitats in which *L. melissa samuelis* was found (Smallidge et al. 1996). Efforts to actively develop more appropriate habitat were undertaken, including the planting of preferred nectaring and larval plants, as well as instituting landscape management practices that support these floral species (Lowell & Loundsbury 2000). Research into the response of key floral species to management practices, including targeted mowing and isolated herbicide treatment, resulted in the development of a landscape more suitable for *L. melissa samuelis*. Together these studies show that less woody landscapes have greater occurrence of *L. melissa samuelis* (Forrester et al. 2005) and that microclimate also matters, with warmer areas having higher abundance (Smallidge et al. 1996). Today these electrical utility landscapes act as habitat reserves, and there have been significant improvements in local species occurrence.

A similar targeted management study conducted in the Neatherlands examined the success of species reintroductions into roadside habitats that, like in the case of the Karner Blue Butterfly, represented the only remaining areas of suitable habitat (Wynhoff et al. 2011). *Phengaris (Maculinea) teleius* and *P. nausithous* are species that have positive associations with local native ant species, and there are strong correlations between their occurrence and ant colony presence. In 2009 these species were reintroduced into roadside habitats as part of a management study to increase native ant presence and subsequently increase the occurrence of *P. (Maculinea) teleius* and *P. nausithous*. Tested management practices include early season mowing and summer mowing, as well as creating areas of open vegetation that increase ground temperature, which increases the productivity of the associated ants. Heavy disturbance was not positive for the ants, and therefore also reduced butterfly occurrence. The resultant prescribed management practice that will be used includes bi-seasonal mowing and residue removal to increase bare ground in favour of the ants, and butterfly numbers are expected to continue increasing (Wynhoff et al. 2011).

More complicated total ecological studies of management impacts and subsequent pollinator occurrence on rights-of-way are not available. Although general conclusions about best management practices for the pollinator community can be drawn from the existing literature, it is clear that conservation and management relies on a deeper understanding of the ecology of these landscapes and the local biology of their resident species.

Negative interactions between rights-of-way and pollinators

Accounts of the potentially negative impacts of roadsides focus on butterfly and moth mortality associated with traffic. Seibert & Conover (1991) conducted a study of general roadside mortality and noted the number of insects, especially butterflies and moths, to be higher than expected, and greater than the observed mammalian road kill that is more easily observed. Direct mortality related to vehicular traffic was investigated by McKenna et al. (2001), and indicated that there was an association with increased travel velocity and traffic volume and the number of dead Lepidoptera found lying on shoulders. This study also provided an estimate of the number of individuals killed along roadways each week in the state of Illinois at 20,000,000. Munguira & Thomas (1992) has contrasting results from the work they conducted in England; they recorded overall low and insignificant mortality rates (0.6 to 1.9%) attributed to vehicular traffic as compared to natural factors. Roads were not found to be barriers to the movement of butterflies, nor where populations of butterflies separated by even busy roadways genetically isolated from each other, again indicating fluid species movement (Munguira & Thomas 1992). What is missing from these studies is whether roadsides that are managed for native plant species or to promote wildlife habitat cause more kills than unmanaged roadsides and if this has any impact on local populations. We are aware of a few new studies that are underway in Britain, the Netherlands, and the United States to better quantify the direct threats of vehicular traffic to local insect populations.

We do not have an idea of the magnitude of direct vehicular impacts on keystone pollinators including bees, flies, and wasps. The studies reviewed have not indicated any direct bee mortality, even along roadsides that have been designed to attract bees (Hopwood et al. 2010). We also do not have a clear understanding of at what heights most insects fly across roads. There are contrasting findings regarding whether roads present barriers to bee movement. Bhattacharya et al. (2003) conducted a mark-recapture study of bees along suburban roadsides and did not find any evidence of movement across roadways. Our personal observation have documented bees crossing roads in cities and in suburbs (V. Wojcik, pers. obs.); Hopwood et al. (2010) also observed that bumble bees do in fact cross roads when foraging along roadsides, although they tend to preferentially forage along the linear transects created by wildflower seeding efforts. Road width, usage, traffic, and local floral resource presence are factors that might influence bee foraging movement, in which case site-specific characteristics would be determinants of flight patterns.

A handful of studies addressing pollinator health examined the impacts of high voltage wires and electromagnetic fields on honey bees, and indicated that lower productivity and behavioural changes (aggression) occurred (Wellenstein 1973; Rogers et al. 1982; Lee & Reiner 1983). Because right-of-way habitats are beginning to be seen as reserves for wild bees understanding the impacts of high voltage and electromagnetic fields on the native bee community will be of value. Parallel declines in productivity could be expected in native bees, but there is no data available. We recommend a comparative study of trap nest occupancy, nest productivity, brood development, and fecundity in natural settings and underneath high voltage lines.

In regard to general pollinator health, we believe that electrical utility rights-of-way present potentially healthy habitats, with less "pollution" compared to roadsides. The presence of run-off and contamination from roads can negatively impact bees, especially ground nesting species – but there is no investigation in this area. Plants growing along roadsides could also potentially accumulate toxins in their nectar and pollen, and these could be magnified in bees. A toxicity study of the impacts of roadway run-off on pollinator nesting and development was not found, and there is insufficient research on whether or not environmental toxins and pollutants accumulate in the soil or nectar of plants and if there are subsequent impacts to pollinator health.

Deficits in the literature

Nesting opportunities for bees in particular have been given little critical attention, although studies allude to the importance of available open space (Hopwood 2008) and guidelines produced by conservation organizations (NAPPC 2010) promote the development of open space for nesting. Furthermore, attempts at increasing nest site availability through either vegetation management or nest installation have not been investigated, yet Potts et al. (2003) indicated that the most important factors impacting the bee community in natural systems were the floral community, nesting site availability and time after fire (also a measure of disturbance). Parallels likely exist in managed electrical utility landscapes and the role of available nesting habitat as well as increasing nest site availability should be investigated.

No studies of flower beetles, hummingbirds, or bats were recovered in our investigation. While not key agricultural pollinators, many beetles, flies, hummingbirds, and bats play essential roles in maintaining local ecosystems. Some hummingbirds and bats are also species of concern that have suffered declines in their natural migratory habitat. Rights-of-way could be providing a potential network of habitat reserves for these species, but there is no account of their study or conservation in the current literature.

Many roads and electrical transmission corridors run parallel to agricultural landscapes that could benefit from pollinator populations present in these lands. No study to date has examined the impact that roads and electrical transmission lines can have on nearby agricultural productivity. If planting hedgerows near crops increases yield and if farms that are situated within more "natural" areas produce more crop (Kremen et al. 2000), then it is likely that nearness to correctly managed rights-of-way could have a similar benefit. This is an area that requires further investigation.

The most understudied parts of the world are areas where populations and supporting infrastructure are the most rapidly developing and where wild pollination is still a significant factor in agricultural productivity. Our review did not find any works from Africa, Asia, or South and Central America. Referring back to Table 1, four of the top five countries by road network length are not present in the literature, in fact, only four of the top 20 road networks have been given attention from scientists and land managers, and in some case this is restricted to only a few studies per region.

CONCLUSIONS AND SUMMARIES

Vegetation remnants alongside roads and underneath overhead transmission corridor provide habitat for pollinating species. We believe right-of-way landscapes can play a role in pollinator conservation and management. They are potential linear refuges, but how we manage these vegetative strips makes a difference. When these landscapes are managed appropriately or restored using native vegetation there is a strong positive effect on native pollinator diversity and local abundance.

We recommend increased monitoring and implementing tests of alternative management techniques to develop best management practices that directly promote pollinators. Furthermore, we encourage the investigation of off-site impacts and benefits that might be derived from roads and rights-of-way to agricultural landscapes. The network of managed roadways and electrical utility rights-of-way also mirrors many migratory pathways, especially those of monarch butterflies, some bats, and hummingbirds. Ideas of facilitated migration and the development of managed way-stations along roads and electrical utility crossings could hold potential for the conservation of these species but require a well-reasoned management approach. Successful species management requires more refined information, and significant gaps persist in our understanding of how roads and rights-of-way can benefit all pollinators.

ACKNOWLEDGEMENTS

We would like to thank T. Redick of Wildlife Habitat Council and S. Sanchez of BFCI/Butterfly Conservation Initiative for assistance in resource acquisition. Thank you to L. Adams and M. Rager for reviews and advice on early drafts of the manuscript. Thank you also to our partners in the electrical utility management industry for continued interest in the pollinator issue (UAA, ISA, SMUD, PG&E, Energy United, and Edison Electric). A special acknowledgement is deserved by the International Commission on Plant-Pollinator Relations for providing the venue for the initial presentation of this work at the 10th International Pollination Symposium June 27 - 29, 2011 Universidad de las Américas Puebla, Cholula, Mexico.

APPENDIX

Additional supporting information may be found in the online version of this article:

Appendix I. Catalogue of studies examining pollinators on right-of-way landscapes

REFERENCES

Angold PG, Sadler JP, Hill MO, Pullin A, Rushton S, Austin K, Small E, Wood B, Wadsworth R, Sanderson R, Thompson K (2006) Biodiversity in urban habitat patches. Science of the Total Environment 360:196-204.

Ashman TL, King EA (2005) Are flower-visiting ants mutualists or antagonists? A study in a gynodioecious wild strawberry. American Journal of Botany 92(5):891-895.

Berlanga H, Kennedy JA, Rich TD, Arizmendi MC, Beardmore CJ, Blancher PJ, Butcher GS, Couturier AR, Dayer AA, Demarest DW, Easton WE, Gustafson M, Iñigo-Elias E, Krebs EA, Panjabi AO, Rodriguez Contreras V, Rosenberg KV, Ruth JM, Santana Castell E, Ma Vidal R, Will T (2010) Saving Our Shared Birds: Partners in Flight Tri-National Vision for Landbird Conservation Cornell Lab of Ornithology: Ithaca, NY.

Bhattacharya M, Primack RB, Gerwein J (2003) Are roads and railroads barriers to bumblebee movement in a temperate suburban conservation area? Biological Conservation 109:37-45.

Bramble WC, Yahner RH, Byrnes WR (1999) Effect of herbicide maintenance of an electric transmission line right-of-way on butterfly populations. Journal of Arboriculture 25(6):302-310.

Brower LP, Fink LS, Walford P (2006) Fueling the fall migration of the monarch butterfly. Integrative and Comparative Biology 46:1123-1142.

Buchmann S, Nabhan GP (1996) The Forgotten Pollinators. Island Press, New York.

Buchmann S, Kunzmann MR, Donovan AJ, Hobbs RJ. 1999. Gap Analysis of Pollinator (bats, bees, hummingbirds) Species Richness in Arizona: Implications for Conservation Biology. [online] URL: http://proceedings.esri.com/library/userconf/proc99/proceed/papers/pap530/p530.htm (accessed December 2011).

Buchmann SL, Ascher JS (2005) The plight of pollinating bees. Bee World 86: 71-74.

Cane JH, Minckley RL, Kervin L, Roulston TH (2005) Temporally persistent patters of incidence and abundance in a pollinator guild at annual and decadal scales: The bees of *Larrea tridentate*. Biological Journal of the Linnean Society 85:319-329.

Carter CI, Anderson MA (1987) Research information note 126: Enhancement of lowland forest ridesides and roadsides to benefit wild lands and butterflies. UK Forestry Commission Research Division, Surrey. *Booklet.*

Champagne N, Bourassa JP (2000) Entomological communities of three landscapes along highway rights-of-way of southern Quebec: effect of mowing frequency on insects abundance and diversity 7th International Symposium on Environmental Concerns in Rights-of-Way Management [online] URL: http://www.rights-of-way.org/17symp.htm (accessed September 2011).

Dramstad WE, Olson JD, Forman RTT (1996) Landscape Ecology Principles in Landscape Architecture and Land-Use Planning. Island Press, Cambridge.

Fahrig L (2003) Effects of habitat fragmentation on biodiversity. Annual Reviews of Ecology Evolution and Systematics 34:487-515.

Forman RTT (1995) Land Mosaics: The Ecology of Landscapes and Regions Cambridge University Press, Boston.

Forman RTT, Sperling D, Bissonette JA, Clevenger AP, Cutshall CD, Dale VH, Fahrig L, France RL, Goldman CR, Heanue K, Jones J, Swanson F, Turrentine T, Winter TC (2002) Road Ecology: Science and Solutions. Island Press, Washington D.C.

Forrester J, Leopold DJ, Hafner SD (2005) Maintaining critical habitat in a heavily managed landscape: effects of power line corridor management on Karner Blue Butterfly (*Lycaeides melissa samuelis*) habitat. Restoration Ecology 13(3):488-498.

Free JB, Gennard D, Stevenson JH, Williams IN (1975) Beneficial insects present on a motorway verge. Biological Conservation 8:61-72.

Garland T, Bradley WG (1984) Effects of a highway on Mojave Desert rodent populations. The American Midland Naturalist 111(1):47-56.

Gess SK, Gess FW (2004) Distributions of flower associations of pollen wasps (Vespidae: Masarinae) in southern Africa. Journal of Arid Environments 57(1):17-44.

Getz LL, Cole FR, Gates DL (1978) Interstate roadsides as dispersal routes for *Microtus pensylvanicus*. Journal of Mammology 59:208-212.

Goodrich-Mahoney JW, Abrahamson L, Ballard J, Tikalsky S (2008) Environment Concerns in Rights-of-Way Management 8th International Symposium. Elsevier Science, New York.

Greenberg B, Bindokas VP, Gauger JR (1981) Biological effects of a 756-kV transmission line: exposure and thresholds in honeybee colonies. Bioelectromagnetics 2(4):315-328.

Greenleaf SS, Williams N, Winfree R, Kremen C (2007) Bee foraging ranges and their relationship to body size. Oecologia 153:589-596.

Harper-Lore B, Wilson M (eds) (2000) Roadside Use of Native Plants. Washington DC: Island Press.

Heinrich B (1979) Bumble bee Economics. Harvard University Press, Cambridge.

Hilton B, Miller MW (2003) Annual survival and recruitment in a Ruby-throated Hummingbird populations, excluding the effect of transient individuals. Condor 105:54-62.

Hopwood JL, Winkler L, Deal B, Chivvis M (2010) The use of roadside prairie plantings by native bees. Living Roadway Trust Fund [online] URL: http://www.iowalivingroadway.com/ResearchProjects/90-00-LRTF-011.pdf (accessed November 2011).

Hopwood JL (2008) The contribution of roadside grassland restorations to native bee conservation. Biological Conservation 141 (10):2632-2640.

Kearns CA, Inouye DW (1997) Pollinators, Flowering Plants, and Conservation Biology. Bioscience 47(5):297-307.

Kevan PG (2001) Pollination: a plinth, pedestal, and pillar for terrestrial productivity. The why, how, and where of pollination protection, conservation, and promotion. In: Stubbs C and F Drummond (eds) Bees and Crop Pollination—Crisis, Crossroads, and Conservation. Lanham, MD: Entomological Society of America, pp 7-68.

Kevan PG (2002) Flowers, pollination, and the associated diversity of flies. In: JH Skevington and PT Dang (eds) Exploring the Biological and Ecological Diversity of Flies (Diptera). Biodiversity 3(4):16-18.

Kevan PG, Baker HG (1998) Insects on Flowers: Pollination and Floral Visitation. In: CB Huffaker and AP Gutierrez (eds) Ecological Entomology (2nd Edition) Wiley, New York, pp 407-637.

King DI, Byers BE (2002) An evaluation of powerline rights-of-way as habitat for early-successional shrubland birds. Wildlife Society Bulletin 30(3):868-874.

Kremen C, Williams NM, Bugg RL, Fay JP, Thorp RW (2004) The area requirements of an ecosystem service: crop pollination by native bee communities in California. Ecology Letters 7:1109-1119.

Kremen C, Williams N, Thorp RW (2002) Crop pollination from native bees at risk from agricultural intensification. PNAS 99(26):16812-16816.

Lanham JD, Nichols MJ (2000) Butterflies and skippers in utility rights-of-way in the Upper Piedmont of South Carolina 7th International Symposium on Environmental Concerns in Rights-of-Way Management [online] URL: http://www.rights-of-way.org/17symp.htm (accessed September 2011).

Larsen KJ (2010) Impact of roadside prairie plantings on plant and insect communities LRTF Research Final Reports [online] URL: http://www.iowalivingroadway.com/ResearchProjects/90-00-LRTF-800,%20910.pdf (accessed December 2011).

Lee JM, Reiner GL (1983) Transmission line eclectic fields and the possible effects on livestock and honeybees. Transactions of the American Society of Agricultural Engineers. 83:279-286.

Lowell F, Lounsbury S (2000) Karner Blue Butterfly habitat restoration on pipeline right-of-way in Wisconsin 7th International Symposium on Environmental Concerns in Rights-of-Way Management [online] URL: http://www.rights-of-way.org/17symp.htm (accessed September 2011).

MacArthur RH, Wilson EO (1967) The Theory of Island Biogeography. Princeton University Press, Princeton.

McFrederick QS, LeBuhn G (2006) Are urban parks refuges for bumblebees *Bombus* (Hymenoptera: Apidea)? Biological Conservation 123:372-382.

McGuiness KA (1984) Equations and explanations in the study of species-area curves. Biological Reviews 59:423-440.

McHarg I (1969) Design with Nature Natural History Press, Garden City, New York.

McKenna DD, McKenna KM, Berenbaum MR (2001) Mortality of Lepidoptera along roadways in central Illinois. Journal of the Lepidopterists' Society 55(2):63-65.

Molina-Freaner F, Rojas-Martinez A, Fleming TH, Valiente-Banuet A (2004) Pollination biology of the columnar cactus *Pachycereus pecten-aboriginum* in north-western Mexico. Journal of Arid Environments 56(1):117-127.

Munguria ML, Thomas JA (1992) Use of road verges by butterflies and burnet populations, and the effect of roads on adult dispersal and mortality. Journal of Applied Ecology 29:316-329.

Murray TE, Kuhlmann M, Potts SG (2009) Conservation ecology of bees: populations, species and communities. Apidologie 40:211-236.

Nabhan G (2001) Nectar Trails of Migratory Pollinators. Conservation Magazine 2(1) [online] URL: http://wwwconservationmagazineorg/2008/07/nectar-trails-of-migratory-pollinators/ (accessed October 2011).

Ney-Nifle M, Mangel M (2000) Habitat loss and changes in the species-area relationship. Conservation Biology 14(3): 893-898.

Noordijk J, Delille K, Schaffers AP, Sýkora KV (2009) Optimizing grassland management for flower-visiting insects in roadside verges. Biological Conservation 14:2097-2103.

North American Electrical Reliability Corporation (NERC) – (2009) Reliability Standards for the Bulk Electric Systems of North America [online] URL: http://wwwnerccom/files/Reliability_Standards_Complete_Setpdf (accessed October 2011).

North American Pollinator Protection Campaign (NAPPC) Rights of Way Task Force of (2010) Plight of the Pollinator: Save money, Time and Energy with IVM and Energy Rights-of-way for Wild Pollinators. *Brochure.*

Ollerton J, Winfree R, Tarrant S (2011) How many flowering plants are pollinated by animals? Oikos 120:321-326.

Potts SG, Vulliamy B, Dafni A, Ne'eman G, Willmer P (2003) Linking bees and flowers: how do floral communities structure pollinator communities? Ecology 84(10)2628–2642.

Raemakers IPS, André P, Sýkora KV, Heijerman T (2001) The importance of plant communities in road verges as a habitat for insects. Proceedings of Experimental and Applied Entomology, Nederlandse Entomologische Vereniging, Amsterdam 12:101-106.

Ries L, Debinski DM, Wieland ML (2001) Conservation value of roadside prairie restoration to butterfly communities. Conservation Biology 15(2): 401-411.

Rogers LE, Warren JT, Hinds NR, Gano KA, Fitzner RE, Piepel GF (1982) Environmental studies of a 1100-kV prototype transmission line: an annual report for the 1981 study period. Battelle Pacific Northwest Laboratories, Richland, Washington.

Russell KN, Ikerd H, Droege S (2005) The potential conservation value of unmowed powerline strips for native bees. Biological Conservation 124:133-148.

Saarinen K, Valtonen A, Jantunen J, Saarnio S (2005) Butterflies and diurnal moths along road verges: Does road type affect diversity and abundance? Biological Conservation 123:403-412.

Schmidt L, Schmidt J, Rao H, Wang W, Xu L (1995) Feeding preference and survival of young worker honey bees (Hymenoptera: Apidae) fed rape, sesame, and sunflower pollen. Journal of Economic Entomology 88(6):1591-1595.

Seibert HC, Conover JH (1991) Mortality of vertebrates and invertebrates on an Athens County, Ohio, Highway. Ohio Journal of Science 91(4):163-166.

Sheperd S, Debinski DM(2005) Evaluation of isolated and integrated prairie reconstructions as habitat for prairie butteries. Biological Conservation 126:51-61.

Shepherd M, Buchmann SL, Vaughan M, Black SH (2003) Pollinator Conservation Handbook. Portland: The Xerces Society. [online] URL: www.xerces.org/bookspollinator-conservationhandbook/ (accessed December 2011).

Smallidge PJ, Leopold DJ, Allen CM (1996) Community characteristics and vegetation management of Karner Blue Butterfly (*Lycaeidesmelissa samuelis*) habitats on rights-of-way in East-Central New York, USA. British Ecological Society 33:1405-1419.

Smith DE (1993) Ecology of Greenways: Design and Function of Linear Conservation Areas: University of Minnesota Press, Minneapolis.

Soulé ME (1991) Land use planning and wildlife maintenance: Guidelines for conserving wildlife in an urban landscape. Journal of the American Planning Association 57(3):313-323.

Sparks TH, Parish T (1995) Factors affecting the abundance of butterflies in field boundaries in swavesey fens. Cambridgeshire, UK Biological Conservation 73:221-227.

Valtonen A, Jantunen J, Saarinen K (2006) Flora and Lepidoptera fauna adversely affected by invasive *Lupinus polyphyllus* along road verges. Biological Conservation 133: 389-396.

Von Der Lippe M, Kowarik I (2007) Long-distance dispersal of plants by vehicles as a driver of plant invasions. Conservation Biology 21(4):986-996.

Way JM (1969) Road verges - research on management for amenity and wildlife. In: Way JM (ed) Road verges – their function and management. Abbots Ripton, The Nature Conservancy, pp 34-40.

Way JM (1977) Roadside verges and conservation in Britain: A review. Biological Conservation 12:65-74.

Weber TC, Allen WL (2010) Beyond on-site mitigation: An integrated, multi-scale approach to environmental mitigation and stewardship for transportation projects. Landscape and Urban Planning 96(4):240-256.

Wellenstein G (1973) The influence of high-tension lines on honeybee colonies (translation from the original German). Zeitschrift Fur Angewandte Entomologie 74:86-94.

Wildlife Habitat Council (2005) Wildlife on Rights-of-Way: suggested best management practices. *Brochure.*

Wildlife Habitat Council and USDA Natural Resource Conservation Service (2005) A guide to managing rights-of-way for wildlife habitat. *Booklet.*

Williams MR (2008) Assessing diversity of diurnal Lepidoptera in habitat fragments: Testing the efficiency of strip transects. Ecological Entomology 37(5):1313-1322.

Williams NM, Minckley RL, Silveira FA (2001) Variation in native bee faunas and its implications for detecting community changes. Conservation Ecology 5(1):7 [online] URL: www.consecol.org/vol5/iss1/art7 (accessed November 2011).

Willyard CJ, Tikalsky SM, Mullins PA (2004) Ecological effects of fragmentation related to transmission line rights-of-way: A review of the state of the science. Wisconsin Focus on Energy Environmental Research Program 1-66.

Winston ML (1992) The biology and management of Africanized honey bees. Annual Reviews of Entomology 37:173-93.

Wynhoff I, van Gestel R, van Swaay C, van Langevelde F (2011) Not only the butterflies: managing ants on road verges to benefit *Phengaris* (*Maculinea*) *butterflies.* Journal of Insect Conservation 15:189-206.

Xerces Society (2010) Invertebrate Fact Sheet: Pollinators and Roadsides Booklet. *Booklet.*

Yahner RH (2004) Wildlife response to more then 50 years of vegetation maintenance on a Pennsylvania US, right-of-way. Journal of Arboriculture 30(2):12.

7

NECTAR PLANT PREFERENCES OF *HEMARIS* (SPHINGIDAE: LEPIDOPTERA) ON CO-OCCURRING NATIVE *CIRSIUM* AND NON-NATIVE *CENTAUREA* (ASTERACEAE) INFLORESCENCES

Elena S. Tartaglia* and Steven N. Handel

Department of Ecology, Evolution and Natural Resources, Rutgers, The State University of New Jersey, 1 College Farm Road, New Brunswick, New Jersey, 08901-1582

Abstract—Preferences for certain nectar plants by moth pollinators have not been extensively studied, particularly seasonal switches in nectar diet. The genus *Hemaris,* found throughout North America and parts of other continents, is a useful new system to address questions of flower-insect interactions as the moths are relatively common and day-flying. In this study we address the following questions with a group of three *Hemaris* species. (1) Does *Hemaris* display a preference for foraging on the native swamp thistle, *Cirsium discolor,* over the non-native invasive knapweeds, *Centaurea* spp.? (2) What factors might be driving *Hemaris'* preferences? The bloom periods of *C. discolor* and *Centaurea* overlap in the northeastern United States, with *Centaurea* blooming from approximately June through September and *C. discolor* blooming for approximately the first three weeks of August. *Hemaris* typically ecloses in the first week of July and fed on *Centaurea* until *C. discolor* bloomed. We tracked and recorded *Hemaris* visits to both plant species. During the co-blooming period of *C. discolor* and *Centaurea, Hemaris* visits significantly more *C. discolor* inflorescences, demonstrating a possible preference for the native species. *Hemaris* reverts to nectaring at the non-native *Centaurea* after native *C. discolor's* bloom period ends. The nectar of *C. discolor* has a significantly higher sugar concentration than does the nectar of *Centaurea*; this difference may account for part of the moths' preference for *C. discolor.*

Keywords: Centaurea; Cirsium discolor; Hemaris; *nectar plant preferences; non-native species*

INTRODUCTION

Preferences for certain nectar plants by pollinators has been examined in bees (Wykes 1952; Roubik & Buchmann 1984; Alm et al. 1990; Leong & Thorp 1999; Schemske & Bradshaw 1999), butterflies (Wiklund et al. 1979; Alm et al 1990; Grundel et al. 2000), birds (Bolten & Feinsinger 1978; Roberts, 1996; Schondube & Del Rio 2003) and, more rarely, in moths (Riffell et al. 2008). The relative abundances of floral resources can change throughout a pollinator's life, necessitating seasonal switches in nectar diets and many types of pollinators have displayed these nectar diet shifts (Feinsinger & Swarm 1982).

Food preference may also change based on a pollinator's capacity to learn and seasonal resource availability. Ensuing shifts in diet are commonly described for a variety of organisms, though rarely in moths (e.g. Tinbergen 1960; Persson & Hansson, 1999; Sydeman et al. 2001). An emerging issue in ecology is the role of non-native plant species in the diets of native pollinators. Interactions between pollinators and non-native plants vary greatly among invaded systems (Parker & Haubensack 2002; Mitchell et al. 2009). In many cases non-natives have been shown to directly compete with natives and draw pollinators away from native plants thereby reducing native seed set (Mosquin 1971; Brown & Mitchell 2001; Brown et al. 2002; Bell et al. 2005; Kandori et al. 2009) while in other cases, the presence of non-natives has no effect on or may even facilitate native pollination (Laverty & Hiemstra 1998; Moragues & Traveset 2005; Bartomeus et al. 2008).

This study focuses on the nectar feeding behaviour of *Hemaris,* a genus of diurnal moths (Lepidoptera) of the family Sphingidae, at native and non-native Asteraceae species. Although they are relatively common organisms, little is known of the behaviour or function of *Hemaris* as a flower visitor. As caterpillars, the niches of *Hemaris* species overlap. *Hemaris thysbe* and *H. diffinis* feed on woody vines such as *Vitis* spp., *Lonicera sempervirens,* shrubs such as *Symphoricarpos* spp. and forbs such as *Apocynum cannabinum. Hemaris gracilis* larval host plants include shrubs of the heath family (Ericaceae) such as *Kalmia* spp. and *Vaccinium vaccilans* (Butterflies and Moths of North America 2013; Tuttle 2007). These ericaceous hosts are not found in abundance at our site, so the lack of larval hosts may explain the dearth of *H. gracilis* individuals encountered. Adult nectar resources overlap broadly. Adults have been observed nectaring from a wide variety of native and horticultural flowers including *Cirsium discolor, Monarda* spp., *Petunia x hybrida, Kalmia latifolia,* and *Ascplepias syriaca* (Tartaglia 2013).

*Corresponding author; email: elena.tartaglia@gmail.com

In this study, we address the following questions: (1) Does *Hemaris* display a preference for foraging on the native swamp thistle, *Cirsium discolor,* over the non-native invasive knapweeds, *Centaurea* spp.? (2) What factors might be driving *Hemaris'* preferences? We expected that our findings would improve our understanding of community pollination dynamics with this little-studied group of pollinators.

MATERIALS AND METHODS

Study system

Hutcheson Memorial Forest (HMF), in Franklin, Somerset County, New Jersey, is a habitat consisting of several early successional meadows interspersed between forest patches of various age and species composition. HMF consists of 226 ha of meadows and upland temperate forest surrounded by various landscape matrix types including urban and suburban areas, as well as being connected to a corridor network of open space throughout central NJ. We made our observations in one of HMF's early-successional meadows. This meadow is permanently maintained in an early successional sere and is surrounded by stands of *Juniperus virginiana.*

We made observations from 16-July until 28-August, 2012. Throughout the study period, the most abundantly blooming plants were *Centaurea* spp., knapweeds native to Eurasia. *Centaurea* was first reported in North America in the early 1890's (Morisawa 1999). There are two species of non-native *Centaurea* at HMF, *C. stoebe* spp. *micranthos* (Gugler) Hayek and *C. nigrescens* Willd (ITIS 2014). As the two species are nearly indistinguishable, we did not differentiate between them in this study. *Cirsium discolor,* a thistle native to the northeast blooms abundantly for a short period of time during the summer, so the bloom periods of these plants overlap for only approximately the first three weeks of August. Thistle grows to an average height of 1.8 m at HMF while knapweed is shorter (average height 1.2 m). Both plants belong to the Asteraceae and have similar floral forms consisting of purple/lavender inflorescences of densely packed discoid florets. However, knapweed also possesses ray florets, which are lacking in thistle. We had established previously that *Hemaris* is attracted to purple flowers over other colours (Tartaglia 2013).

Data Collection

We netted *Hemaris* of all three species on two 20 m unidirectional transects, one running through a patch of *Cirsium discolor* plants and one running through a patch of *Centaurea* plants. Upon capture, each moth was manually dusted with fluorescent dye powder (Shannon Luminous Materials, Inc., Santa Ana, California *via* BioQuip, Rancho Dominguez, California) to track which flowers they landed on (Kearns & Inouye 1993). Care was taken to apply dye to both dorsal and ventral surfaces of the moths and to remove any powder that might obstruct their vision or antennae. Moths were released immediately after dye application. *Hemaris thysbe* individuals were dusted with orange dye powder, *H. diffinis* individuals were dusted with blue dye powder and *H. gracilis* individuals were dusted with yellow dye powder. At dusk, we returned to the site with blacklight flashlights to track moth landings and recorded the number and species of inflorescences in the transects with orange, yellow or blue dye on them. We also examined other floral resources present in the field to determine if *Hemaris* utilises species besides *C. discolor* and *Centaurea* in this study system. Since the dye faded overnight to the degree that it was nearly indistinguishable the following day (particularly if rainfall occurred), we were able to make a distinction newly-visited inflorescences (i.e. visited during the course of that day) as discernibly brighter than previously visited inflorescences (i.e. dye was very faint or had disappeared totally).

We also counted thistle and knapweed inflorescences in our transects each day. Data for inflorescence number and *Hemaris* visitation was pooled for the two transects. We used t-tests to assess whether there were significant differences in *Hemaris* visitation to thistle and knapweed inflorescences.

To determine if sugar concentrations in nectar might be driving any observed preferences for thistle or knapweed, we sampled 20 disc florets from 10 inflorescences each for thistle and knapweed plants growing adjacent to our transect. We extracted nectar by separating individual disc florets, removing the bottom of the tube and squeezing nectar onto a Reichert 10431 hand refractometer surface. Nectar extracted from individual florets from separate inflorescences was pooled. We sampled nectar once at mid-day (between 1200 and 1300h) and once in afternoon (between 1500 and 1600h). We used t-tests to assess whether there were significant differences in mean nectar concentrations of thistle and knapweed.

RESULTS

In the northeastern United States, the bloom periods of knapweed and thistles coincide for only a few weeks (Fig. 1). Knapweed blooms for nearly the entire summer (from June through September) and has a much longer bloom period than *Cirsium discolor,* which blooms for approximately the first three weeks of August. Knapweed is also more abundant than thistle. During the overlapping bloom period, there were on average 45.3 thistle inflorescences (115 at peak) and 157 on average (175 at peak) knapweed inflorescences present in our transects. Prior to thistle blooming at the study site, *Hemaris* (N = 50) was netted from knapweed and fluorescent dye tracking revealed that they nectared exclusively at knapweed during this time. When thistle bloom began, *Hemaris* (N = 64) was netted from both thistle and knapweed and fluorescent dye tracking revealed that they nectared at both thistle and knapweed. However, the moths visited significantly more thistle inflorescences ($t = 2.82$, $df = 24$, $P = 0.007$) during overlapping bloom times, despite the difference in abundance between the two plant species at the site. During the co-blooming period, *Hemaris* made 351 total visits to *Cirsium* inflorescences vs. 78 visits to *Centaurea* inflorescences (Fig. 1). When thistles ceased blooming, fluorescent dye tracking revealed that *Hemaris* (N = 11) reverted to nectaring at knapweed inflorescences (58 visits).

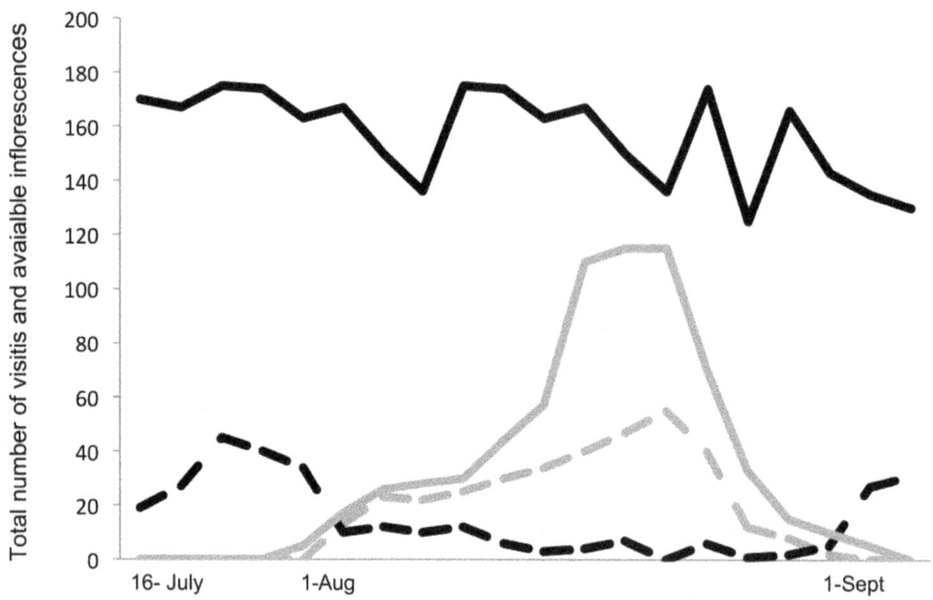

FIGURE I: Total visits of *Hemaris* to thistle and knapweed inflorescences vs available thistle and knapweed inflorescences in two 20m transects over the course of the study. Solid black line represents available knapweed inflorescences; solid gray line represents available thistle inflorescences; dotted black line represents visits to knapweed; dashed gray line represents visits to thistle.

Three *Hemaris* species are present at HMF. Of the 114 individuals dusted with fluorescent powder in this study, 65 were *H. diffinis,* 43 were *H. thysbe,* and 6 were *H. gracilis.* Among *Hemaris* species there were no significant differences ($F_{2,34}$ = 0.395, P = 0.68) in visitation to thistle and knapweed. In a previous study specifically examining *Hemaris* foraging behaviours (Tartaglia 2013), we also found no differences between the three *Hemaris* species, so we have treated them at the genus level for our study. Other floral resources (*Asclepias syriaca, Convolvulus* spp., *Erigeron philadelphicus, Leucanthemum vulgare, Penstemon hirsutus, Solidago* spp.) existed in the field but were ignored by the moths.

Refractometry to determine sugar concentration differences between the two preferred nectar resources revealed that thistle (56.2% sugar) has significantly (t = 7.22, df = 38, $P < 0.01$) more sugar-rich nectar than knapweed (47.1% sugar).

DISCUSSION

Our results add to the growing body of literature regarding the role of non-native plant species in plant-pollinator relationships. The diurnal *Hemaris* genus is an active and selective visitor to floral communities. It seems that the non-native *Centaurea* serves as a "place-holder" food resource for *Hemaris* until the preferred nectar source, *C. discolor,* blooms, creating an apparent sequential mutualism between the moths, the non-native *Centaurea,* and the native *C. discolor. Centaurea* may help sustain the *Hemaris* population, allowing sufficient numbers of the moths to survive until their preferred food source becomes available. Our data set did not address the parallel question of whether *C. discolor* benefits as well. Waser & Real (1979) were among the first to consider in detail the concept that sequentially flowering species may sustain pollinator populations throughout seasons, ensuring reproductive success for both plants and pollinators and our data lends support to this concept as *Hemaris* heavily utilises *Centaurea* until *C. discolor* becomes available.

No specific research has been done addressing colour preference in *Hemaris,* but it has been established that naive *Macroglossum stellatarum* is able to learn and retain visual colour cues (Kelber 1996). Comparisons between diurnal and nocturnal Sphingidae have revealed that diurnal hawkmoths do rely more heavily on visual cues (colour and light intensity) to detect resources, while nocturnal species relied on odor cues (Balkenius et al.,2006) so colour may indeed be a critical factor in these moths' selection preferences.

Though the reason why *Hemaris* seems to prefer *C. discolor* may be due to the higher sugar concentration in the native thistle, other factors influence the differential attraction. Hawkmoths rely on multiple sensory inputs to detect nectar resources (Raguso 2004; Balkenius et al. 2006), and Sphingidae have keen olfactory capacities in addition to their visual capacities. In fact, resource switching in Sphingidae has been documented by Riffell et al. (2008) when *Manduca sexta* switched from flowers of *Agave* flowers to those of *Datura* in their study system. Though *Datura* flowers have less nectar than *Agave,* the moths utilised them due to an innate preference for *Datura* (Riffell et al. 2008). Although the plants seem odorless to humans, *Hemaris* may be able to detect different more- or less-attractive volatiles from the two plants. Additionally, *C. discolor* does not possess ray florets, so differences in ray floret reflectance, inflorescence shape or inflorescence size may drive differential visitation to the flowers. Diurnal Sphingidae have been shown to rely heavily on visual cues when foraging (Kelber 1996; Balkenius et al. 2006) and Dafni et al. concluded that a wide variety of floral spatial factors contribute to insects' foraging behaviours (1997).

ACKNOWLEDGEMENTS

The authors thank the Hutcheson Memorial Forest Center for funding. C. Holzapfel, R.A. Raguso and L. Struwe provided valuable feedback on this manuscript. S. Wehman and E. Golan assisted with fieldwork.

REFERENCES

Alm J, Ohnmeiss TE, Varza J, Vriesenga L (1990) The preference for cabbage white butterflies and honeybees for nectar that contains amino acids. Oecologia 84:53-57.

Balkenius A, Rosen W, Kelber A (2006) The relative importance of olfaction and vision in a diurnal and a nocturnal hawkmoth. Journal of Comparative Physiology A. 192:431-437.

Bartomeus I, Vilà M, Santamaría L (2008) Contrasting effects of invasive plants in plant–pollinator networks. Oecologia 155:761-770.

Bell JM, Karron JD, Mitchell RJ (2005) Interspecific competition for pollination lowers seed production and outcrossing in *Mimulus ringens*. Ecology 86:762-771.

Bolten AB, Feinsinger P (1978) Why do hummingbird flowers secrete dilute nectar? Biotropica 10:307-309.

Brown BJ, Mitchell RJ (2001) Competition for pollination: effects of pollen of an invasive plant on seed set of a native congener. Oecologia 129:43-49.

Brown BJ, Mitchell RJ, Graham SA (2002) Competition for pollination between invasive species (purple loosestrife) and a native congener. Ecology 83:2328-2336.

Butterflies and Moths of North America. 2013. (http://www.butterfliesandmoths.org/)

Cartar RV (1991) A test of risk-sensitive foraging in wild bumblebees. Ecology 72:888-895.

Dafni A, Lehrer M, Kevan PG (1997) Spatial flower parameters and insect spatial vision. *Biological Reviews* 72:239-282.

Feinsinger P, Swarm LA (1982) Ecogical release, seasonal variation in food supply and the hummingbird *Amazilia tobaci* on Trinidad and Tobago. *Ecology* 63:1574-1587.

Grundel R, Pavlovic NB, Sulzman CL (2000) Nectar plant selection by the Karner blue butterfly (*Lycaedes Melissa samuelis*) at the Indiana Dunes national lakeshore. The American Midland Naturalist 144:1-10.

Heinrich B (1979a) "Majoring" and "minoring" by foraging bumblebees, *Bombus vagans*: An experimental analysis. Ecology 60:245-255.

Heinrich B (1979b) Resource heterogeneity and patterns of movement in foraging bumblebees. Oecologia 40: 235-245.

Integrated Taxonomic Information System (ITIS). Retrieved Jan 31, 2014 from (http://www.itis.gov).

Kandori I, Hirao T, Matsunaga S, Kurosaki T (2009) An invasive dandelion unilaterally reduces the reproduction of a native congener through competition for pollination. Oecologia 159:559-569.

Kearns CA, Inouye DW (1993) Animals. In: Kearns CA and Inouye (eds) Techniques for pollination biologists. University Press of Colorado, Niwot, CO. pp 263-392.

Kelber A (1996) Colour learning in the hawkmoth *Macroglossum stellatarum*. Journal of Experimental Biology 199:1127-1131.

Kelber A (1997) Innate preferences for flower features in the hawkmoth *Macroglossum stellatarum*. Journal of Experimental Biology 200:827-836.

Laverty T, Hiemstra H (1998) Effect of flowering dandelion as a competitor to flowers of fruit trees for pollen-collecting honey bees in Ontario. Proceedings of the Entomological Society of Ontario 129:3-8.

Leong JM, Thorp RW (1999) Colour-coded sampling: the pan trap colour preferences of oligolectic and nonoligolectic bees associated with a vernal pool plant. Ecological Entomology 24:329-335.

Mitchell RJ, Flanagan RJ, Brown BJ, Waser NM, Karron JD (2009) New frontiers in competition for pollination. Annals of Botany 103:1403-1413.

Moragues E, Traveset A (2005) Effect of *Carpobrotus* spp. on the pollination success of native plant species of the Balearic Islands. Biological Conservation 122:611-619.

Morisawa T-L (1999) Weed Notes: Centaurea maculosa. The Nature Conservancy Wildland Weeds Management and Research. http://tncweeds.ucdavis.edu.

Mosquin T (1971) Competition for pollinators as a stimulus foot the evolution of flowering time. Oikos 22:398–402.

Parker IM, Haubensak KA (2002) Comparative pollinator limitation of two non-native shrubs: do mutualisms influence invasions? Oecologia 130:250-258.

Persson A, Hansson LA (1999) Diet shift in fish following competitive release. Canadian Journal of Fisheries and Aquatic Sciences 56:70-78.

Raguso RA (2004) Flowers as sensory billboards: progress towards an integrated understanding of floral advertisement. Current Opinions in Plant Biology 7:434-440

Riffell JA, Alarcón R, Abrell L, Davidowitz G, Bronstein JL, Hildebrand JG (2008) Behavioral consequences of innate preferences and olfactory learning in hawkmoth–flower interactions. Proceedings of the National Academy of Sciences 105:3404-3409.

Roberts MW (1996) Hummingbirds' nectar concentration preferences at low volume: the importance of time scale. Animal Behaviour 52:361-370.

Roubik DW, Buchmann SL (1984) Nectar selection by *Melipona* and *Apis mellifera* (Hymenoptera: Apidae) and the ecology of nectar intake by bee colonies in a tropical forest. Oecologia 61:1-10.

Schemske DW, Bradshaw HD (1999) Pollinator preference and the evolution of floral traits in monkeyflowers (*Mimulus*). Proceedings of the National Academy of Sciences 96:11910-11915.

Schondube JE, Del Rio CM (2003) Concentration-dependent sugar preferences in nectar-feeding birds: mechanisms and consequences. Functional Ecology 17:445-453.

Sydeman WJ, Hester MM, Thayer JA, Gres F, Martin P, Buffa J (2001) Climate change, reproductive performance and diet composition of marine birds in the southern California Current system, 1969–1997. Progress in Oceanography 49:309-329.

Tartaglia ES (2013) Hawkmoth-Flower interactions in the urban landscape: Sphingidae ecology with a focus on the genus *Hemaris*. Ph.D. dissertation, Rutgers, The State University of New Jersey, New Brunswick, New Jersey, USA.

Tuttle JP (2007) The Hawkmoths of North America. Washington, D.C.: The Wedge Entomological Society.

Waser NM, Real LA (1979) Effective mutualism between sequentially flowering plant species. Nature 281:671-672.

Wiklund C, Eriksson T, Lundberg H (1979) The wood white butterfly *Leptidea sinapis* and its nectar plants: A case of mutualism or parasitism? Oikos 33:358-362.

Wykes GR (1952) The preferences of honeybees for solutions of various sugars which occur in nectar. Journal of Experimental Biology 29:511-519

8

EXPERIMENTAL POLLINATOR DECLINE AFFECTS PLANT REPRODUCTION AND IS MEDIATED BY PLANT MATING SYSTEM

Rebekka Lundgren[1,*], Amparo Lázaro[1,2] and Ørjan Totland[1]

[1]*Norwegian University of Life Sciences, Department of Ecology and Natural Resource Management, P. O. Box 5003, Ås, Norway*
[2]*Present address: Mediterranean Institute for Advanced Studies, Miquel Marques 21, 07190 Esporles, Balearic Islands, Spain.*

Abstract—There is growing concern that current pollinator decline will affect the reproduction of plant species, potentially driving a decline in plant population densities. We experimentally tested whether a reduction in flower visitation caused a reduction in fertilization rate in several species, and whether any reduction in fecundity of species depends on their degree of reproductive dependence on pollinators and their attractiveness for pollinators. We recorded visitation rate, fertilization rate, seed weight, flower size and density of nineteen insect-pollinated perennial herbs inside thirty 2 x 2 m dome-shaped cages covered with fishnet (experimental plots) and in thirty control plots in a Norwegian hay meadow. We used a bagging experiment to estimate the ability of the study species to produce seeds in the absence of pollinators. The visitation rate for fifteen of nineteen study species was lower inside cages than outside and only three of the fifteen species showed significantly reduced fertilization rates in the experimental plots. The magnitude of reduction in fertilization rate was positively related to the degree of pollinator dependence, but not to their attractiveness for pollinators or to the reduction in visitation rate. Seed weight was not affected by the experiment. The lack of an overall effect of reduced pollinator visitation on fertilization rate suggests that some species may be robust to a pollinator decline that could increase pollen limitation on plant reproduction. Our results suggest that species with greater pollinator dependence are more vulnerable to pollinator loss.

Keywords: blossom cover, degree of pollinator dependence, fertilization rate, pollinator decline, seed weight, visitation rate

INTRODUCTION

The majority of plant species are dependent on animal pollination (Burd 1994; Dauber et al. 2010) and many of them experience pollen limitation on their reproductive success (Knight et al. 2005). Both low quantity and low quality of pollen may result in pollen limitation (Ashman et al. 2004; Wilcock & Neiland 2002). One component of quantitative pollen limitation is pollinator limitation (Aizen & Harder 2007) and this particular component has received considerable attention in relation to the recently documented pollinator decline (Biesmeijer et al. 2006; Goulson et al. 2008; Pauw 2007). Pollinator decline is caused by many factors, such as degradation of natural and semi-natural habitats, climate change, and changes in plant and pollinator distributions (Aguilar et al. 2006; Hegland et al. 2009; Kearns et al. 1998; Knight et al. 2005). These threats may all disrupt plant-pollinator interactions, leading to pollinator and ultimately pollen limitation, with potential large effects on plant population dynamics and plant community composition (Ashman et al. 2004). Even so, there is a lack of experimental studies of the effect of community-wide reductions in pollinator visitation rate and their effects on plant fecundity through pollen limitation (but see Lázaro et al. *submitted*).

*Corresponding author; email: rebekka.lundgren@umb.no

In order to maintain seed production plant species that experience pollen limitation due to low visitation rates can either evolve higher attractiveness and/or alternative pollination modes (Ashman et al. 2004; Eckert et al. 2009). In particular, plant species which can use different pollination modes (e.g. selfing and wind pollination) may be reproductively assured against reductions in pollinator availability (Culley et al. 2002; Kennedy & Elle 2008). In addition to life history traits, such as breeding system, ecological factors may affect plant reproductive responses to pollinator decline. For instance, large individual flower displays and dense patches of plants often receive a higher number of visits (Ebeling et al. 2008; Grindeland et al. 2005; Hegland & Totland 2005) due to a higher attraction (Dreisig 1995) and/or foraging economy of pollinators (Goulson 2003).

A reduction in pollinator availability may not only affect the quantity of seeds produced but also their quality (Aizen & Harder 2007), since selfing in self-compatible plant species may result in reduced seed quality due to increased inbreeding (Dudash & Fenster 2000; Grindeland 2008). The seed quality in purely outcrossing plant species may also suffer from reduced pollinator availability if pollinators forage within restricted areas instead of moving long distances to find rewarding flowers (Dukas & Real 1993; Thomson et al. 1982). Such spatially restricted foraging may increase biparental inbreeding in self-incompatible plant species, since closely situated plants are often related (Waser & Price 1994).

Here we use an experimental approach to study whether a reduction in flower visitation affects reproductive success in nineteen plant species in a mown hay meadow in south Norway. Furthermore, we examine whether the magnitude of reduction in fecundity of these nineteen species is related to the reduction in pollinator visitation rate, to their dependence on pollinators for fertilization and to their attractiveness for pollinators. We simulated a decline in pollinator visitation by using dome-shaped cages covered with fishnet (experimental plots) that effectively reduced flower visitation rates and estimated the reduction in plant fecundity (fertilization rate and seed weight) and visitation rates as the difference between the values obtained in experimental versus control plots. Our specific questions were: 1) does a reduction in visitation rate cause a reduction in plant fecundity?; and 2) is there a relationship between the reduction in visitation rate, the degree of pollinator dependence, the attractiveness of plant species for pollinators and the reduction in plant fecundity?

MATERIALS AND METHODS

Study area and study species

The study site is located on a species-rich hay meadow at Ryghsetra (59°44`03``N, 10°02`48``E), in Buskerud county in Norway. The blooming season in this species rich community begins in early May and ends in mid-late August, and approximately 55 insect-visited plant species bloom during this period. We studied the nineteen species with the highest number of flowering individuals among the insect-pollinated perennial herbs occurring at Ryghsetra. However, the experiment reduced pollinator visitation in only fifteen species and only these species are used in analyses. From literature we know that most of the fifteen species, which we studied in our study site, are probably able to produce seeds to a greater or lesser degree without pollinator visitation, with the exception of *Centaurea jacea, Primula veris, Trifolium pratense* and most likely also *Lathyrus linifolius* (Table 1). From flower visitation observations conducted in control plots in 2006, we found that the flower visitor assemblage consisted of 72.7% bumblebees, 11.0% muscoid flies, 5.5% solitary bees, 4.7% hover flies, 2.4% ants, 1.6% butterflies, 0.9% honeybees, 0.5% beetles and 0.5% bee flies.

Experimental manipulation of pollinator visitation

The hay meadow was large and relatively homogeneous, and we therefore placed thirty pairs of permanent plots (2 × 2 m) systematically along two parallel rows (separated by ca. 5 m) across the study site and marked the inner square of 1 × 1 m of each plot, at the beginning of the field season in 2006. In order to reduce any effect of the experimental setup (see below) we only conducted flower visitation observations inside this inner square meter. Plot pairs were separated by at least 3 m and plots within a pair were separated by 2 m. We randomly selected one plot of each pair for the experimental reduction of pollinator visitation (experimental plots hereafter), whereas the other plot was left unmanipulated and open for natural pollination (control plots hereafter). To obtain a reduction in pollinator visitation, we placed dome-shaped cages made of two 4 m long PVC-tubes bent diagonally over the experimental plots, and covered the domes with transparent nylon fishnet with a mesh width of 1.05 × 1.05 cm. The size of the cages was 2 × 2 × 1 m (l, w, h). To allow pollinators entering the cages an easy exit, we left the lowest approximately 10 cm without mesh, and had in addition another opening of 0.5 × 0.5 m on the top.

TABLE 1. The fifteen study species and information from references about their plant family, compatibility system and the presence of apomixis in these.

Species	Family	Compatibility system	Apomixis	References
Centurea jacea	Asteraceae	Self-incompatible	No	Hardy et al. (2001), Noyes (2007)
Fragaria vesca	Rosaceae	Self-compatible	No	Schulze et al. (2012), Nosrati et al. (2010)
Galium mollugo	Rubiaceae			
Geranium sylvaticum	Geraniaceae	Self-compatible		Ramula et al. (2007)
Hieracium cymosum	Asteraceae			
Hieracium lactucella	Asteraceae	Self-incompatible	Facultative apomictic	Krahulec et al. (2008), Koltunow et al. (1998)
Hieracium pilosella	Asteraceae	Self-incompatible	Facultative apomictic	Koltunow et al. (1998), Houliston & Chapman (2004)
Knautia arvensis	Dipsacaceae	Self-compatible		Vange (2002)
Lathyrus linifolius	Fabaceae			
Leucanthemum vulgare	Asteraceae	Self-incompatible	Not demonstrated	Andersson (2008), Noyes (2007)
Linum catharticum	Linaceae			
Polygala vulgaris	Polygalaceae	Self-compatible		Norderhaug (1995)
Potentilla thuringiaca	Rosaceae		Facultative pseudogamous apomictic	Dobeš et al. (2013)
Primula veris	Primulaceae	Self-incompatible		Wedderburn & Richards (1990)
Trifolium pratense	Fabaceae	Self-incompatible		Leduc et al. (1990)

In a pilot-study conducted in 2005, we checked potential biotic and abiotic side-effects of the cages, such as changes in herbivory, seed predation, photosynthetically active radiation and wind speed. No significant side-effects of the cages were found (all P>>0.05; results not shown), except for a marginally significant reduction in wind speed inside the cages (19.4 %; $P = 0.066$), that did not seem to affect the flying insects. The experimental set up and/or the reduced density of simultaneous foragers (Inouye 1978; Lázaro et al. 2011; Valido et al. 2002) might also affect the foraging behaviour of individual pollinators. If so, we would not know whether the results of our study were due to a reduction in pollinator availability exclusively or to related changes caused by the experimental set up itself. In another study (Lázaro et al. *Submitted*) we therefore checked whether the experimental setup altered the behaviour and composition of pollinators of *Centaurea jacea, Galium mollugo* and *Leucanthemum vulgare*, three of the most abundant and visited species in this community. We found no difference in the number of flowers contacted per individual visitor or visit duration inside the plots. There was no significant difference either in the composition of pollinators visiting these species in control and experimental plots, except for *Centaurea jacea,* which showed a minor decrease of bumblebees inside the experimental plots (72.2% vs. 82.8% of total pollinators for experimental and control plots, respectively; Lázaro et al. *submitted*). These results indicate that our experiment reduced the number of pollinator visits without affecting the biotic and abiotic environment of plants or the behaviour and composition of their pollinators. Our experimental set up therefore seems appropriate to test the effects of pollinator reduction on plant fecundity.

In four of the nineteen species (*Anthyllis vulneraria, Lotus corniculatus, Potentilla erecta* and *Vicia cracca*) the visitation rate was higher in experimental than control plots. We therefore excluded these four species from analyses since we focus on effects of reduced visitation rate.

Visitation rate and plant fecundity

We observed flower visitation at Ryghsetra from 27 May to 19 July, 2006, covering the entire blooming season of the study species. Observations were done between 0800 and 1800 h on days without rain and when winds were calm. Both experimental and control plots of the same pair were observed simultaneously (or immediately after each other) using 20 minute observation periods (N = 247). The order of observation of pairs was random, but we never observed the same pair more than once per day. We observed pollinator visitation to each flower or inflorescence (depending on the species; flowers hereafter) of all flowering species occurring within the inner square of the plots. Since pollinator observations were plot based, the observation time to each individual study plant species ranged from 1.7 hours to 15.3 hours (Table 2). A visit was defined to have occurred when there was contact between the visitor's body and the reproductive organs of the flower. We categorized each visitor into bumblebees (*Bombus* spp.), solitary bees (e.g. Colletidae, Andrenidae, Halictidae; all of them of similar small size), honeybees (*Apis mellifera*), ants (Formicidae), hover flies (Syrphidae), muscoid flies (mainly Muscidae, Anthomyiidae), beetles (Coleoptera), butterflies (Lepidoptera) and beeflies (*Bombylius minor*). After each observation period we counted the number of open flowers of all plant species occurring within both the inner square and the entire plot.

We obtained the average visitation rate of each pollinator groups to all fifteen study plant species in each observation period by dividing the number of pollinator visits to each plant species by the number of open flowers of that plant species in the inner squares of each plot (Table 3). We also obtained the total visitation rate per flower and plot for all study plant species in each observation period, by dividing the total number of pollinator visits to each plant species by the number of open flowers of that plant species in the inner squares of the plots. Previous studies suggest that seed set may be related to the abundance of specific flower-visitor groups, rather than to the total abundance of all pollinators (Franzen & Larsson 2009; Steffan-Dewenter & Tscharntke 1999). However, total visitation by all pollinator groups and visitation by the most abundant pollinator group to each plant species were significantly correlated ($r = 0.788$, $n = 15$, $P = 0.000$). Therefore, we only used the total visitation rate from all pollinators groups (visitation rate, hereafter) in the analyses where we study the relationships between the magnitude reduction in plant fecundity and the reduction in flower visitation.

To estimate fertilization rate, we haphazardly selected and marked one flower in three individuals per species and plot (unless fewer individuals were present) in both control and experimental plots. We marked flowers while they were still blooming to avoid bias due to subconscious selection of fruits. We collected the fruits or infructescences (depending on the species; fruits hereafter) of marked flowers, when fruits were dry and immediately before dispersal. Fruits were stored in paper bags until they were dissected in the laboratory and the number of undeveloped seeds and fully developed seeds was counted. The extent of seed predation was very low and only occurred in three species. In these species (*Centaurea jacea, Hieracium pilosella* and *Lathyrus linifolius*) we counted the number of seed as accurately as possible and only weighted undamaged seeds. We weighed the dry weight of all undamaged developed seeds per fruit together to the nearest mg. Fertilization rates were obtained by dividing the number of fully developed and aborted seeds by the total number of ovules. In *Primula veris,* we could not reliably determine the number of ovules, and therefore we calculated the fertilization rates of this species by dividing the number of fertilized seeds by the average number of ovules per flower (50 ovules; Wissman (2006)). When plant species had single-seeded fruits (e.g. *Potentilla*), an average was calculated for each infructescence to obtain a single measure of fertilization rate per individual plant.

TABLE 2 The fifteen study species and the reduction in visitation rate (VR), their degree of pollinator dependence (DPD) and log transformed blossom cover (logBC). For control (C) and experimental plots (E): the number of hours each species was observed (Obs. hour), and the number of plots (P) and number of individuals (I) that were used for the analyses of fertilization rate (FR) and seed weight (SW).

						FR	SW
Species	DPD	logBC	VR	Plot	Obs. hour	P (I)	P (I)
Centaurea jacea	0.05	3.82	0.38	C	9.7	20 (72)	20 (71)
				E	11	18 (57)	18 (57)
Fragaria vesca	0.22	2.04	0.64	C	4.7	15 (45)	15 (45)
				E	5	13 (40)	13 (40)
Galium mollugo	0.34	1.85	0.15	C	16	30 (111)	30 (108)
				E	15.3	30 (114)	30 (107)
Geranium sylvaticum	0.68	2.49	0.7	C	3	6 (13)	4 (4)
				E	1.7	2 (4)	2 (2)
Hieracium cymosum	0	2.40	0.82	C	4	16 (36)	15 (28)
				E	3.7	16 (33)	14 (26)
Hieracium lactucella	0.22	2.30	0.52	C	5	20 (45)	19 (36)
				E	4.7	19 (35)	13 (19)
Hieracium pilosella	0	2.71	0.56	C	7	21 (47)	19 (36)
				E	7.3	21 (53)	21 (44)
Knautia arvensis	0.9	2.40	0.84	C	2.3	6 (13)	6 (13)
				E	3.7	12 (23)	12 (21)
Lathyrus linifolius	1	2.04	1	C	1.7	4 (7)	2 (2)
				E	1.7	6 (11)	2 (2)
Leucanthemum vulgare	0.61	3.33	0.61	C	9.7	28 (75)	28 (74)
				E	10	27 (63)	27 (63)
Linum catharticum	0.16	2.07	0.62	C	8	26 (69)	26 (69)
				E	9.3	27 (80)	27 (80)
Polygala vulgaris	0.57	2.54	1	C	10.3	24 (48)	23 (47)
				E	10.7	25 (53)	25 (53)
Potentilla thuringiaca	0.22	2.76	0.62	C	3.7	15 (35)	15 (35)
				E	4.3	16 (38)	16 (38)
Primula veris	1	3.70	0.64	C	5.7	8 (10)	5 (6)
				E	6.3	11 (11)	4 (4)
Trifolium pratense	1	3.42	0.12	C	3.7	9 (21)	8 (18)
				E	5.7	11 (26)	10 (18)

Plant species characteristics: degree of pollinator dependence and blossom cover

To test the ability of the study species to produce seeds in the absence of pollinators we conducted a bagging experiment where we used fifteen pairs of individuals outside the plots at Ryghsetra. Individuals in a pair occurred close together in space and were as similar in size as possible. One individual of each pair was haphazardly selected for the bagging treatment, i.e. all or part of its flowers were bagged with mosquito net before anthesis, whereas the other was left unmanipulated and open for natural pollination. For plant species with many flowers and branches (e.g. *Potentilla* or *Trifolium*), we used pairs of branches instead of pairs of individuals to perform this experiment. We collected their seeds when ripe and calculated fertilization rates per treatment as explained above. We estimated the degree of pollinator dependence for each plant species as DPD = 1-

B/C, where B is the average fertilization rate of all fifteen bagged individuals (or branches) and C is the average fertilization rate of all fifteen unmanipulated individuals (or branches). A value of 0 indicate no differences in fertilization rate between bagged and unmanipulated flowers, and thus no dependency of pollinator visitation for fertilization, whereas a value of 1 indicates a complete dependency on pollinator visitation for fertilization. This experiment was performed for all the study species, except *Primula veris*. However, we set the degree of pollinator dependence of this species to 1, since this species is an obligate outbreeder (Wedderburn & Richards 1990). Table 2 shows DPD for each study species.

Flower density on its own might not be an appropriate measure of differences in pollinator attractiveness among plant species, since the attractiveness of a plant species also could be influenced by the size of its reproductive units (Bell 1985; Dudash et al. 2011; Eckhart 1991; Galen & Newport 1987). We therefore followed the procedure by Hegland and Boeke (2006) and combined flower size and flower density for each species into one standardized measure to obtain a variable that more accurately describes the attractiveness of plant species for pollinators (blossom cover, hereafter). Thus, we calculated blossom cover as the product of the average number of flowers per m^2 and the average area of one flower (see values for each species in Table 2). We used data from the entire 2 x 2 m plots to calculate the density of flowers. To calculate the area occupied by a flower, we measured the size of the flowers in the field on thirty hazarphardly selected individuals per species with a digital calliper. For flowers with circular outline (e.g. *Leucanthemum vulgare*) we used the formula of πr^2, where r = radius. For flowers with a depth dimension (e.g. *Lathyrus linifolius*) we used the formula $2\pi rd + \pi r^2$, where r = radius and d = depth (Hegland & Totland 2005).

Data analyses

We used Generalized Linear Models (IBM SPSS Statistics for Windows, Version 20.0. Armonk, NY: IBM Corp) to test whether the experiment significantly reduced the overall visitation rate as well as the visitation rate for each pollinator group to the study species. We performed separate analyses for overall visitation rate to each plant species as well as separate analyses for each pollinator group and study plant species. Due to the nature of the data, we used gamma distributions with log link functions for all the analysis except for visitation rate of bumblebees in *Centaurea jacea* and *Knautia arvensis* where we used normal distributions with link identity. We used individual observation periods as sample units and the treatment (experimental vs. control) as a fixed factor. Due to the very low visitation rate for some species, we only analysed the difference in visitation rate between control and experimental plots in species with more than 10 visits in total from a specific pollinator group to a study species.

To test whether the experiment significantly reduced fecundity (fertilization rate and seed weight) in the study species we used Mixed Effects Models (R 2.10.1; R Development Core Team 2009). We used Gaussian distributions since data fulfilled the assumptions of normality. We performed separate analyses for each response variable and study species. In these analyses we used individual plants as sample units, plot as a random factor and the treatment (experimental vs. control) as a fixed factor.

To study the relationships between the magnitude reduction in plant fecundity, the reduction in flower visitation, blossom cover, and degree of pollinator dependence we used Multiple Linear Regression (IBM SPSS Statistics for Windows, Version 20.0. Armonk, NY: IBM Corp). We used Gaussian distributions since data fulfilled the assumptions of normality and log-transformed the 'blossom cover' predictor variable for a better fit of the models. In order to obtain a single measurement per treatment and species and avoid potential problems of pseudo replication, we averaged the data of visitation rate (VR), fertilization rate (FR) and seed weight (SW) first per plot and then by species. For all three variables we calculated the magnitude of reduction in the experimental plots as 1-E/C, where E is the average VR, FR or SW for the experimental plots and C is the average VR, FR or SW for the control plots. Negative values indicate an increase of VR, FR or SW in the experimental plots compared to the control plots, whereas positive values indicate a reduction in these variables in the experimental plots compared to the control plots. We conducted separate models for each fecundity variable (reduction in fertilization rate and seed weight). The fifteen plant species were the sample units in both models. We ran the full models containing the three continuous variables: 'visitation rate', 'log blossom cover', 'degree of pollinator dependence'. We did not include any interactions in the analysis due to the low number of samples (fifteen samples) and therefore few degrees of freedom.

RESULTS

Experimental reduction in visitation rates and plant fecundity

In the fifteen study species the overall visitation rate was on average 61% lower inside than outside the experimental plots (ranging from 12% to 100%), and 80 % of the species showed a more than 50% lower visitation rate inside cages (Table 2). The overall visitation rate was significantly lower in the experimental plot compared to the control plots (Table 3) for *Centaurea jacea* (χ^2_1 = 13.7, P = 0.000), *Hieracium cymosum* (χ^2_1 = 4.0, P = 0.045), *Knautia arvensis* (χ^2_1 = 15.9, P = 0.000) and for *Leucanthemum vulgare* (χ^2_1 = 14.8, P = 0.000). From the 20 cases where the specific pollinator group visited a specific study species more than 10 times, the difference between control and experimental plots were significant in 10 cases (Table 3). The visitation rate of bumblebees (χ^2_1 = 13.2, P = 0.000), butterflies (χ^2_1 = 10.1, P = 0.002) and hover flies (χ^2_1 = 6.2, P = 0.012) was significantly lower in the experimental plot for *Centaurea jacea*. The visitation rate of flies (χ^2_1 = 3.9, P = 0.046) was significantly lower in the experimental plot for *Hieracium cymosum*. The visitation rate of flies (χ^2_1 = 5.5, P = 0.019) was significantly lower in the experimental plot for *Knautia arvensis*, and finally the visitation rate of flies (χ^2_1 = 7.3, P = 0.007), butterflies (χ^2_1 = 8.5, P = 0.004) and hover flies (χ^2_1 = 14.2, P = 0.000)

TABLE 3 The fifteen study plant species and the average visitation rate (Av. VR) measured as number of visits/flower/observation period and the total number of visits (# Visits) from all pollinator groups in control plots (C) and experimental plots (E). Significant different average visitation rates in control and experimental plots for each plant species and overall visitation rate as well as pollinator group are shown in bold.

		Overall		Ants		Bee flies		Beetles		Bumblebees		Butterflies		Flies		Honeybees		Hover flies		Solitary bees	
Species	Plot	Av. VR	# Visits	Av. VR	# Visits	Av. VR	# Visits	Av. VR	# Visits	Av. VR	# Visits	Av. VR	# Visits	Av. VR	# Visits	Av. VR	# Visits	Av. VR	# Visits	Av. VR	# Visits
Centaurea jacea	C	**4.966**	**1784**	0	0	0	0	0	0	**3.986**	**1607**	**0.12**	**22**	0.317	47	0.018	4	**0.123**	**67**	0.221	37
	E	**3.055**	**953**	0	0	0	0	0	0	**2.142**	**811**	**0.005**	**3**	0.328	83	0.02	5	**0.034**	**12**	0.228	39
Fragaria vesca	C	0.492	18	0.229	8	0.036	1	0	0	0	0	0.048	2	0.052	3	0	0	0	0	0.119	4
	E	0.176	21	0.067	5	0	0	0.053	4	0	0	0	0	0	0	0	0	0.027	2	0.109	10
Galium mollugo	C	0.203	75	0.011	4	0	0	0.004	2	0	0	0	0	0.155	54	0	0	0.016	13	0.2	2
	E	0.173	81	0.001	1	0	0	0.006	2	0	0	0	0	0.108	53	0	0	0.031	23	0.005	2
Geranium sylvaticum	C	0.452	24	0	0	0.029	1	0	0	0.143	1	0	0	0.043	3	0.195	15	0	0	0.249	4
	E	0.133	2	0	0	0	0	0	0	0	0	0	0	0	0	0	0	0	0	0.222	2
Hieracium cymosum	C	**0.389**	**24**	0	0	0	0	0	0	0	0	0	0	**0.385**	**22**	0	0	0	0	0.011	2
	E	**0.071**	**6**	0	0	0	0	0	0	0	0	0	0	**0.082**	**5**	0	0	0	0	0.004	1
Hieracium lactucella	C	0.161	14	0.006	1	0	0	0	0	0	0	0	0	0.154	13	0	0	0	0	0	0
	E	0.078	15	0.007	1	0	0	0.008	1	0	0	0	0	0.072	12	0	0	0	0	0.003	1
Hieracium pilosella	C	0.253	9	0	0	0	0	0	0	0.007	1	0	0	0.219	7	0	0	0	0	0.048	1
	E	0.112	20	0.011	1	0	0	0.008	1	0	0	0	0	0.142	17	0	0	0	0	0.002	1
Knautia arvensis	C	**4.75**	**70**	0	0	0	0	0	0	**3.188**	**40**	**1.063**	**11**	**0.875**	**14**	0.125	1	0.125	1	0.188	3
	E	**0.738**	**19**	0	0	0	0	0	0	**0.703**	**17**	**0**	**0**	**0.061**	**2**	0	0	0	0	0	0
Lathyrus linifolius	C	0.1	20	0	0	0.02	4	0	0	0.08	16	0	0	0	0	0	0	0	0	0	0
	E	0	0	0	0	0	0	0	0	0	0	0	0	0	0	0	0	0	0	0	0
Leucanthemum vulgare	C	**0.784**	**88**	0	0	0	0	0.06	9	0	0	0	0	**0.439**	**54**	0	0	**0.198**	**22**	0.021	3
	E	**0.312**	**29**	0	0	0	0	0.042	3	0	0	0	0	**0.162**	**21**	0	0	**0.011**	**2**	0.016	3
Linum catharticum	C	1.575	4	0	0	0	0	0	0	0	0	0	0	0.04	1	0	0	0.007	1	0.001	2
	E	0.6	3	0	0	0	0	0	0	0	0	0	0	0.007	2	0	0	0.009	1	0	0
Polygala vulgaris	C	0.008	3	0	0	0	0	0	0	0.002	1	0	0	0.007	2	0	0	0	0	0	0
	E	0	0	0	0	0	0	0	0	0	0	0	0	0	0	0	0	0	0	0	0
Potentilla thuringiaca	C	0.608	52	0.141	27	0.023	1	0	0	0	0	0	0	0.106	7	0.047	6	0	0	0.135	11
	E	0.232	74	0.045	30	0	0	0	0	0	0	0	0	0	0	0.02	9	0.003	1	0.103	34
Primula veris	C	0.018	15	0	0	0.001	1	0.004	1	0	0	0	0	0.001	1	0	0	0.007	5	0.007	7
	E	0.006	3	0	0	0	0	0	0	0	0	0	0	0	0	0	0	0.003	1	0.005	2
Trifolium pratense	C	0.151	15	0	0	0.01	1	0	0	0.125	12	0	0	0.006	1	0	0	0.008	1	0	0
	E	0.133	6	0	0	0	0	0	0	0.123	6	0	0	0	0	0	0	0	0	0	0

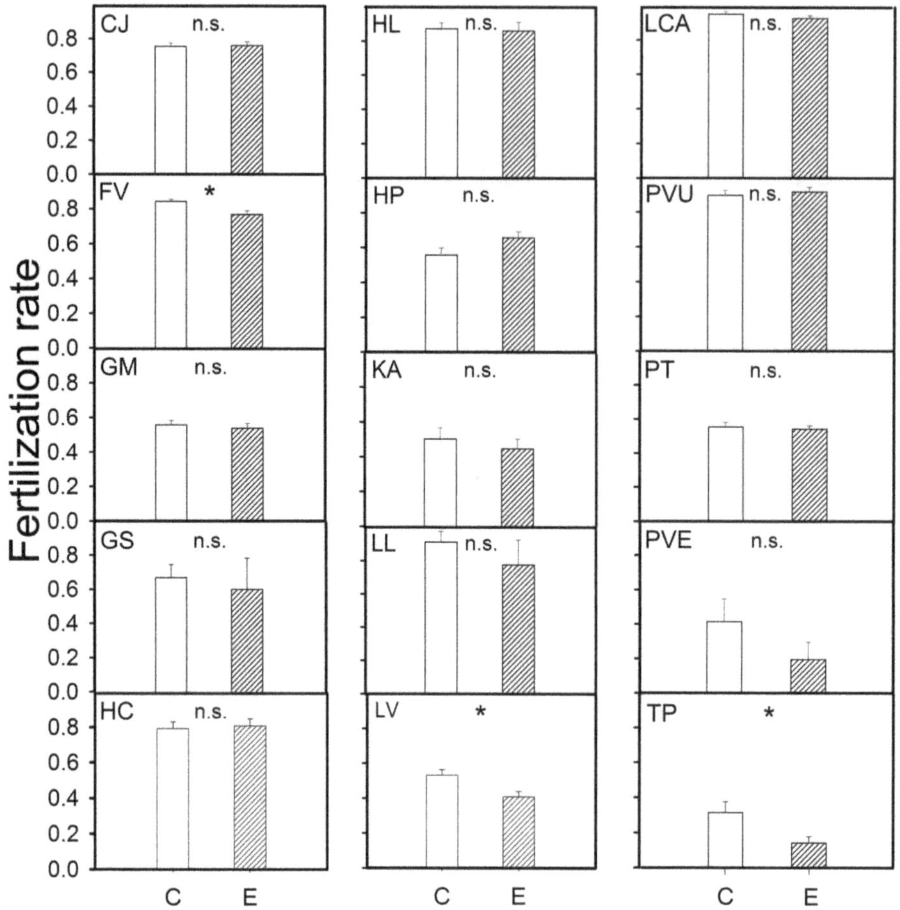

FIG. 1 Mean fertilization rate ± SE in experimental (E) and control (C) plots for all study species. *Centaurea jacea* (CJ), *Fragaria vesca* (FV), *Galium mollugo* (GM), *Geranium sylvaticum* (GS), *Hieracium cymosum* (HC), *Hieracium lactucella* (HL), *Hieracium pilosella* (HP), *Knautia arvensis* (KA), *Lathyrus linifolius* (LL), *Leucanthemum vulgare* (LV), *Linum catharticum* (LCA), *Polygala vulgaris* (PVU), *Potentilla thuringiaca* (PT), *Primula veris* (PVE) and *Trifolium pratense* (TP). 'n.s.' indicates non–significant differences between treatments ($P > 0.1$), whereas '*' indicates significant statistical differences between treatments ($P < 0.05$).

was significantly lower in the experimental plot for *Leucanthemum vulgare.*

Despite the relatively large reductions in visitation rates, the effects on plant fecundity were modest and most of the species showed no statistically significant differences in fecundity between experimental and control plots (Fig. 1). Nevertheless, the experiment significantly affected the fertilization rate of *Fragaria vesca* ($F_{1,26} = 7.46$, $P = 0.01$), *Leucanthemum vulgare* ($F_{1,53} = 6.49$, $P = 0.01$) and *Trifolium pratense* ($F_{1,18} = 4.91$, $P = 0.04$), which all had significantly lower fertilization rate in the experimental plots (Fig.1). Other species, also showed lower fertilization rate in the experimental plots than in control plots, but the differences between treatments were not statistically significant (Fig.1). For some of them, such as *Geranium sylvaticum, Knautia arvenis, Lathyrus linifolius* and *Primula veris* (from $F_{1,8} = 1.57$, $P > 0.25$ to $F_{1,36} = 0.07$, $P > 0.79$), the lack of significance is most likely due to low sample sizes since the differences between treatments seems large and the 95% confidence interval are wide (Fig. 1; Table 2; Figure S1 in Electronic Supplemental Material). The experimental reduction in visitation rate did not cause a significant reduction in the seed weight of any of the study species (from $F_{1,37} = 3.38$, $P = 0.07$ to $F_{1,39} = 0.01$, $P = 0.93$).

Relationships between reduction in plant fecundity, reduction in visitation rates, degree of pollinator dependence and blossom cover

The degree of pollinator dependence was significantly and positively related to the magnitude of reduction in fertilization rate ($F_{1,11} = 21.1$, $P = 0.001$; Fig 2A). The magnitude of reduction in fertilization rate was neither related to blossom cover ($F_{1,11} = 1.99$, $P = 0.186$; Fig 2B) nor to the reduction in visitation rate ($F_{1,11} = 2.52$, $P = 0.141$; Fig 2C). The reduction in seed weight was not significantly related to any of the predictor variables ($P > 0.61$ in all cases).

DISCUSSION

Pollinator decline and its potential effects on ecosystem services have recently received considerable attention (for review see Potts et al. 2010). Although several studies on single plant species have used correlative approaches to study how differences in pollinator visitation lead to variation in pollen limitation (Gomez et al. 2010; Moeller et al. 2012) and seed set (Jennersten 1988; Pauw 2007), very few studies have experimentally manipulated pollinator availability directly (Fontaine et al. 2008; Fontaine et al. 2006). We partially reduced pollinator availability to an entire plant community to more realistically simulate potential effects of pollinator decline on plant fecundity. Our results show that even large experimental reductions in pollinator visitation rates do not necessarily reduce plant fecundity. We found that the effects of a reduction in pollinator visitation on plant fecundity varied among species and these differences appear to be driven by differences in their degree of pollinator dependence.

As expected, the effects of reduced pollinator visitation on fecundity increased with the plants' dependence on pollinators. This result supports the hypothesis that the ability to use different pollination modes, such as the capability of both selfing (Kalisz & Vogler 2003) and/or wind pollination (Totland & Sottocornola 2001), can serve as reproductive assurance when pollinator availability is low (Culley et al. 2002). In our study species, selfing is the most likely alternative to insect pollination due to their showy flowers and morphological features (e.g. closed flowers) which makes wind pollination improbable. Previous studies have also found that obligate outcrossing animal-pollinated species are more prone to pollen limitation (Knight et al. 2005), more affected by disturbances (e.g. habitat fragmentation) which reduce pollinator availability (Aguilar et al. 2006) and have recently declined more than plant species capable of selfing (Biesmeijer et al. 2006). Nevertheless, the recent review by Potts et al. (2010) highlights the need for more mechanistic evidence for the consequences of pollinator loss. We believe that our experiment and results provide such mechanistic evidence and support the hypothesis that plant species with a high degree of pollinator dependence are more vulnerable to declines in pollination services than predominately selfing species (Potts et al. 2010). However, if a reduction in visitation rate leads to an increase of the self-fertilization rate, then plant species with low pollinator dependence may also suffer from reduced pollinator availability since inbreeding depression can occur in later stages of recruitment (Husband & Schemske 1996). Inbred seeds are often smaller than outcrossed seeds (Lienert & Fischer 2004; Olesen et al. 1998) and larger seeds often have better germination, greater survival, and give rise to larger seedlings than smaller seeds (Baker et al. 1994). We, therefore, used seed weight as an estimate of seed quality, but did not find any effect of the reduction in visitation rate on seed weight. However, an effect on germination and survival (Husband & Schemske 1996) or decreased genetic diversity over time (Rusterholz & Baur 2010) cannot be discarded for species with low dependence of pollinators. For instance, a reduction in visitation rate may not affect the number and weight of seeds in aposporous apomicts species, such as *Hieracium* spp.. However, an increased number of asexually produced seeds in facultative apomictic species may reduce the amount of genetic variation within populations (Houliston & Chapman 2004). Future studies on population dynamics, outcrossing rate and genetic variability will hopefully shed light on the effect of pollinator decline on self-compatible and facultative apomictic species.

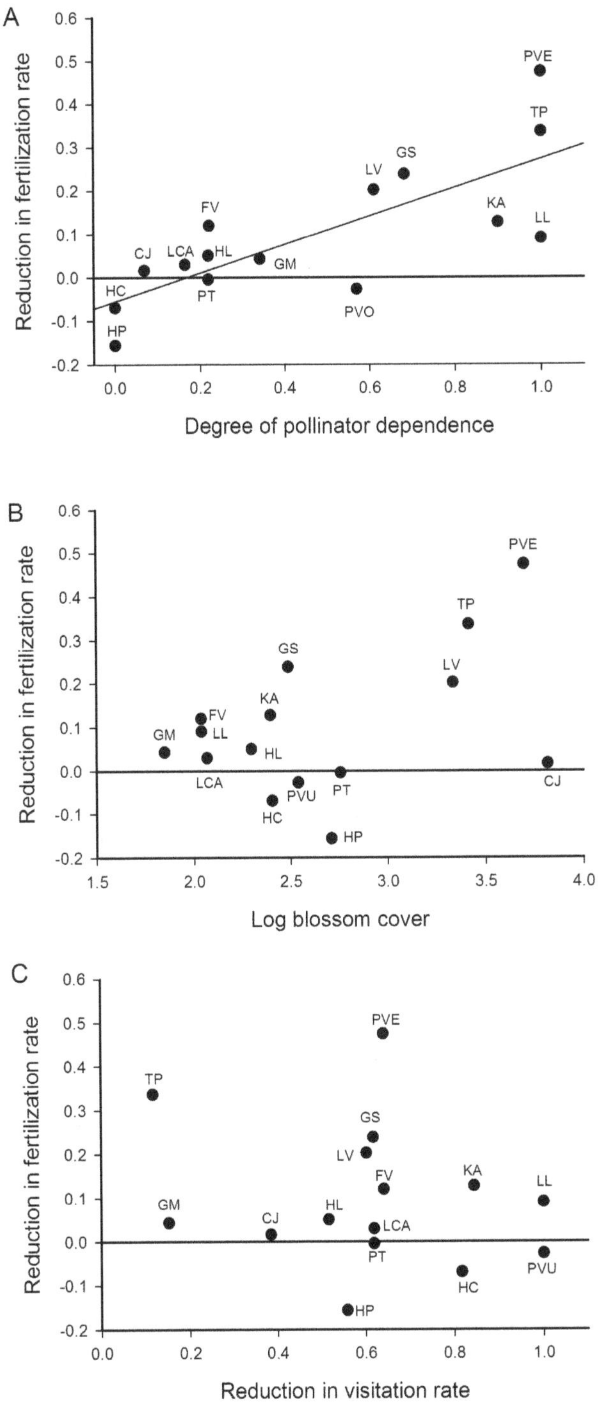

FIG. 2 Relationship between the reduction in fertilization rate and: A) the degree of pollinator dependence, B) blossom cover and C) reduction in visitation rate. *Centaurea jacea* (CJ), *Fragaria vesca* (FV), *Galium mollugo* (GM), *Geranium sylvaticum* (GS), *Hieracium cymosum* (HC), *Hieracium lactucella* (HL), *Hieracium pilosella* (HP), *Knautia arvensis* (KA), *Lathyrus linifolius* (LL), *Leucanthemum vulgare* (LV), *Linum catharticum* (LCA), *Polygala vulgaris* (PVU), *Potentilla thuringiaca* (PT), *Primula veris* (PVE) and *Trifolium pratense* (TP).

Plants in small populations and flower patches of low density often have lower reproductive success than plants in dense and large populations (Dauber et al. 2010; Fischer & Matthies 1998; Hensen & Wesche 2006; Jennersten 1988; Jennersten & Nilsson 1993; Kunin 1997; Oostermeijer et al. 1998). A recent inter-specific study has shown that seed set of species also increase with their conspecific flower density (Lázaro et al. 2013). This is because plants at low density may be less attractive to pollinators (Dauber et al. 2010; Hegland & Totland 2005; Kunin 1997) and/or because there is a decreased likelihood of receiving outcrossed and compatible pollen when the abundance of conspecifics is low (Jennersten & Nilsson 1993; Karron et al. 1995; Kunin 1997). Therefore, we expected species with lower blossom cover to be more affected by a reduction in pollinator availability, because when pollinator availability is low competition theory predicts low competition among pollinators for flowers (Alley 1982), and pollinators may choose the most abundant species in order to forage economically (Stephen & Krebs 1986). However, we did not find a relationship between the reduction in fertilization rate and blossom cover. Therefore it seems that attractiveness of plant species to pollinators does not change even if pollinator availability is reduced and thus pollinators compete less for food resources. We also expected the reduction in plant fecundity to increase with the reduction in flower visitation rate, but we found no such relationship between these two variables. At the within-species level, some studies have shown a strong positive relationship between seed production and visitation rate (Jennersten 1988; Pauw 2007), whereas others both at the within-species level (Dauber et al. 2010) and at the inter-specific level (Lázaro et al. 2013) have failed to find this intuitive relationship. However, the relationships present at the within species level might not occur at the among species level, because plant species may require a different visitation intensity to obtain a similar proportion of fertilized seeds due to, for example, large differences in ovule number among species.

The lack of an effect of reduced pollinator visitation on fertilization rate at the community level suggests that some species may be robust to a pollinator decline that could increase pollen limitation on plant reproduction. However, our approach has important limitations. First, the presence of resource limitation in this hay meadow could cancel out any difference in fertilization rate created by differences in visitation rate. However, although this possibility cannot be completely discarded, we used fertilization rate instead of seed set as a response variable in order to minimize an eventual effect of resource limitation. Experimental supplemental pollination, coupled with resource addition, would be required to reveal if pollen limitation occur in our study species. Second, sample size for some study plant species may be not high enough to detect significant effects in fertilization rate. At least four of the study species had very reduced fertilization rate in the experimental plots (Fig.1), but the 95% confidence intervals showed that the sample size was too small to detect a meaningful statistically significant difference in fecundity (Figure S1 in Electronic Supplemental Material). Sampling adequately in studies at the community level in which several variables have to be recorded at time is always challenging, because plant species are not distributed homogeneously and show different abundances. Increasing sample size of less abundant species would help to improve future studies in the line of ours. Third, the relative low number of plant species in the study prevented us from including interactions in the model. However, one could expect a significant interaction between the reduction in visitation rate and the degree of pollinator dependence on fertilization rates, since low visitation rates might reduce more fertilization rates in plant species that are pollinator dependent more than in autogamous species. Studies with a larger number of study species would be needed to confirm whether this interaction exists. Lastly, for four of the nineteen study species the experiment increased the visitation rate in the experimental plots compared to the control plot. The sample size of three of them (*Anthyllis vulneraria, Lotus corniculatus* and *Vicia cracca*) was rather low and this might explain the increase in visitation rate. However, *Potentilla erecta* was the study species which was observed for the longest time (16.7 hours in control and 17.3 hours in experimental plots) and showed a higher number of visits in the experimental plots compared to the control plots. It seems that the main reason for the increase in visitation rate in the experimental plots is a strong relationship between the number of visits and the number of *Potentilla erecta* flowers in the plots ($\chi^2 = 246.28$, $P < 0.0001$). Unfortunately, the number of *Potentilla erecta* flowers was significantly higher in the experimental plots compared to control plots ($\chi^2 = 30.42$, $P < 0.0001$) and this is a likely cause of the increased amount of visits in the experimental plots compared to the control plots.

Conclusion

A large decline in pollinator availability, such as predicted under the global pollination crisis (Allen-Wardell et al. 1998; Biesmeijer et al. 2006; Kearns et al. 1998), may not necessarily cause a uniform large reduction in plant fecundity. Our experimental reduction of pollinator visitation did not imply a significant reduction in fecundity which indicates that some plant species may be quite robust to a pollinator decline. Also, plant species varied in their response to a pollinator decline, where the species most dependent on pollinators for reproduction were most strongly affected. However, the accumulated effects of annually small reductions in fecundity on population dynamics and persistence, and whether such differences in response to pollinator decline may translate into community-wide changes in species composition, are largely unknown. Experimental studies like ours are crucial to understand how pollinator declines may influence plant fecundity, population dynamics and ultimately the structure of plant communities.

Acknowledgements

We would like to thank Manuel Hidalgo, Kristian Kyed, Kirsten Marthinsen, Mari Steinert, Siril Stenerud, Hartvig Velund, Silje Wang, and Magnus Øye for their enthusiastic help in the field. We are also very grateful to Paul Aakerøy, Kari Astrid Ehrlinger, Astrid Haarik, Kristian Kyed, Antón Pérez, Anna Karina Schmidt, Jorunn Vallestad, Silje Wang and Magnus Øye for their help in the lab. Jaime Otero provided statistical advice. We thank two anonymous referees for comments on the manuscript. This work was supported by the project 170532/V40, financed by the Norwegian Research Council.

References

Aguilar R, Ashworth L, Galetto L, Aizen MA (2006) Plant reproductive susceptibility to habitat fragmentation: review and synthesis through a meta-analysis. Ecology Letters 9:968-980

Aizen MA, Harder LD (2007) Expanding the limits of the pollen-limitation concept: effects of pollen quantity and quality. Ecology 88:271-281

Allen-Wardell G et al. (1998) The potential consequences of pollinator declines on the conservation of biodiversity and stability of food crop yields. Conservation Biology 12:8-17

Alley T (1982) Competition theory, evolution, and the concept of an ecological niche. Acta Biotheoretica 31:165-179

Andersson S (2008) Pollinator and nonpollinator selection on ray morphology in *Leucanthemum vulgare* (oxeye daisy, Asteraceae). American Journal of Botany 95:1072-1078

Ashman T-L et al. (2004) Pollen limitation of plant reproduction: ecological and evolutionary causes and consequences. Ecology 85:2408-2421

Baker K, Richards AJ, Tremayne M (1994) Fitness constraints on flower number, seed number and seed size in the dimorphic species *Primula farinosa* L. and *Armeria maritima* (Miller) Willd. New Phytologist 128:563-570

Bell G (1985) On the function of flowers. Proceedings of the Royal Society of London. Series B, Biological Sciences 224:223-266

Biesmeijer JC et al. (2006) Parallel declines in pollinators and insect-pollinated plants in Britain and the Netherlands. Science 313:351-354

Burd M (1994) Bateman's principle and plant reproduction: the role of pollen limitation in fruit and seed set. Botanical Review 60:83-139

Culley TM, Weller SG, Sakai AK (2002) The evolution of wind pollination in angiosperms. Trends in Ecology & Evolution 17:361-369

Dauber J et al. (2010) Effects of patch size and density on flower visitation and seed set of wild plants: a pan-European approach. Journal of Ecology 98:188-196

Dobeš C, Lückl A, Hülber K, Paule J (2013) Prospects and limits of the flow cytometric seed screen – insights from Potentilla sensu lato (Potentilleae, Rosaceae). New Phytologist 198:605-616

Dreisig H (1995) Ideal free distributions of nectar foraging bumblebees. Oikos 72:161-172

Dudash MR, Fenster CB (2000) Inbreeding and outbreeding depression in fragmented populations. In: A. Y, G. C (eds) Genetics, demography and viability of fragmented populations. Cambridge University Press, Cambridge, United Kingdom, pp 35-53

Dudash MR, Hassler C, Stevens PM, Fenster CB (2011) Experimental floral and inflorescence trait manipulations affect pollinator preferences and function in a hummingbird-pollinated plant. American Journal of Botany 98:275-282

Dukas R, Real LA (1993) Effects of recent experience on foraging decisions by bumble bees. Oecologia 94:244-246

Ebeling A, Klein AM, Schumacher J, Weisser WW, Tscharntke T (2008) How does plant richness affect pollinator richness and temporal stability of flower visits? Oikos 117:1808-1815

Eckert CG et al. (2009) Plant mating systems in a changing world. Trends in Ecology & Evolution 25:35-43

Eckhart V (1991) The effects of floral display on pollinator visitation vary among populations of *Phacelia linearis* (Hydrophyllaceae). Evolutionary Ecology 5:370-384

Fischer M, Matthies D (1998) Effects of population size on performance in the rare plant *Gentianella germanica*. Journal of Ecology 86:195-204

Fontaine C, Collin CL, Dajoz I (2008) Generalist foraging of pollinators: diet expansion at high density. Journal of Ecology 96:1002-1010

Fontaine C, Dajoz I, Meriguet J, Loreau M (2006) Functional diversity of plant–pollinator interaction webs enhances the persistence of plant communities PLoS Biol, vol. 4, *P* e1

Franzen M, Larsson M (2009) Seed set differs in relation to pollen and nectar foraging flower visitors in an insect-pollinated herb. Nordic Journal of Botany 27:274-283

Galen C, Newport MEA (1987) Bumble bee behavior and selection on flower size in the sky pilot, *Polemonium viscosum*. Oecologia 74:20-23

Gomez JM, Abdelaziz M, Lorite J, Munoz-Pajares AJ, Perfectti F (2010) Changes in pollinator fauna cause spatial variation in pollen limitation. Journal of Ecology 98:1243-1252

Goulson D (2003) Bumblebees - their behaviour and ecology. Oxford University Press

Goulson D, Lye GC, Darvill B (2008) Decline and conservation of bumble bees. Annual Review of Entomology 53:191-208

Grindeland JM (2008) Inbreeding depression and outbreeding depression in *Digitalis purpurea*: optimal outcrossing distance in a tetraploid. Journal of Evolutionary Biology 21:716-726

Grindeland JM, Sletvold N, Ims RA (2005) Effects of floral display size and plant density on pollinator visitation rate in a natural population of *Digitalis purpurea*. Functional Ecology 19:383-390

Hardy OJ, De Loose M, Vekemans X, Meerts *P* (2001) Allozyme segregation and inter-cytotype reproductive barriers in the polyploid complex *Centaurea jacea*. Heredity 87:136-145

Hegland SJ, Boeke L (2006) Relationships between the density and diversity of floral resources and flower visitor activity in a temperate grassland community. Ecological Entomology 31:532-538

Hegland SJ, Nielsen A, Lázaro A, Bjerknes AL, Totland O (2009) How does climate warming affect plant-pollinator interactions? Ecology Letters 12:184-195

Hegland SJ, Totland O (2005) Relationships between species' floral traits and pollinator visitation in a temperate grassland. Oecologia 145:586-594

Hensen I, Wesche K (2006) Relationships between population size, genetic diversity and fitness components in the rare plant *Dictamnus albus* in Central Germany. Biodiversity and Conservation 15:2249-2261

Houliston GJ, Chapman HM (2004) Reproductive strategy and population variability in the facultative apomict *Hieracium pilosella* (Asteraceae). American Journal of Botany 91:37-44

Husband BC, Schemske DW (1996) Evolution of the magnitude and timing of inbreeding depression in plants. Evolution 50:54-70

Inouye DW (1978) Resource partitioning in bumblebees: experimental studies of foraging behavior. Ecology 59:672-678

Jennersten O (1988) Pollination in *Dianthus deltoides* (Caryophyllaceae): effects of habitat fragmentation on visitation and seed set. Conservation Biology 2:359-366

Jennersten O, Nilsson SG (1993) Insect flower visitation frequency and seed production in relation to patch size of *Viscaria vulgaris* (Caryophyllaceae). Oikos 68:283-292

Kalisz S, Vogler DW (2003) Benefits of autonomous selfing under unpredictable pollinator environments. Ecology 84:2928-2942

Karron JD, Thumser NN, Tucker R, Hessenauer AJ (1995) The influence of population density on outcrossing rates in *Mimulus ringens*. Heredity 75:175-180

Kearns CA, Inouye DW, Waser NM (1998) Endangered mutualisms: the conservation of plant-pollinator interactions. Annual Review of Ecology and Systematics 29:83-112

Kennedy BF, Elle E (2008) The reproductive assurance benefit of selfing: importance of flower size and population size. Oecologia 155:469-477

Knight TM et al. (2005) Pollen limitation of plant reproduction: pattern and process. Annual Review of Ecology, Evolution, and Systematics 36:467-497

Koltunow AM, Johnson SD, Bicknell RA (1998) Sexual and apomictic development in Hieracium. Sexual Plant Reproduction 11:213-230

Krahulec F, Krahulcová A, Fehrer J, Brautigam S, Schuhwerk F (2008) The structure of the agamic complex of Hieracium subgen. Pilosella in the Sumava Mts and its comparison with other regions in Central Europe. PRESILA-PRAHA- 80:1

Kunin WE (1997) Population size and density effects in pollination: pollinator foraging and plant reproductive success in experimental arrays of *Brassica kaber*. Journal of Ecology 85:225-234

Lázaro A, Aase ALTO, Totland O (2011) Relationships between densities of previous and simultaneous foragers and the foraging behaviour of three bumblebee species. Ecological Entomology 36:221-230

Lázaro A, Jakobsson A, Totland O (2013) How do pollinator visitation rate and seed set relate to species' floral traits and community context? Oecologia DOI 10.1007/s00442-013-2652-5

Lázaro A, Lundgren R, Totland O (*Submitted*) Experimental reduction of pollinator visitation modifies plant-plant interactions for pollination.

Leduc N, Douglas GC, Monnier M, Connolly V (1990) Pollination in vitro: effects on the growth of pollen tubes, seed set and gametophytic self-incompatibility in *Trifolium pratense* L. and *T. repens* L. Theoretical and Applied Genetics 80:657-664

Lienert J, Fischer M (2004) Experimental inbreeding reduces seed production and germination independent of fragmentation of populations of *Swertia perennis*. Basic and Applied Ecology 5:43-52

Moeller DA, Geber MA, Eckhart VM, Tiffin P (2012) Reduced pollinator service and elevated pollen limitation at the geographic range limit of an annual plant. Ecology 93:1036-1048

Norderhaug A (1995) Mating systems of three meadow plant species. Nordic Journal of Botany 15:243-250

Nosrati H, Price AH, Wilcock CC (2010) No evidence of apomixis in matroclinal progeny from experimental crosses in the genus Fragaria (strawberry) based on RAPDs. Euphytica 171:193-202

Noyes RD (2007) Apomixis in the Asteraceae: diamonds in the rough. Functional Plant Science and Biotechnology 1:207-222

Olesen JM, Forfang AS, Baez M (1998) Stress-induced male sterility and mixed mating in the island plant *Cedronella canariensis* (Lamiaceae). Plant Systematics and Evolution 212:159-176

Oostermeijer JGB, Luijten SH, Krenova ZV, Den Nijs HCM (1998) Relationships between population and habitat characteristics and reproduction of the rare *Gentiana pneumonanthe* L. Conservation Biology 12:1042-1053

Pauw A (2007) Collapse of a pollination web in small conservation areas. Ecology 88:1759-1769

Potts SG, Biesmeijer JC, Kremen C, Neumann P, Schweiger O, Kunin WE (2010) Global pollinator declines: trends, impacts and drivers. Trends in Ecology & Evolution 25:345-353

Ramula S, Toivonen E, Mutikainen P (2007) Demographic Consequences of Pollen Limitation and Inbreeding Depression in a Gynodioecious Herb. International Journal of Plant Sciences 168:443-453

Rusterholz HP, Baur B (2010) Delayed response in a plant-pollinator system to experimental grassland fragmentation. Oecologia 163:141-152

Schulze J, Rufener R, Erhardt A, Stoll P (2012) The relative importance of sexual and clonal reproduction for population growth in the perennial herb *Fragaria vesca*. Population Ecology 54:369-380

Steffan-Dewenter I, Tscharntke T (1999) Effects of habitat isolation on pollinator communities and seed set. Oecologia 121:432-440

Stephen D, Krebs J (1986) Foraging theory. Princeton University Press

Thomson JD, Maddison WP, Plowright RC (1982) Behavior of bumble bee pollinators of *Aralia hispida* Vent. (Araliaceae). Oecologia 54:326-336

Totland O, Sottocornola M (2001) Pollen limitation of reproductive success in two sympatric alpine willows (Salicaceae) with contrasting pollination strategies. American Journal of Botany 88:1011-1015

Valido A, Dupont YL, Hansen DM (2002) Native birds and insects, and introduced honey bees visiting *Echium wildpretii* (Boraginaceae) in the Canary Islands. Acta Oecologica 23:413-419

Vange V (2002) Breeding system and inbreeding depression in the clonal plant species *Knautia arvensis* (Dipsacaceae): implications for survival in abandoned grassland. Biological Conservation 108:59-67

Waser NM, Price MV (1994) Crossing-distance effects in *Delphinium nelsonii*: outbreeding and inbreeding depression in progeny fitness. Evolution 48:842-852

Wedderburn F, Richards AJ (1990) Variation in within-morph incompatibility inhibition sites in heteromorphic *Primula* L. New Phytologist 116:149-162

Wilcock C, Neiland R (2002) Pollination failure in plants: why it happens and when it matters. Trends in Plant Science 7:270-277

Wissman J (2006) Reproduction and population structure of *Primula veris* L. Primulaceae) in semi-natural grasslands – a landscape and management perspective. Dept. of Conservation Biology. SLU

9

RELATIVE FLORAL DENSITY OF AN INVASIVE PLANT AFFECTS POLLINATOR FORAGING BEHAVIOUR ON A NATIVE PLANT

Amy M. Iler[1,3]* & Karen Goodell[2]

[1]*Department of Evolution, Ecology, and Organismal Biology, The Ohio State University, 318 W 12th Ave, Columbus, OH 43210 USA*
[2]*Department of Evolution, Ecology, and Organismal Biology, The Ohio State University, 1179 University Dr, Newark, OH 43055 USA*
[3]*Current address: Department of Biology, University of Maryland, College Park, Maryland 20742 USA*

Abstract—Interactions between invasive and native plants for pollinators vary from competition to facilitation of pollination of native plants. Theory predicts that relative floral densities should account for some of this variation in outcomes, with facilitation at low floral densities and competition at high floral densities of the invader. We tested this prediction by quantifying pollination and female reproductive success of a native herb, *Geranium maculatum*, in three experimental arrays that varied in floral density of the invasive shrub *Lonicera maackii*: control (no *L. maackii*), low floral density of *L. maackii*, and high floral density of *L. maackii*. A low density of *L. maackii* flowers was associated with an increase in pollinator visitation rate to *G. maculatum* flowers and an increase in conspecific pollen deposition compared to controls and high density arrays. Increased visitation rates were not associated with an increase in the number of visitors to low density arrays, suggesting instead that a behavioural switch in visitation within the array accounted for increased pollen deposition. In contrast, the only evidence of competition in high density arrays was a shorter duration of visits to *G. maculatum* flowers relative to the other treatments. The number of seeds per flower did not vary among treatments, although trends in seeds per flower were consistent with patterns of pollinator foraging behaviour. Given increased pollinator visits and pollen deposition at a low density of the invader, our study indicates that complete eradication of invasives as a management or restoration technique may have unintended negative consequences for pollination of native plants.

Keywords: biological invasion; density-dependence; experimental array; indirect interaction; Lonicera; *pollination*

INTRODUCTION

Invasive plant species can directly out-compete native plant species for space and other abiotic resources, depending on resource availability and past disturbance regimes (Daehler 2003; Levine et al. 2003). Despite increasing interest in indirect effects of invasive plants on the biotic interactions of native plants, such as those mediated through pollinators, indirect effects are still not as well understood as direct effects (Strauss 1991; White et al. 2006). Establishment of mutualisms in the introduced range can be a crucial component of successful plant invasions (Mitchell et al. 2006). Mutualists are resources for which plants can compete (Waser 1983), and invasive plants are likely to be particularly strong competitors for mutualists because they occur at high densities (Traveset & Richardson 2006). Indeed, invasive plants compete for pollinators with native plants and reduce native plant reproductive success, an effect that is more pronounced as invasive plant density increases (Chittka & Schürkens 2001; Morales & Travesest 2009; Kandori et al. 2009; Takakura et al. 2009; Matsumoto et al. 2010; Dietzsch et al. 2011). On the other hand, facilitation of pollination and native plant seed set by invasive plants has also been observed (Muñoz & Cavieres 2008; McKinney & Goodell 2011). Whether native plants experience facilitation, no effect, or competition for pollination with an invasive plant species seems to depend on relative floral densities and the spatial scale over which interactions are studied (Muñoz & Cavieres 2008; Cariveau & Norton 2009; McKinney 2010). Because invasive plant densities increase as plant invasions progress, studying the role of invasive-native relative floral density in mediating pollination services can serve as a model for understanding ecological processes at different stages of invasion.

Pollinator foraging behaviour depends on processes operating at the level of patches of floral resources within a broader floral landscape and processes within the patch (Fig. 1). Total floral density and floral diversity affect pollinator choice among patches, whereas relative floral density should affect pollinator foraging decisions among floral resources within a patch (Ghazoul 2006; Yang et al. 2011). If attraction to larger patches of floral resources is an accelerating function of floral density, facilitation of pollinator visits to plant species within the patch is possible (sensu Feinsinger et al. 1991; Rathcke 1983; Feldman 2004; Fig. 1A & B). However, facilitation is not expected at any floral density if pollinator attraction increases linearly with patch-level floral density (Feldman 2004). After initial pollinator attraction to an invaded patch, facilitation is more likely at low relative floral density of the invader (Fig. 1B vs. C). As a plant invasion progresses, floral availability is likely

*Corresponding author; email: amy.marie.iler@gmail.com

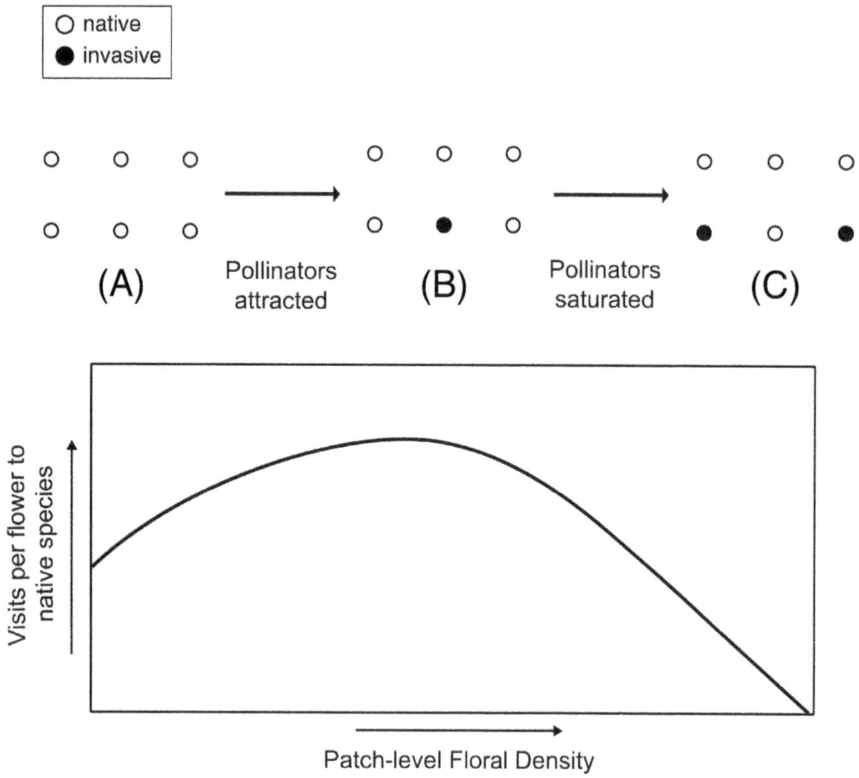

FIGURE 1. Conceptual diagram of pollinator foraging responses to increasing floral density in patches of flowers. Total patch-level floral density increases from (A) to (C), as patches become increasingly invaded by an invasive plant species. Circles represent individual plants. As pollinators are attracted to patches with a higher density of floral resources, facilitation of visits to the native species is possible (B), until pollinators become saturated and competition for visits is likely (C). In this study, we use plant arrays in which total floral density increases in invaded arrays, similar to those portrayed here (see Table 1 for experimental design).

to saturate the resident pollinator community, and competition for pollinators within the patch is more likely (Feinsinger 1987; Rathcke 1983; Fig. 1C).

The reproductive consequences of altered pollinator foraging behaviour to native plants depend on subsequent patterns of pollen removal and deposition, in addition to whether conspecific pollen receipt limits native plant reproduction (McKinney & Goodell 2010; Tscheulin & Petanidou 2013). For example, heterospecific pollen deposition can offset increased visitation rates to native plants and reduce native plant reproductive success (Grabas & Laverty 1999; Brown & Mitchell 2001; Lopezaraiza-Mikel et al. 2007; Jakobsson et al. 2008; but see Bartomeus et al. 2008). Transfer of heterospecific pollen can counteract any positive effects of increased pollinator visitation if pollen clogs plant stigmas or styles or if pollen allelopathy interferes in plant reproduction (among native plants: Waser 1978; Galen & Gregory 1989; Murphy & Aarssen 1995). Sequential visits to flowers of the invader may also result in loss of native plant pollen and lower seed production for the native plant, even when more pollinators are attracted to invaded patches (Flanagan et al. 2010). Finally, changes in pollinator foraging behaviour may have no effect on native plant reproduction if seed production is not pollen-limited (Ghazoul 2004; Totland et al. 2006; Kaiser-Bunbury & Muller 2009; Bartomeus et al. 2010).

Here we focus on pollination services to a native plant species within invaded and uninvaded artificial patches of floral resources. We use experimental arrays that vary in relative floral densities of the invasive shrub *Lonicera maackii* and the native herb *Geranium maculatum*, two species that have been shown to interact for pollinators in our previous work (McKinney & Goodell 2010). The goal of the treatments is to emulate different stages of invasion. In a previous experiment at this site, we detected evidence of facilitation of *G. maculatum* pollination at an approximate 20:1 floral ratio (and at a distance up to 20 m away from *L. maackii*; McKinney 2010). Here we investigate whether increasing the floral density of the invader will reduce the facilitative effect or even result in competition with the native plant (McKinney 2010). We specifically ask how a low (20:1) and high (60:1) relative floral density of the invader affects pollinator foraging (number of visitors, visitation rate, and duration of visits), heterospecific and conspecific pollen deposition, and seed set of *G. maculatum* relative to control plots containing only *G. maculatum* flowers.

MATERIALS AND METHODS

Study species

Lonicera maackii (Rupr.) Herder was introduced to North America from China in the late 19th century as an ornamental shrub and is now invasive in forests and fields of the eastern USA (Luken & Thieret 1996). Mature plants, such as those in our study site, produce thousands of medium-sized (2-2.5 cm wide), zygomorphic, white to light pink flowers in early May. Non-native honeybees (*Apis mellifera*) and native solitary bees forage on *L. maackii* flowers in central Ohio (Goodell & Iler 2007). *Lonicera maackii* invasions seem to negatively affect native flora and fauna through direct pathways, such as competition for abiotic resources and by increasing nest predation of native birds (Hutchinson & Vankat 1997; Gould & Gorchov 2000; Collier et al. 2002; Borgmann & Rodewald 2004; Miller & Gorchov 2004) and through indirect pathways, such as increasing human disease risk via altering white-tailed deer host dynamics (Allan et al. 2010; Mattos & Orrock 2010).

Geranium maculatum L. is a common, widespread, perennial herb of eastern North American deciduous forests and open fields (Martin 1965). It was selected for this study because it shares pollinators and co-occurs with *L. maackii*, and its geographic range, habitat, and flowering phenology overlap with *L. maackii*. *Geranium maculatum* relies on sexual reproduction as well as vegetative reproduction, but seeds are required for dispersal and population expansion (Martin 1965). Flowering plants bear 2-30 large (2.5-4 cm wide), actinomorphic, purple-pink flowers and produce more flowers in full sun than plants in shaded habitat (Martin 1965). Some populations are gynodioecious, with a small proportion of female plants relative to hermaphrodites (Agren & Willson 1991). The hermaphroditic flowers of *G. maculatum* are protandrous (Bertin & Sholes 1993). Plants are self-compatible, but pollinators are required for movement of pollen between flowers (Martin 1965; Willson et al. 1979). At our study site, the most common visitors to *G. maculatum* are native solitary bees that also visit *L. maackii*; Lepidopterans and syrphid flies also occasionally visit *G. maculatum* (McKinney 2010). Reproduction of potted hermaphroditic *G. maculatum* plants at our study site was pollen limited in spring 2008, suggesting that pollinators generally play an important role in seed production of potted *G. maculatum* plants at this location (McKinney 2010).

Study site

We conducted this study during May 2009 in an old field habitat in Three Creeks Metro Park, Groveport, Ohio, USA. The area of the park in which our experiment was conducted was used for agriculture until the late 1960's, when it was permitted to re-grow naturally and was invaded by *Lonicera maackii* (J. Snyder, park naturalist, personal communication). The old field was approximately 160 m (E to W) × 350 m (N to S) and was bordered to the west and north by a mixed white pine and deciduous forest that was invaded by *L. maackii* and to the east by a restored prairie that was separated from the old field site by a paved bicycle path. A park entrance road bordered the site to the south, approximately 200 m from the experiment. *Lonicera maackii* density in the forest adjacent to the old field was 0.14 ± 0.0086 shrubs per m^2 (mean ± 1 SE here and throughout). The prairie was 40 m away from the nearest experimental plants and was mainly composed of *Andropogon gerardii* and *Sorghastrum nutans*; *Chrysanthemum leucanthemum* was also present and co-flowered with *L. maackii* and *G. maculatum* from 15-25 May (the latter 2/3 of the experiment). To deter the spread of weedy species, park management mowed the old field in the spring. It was not mowed during spring 2008 or 2009 because of other experiments, and small (mostly non-flowering) *L. maackii* and *Eleagnus umbellata* shrubs started to colonize the field. There were no naturally occurring *G. maculatum* plants at our study site, probably because of clearing of natural vegetation for agriculture, so that natural variation in *G. maculatum* plant density was not a confounding factor in this study. There were no other flowering plants in the old field during the study.

Experimental design

We established 24 1 m × 0.5 m plots within the old field that were at least 40 m away from invaded forest edge habitat and ca. 20 m apart from one another. Plots were randomly assigned to one of three array treatments: control, low floral density invasion, or high floral density invasion (N = 8 plots per treatment; Table 1). We arranged potted *G. maculatum* plants in rectangular arrays in which each plant was separated by 0.5 m (3 × 2 plants in controls; Table 1). In low and high density invasion arrays, *L. maackii* branches replaced one and two *G. maculatum* plants, respectively (Table 1). *Lonicera* branches were cut each morning and placed in brown 1L pitchers containing water, which were attached to a fence post at ca. 1.25 m from the ground (vertical separation of ca. 0.75 m between flowers of each species). *Geranium maculatum* rhizomes were purchased from Prairie Moon Nursery (Minnesota, USA), planted in 2.8 L plastic pots containing ProMix potting soil, fertilized with Osmocote smart-release ® plant food to encourage flowering, and watered as necessary. Plants were maintained in a screen house at The Ohio State University Newark until transportation to the field. All plants used in the study were hermaphrodites. The low density ratio of 20:1 invasive:native flowers was based on a previous experiment at this site in which we detected evidence of facilitation (McKinney 2010). We aimed to keep relative floral densities relevant to densities observed in the field by setting the high floral density treatment at 60:1, based on floral densities of *L. maackii* and other native herbs within the forest at this site (A. Iler, personal observation). While the total number of *G. maculatum* plants varied across treatments, the average number of *G. maculatum* flowers open per day per plot did not significantly differ across treatments (ANOVA, $F_{2,21}$ = 0.86, P = 0.44; control: 11.5 ± 2.3 flowers, low: 8.7 ± 1.3 flowers, high: 7.4 ± 0.96 flowers).

TABLE 1. Experimental arrays of native *Geranium maculatum* and invasive *Lonicera maackii* used to simulate effects of plant invasion on pollination services to native plants. *Lonicera maackii* branches were placed in containers of water daily to create target floral ratios, and *G. maculatum* plants were permanently contained in pots. Plants were arranged in two rows of three plants in each treatment. N = 8 arrays per treatment.

	Control	Low density invasion	High density invasion
# Native plants	6	5	4
# Invasive branches	0	1	2
Invasive:native floral ratio	–	20:1	60:1

The spatial scale of this experiment is within the foraging ranges of most native bee species (Gathmann & Tscharntke 2002; Greenleaf et al. 2007), yet relevant to foraging decisions of bees within and among patches of floral resources (Ghazoul, 2006). It is a reasonable representation of plant densities in natural systems because flowers are often patchily distributed (e.g. Sih & Baltus, 1987). Our experimental design intentionally confounds pollinator responses to presence of *L. maackii* with higher overall flower density, because flowering *L. maackii* shrubs increase community-level floral density relative to uninvaded areas (A. Iler, personal observation). Furthermore, from an invasion perspective, *L. maackii* is likely to replace herbs as the invasion progresses (Hutchinson & Vankat 1997), a pattern with copious support from studies of other invasive species (e.g., Woods 1993; Tilman 1997; D'Antonio et al. 1998; Christian & Wilson 1999).

Arrays were set up on all sunny or partly cloudy days (days when pollinators were active), commencing when *G. maculatum* began to flower on 10 May and ending when *L. maackii* stopped flowering on 25 May. We removed the few remaining *G. maculatum* flower buds after this date; the few plants with remaining flowers were equally distributed across treatments and should therefore not affect our ability to detect differences in seed production across treatments. Out of a total of 16 days of co-flowering between our study species, experimental arrays were set up on 13 days, 10 of which included pollinator observations; the remaining three days were rainy and overcast. New *L. maackii* branches were cut each morning before pollinators became active (prior to 10:00am) and flowers removed as needed to create target floral ratios of 20:1 and 60:1. The average floral ratios were 20.4 ± 0.7:1 and 57.6 ± 2.3:1, respectively.

Data collection & analysis

To measure pollinator services to *G. maculatum*, we quantified pollinator visitation and pollen deposition. We conducted multiple 10 min pollinator observation sessions per plot, on all open *G. maculatum* flowers within arrays on a total of ten sunny to partly cloudy days when air temperature was above 15°C. Conditions of maximum pollinator activity are based on previous *L. maackii* observations in central Ohio (Goodell et al. 2010). Observations were conducted within the hours of 10:30am to 4:00pm, and all plots were observed in a random order on each observation date. Although we were unable to observe all 24 plots each time pollinator observations were conducted, we observed an equal number of randomly selected arrays from each treatment on each day. The number of open *G. maculatum* flowers, number of visitors and visits to all open *G. maculatum* flowers in the array (measures of visitation quantity), visitor identification, and visit duration per flower (a measure of visit quality, following Muñoz & Cavieres 2008) were recorded during each observation session. A visit was defined as contact with the stigma or anthers. We did not track visits to individual plants during observations. We followed individual bees foraging on *G. maculatum* until they left the plot; bees were not collected and therefore could have returned to the same plot in a later foraging bout. Bees were identified on the wing to genus with the use of a reference collection, and all other visitors were identified to family. If too many visitors were present during a 10-min session to track both the number and duration of visits, we conducted a separate 10-min observation to record duration of visits. Because of time constraints, we have fewer observations of the duration of individual visits (duration data from a total of 106 10-min observation sessions vs. a total of 161 observation sessions for quantity of visits). Pollinator observations were not conducted on *L. maackii* branches because of time constraints, but pollinators were observed foraging on these flowers.

Ten post-receptive *G. maculatum* stigmas were collected haphazardly from each plot on six separate days throughout the flowering period and placed in microcentrifuge tubes containing 75% ethanol. Post-receptive stigmas are easily identifiable because they begin to close and dry out. Stigmas are receptive only during the last day of the lifetime of a flower (4-6 days), and pollen tubes reach the ovary in less than 2 h 30 min (Mulcahy et al. 1983). Therefore, pollen tubes should have reached the ovary when stigmas were collected. Stigma removal did not appear to affect fruit set because several mature fruits were missing stigmas (A. Iler, personal observation). Pollen grains were stained with fuchsin dye and mounted on slides with glycerin gel (Kearns & Inouye 1993). We counted pollen grains adhered to stigmas and classified them as *G. maculatum*, *L. maackii*, or unknown pollen grains using a compound microscope. Our protocol may underestimate the amount of *L. maackii* pollen that arrives on *G. maculatum* stigmas because we placed stigmas in ethanol after collection, and any pollen that had not adhered to the sigma may have washed away. Nevertheless, we observed similar numbers of *L. maackii* pollen grains on *G. maculatum* stigmas in another study within the old field in which we mounted stigmas immediately after collection (McKinney 2010). We collected *G. maculatum* fruits from experimental arrays as they matured (1-3 weeks after all flowering ended) and counted the number of seeds.

We used one-way ANOVAs to compare mean responses across treatments (except for the number of unknown pollen grains, which was analyzed with a nonparametric ANOVA). When ANOVAs revealed significant differences among treatments, we used Tukey-Kramer HSD (honest significant difference) tests to compare means between each pair of treatments. The number of visitors to *G. maculatum* per plot, visitation rate (number of visits per *G. maculatum* flower), and duration of visits were each averaged across all 10-min observation sessions for each plot, the unit of replication ($N = 8$ plots). We obtained visit durations in all but one plot in the high-density treatment, so $N = 7$ for this treatment. The number of conspecific, *L. maackii*, and unknown pollen grains per flower were first averaged by plant then by plot. We calculated reproductive output as the number of seeds per flower instead of seeds per plant because the number of flowers per plant differed across treatments ($F_{2,21} = 6.13$, $P = 0.008$). Plants in the low density treatment had significantly fewer total flowers per plant than plants in the control ($P = 0.0058$), and neither the control nor the low density treatment differed from the high density treatment ($P = 0.20$, $P = 0.22$, respectively; all *P*-values for comparisons between treatments are from Tukey's HSD tests). Therefore, seed production per plant could reflect this pattern of flower production rather than treatment effects. The mean number of seeds per flower for each plant was averaged by plot and was log transformed to meet normality assumptions. All analyses were performed in R v 2.15.3 (R Development Core Team 2013, http://www.r-project.org/).

Pollinator foraging choices across plots could reflect a response to relative density of flowers or absolute abundance of flowers, both of which varied across plots. We conducted a regression analysis to investigate evidence for effects of total plot-level *G. maculatum* floral abundance on pollinator foraging responses to *G. maculatum*, focusing on variation in total flower number across control arrays. We used simple linear regression to analyze the relationship between the number of *G. maculatum* flowers open in control arrays and pollinator responses: number of visitors (patch-level attraction), number of visits, and duration of visits. Daily means were calculated for each pollinator response from 10-min observation sessions ($N = 10$ days). Response variables were log transformed as necessary to account for unequal variances. One extreme outlier was removed from the analysis of the number of visits (Bonferonni $P = 0.0002$); removing the outlier made the trend weaker but does not affect any of our conclusions. If *G. maculatum* floral abundance determines pollinator choice among arrays, as opposed to relative floral density, we expect the number of visitors per plot to increase with increasing *G. maculatum* floral abundance.

RESULTS

The ratio of *G. maculatum* to *L. maackii* flowers remained constant within treatments throughout the experiment, but the absolute number of flowers changed through time, depending on how many *G. maculatum* flowers were open each day. On average, the low density arrays contained 163.8 ± 19.0 *L. maackii* flowers, and the high density arrays contained 411.8 ± 48.1 *L. maackii* flowers.

Eight bee genera were recorded visiting *G. maculatum* flowers in over 28 hours of observation (approximately 9 hours per treatment). The most common visitors were native bees in the genera *Ceratina*, *Halictus*, and *Osmia*, and we observed only one honey bee visit (*Apis melifera*) and no bumble bees (*Bombus* spp.) during our observations (see McKinney 2010 for a list of visitors). This result is consistent with previously published records of bee communities in old field habitat (Ginsberg 1983). The average number of individual insect visitors per 10 min observation session observed on *G. maculatum* flowers differed among treatments ($F_{2,21} = 4.48$, $P = 0.024$, Fig. 2a). The number of visitors to *G. maculatum* did not differ between the control and either invaded array (low density: $P = 0.52$, high density: $P = 0.18$), but *G. maculatum* in low density arrays had more visitors than in high density arrays ($P = 0.02$; Fig. 2A). Mean pollinator visitation rate to *G. maculatum* differed among treatments ($F_{2,21} = 6.58$, $P =$

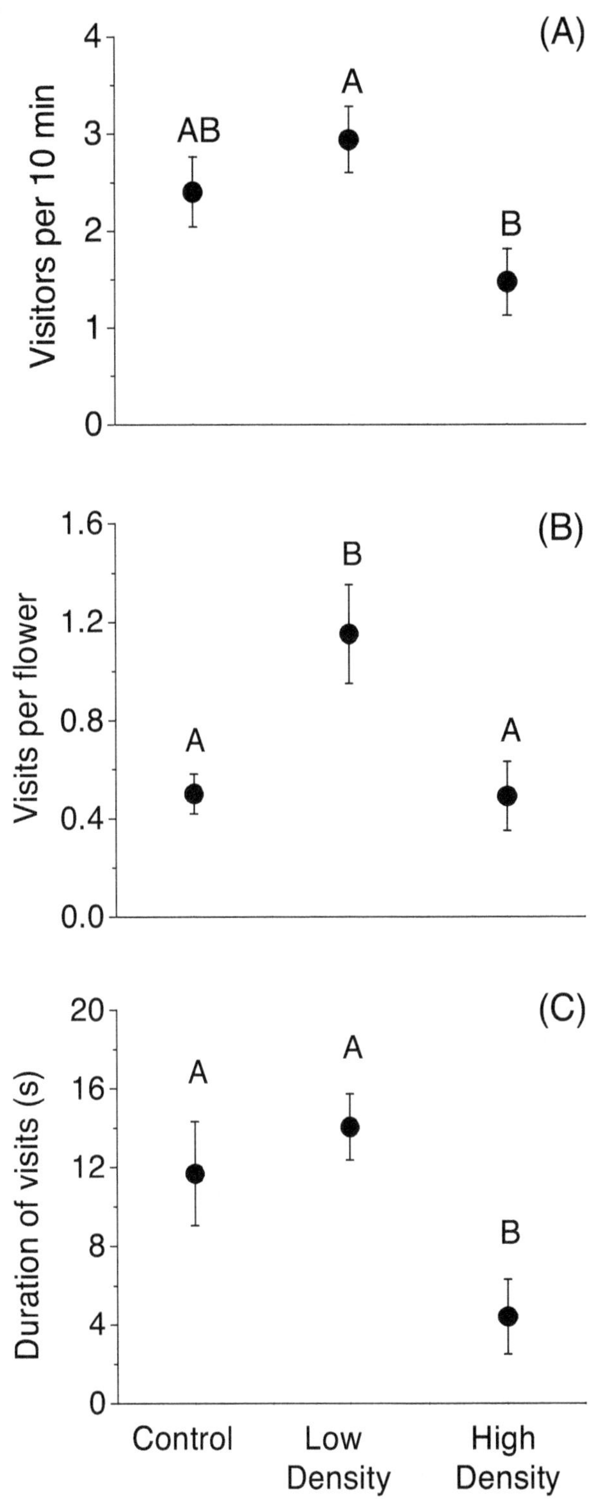

FIGURE 2. Pollinator foraging on *G. maculatum* flowers in three array treatments varying in relative floral density of the invasive shrub *L. maackii* (control: no *L. maackii*). (A) Number of visitors. (B & C) Pollinator foraging behaviour. Bars are mean values and error bars ± 1 SE. Capital letters represent significant differences at $P < 0.05$ from Tukey's multiple comparisons.

0.0061); flowers in low density arrays received more than twice as many pollinator visits as those in control and high density arrays ($P = 0.013$, $P = 0.013$; Fig 2B). In contrast, visitation rates in high density arrays did not differ from

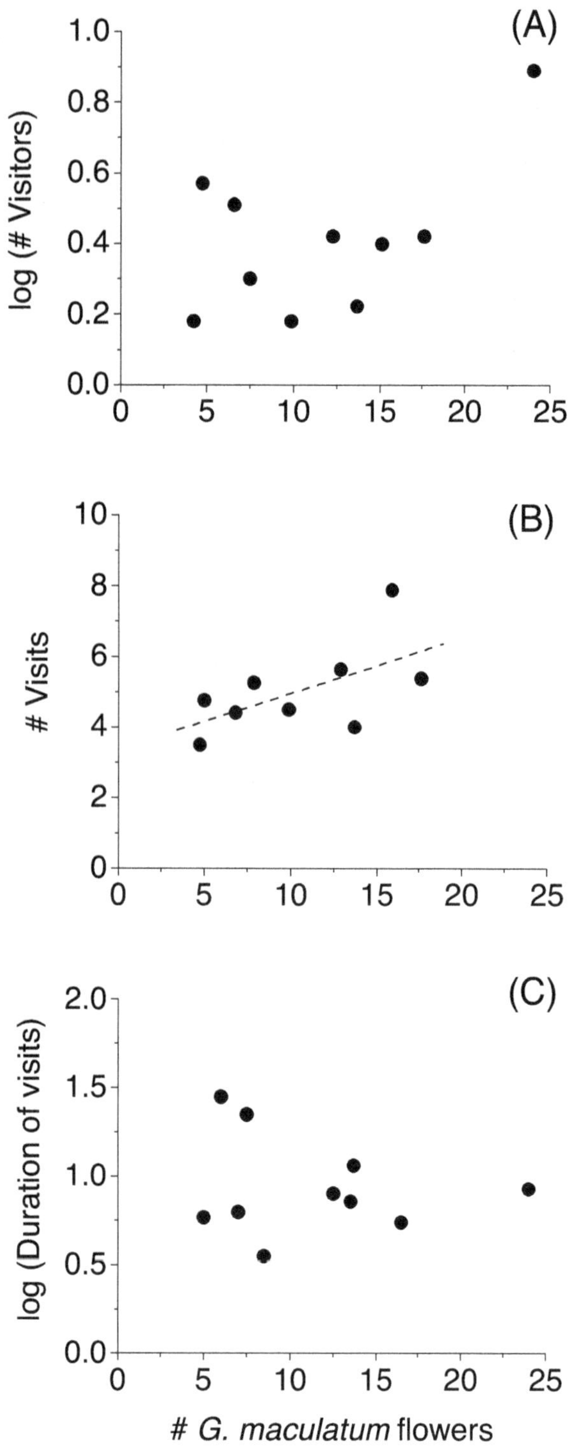

FIGURE 3. Relationship between pollinator foraging behaviour and total number of conspecific flowers in control arrays (only *G. maculatum* flowers). Pollinator foraging was quantified as (A) visitor attraction at the patch level (log transformed number of visitors), (B) pollinator visits, and (C) log transformed duration of visits (seconds). Each point represents mean daily values calculated for each 10-min observation session in control arrays (each plot was observed on each day, $N = 10$ days). Dashed line is from simple linear regression and represents a marginally significant relationship ($0.05 < P < 0.10$).

control arrays ($P = 0.99$, Fig. 2B). The average duration of pollinator visits (log transformed) also differed among treatments ($F_{2,20} = 10.86$, $P = 0.0006$). Visitors spent less than half as much time on *G. maculatum* flowers in high density compared to control and low density arrays ($P = 0.0047$; $P = 0.0007$, respectively; Fig 2C), and there was no difference in the duration of visits in low density compared to control arrays ($P = 0.69$, Fig. 2C).

There was no effect of *G. maculatum* floral abundance in control arrays on the number of visitors attracted to these patches of flowers (log transformed), the number of visits observed to these flowers, or in the duration of visits (log transformed) ($R^2 = 0.29$, $F_{1,8} = 3.31$, $P = 0.11$, Figure 3A; $R^2 = 0.35$, $F_{1,7} = 3.76$, $P = 0.094$, Figure 3B; $R^2 = 0.03$, $F_{1,8} = 0.25$, $P = 0.63$, Figure 3C). These results do not support the hypothesis that *G. maculatum* flower abundance, at least across the range used in this experiment, determined attraction of visitors and visitor foraging behaviour and instead support the hypothesis that relative floral densities had an effect on pollinator foraging behaviour. Although there is a marginally significant trend of increased visits at higher *G. maculatum* floral abundances, this trend is in the opposite direction expected if *G. maculatum* floral abundance was accounting for increased visitation rates to low density arrays.

Conspecific pollen grains accounted for 78.55%, *L. maackii* grains for 14.20%, and unknown pollen grains for 7.43% of total pollen deposition to *G. maculatum* stigmas. The number of *G. maculatum* pollen grains differed significantly among treatments ($F_{2,21} = 5.38$, $P = 0.013$). Consistent with pollinator visitation data, conspecific pollen deposition was higher to flowers in low density arrays compared to control and high density arrays ($P = 0.048$, $P = 0.015$, respectively), and there was no difference in conspecific pollen deposition between high density arrays compared to controls ($P = 0.86$, Fig. 4A). The numbers of *L. maackii* (Lm) and log-transformed unknown (un) pollen grains on *G. maculatum* stigmas were relatively consistent across treatments ($F_{2,21} = 1.22$, $P = 0.31$, $F_{2,20} = 2.18$, $P = 0.14$, respectively; control: 4.84 ± 0.64 Lm grains, 2.96 ± 0.69 un grains; low: 3.52 ± 0.42 Lm grains, 2.75 ± 1.04 un grains; high: 3.93 ± 0.73 Lm grains, 1.27 ± 0.53 un grains). *Geranium maculatum* seeds per flower (log transformed) did not differ among treatments ($F_{2,21} = 0.056$, $P = 0.95$, Fig. 4B).

DISCUSSION

Our results highlight the importance of relative floral density within patches of floral resources in shaping outcomes of interactions for pollinators between co-flowering plant species. Although the total number of flowers in each array varied across days, relative floral density remained constant, and pollinator foraging responses were consistent with our hypotheses about the effects of relative floral densities. We detected increased visitation rates to *G. maculatum* flowers at a low relative density of *L. maackii* flowers compared to our other treatments. Higher densities of *L. maackii* flowers were associated with a lower quality of visits, evidenced by shorter visits to *G. maculatum* flowers in high density arrays relative to other treatments. Bjerknes et al. (2007) emphasize that variation in spatial and temporal

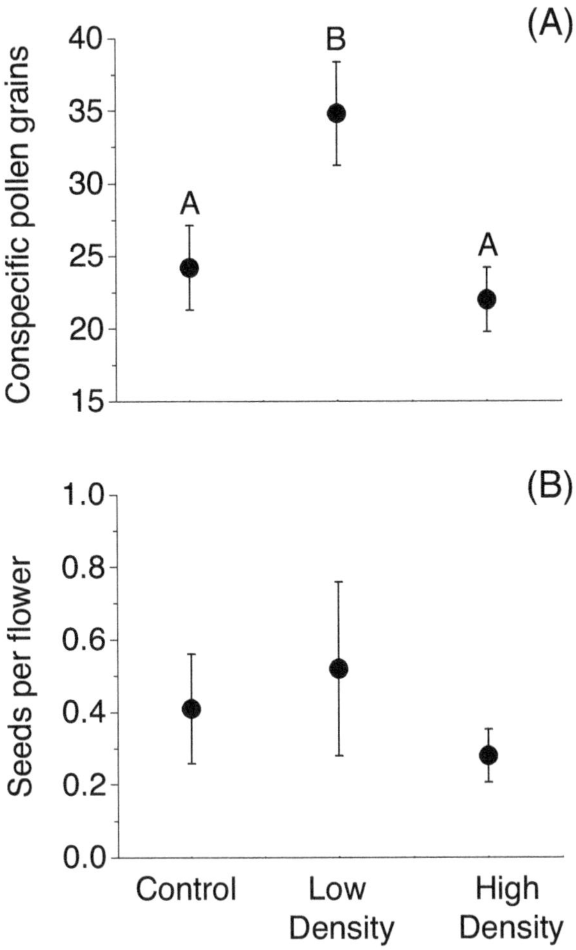

FIGURE 4. *Geranium maculatum* female reproductive success, averaged per plant then by plot, in three arrays varying in relative floral density of *L. maackii* (control: no *L. maackii*). (A) Conspecific pollen deposition to *G. maculatum* flowers (B) Mean values of seeds per flower are shown, although they were log-transformed for analysis. Error bars ± 1 SE. Capital letters represent significant differences at $P < 0.05$ from Tukey's multiple comparisons.

scales may explain variation among outcomes of studies of pollinator-mediated interactions between invasive and native plant species. Theoretical work and field studies have also confirmed the role of spatial scale in mediating interactions for pollinators between co-flowering plant species (Cariveau & Norton 2009; Hanoteaux et al. 2013). Our study supports the idea that relative floral density within patches of floral resources is also likely to play an important role in mediating these outcomes, perhaps as an explanation for why interactions vary through time, both within seasons as an invasive plant species comes into bloom and across years as an invader increases in density. Our results also suggest that facilitation of pollination is possible in incipient invasions, but that this effect may be neutralized as relative floral density of the invader increases.

The particular floral density at which facilitation or competition for pollination occurs is likely to vary considerably among plant species. We detected density-dependent pollinator foraging behaviour at relative floral densities of 20:1 (low invader) and 60:1 (high invader), compared to 1:1 and 5:1 in another study (Muñoz & Cavieres 2008). While strong interaction effects for pollination are expected for species of similar growth form and floral morphology (Morales & Traveset 2009), we find that interaction effects, in this case facilitation, can also be quite strong for pairs of species with non-similar morphologies (Ghazoul 2006). In addition to floral morphology, floral reward should also influence the relative floral densities at which facilitation and competition for pollinators occur. Native species with less rewarding nectar than co-flowering invasive species may experience competition at lower floral densities of the invader compared to native plants with more rewarding nectar or a higher rate of nectar production (Chittka & Schürkens 2001). Indeed, *G. maculatum* nectar contains on average 40% more total sugar than *L. maackii* nectar, but differences in nectar volume are unknown (McKinney & Goodell 2010). The higher nectar sugar content in the native species in our study may explain why we detected facilitation of pollination at a relatively high floral density of the invader compared to the findings of Muñoz & Cavieres (2008). Our results suggest that even higher *L. maackii* densities are necessary to observe negative effects on *G. maculatum* pollination.

Increased *G. maculatum* pollen receipt at a low density of the invader seems to be a result of a shift in the behaviour of individual pollinators relative to uninvaded patches. *Geranium maculatum* flowers received more than twice as many visits in low density arrays compared to controls, but not significantly more visitors, indicating that visitation rate increased because individual pollinators made more visits to *G. maculatum*. This is in contrast to the more common trend of invaders acting as a 'magnet species' by attracting more pollinators to mixed patches of floral resources, whose visitors then spill over onto nearby plants (Fig. 1b; Moragues & Traveset 2005; Muñoz & Cavieres 2008; Molina-Montenegro et al. 2010; Flanagan et al. 2010; McKinney & Goodell 2011). Additionally, the non-significant trend towards more visitors in low density arrays relative to controls cannot account for a doubling of visits in low density arrays. These results suggest that intraspecific competition for pollinators between native plants may decrease in the presence of a low floral density of the invader when the number of visitors remains constant. This behavioural switch in the presence of a low density of invasive flowers could reflect perceived reward per unit of foraging effort within the patch of floral resources. Bees that encounter *G. maculatum* flowers after visiting less rewarding *L. maackii* flowers may have a higher 'departure' threshold for leaving the patch, compared to bees that only have the option of intraspecific floral transitions in control arrays. Bees that opt to continue foraging on *G. maculatum* at low densities of *L. maackii* may not do so at high densities merely because of the surplus of *L. maackii* flowers relative to *G. maculatum* in high density arrays. Our results are consistent with this hypothesis because visitors spent less time on *G. maculatum* flowers in high density arrays compared to controls and low density arrays.

Behavioural shifts in pollinator foraging to *G. maculatum* did not translate into differences in *G. maculatum* reproductive success (seeds per flower) among treatments. This result may reflect numerous issues: pollen quality, variation in maternal resources across treatments, and/or low statistical power. The aim of our experimental design was to represent different *L. maackii* invasion scenarios. However, placing fewer *G. maculatum* plants (i.e. fewer nearby pollen donors) in low and high density arrays compared to controls could have influenced reproduction via pollen quality because *G. maculatum* reproduction shows evidence of inbreeding depression (Ågren & Willson 1991); receipt of self pollen results in a lower probability of seed set in this species (Chang 2007). For our design to affect *G. maculatum* reproduction in this way, bees have to make more intra-plant movements and transfer more self pollen (and insufficient outcross pollen) in mixed arrays compared to controls, a pattern that has empirical support (Yang et al. 2011). We did not distinguish between inter- and intra-plant pollinator movements in our observations and therefore cannot conclusively evaluate of the role of pollen quality vs. pollen quantity. Because all of our plants were hermaphrodites, however, they may be more prone to inbreeding depression than female plants that require outcross pollen to set seed. At the same time, we expect female plants to respond more strongly to changes in pollination because they require outcross pollen; in this sense our experiment is a conservative test for effects of plant invasion on pollination and reproductive success of *G. maculatum*. Heterospecific pollen transfer (HPT) also may have countered positive effects of increased pollinator visitation in low density arrays (sensu Brown & Mitchell 2001), but in our study, HPT was consistently low across arrays (consistent with Bartomeus et al. 2008). We may detect little HPT because pollinators do not frequently switch between species (Chittka et al. 1999; Feldman 2008) or because dissimilar floral morphologies lead to differential pollen placement on pollinator bodies (Harder & Barrett 1993). Finally, if flower production is an indicator of maternal resources, it is possible that *G. maculatum* plants in low density arrays were unable to take full advantage of increased conspecific pollen deposition, because plants in low density arrays had significantly fewer flowers than plants in control arrays. Indeed, conspecific pollen deposition was above that required for maximum seed production in *G. maculatum* (Mulchahy et al. 1983), suggesting resource rather than pollen limitation.

The net effect of invasive plants on pollination of neighboring plants will ultimately depend on how pollinator populations and communities respond to invasion, in addition to pollinator attraction to native plants at local spatial scales (Tepedino et al. 2008; Goodell 2008; Mitchell et al. 2009; Moron et al. 2009; Bartomeus et al. 2010; Yang et al. 2011). Here we show that relative floral density within patches of floral resources may account for some of the variation in the direction of pollinator-mediated interactions between invasive and native plants at local spatial scales (i.e. competition, neutral effects, and facilitation) (Traveset & Richardson 2006; Bjerknes et al. 2007). Plant-pollinator interactions fluctuate across years and between months within a year (Basilio et al. 2006; Petanidou et al. 2008), and much of this variation may relate to changes in absolute or relative floral density of plant species within a plant-pollinator community. In sites severely degraded from initial or target conditions, like our study site, invasive plants may be important for maintaining or developing plant-pollinator communities (Ewel & Putz 2004). Invaders may serve as facilitators of pollination in early stages of invasion and in some restoration scenarios, at least when HPT does not negatively affect native plant reproduction and when invaders act as magnet species. In this study, we provide novel evidence for facilitation as a result of a behavioural shift in pollinator foraging, independent of a magnet effect. Although this may be another mechanism by which invasive plants can facilitate pollination of native plants, facilitation is still the exception to the more common trend of competition for pollination (Larson 2006; Morales & Traveset 2009).

ACKNOWLEDGEMENTS

The authors thank S. Collins, S. McKinney, S. Seibert, and D. Shields for assistance with field work and Three Creeks Metro Park for access to field sites. Thoughtful and constructive comments from I. Bartomeus, A. Snow, C. Lin, S. Cusser, and anonymous reviewers greatly improved earlier drafts of this paper. We also thank P. CaraDonna, N. Hass, N. Waser, and M. Price for valuable feedback on the conceptual figure. An Alumni Grant for Graduate Student Research and a Supplemental Research Grant (awarded to AMI), both from The Ohio State University, funded this research.

REFERENCES

Ågren J, Willson MF (1991) Variation and sexual differences in reproductive characters and seed production in gynodioecious *Geranium maculatum*. American Journal of Botany 78: 470–480.

Allan BF, Dutra HP, Goessling LS, Barnett K, Chase JM, Marquis RJ, Pang G, Storch GA, Thach RE, J.L. Orrock JL. (2010) Invasive honeysuckle eradication reduces tick-borne disease risk by altering host dynamics. Proceedings of the National Academy of Sciences USA 107: 18523–18527.

Bartomeus I, Bosch J, Vilà M (2008) High invasive pollen transfer, yet low deposition on native stigmas in a *Carpobrotus*-invaded community. Annals of Botany 102: 417–424.

Bartomeus I, Vila M, Steffan-Dewenter I (2010) Combined effects of *Impatiens glandulifera* invasion and landscape structure on native plant pollination. Journal of Ecology 98:440–450.

Basilio AM, Medan D, Torretta JP, Bartoloni NJ (2006) A year-long plant-pollinator network. Austral Ecology 31:975–983.

Bertin RI, Sholes ODV (1993) Weather, pollination, and the phenology of *Geranium maculatum*. American Midland Naturalist 129:52–66.

Bjerknes AL, Totland O, Hegland SJ, Nielsen A (2007) Do alien plant invasions really affect pollination success in native plant species? Biological Conservation 138:1–12.

Borgmann KL, Rodewald AD (2004) Nest predation in an urbanizing landscape: The role of exotic shrubs. Ecological Applications 14:1757–1765.

Brown BJ, Mitchell RJ (2001) Competition for pollination: effects of pollen of an invasive plant on seed set of a native congener. Oecologia 129: 43–49.

Cariveau DP, Norton AP (2009) Spatially contingent interactions between an exotic and native plant mediated through flower visitors. Oikos 118: 107–114.

Chang SM (2007) Gender-specific inbreeding depression in a gynodioecious plant, *Geranium maculatum* (Geraniaceae). American Journal of Botany 94: 1193–1204.

Chittka L, Thomson JD, Waser NM (1999) Flower constancy, insect psychology, and plant evolution. Naturwissenschaften 86:361–377.

Chittka L, Schürkens S (2001) Successful invasion of a floral market - An exotic Asian plant has moved in on Europe's river-banks by bribing pollinators. Nature 411:653–653.

Christian JM, Wilson SD (1999) Long-term ecosystem impacts of an introduced grass in the northern Great Plains. Ecology 80:2397–2407.

Collier MH, Vankat JL, Hughes MR (2002) Diminished plant richness and abundance below *Lonicera maackii*, an invasive shrub. American Midland Naturalist 147:60–71.

D'Antonio CM, Hughes RF, Mack M, Hitchcock D, Vitousek PM (1998) The response of native species to removal of invasive exotic grasses in a seasonally dry Hawaiian woodland. Journal of Vegetation Science 9:699–712.

Daehler CC (2003) Performance comparisons of co-occurring native and alien invasive plants: Implications for conservation and restoration. Annual Review of Ecology Evolution and Systematics 34:183–211.

Dietzsch AC, Stanley DA, Stout JC (2011) Relative abundance of an invasive alien plant affects native pollination processes. Oecologia 167:469–479.

Ewel JJ, Putz FE (2004) A place for alien species in ecosystem restoration. Frontiers of Ecology and the Environment 2:354–360.

Feinsinger P (1987) Effects of plant species on each other's pollination: is community structure influenced? Trends in Ecology and Evolution 2: 123–126.

Feinsinger P, Tiebout HM, Young BE (1991) Do tropical bird-pollinated plants exhibit density-dependent interactions? Field experiments. Ecology 72: 1953–1963.

Feldman TS, Morris WF, Wilson WG (2004) When can two plant species facilitate each other's pollination? Oikos 105: 97–207.

Feldman TS (2008) The plot thickens: does low density affect visitation and reproductive success in a perennial herb, and are these effects altered in the presence of a co-flowering species? Oecologia 156: 807–817.

Flanagan RJ, Mitchell RJ, Karron JD (2010) Increased relative abundance of an invasive competitor for pollination, *Lythrum salicaria*, reduces seed number in *Mimulus ringens*. Oecologia 164: 445–454.

Galen C, Gregory T (1989) Interspecific pollen transfer as a mechanism of competition: consequences of foreign pollen contamination for seed set in the alpine wildflower, *Polemonium viscosum*. Oecologia 81:120–123.

Gathmann A, Tscharntke T (2002) Foraging ranges of solitary bees. Journal of Animal Ecology 71:757–764.

Ghazoul J (2004) Alien abduction: Disruption of native plant-pollinator interactions by invasive species. Biotropica 36:156–164.

Ghazoul J (2006) Floral diversity and the facilitation of pollination. Journal of Ecology 94:295-304.

Ginsberg HS (1983) Foraging ecology of bees in an old field. Ecology 64: 165-175.

Goodell K (2008) Invasive exotic plant-bee interactions. In: James RR, Pitts-Singer TL (eds) Bee Pollination in Agroecosystems. Oxford University Press, Oxford, pp 166-183.

Goodell K, Iler AM (2007) Reproduction and habitat-dependent pollinator services in the invasive shrub *Lonicera maackii*. In: Ohio Invasive Plants Research Conference. The Ohio Biological Survey, Delaware, OH.

Gould AMA, Gorchov DL (2000) Effects of the exotic invasive shrub *Lonicera maackii*, Amur honeysuckle, on survival and fecundity of native forest annual herbs. American Midland Naturalist 144:36-50.

Grabas GP, Laverty TM (1999) The effect of purple loosestrife (*Lythrum salicaria* L,; Lythraceae) on the pollination and reproductive success of sympatric co-flowering wetland plants. Ecoscience 6:230-242.

Greenleaf SS, Williams NM, Winfree R, Kremen C (2007) Bee foraging ranges and their relationship to body size. Oecologia 153:589-596.

Hanoteaux S, Tielbörger K, Seifan M (2013) Effects of spatial patterns on the pollination success of a less attractive species. Oikos 122: 867–880.

Harder LD, Barrett SCH (1993) Pollen removal from tristylous *Pontederia cordata* - Effects of anther position and pollinator specialization. Ecology 74:1059-1072.

Hutchinson TF, Vankat JL (1997) Invasibility and effects of Amur honeysuckle in southwestern Ohio forests. Conservation Biology 11:1117-1124.

Jakobsson A, Padron B, Traveset A (2008) Pollen transfer from invasive *Carpobrotus* spp. to natives - A study of pollinator behaviour and reproduction success. Biological Conservation 141:136-145.

Kaiser-Bunbury C, Muller C (2009) Indirect interactions between invasive and native plants via pollinators. Naturwissenschaften 96:339-346.

Kandori I, Hirao T, Matsunaga S, Kurosaki T (2009) An invasive dandelion unilaterally reduces the reproduction of a native congener through competition for pollination. Oecologia 159:559-569.

Kearns CA, Inouye DW (1993) Techniques for pollination biologists. University Press of Colorado. Niwot, Colorado, USA.

Larson DL, Royer RA, Royer MR (2006) Insect visitation and pollen deposition in an invaded prairie plant community. Biological Conservation 130:148-159.

Levine JM, Vila M, D'Antonio CM, Dukes JS, Grigulis K, Lavorel S (2003) Mechanisms underlying the impacts of exotic plant invasions. Proceedings of the Royal Society of London Series B 270:775-781.

Lopezaraiza-Mikel ME, Hayes RB, Whalley MR, Memmott J (2007) The impact of an alien plant on a native plant–pollinator network: an experimental approach. Ecology Letters 10:539-550.

Luken JO, Thieret JW (1996) Amur honeysuckle, its fall from grace. Bioscience 46:18-24.

Martin MC (1965) An ecological life history of *Geranium maculatum*. American Midland Naturalist 73:111-149.

Matsumoto T, Takakura KI, Nishida T (2010) Alien pollen grains interfere with the reproductive success of native congener. Biological Invasions 12:1617-1626.

Mattos KJ, Orrock JL (2010) Behavioral consequences of plant invasion: an invasive plant alters rodent antipredator behavior. Behavioral Ecology 21: 556-561.

McKinney AM (2010) Pollinator-mediated interactions between the invasive shrub *Lonicera maackii* and native herbs: the roles of shade, flowering phenology, spatial scale, and floral density. PhD Dissertation, The Ohio State University, Columbus.

McKinney AM, Goodell K (2010) Shading by invasive shrub reduces seed production and pollinator services in a native herb. Biological Invasions 12:2751-2763.

McKinney AM, Goodell K (2011) Plant-pollinator interactions between an invasive and native plant vary between sites with different flowering phenology. Plant Ecology 212:1025-1035.

Miller KE, Gorchov GL (2004) The invasive shrub, *Lonicera maackii*, reduces growth and fecundity of perennial forest herbs. Oecologia 139:359-375.

Mitchell CE, Agrawal AA, Bever JD, et al. (2006) Biotic interactions and plant invasions. Ecology Letters 9:726-740.

Mitchell RJ, Flanagan RJ, Brown BJ, Waser NM, Karron JD (2009) New frontiers in competition for pollination. Annals of Botany 103:1403-1413.

Molina-Montenegro MA, Badano EI, Cavieres LA (2010) Positive interactions among plant species for pollinator service: assessing the 'magnet species' concept with invasive species. Oikos 117:1833-1839.

Moragues E, Traveset A (2005) Effect of *Carpobrotus* spp. on the pollination success of native plant species of the Balearic Islands. Biological Conservation 122:611-619.

Morales CL, Traveset A (2009) A meta-analysis of alien vs. native plants on pollinator visitation and reproductive success of co-flowering native plants. Ecology Letters 12:716-728.

Mulcahy DL, Curtis PS, Snow AA (1983) Pollen competition in a natural population. In: Jones, C.E., and R.J. Little (eds) Handbook of Experimental Pollination Biology. Academic Press, New York, pp 330-337.

Muñoz, A.A., and L.A. Cavieres LA (2008) The presence of a showy invasive plant disrupts pollinator service and reproductive output in native alpine species only at high densities. Journal of Ecology 96:459-467.

Murphy SD, Aarssen LD (1995) Reduced seed set in *Elytrigia repens* caused by allelopathic pollen from *Phleum pratense*. Canadian Journal of Botany 73:1417-1422.

Olesen JM, Eskildsen LI, Venkatasamy S (2002) Invasion of pollination networks on oceanic islands: importance of invader complexes and endemic super generalists. Diversity and Distributions 8:181-192.

Petanidou T, Kallimanis AS, Tzanopoulos J, Sgardelis SP, Pantis JD (2008) Long-term observation of a pollination network: fluctuation in species and interactions, relative invariance of network structure and implications for estimates of specialization. Ecology Letters 11:564-575.

R Development Core Team (2013) R: A language and environment for statistical computing. R Foundation for Statistical Computing, Vienna, Austria. ISBN 3-900051-07-0, URL http://www.R-project.org.

Rathcke B (1983) Competition and facilitation among plants for pollination. In: Real, L. (ed) Pollination Biology. Academic Press, New York, pp 305-329.

Sih A, Baltus MS (1987) Patch size, pollinator behavior, and pollinator limitation in Catnip. Ecology 68:1679-1690.

Strauss SY (1991) Indirect effects in community ecology - their definition, study and importance. Trends in Ecology & Evolution 6:206-210.

Takakura K, Nishida T, Matsumoto T, Nishida S (2009) Alien dandelion reduces the seed-set of a native congener through frequency-dependent and one-sided effects. Biological Invasions 11:973-981.

Tepedino V, Bradley B, Griswold T (2008) Might flowers of invasive plants increase native bee carrying capacity? Intimations from Capitol Reef National Park, Utah. Natural Areas Journal 28:44-50.

Tilman D (1997) Community invasibility, recruitment limitation, and grassland biodiversity. Ecology 78:81-92.

Totland O, Nielsen A, Bjerknes AL, Ohlson M (2006) Effects of an exotic plant and habitat disturbance on pollinator visitation and reproduction in a boreal forest herb. American Journal of Botany 93:868-873.

Traveset A, Richardson DM (2006) Biological invasions as disruptors of plant reproductive mutualisms. Trends in Ecology & Evolution 21:208-216.

Tscheulin T, Petanidou T (2013) The presence of the invasive plant *Solanum elaeagnifolium* deters honeybees and increases pollen limitation in the native co-flowering species *Glaucium flavum*. Biological Invasions 15: 385-393.

Waser NM (1978) Competition for hummingbird pollination and sequential flowering in two Colorado wildflowers. Ecology 59:934-944.

Waser NM (1983) Competition for pollination and floral character differences among sympatric plant species: a review of the evidence. In: Jones, C.E., and Little, R.J. (eds) Handbook of Experimental Pollination Biology. Academic Press, New York, pp 277-293.

White EM, Wilson JC, Clarke AR (2006) Biotic indirect effects: a neglected concept in invasion biology. Diversity and Distributions 12:443-455.

Willson MF, Miller LJ, Rathcke BJ (1979) Floral display in *Phlox* and *Geranium*: adaptive aspects. Evolution 33:52-63.

Woods KD (1993) Effects of invasion by *Lonicera tatarica* L. on herbs and tree seedlings in four New England forests. American Midland Naturalist 130:62-74.

Yang S, Ferrari MJ, Shea K (2011) Pollinator behavior mediates negative interactions between two congeneric invasive plant species. American Naturalist 177: 110-118.

10

SPATIAL VARIATION IN POLLINATOR COMMUNITIES AND REPRODUCTIVE PERFORMANCE OF *PROSOPIS JULIFLORA* (FABACEAE)

Asif Sajjad*, Shafqat Saeed and Muhammad Amjad Bashir

Department of Entomology, University College of Agriculture, Bahauddin Zakariya University Multan, Pakistan

Abstract— This study was conducted in an effort to understand the effects of spatial variations in pollinator assemblage due to habitat isolation on the reproductive performance of perennial plant species. Variations in pollinator assemblage structure (abundance, diversity and Shannon-Wiener index) were studied at three widely isolated (100 to 200 km apart) nature reserves of Southern Punjab, Pakistan, in order to explore its effects on reproductive performance of *Prosopis juliflora*. Species richness and abundance were highest in Pirowal Sanctuary followed by Chichawatni Sanctuary and Chak Katora forest reserve. The pollination system of *P. juliflora* was highly generalized with 77 insect visitor species in four orders among all the three sites. However, pollinator assemblage varied significantly in composition among the sites. Out of the four reproductive parameters considered, the number of pods per raceme and germination varied significantly among the three locations. The reproductive performance of *P. juliflora* in terms of number of pods per raceme and germination improved with abundance of pollinators.

Keywords: pollinator assemblages, Prosopis juliflora, *reproductive success, spatial variation*

INTRODUCTION

Pollinator assemblage structure can have important influences on floral evolution and reproductive interactions among plant species (Moeller 2005). Recent studies of structure of plant-pollinator networks have shown that they exhibit 'nestedness', i.e. specialist pollinators and plants are subsets of more generalised networks (Kallimanis et al. 2009). Following these ideas, an increasing number of community-level studies have concluded that generalization is the rule and specialization is rare (e.g., Olesen & Jordano 2002). However, most studies have not distinguished between pollinators and floral visitors, despite the fact that a visitor may not actually pollinate the flowers.

Pollination systems can be examined in the same general framework used to understand community structure and assembly. For example, the variation of visitor abundance and their spatial predictability within plant populations allow regular floral visitors to be distinguished from occasional or incidental floral visitors (Root 1973). From the ecological point of view, generalization is considered a positive trait that may favour competitive ability, colonization capacity and invasion ability in plants (Richardson et al. 2000).

In a geographical context, the spatial structure of variation in pollinator abundance and community composition can also have important implications for plant reproductive performance and ultimately floral evolution (Gomez et al. 2007). This spatial scale variation in pollinator communities remains poorly documented. Flower visitation activity of pollinators is often strongly affected by climatic conditions (McCall & Primack 1992) and climate varies on both large and small spatial scales. The size and quality (availability of nesting places and floral resources) of any natural habitat also determines the richness and abundance of pollinator species (Cunningham 2000).

When large tracts of forests are subjected to fragmentation due to deforestation, the organisms remaining in discontinuous remnants are exposed to many biotic and abiotic changes (Saunders et al. 1991). The survival of any species under such conditions depends on its life history, ecological characteristics and mutualistic interactions (Rathcke & Jules 1993). Habitat isolation and quality certainly affects pollinator diversity and hence reproductive success of plants, but seed set may also be influenced by other factors such as patch size, density of flowering plants, occurrence of competing alternative flowers (Jennersten & Nilsson, 1993; Kleijn & Langevelde 2006), genetic variability (Van Treuren et al. 1994) and abundance of pollinators (Donaldson et al. 2002).

Previously, many studies assessed the affect of habitat quality and fragmentation or isolation on the diversity and abundance of pollinators and ultimately plant reproductive success (Steffan-Dewenter & Tscharntke 1999; Donaldson et al. 2002; Murren 2002). These studies were done on small scales, i.e. either on islands or in isolated patches of forest in single habitats suffering from deforestation. Our study was conducted on a different scale: The study locations in this research were relatively large (> 1000 acres) and widely separated (100 - 200 km) from each other. Since there is no geographical barrier in the plains of Punjab, we can presume that thousands of years back, the entire study area may have been covered by continuous, similar vegetation and inhabited by a homogenous population of pollinator species. However, the distribution of pollinators may vary in

*Corresponding author; email: asifbinsajjad@gmail.com

space and time (Ollerton & Cranmer 2002). Very few studies have explicitly tested for spatial variation in pollinator richness and diversity (Herrera 2005).

This study focuses on the invasive *Prosopis juliflora* (Sw.) DC, a multipurpose perennial tree native to northern South America. Its fast growth, drought resistance and salt tolerance has led to its successful naturalization in dry tropical and sub-tropical areas of Africa, south-east Asia, the Indian sub-continent and Australia. It was deliberately introduced to the sub-continent in 1857 (Luna 1996) with view point of its qualities and uses, i.e. construction material, charcoal, soil conservation and rehabilitation of degraded and saline soils (Pasiecznik et al. 2001). Due to improper management, however, its potential benefits were undermined, because it becomes a serious invasive tree (Pasiecznik 2001). Its invasive qualities have resulted in multiple negative effects on food security, livelihoods and environment in many parts of the world, e.g. Ethopia (Dubale, 2006), India and Kenya (Mwangi & Swallow 2005) by declining livestock production and productivity due to its competition with palatable native trees.

P. juliflora is self-incompatible (Simpson 1977) and receives visitors from a variety of insect orders (Ward et al. 1977). Many pollination biologists have explored the reproductive biology of *P. juliflora* (DeOliveira & Pires 1990; Iqbal & Shafiq 1997; Zaitoun et al. 2009). About 300-400 flowers are arranged on a raceme (Zaitoun et al. 2009). An individual flower has petals only 3 mm wide while its anthers are 4 mm wide. The small size of the flower ensures contact of even small insect with reproductive organs and minimizes the risks of nectar robbing. Thus, it is fair to assume that all species recorded act as effective pollinators to some degree.

We investigated *P. juliflora* populations in three widely separated wildlife reserves (*ca.* 100 to 200 km apart from each other) of the southern Punjab (Pakistan). The aim was to study (i) variation in the pollinator community structure (species richness, diversity and abundance) among these reserves due to their isolation a hundred years ago; and (ii), whether the variation in pollinator community structure affects the reproductive performance of *P. juliflora*.

MATERIALS AND METHODS

Study area

The study was conducted in the southern Punjab of Pakistan. Climate of the area is sub-tropical with cold winter and hot summer; the mean monthly temperature ranges between 25°C and 30°C, with mean maxima 35°C to 40°C, and mean minima 10°C to 20°C. The extreme maximum temperature of the region varies between 45°C and 48°C, recorded in May and June, while the lowest minimum temperature is 0°C to -2°C, recorded in January. May and June are the hottest months, whereas January is the coldest month of the region (Khan et al. 2010). We selected three natural habitats: Chichawatni Sanctuary (District Sahiwal), Pirowal Sanctuary (District Khanewal) and Chak Katora forest reserve (District Bahawalpur) (Table 1). Chichawatni is 100 km north-east from Pirowal, while Chak Katora is 150 km south-east: between them are either cities or agricultural land. These natural habitats have never been cultivated and according to the definition as wildlife sanctuary, entry is prohibited without official permission. The largest (17 823 acres) is Pirowal, where we selected a fenced area of 1500 acres with very few human activities (wood cutting only). Chichawatni is smaller (11 530 acres) with low level human interventions like grazing and wood cutting, while Chak Katora is the smallest fragment (only 1 323 acres) with more grazing and wood cutting than the other two locations.

Flower visitor census

Flowering of *P. juliflora* and our study took place during the hottest season from April to June in 2008. In each location we chose four experimental plots (*ca.* 2 acres each) with similar population size (*ca.* 100 trees) of *P. juliflora* and selected 25 similar sized trees in each plot, widely separated from each other. On each tree, we tagged one branch ~10ft long originating from the main stem. Once at every fortnight, from first week of April to end of June, the number of individuals of each insect species visiting the flowers on these tagged branches was counted for 60 seconds/branch. We reselected another branch on the same tree if the chosen one was not flowering. Records were taken on sunny days from 08:00 to 11:00 h (local time). This fortnightly data were completed on three consecutive days (one day per site).

All the visitors were morphotyped and counted in free flight, while later identification was done in the laboratory. Identification of Diptera and Coleoptera was accomplished by using standard keys (Borror et al. 1981). Syrphids (Diptera: Syrphidae) and butterflies (Lepidoptera:

TABLE 1. Properties and types of three studied nature reserves at southern Punjab, Pakistan.

Location	Type of reserve	Area (acres)	GPS	Altitude (feet)	Human intervention
Pirowal District Khabewal	Wildlife Sanctuary (fenced)	17 823	30°34′ N 72°03′ E	437±16.5	Very low (wood cutting)
Chichawatni District Sahiwal	Wildlife Sanctuary (unfenced)	115 30	30°54′ N 72°70′ E	205±22	Low (Grazing and wood cutting)
Chak Katora District Bahawalpur	Reserve forest (unfenced)	1 323	29°78′ N 72°55′ E	447±19.4	Intensive (Grazing and wood cutting)

Rhopalocera) were identified to species by experts (see Acknowledgement). The keys of Michener (2000) were used to identify bee genera. The voucher specimens were deposited at the Agricultural Museum of the University College of Agriculture, Bahauddin Zakariya University Multan.

Reproductive success

Since the pod-setting percentage is very low in *P. juliflora* (De Oliveira & Pires 1990), we tagged three branches on each of the 25 sampled trees in each location and harvested all the mature pods fortnightly. Plant reproductive success was measured in terms of number of pods per raceme, pod length, 100-seed weight, number of seeds per pod and percentage germination. Seeds were mixed thoroughly and 500 seeds were randomly selected and subjected to germination test in plastic pots for each location.

Data analysis

Two traits of the flower visitor assemblage visiting *P. juliflora* flowers are analyzed in this study, i.e. abundance (total number of flower visits) and diversity (Magurran 2004). We assessed flower visitor diversity by calculating species richness, diversity, evenness and dominance. Species richness is simply the number of insect visitor species in each location. Dominance was calculated as the relative abundance of the most abundant visitor species. We used pollinator rank-abundance plots as a way to visualize the structure of the pollinator communities (Magurran 2004). Diversity was calculated using Shannon-Wiener index and Hurlbert's Probability of an Interspecific Encounter (PIE) index (Hurlbert 1971). Hurlbert's PIE is an evenness index that combines the two mechanistic factors affecting diversity, i.e. dominance and species abundance which itself is the complement of Simpson's index.

Among-location differences in pollinator abundance, richness, diversity (Shannon-Wiener and Hurlbert indices) and plant reproductive success were analyzed with one-way ANOVA, followed by the multiple-comparison Least Significant Difference (LSD) test. Richness, dominance and diversity (Shannon-Wiener and Hurlbert indices) were compared among the locations using individual-based rarefaction curves generated by EcoSim (Gotelli & Entsminger 2005); the graph was drawn in PAST software (Hammer et al. 2001). Rarefaction allows for estimation of number of species (S) expected in a random samples of N individuals taken from a larger collection made up of N individuals and S species (Gotelli & Entsminger 2005).

The effect of pollinator diversity on plant reproductive success was explored with simple linear regression analysis between species richness, abundance and Shannon-Wiener index as predictors and components of reproductive success as responses.

RESULTS

A total of 800 insects belonging to 77 species in four orders were observed visiting the flowers of *P. juliflora* in the three isolated locations. The majority of species were Hymenoptera (41) and Diptera (17; Fig. 1). Only four species made up more than 6% of the total visits across all sites. These species were two long-tongued large bees (*Apis dorsata* and *A. florea*) and two syrphid flies (*Eristalinus aeneus* and *Ischiodon scutellaris*). Considering the seven most abundant floral visitors across all sites (Fig. 2), four species (*A. florea, A. dorsata, Megachile* sp. and *E. aeneus*) had long mouth parts (3.44, 6.14, 5.22 and 4 mm, respectively; Somanathan et al. 2009; Chaiyawong et al. 2004) and three species (*I. scutellaris, Episyrphus balteatus* and *Agapostemon* sp.) had shorter mouth parts (2-3 mm; Gilbert, 1981). Among these four most abundant floral visitors, *E. balteatus* and *A. dorsata* were the most abundant at Chichawatni (Appendix 1) while *E. aeneus, I. scutellaris* and *A. florea* at Khanewal. On the other hand, *E. aeneus* and *I. scutellaris* were not observed in Chak Katora, which was dominated by two bees *Megachile* sp. 2 and *Agapostemon* sp.

Pollinator assemblage structure was similar in three locations with few abundant species and high number of scarce species (Fig. 2), but the assemblage varied significantly in composition among the sites. The most abundant species in three locations also varied (Fig. 2). Eleven species (two fly, two bee, three wasp, two butterfly and two beetle species) were observed solely in Chichawatni (Appendix 1). Five species (one dipteran, one wasp, two butterfly and one beetle species) were restricted to Chak Katora while seventeen species (two flies, ten bees, three wasps, one butterflies and one moth) occurred only in Khanewal. Twelve species were commonly found in the three locations, i.e. four fly, four bee, one wasp and three butterfly species (Appendix 1).

At all three sites Hymenoptera was the largest order in terms of abundance and number of species followed by Diptera (Fig. 1). There was significant variation in pollinator abundance among sites in terms of visitation frequency ($F = 0.037$, $df = 126$, $P = 0.05$). The maximum visitation frequency (5.22 individuals per 30 minutes) was observed in Pirowal followed by Chichawatni and Chak-Katora (3.50 and 1.59 individuals per 30 minutes).

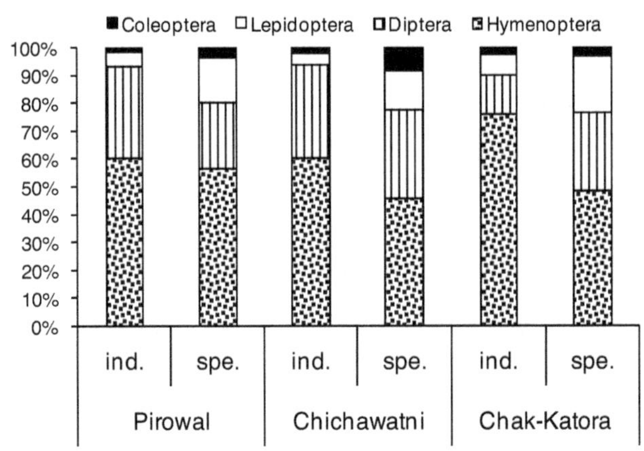

FIGURE 1. Percentage of species (spe.) and individuals (ind.) of related insect orders visiting *Prosopis juliflora* flowers at three locations in southern Punjab during 2008.

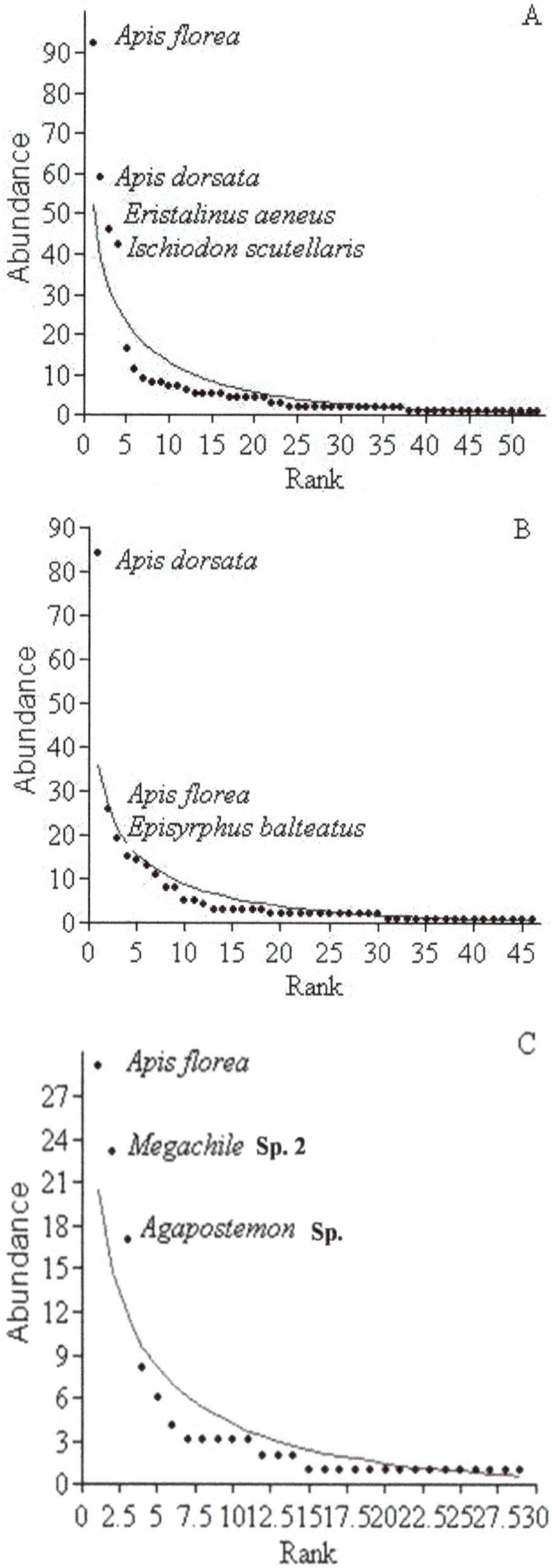

FIGURE 2. Rank-abundance (total numbers over complete study period) curves of insect species visiting flowers of Prosopis juliflora at three locations (A: Pirowal; B: Chichawatni; C: Chak-Katora) in Southern Punjab. The names of the species accounting for at least 10% of the visits at a given location are provided.

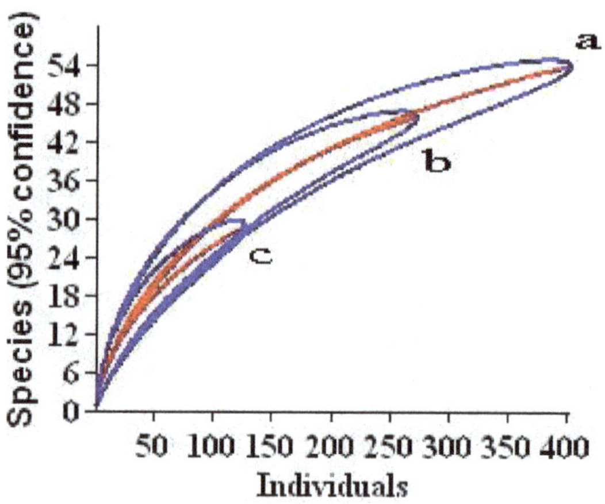

FIGURE 3. Rarefaction curves of three locations, i.e. (a) Pirowal (b) Chichawatni (c) Chak-Katora, based on individual rarefaction method showing the expected number of species as a function of sample size.

The rarefaction curves of three locations based on individual rarefaction method (using the expected number of species as a function of sample size) have not yet reached an asymptotic level so species richness can be expected to increase with sampling effort (Fig. 3). There was a significant difference ($F = 35.79$, df = 82.18, $P = 0.00$) in species richness among the three locations (Table 2) i.e. highest in Pirowal followed by Chichawatni and Chak-Katora. One-way ANOVAs showed that there were no significant differences in seeds per pod and seed weight of *P. juliflora* among the three sites, but there were significant differences in the number of pods per raceme and germination. Significantly more pods per raceme were recorded in Chichawatni and Pirowal than Chak-Katora ($F = 70.33$, df = 75, $P < 0.001$), whereas germination percentage showed a non-significant (df = 27, $F = 2.35$, $P = 0.114$) tendency being higher in Pirowal (51%) than Chichawatni (41.2) and Chak-Katora (40.8%). The number of pods per raceme and germination appeared pollinator limited (Table 3).

Since there were no significant differences in seeds per pod and seed weight among the three locations, we did not relate them to species abundance, richness and Shannon-Wiener index. Only significantly varying components were selected for that purpose, i.e. number of pods per raceme and germination. Abundance was positively related to the number of pods per flower and germination. Species richness and the Shannon-Wiener index (Fig. 4) were not related.

DISCUSSION

The pollination system of *P. juliflora* is extremely generalized. Its brush-shaped flowers were visited by 77 insect species with different morphology, size and behaviour from a variety of functional groups, i.e. short-and long-tongued bees, butterflies, flies and beetles. Previous studies

Locations	Abundance	Dominance	Shannon-Wiener	Species Richness	Hulbert
Pirowal	5.22 a	0.24 b	2.68 a	36.42 a	0.88 a
Chichawatni	3.50 ab	0.32 a	2.58 a	30.35 b	0.87 a
Chak-Katora	1.59 b	0.27 ab	2.31 b	18.95 c	0.89 a

TABLE 2. Among-location differences in pollinator abundance and diversity.

	No of pods/raceme	No of seed/pod	100-Seed weight (gm)	Germination (%)
Pirowal	5.829 a	10.77 a	2.767 a	51.00 a
Chichawatni	6.413 a	10.95 a	2.778 a	41.20 b
Chak-Katora	3.241 b	10.90 a	2.787 a	40.80 b

TABLE 3. Comparison of reproductive success of three locations.

Mean values sharing similar letters in respective columns show non-significant differences ($P < 0.05$) after LSD test.

have also failed to relate any particular functional group of insects to the evolution of brush-shaped flowers (Hingston & McQuillan 2000; Devy & Davidar 2003). Given the morphology of *P. juliflora* flowers, contact of insect visitors with flower's reproductive organs is unavoidable. Muzaffar and Ahmad (1991) noted *P. juliflora* as a major nectar and pollen resource for bees in Pakistan.

Prosopis juliflora is not native to the sub-continent, yet it received visits by a large array of potentially pollinating insects. Pollination success in *P. juliflora* is always low and very few legumes are produced despite large numbers of flowers per tree: DeOliveira & Pires (1990) reported a pollination efficiency of 29% in *P. juliflora* based on the number of inflorescences per tree. However, when related to the number of flowers, it dropped to 1.48%

The low fruiting success under open pollination as well as under exclusion of pollinators indicates that *P. juliflora* is more or less incapable of autonomous selfing (DeOliveira & Pires 1990). Another reason could be that many plant species mass-flower to increase floral display and abort most of their flowers and may not have enough resources to develop fruits from all these flowers (Trueman & Wallace 1999). The total number of seeds produced by one individual may, however, still be high enough for the species to survive (Karron & Mitchell 2011). Several reasons have been documented for the low reproductive success by DeOliveira & Pires (1990): poor pollen viability, short period of pollen release or stigma receptivity, lack of synchronization between pollen release and pollen reception, flower sterility or high rates of ovary abortion, few pollinating insects at the time of maximum receptivity. Goel & Behl (1995) found maximum pollen production in *P. juliflora* at midday, but insects are less mobile during high temperatures at this time (ca 40°C in this study).

The pollinator assemblage in rank-abundance curves revealed *A. dorsata* and *A. florea* as the most abundant species in all three locations. A single colony of *Apis* may have thousands of workers and trees provide ideal nesting locations, making them the most dominant visitors, as in case of this study. Both species have already been reported as the most dominant floral visitors of various crops (Sajjad et al. 2008; Sajjad et al. 2009; Ali et al. 2011) in the study location.

We also found a significant spatial variation among locations in pollinator assemblages in terms of species richness, diversity and dominance. The abundance-richness relationship is frequent in pollinator assemblages (Steffan-Dewenter et al. 2002), i.e. few abundant species and high number of scarce species in this study. The observed species richness ranged from 19 to 34 species. The highest species richness was recorded in the largest and least disturbed site, at Pirowal. Habitat quality includes factors that directly influence the life history of a pollinator species, i.e. availability of breeding and shelter places and host plants (Winfree et al. 2011). The surrounding quantity and quality of habitats is usually positively related to insect species richness and abundance (Oeckinger & Smith. 2007) but the discovery of these relationships may depend upon the scale of investigation.

Perennial plant species respond in different ways towards habitat fragmentation and it is difficult to find a clear rule in general (Donaldson et al. 2002), i.e. populations in small fragments do not always experience pollination deficits. However, many other studies suggest a better plant reproductive performance in larger reserves or mainland than smaller ones or on islands (Jennersten 1988; Cunningham 2000; Donaldson et al. 2002; Murren 2002).

Our data clearly suggests that for *P. juliflora,* the number of pods and germination rate of seeds rise with abundance of its flower visitors. Further, flower visitor abundance was a better predictor of plant reproductive performance than species richness and the Shannon-Wiener index. This might be because of less sensitivity of Shannon-Wiener index to actual site differences, i.e. the Shannon-Wiener index can be similar even if two sites are different in species richness. Only few studies give empirical evidence regarding consequences of pollinator assemblage on plant reproductive success concerning pollen limitation. Diversity and identity (Gomez et al. 2010), and relative abundance of floral visitors (Sahli and Conner 2006) have been reported as the important predictors of plant reproductive success. The sensitivity of pollinator assemblages to predict plant reproductive success

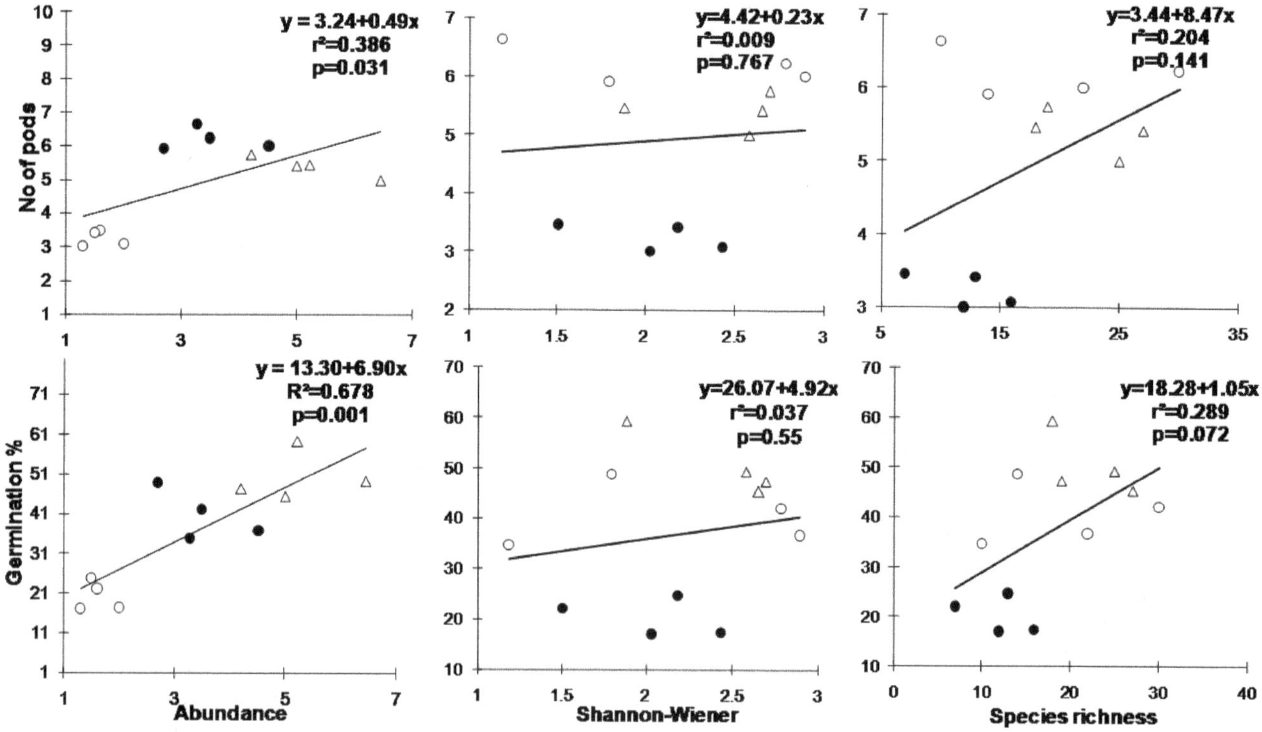

FIGURE 4. Relationship between flower visitor abundance, species richness and diversity (Shannon-Wiener) and plant reproductive performance measured as germination % (lower graphs) and fructification (number of pods, upper graphs) in *P. juliflora* at southern Punjab, Pakistan. Open triangles: Pirowal; Open circles: Chichawatni; Solid circles: Chak-Katora.

may vary with plant species and the pollination effectiveness of available pollinator species (Talavera et al. 2001).

In conclusion, flower visitor spectra of *P. juliflora* at Pirowal and Chichawatni were more species rich and abundant than at Chak-Katora possibly due to the former two sites being larger and least disturbed if not undisturbed compared to the latter. This ultimately affected insect visitation frequency and hence fruit set of *P. juliflora*. In contrast to flower visitor abundance, species richness and the Shannon-Wiener index were no good predictors of reproductive performance of *P. juliflora*.

APPENDIX

Additional supporting information may be found in the online version of this article:

APPENDIX I. Insects visiting *P. juliflora* from April to June 2008 in three selected locations.

ACKNOWLEDGEMENTS

This study was funded by Higher Education Commission of Pakistan. We are grateful to Dr. Claus Claussen (Twedter Holz 12 D-24944 Flensburg, Germany) and Dr. Andrew Whittington (Fly-evidence, Scotland) for their help in identification of many of the Diptera species and Dr. Ather Rafi (National Agriculture Research Centre, Pakistan) for butterfly species.

REFERENCES

Ali M, Saeed S, Sajjad A, Whittington A (2011) In search of the best pollinators for canola (*Brassica napus* L.) production in Pakistan. Applied Entomology and Zoology 46:353-361.

Bingham CT (1987) The Fauna of British India, including Ceylon and Burma, Hymenoptera. Vol. I Wasps and Bees. Taylor and Francies, London.

Borror DJ, Long DM, Triplehorn CA (1981) An introduction to the study of insects. Saunders College Publishing, Philadelphia.

Chaiyawong T, Deowanish S, Wongsiri S, Sylvester HA, Rinderer TE, Guzman LD (2004) Multivariate morphometric study of *Apis florea* in Thailand. Journal of Apicultural Research 43(3): 123-127.

Cunningham SA (2000) Depressed pollination in habitat fragments causes low fruit set. Proceedings of the Royal Society of London 567: 1149-1152.

De Oliveira VR, Pires IE (1990) Pollination efficiency of *P. juliflora* (Sw) DC in Petrolina, Pernambuco. In: The Current State of Knowledge on *P. juliflora*. Habit MA, Saavedra JC (eds), FAO, Rome, Italy, 233-239.

Devy MS, Davidar P (2003) Pollination systems of trees in Kakachi, a mid-elevation wet evergreen forest in Western Ghats, India. American Journal of Botany 90: 650-657.

Donaldson J, Nanni J, Zachariades C, Kemper J (2002) Effects of habitat fragmentation pollinator diversity and plant reproductive success in renosterveld shrublands of South Africa. Conservation Biology 16: 1267-1276.

Dubale A (2006) Impacts of *Prosopis juliflora* invasion and control using charcoal production in Afar National Regional State, Ethiopia. MSc thesis University of Wales, Bangor, UK.

Gilbert FS (1981) Foraging ecology of hoverflies: morphology of the mouthparts in relation to feeding on nectar and pollen in some common urban species. Ecological Entomology. 6: 245-262

Goel VL, Behl HM (1995) Propagation of *Prosopis juliflora* from rooted stem cuttings. International Tree Crops Journal 8: 193-201.

Gomez JM, Bosch J, Perfectti F, Fernandez J, Abdelaziz M (2007) Pollinator diversity affects plant reproduction and recruitments: the tradeoffs of generalization. Oecologia 153: 597-605.

Gomez JM, Abdelaziz M, Lorite J, Pajares AJM, Perfectti F (2010) Changes in pollinator fauna cause spatial variation in pollen limitation. Journal of Ecology 98: 1243-1252.

Gotelli NJ, Entsminger DF (2005) EcoSim, Null models software for ecology, v 7.72 Acquired Intelligence Inc and Kesey-Bearm, at http://www.homepages.together.net/~gentsmin/ecosim.htm

Hammer Ø, Harper DAT, Ryan PD (2001) PAST: Paleontological statistics software for education and data analysis. Paleontologia Electronica 4 (1): 9.

Herrera CM (2005) Plant generalization on pollinators, species property or local phenomenon? American Journal of Botany 92: 13-20.

Hingston AB, McQuillan PB (2000) Are pollination syndromes useful predictors of floral visitors in Tasmania? Austral Ecology 25: 600-609.

Hurlbert SH (1971) The nonconcept of species diversity: a critique and alternative parameters. Ecology 52: 577-586

Iqbal MZ, Shafiq M (1997) Seedling performance of two desert plant species (*Prosopis juliflora* and *Blepharis sindica*) grown under uniform edaphic conditions. Journal of Tropical Forest Science 9: 458-464.

Jennersten O (1988) Pollination in *Dianthus deltoides* (Caryophyllaceae): Effects of habitat fragmentation on visitation and seed set. Conservation Biology 2 (4): 359-366.

Jennersten O, Nilsson SG (1993) Insect flower visitation frequency and seed production in relation to patch size of *Viscaria vulgaris* (Caryophyllaceae). Oikos 68: 283-292.

Kallimanis AS, Petanidou T, Tzanopoulos J, Pantis JD, Sgardelis SP (2009) Do plant-pollinator interaction networks result from stochastic process? Ecological Modelling 220: 684-693.

Karron JD, Mitchell RJ (2011) Effects of floral display size on male and female reproductive success in Mimulus ringens. Annals of Botany doi: 10.1093/aob/mcr193

Khan SU, Hassan M, Khan FK, Bari A. (2010) Climate classification of Pakistan. Balwois 2010 Conference, Ohrid, Republic of Macedonia, http://www.balwois.com/balwois/administration/full_paper/ffp-1295.pdf: 1-47 (Date of access: 20 Aug. 2011)

Kleijn D, Langevelde FV (2006) Interacting effects of landscape context and habitat quality on flower visiting insects in agricultural landscapes. Basic and Applied Ecology 7: 201-214.

Luna R K (1996) *Prosopis juliflora* (Swartz) DC. In: Plantation Trees. International Book Distributors, Delhi,

India.

Magurran AE (2004) Ecological Diversity and Measurements. 2nd edn. Princeton University Press, Princeton, N.J.

McCall C, Primack RB (1992) Influence of flower characteristics, weather, time of day, and season on insect visitation rates in three plant communities. American Journal of Botany 79: 434-442.

Michener CD (2000) The Bees of the World. John Hopkins University Press, Baltimore.

Moeller DA, Tiffin P (2005) Genetic diversity and the evolutionary history of plant immunity genes in two species of *Zea*. Molecular Biology and Evolution 22: 2480-2490.

Murren OJ (2002) Effect of habitat fragmentation on pollination: pollinators, pollinia viability and reproductive success. Journal of Ecology 90: 100-107.

Muzaffar N, Ahmad R (1991) Some honeybee flora for *Apis* spp. in Pakistan. Pakistan Journal of Zoology 23:201-218.

Mwangi E and Swallow B (2005) Invasion of *Prosopis juliflora* and local livelihoods: Case study from the lake Baringo area of Kenya. ICRAF Working Paper -no. 3, Nairobi, World Agroforestry Centre.

Ockinger E, Smith HD (2007) Semi-natural grasslands as population sources for pollination insects in agricultural landscapes. Journal of Applied Ecology 44: 50-59.

Olesen JM, Jordano P (2002) Geographic patterns in plant-pollinator mutualistic networks. Ecology 83: 2416-2424.

Ollerton J, Cranmer L (2002) Latitudinal trends in plant-pollinator interactions: are tropical plants more specialized? Oikos 98: 340-350.

Pasiecznik NM (2001) Managing *Prosopis juliflora* for timber production in arid zones. In: Neelakantan KS *et al.*, eds. Management and Utilisation of *Prosopis juliflora* - Training Manual, pp85-90. Forest College and Research Institute, Tamil Nadu Agricultural University, Mettupalayam, India.

Pasiecznik NM, Felker P, Harris PJC, Harsh LN, Cruz G., Tewari JC, Cadoret K, Maldonado LJ (2001) The *Prosopis juliflora* - *Prosopis pallida* complex: a monograph. HDRA, Coventry, UK. 162pp.

Rathcke BJ, Jules ES (1993) Habitat fragmentation and plant-pollinator interactions. Current Science 65: 273-277.

Richardson DM, Allsopp N, D'Antonio CM, Milton SJ, Rejmanek M (2000) Plant invasions – the role of mutualisms. Biological Reviews 75: 63-99.

Root RB (1973) Organization of a plant-arthropod association in simple and diverse habitats: the fauna of collards (*Brassica oleraceae*). Ecological Monographs 43: 95-124.

Sajjad A, Saeed S, Masood A (2008) Pollinator community of onion (*Allium cepa* L.) and its role in crop reproductive success. Pakistan Journal of Zoology 40(6):451-456.

Sajjad A, Saeed S, Muhammad W, Arif MJ (2009) Role of insects in cross-pollination and yield attributing components of *Susbania sesban*. International Journal of Agriculture and Biology 11(1): 77-80.

Sahli HF, Conner JK (2006) Characterizing ecological generalization in plant-pollinator systems. Oecologia 148:365-372.

Saunders DA, Hobbs RJ, Margules CR (1991) Biological consequences of ecosystem fragmentation: a review. Conservation Biology 5: 18-32.

Simpson BB (1977) Mesquite, its Biology in Two Desert Shrub Ecosystems. Dowden, Hutchinson and Ross, Stroudsberg, Pennsylvania, USA.

Somanathan H, Warrant EJ, Borges RM, Wallen R, Kelber A (2009) Resolution and sensitivity of eyes of the Asian honeybees *Apis florea*, *Apis cerana* and *Apis dorsata*. The Journal of Experimental Biology 212: 2448-2453.

Steffan-Dewenter I, Tscharntke T (1999) Effects of habitat isolation on pollinator communities and seed set. Oecologia 121: 432-440.

Steffan-Dewenter I, Munzenberg U, Burger C, Thies C, Tscharntke T (2002) Scaled-dependent effects of landscape context on three pollinator guilds. Ecology 83: 1421-1432.

Talavera S, Bastida F, Ortiz PL, Arista M (2001) Pollinator attendance and reproductive success in *Cistus libanotis* L. (Cistaceae). International Journal of Plant Sciences 162:343-352.

Truman ST, Wallace HM (1999) Pollination and resource constraints on fruit set and fruit size of *Persoonia rigida* (Proteaceae). Annals of Botany 83: 145-155.

Van Treuren R, Bijlsma R, Ouborg NJ, Kwak MM (1994) Relationship between plant density, outcrossing rates and seed set in natural and experimental populations of *Scabiosa columbaria.* Journal of Evolutionary Biology 7: 287-302.

Ward CR, O'Brien CW, O'Brien LB, Foster DE, Huddleston EW (1977) Annotated Checklist of New World Insects Associated with Prosopis (Mesquite). Technical Bulletin Vol. 1557. United States Department of Agriculture, USA.

Winfree R, Bartomeus I, Cariveau DP (2011) Native pollinators in anthropogenic habitats. Annual Review of Ecology, Evolution and Systematic 42:1-22.

Zaitoun S, Al-Ghzawi AA, Samarah N, Mullen R (2009) Pod production of *Prosopis juliflora* (Sw.) DC. As affected by supplementary and honeybee pollination under arid conditions. Acta Agriculturae Scandinavica 59: 349-356.

11

SIT-AND-WAIT POLLINATION IN THE SPRING FLOWERING WOODLAND PLANT, *TRILLIUM GRANDIFLORUM*

Emily S. Darling[1,]* and Spencer C.H. Barrett[2]

[1]*Department of Biological Sciences, Simon Fraser University, 8888 University Drive, Burnaby, British Columbia, Canada V5A 1S6*
[2]*Department of Ecology and Evolutionary Biology, University of Toronto, 25 Willcocks Street, Toronto, Ontario, Canada M5S 3B2*

Abstract—In animal-pollinated plants, reproductive success is commonly limited by pollen availability, which can occur in environments where pollinator activity is scarce or variable. Extended floral longevity to maximize a plant's access to pollinators may be an adaptation to such uncertain pollination environments. Here, we investigated the effects of flower exposure time to pollinators on female fertility (fruit and seed set) in the bee-pollinated woodland herb *Trillium grandiflorum,* a species with long-lived flowers (~17-21 d) that blooms in early spring when pollinator activity is often variable. We experimentally exposed flowers to pollinators for different amounts of time to determine the extent to which floral longevity influenced reproductive success. The amount of time that flowers were exposed to pollinators significantly increased fruit set and seed set per flower, but not seed set per fruit. Our results provide experimental evidence that long floral life spans may function as a 'sit-and-wait' pollination strategy to increase the amount of exposure time to pollinators and promote seed set in the unpredictable pollination environments often experienced by early spring ephemerals. In large populations with infrequent pollinator visitation, as commonly occurs in *T. grandiflorum,* pollination may be a largely stochastic process.

Keywords: Flower longevity, pollinator exposure, pollen limitation, stochastic pollination environment, bee pollination

INTRODUCTION

The reproductive success of plants is often limited by pollen availability (reviewed in Burd 1994; Larson & Barrett 2000; Ashman et al. 2004; Knight et al. 2005). This occurs when the quantity and/or quality of pollen that a plant receives during pollination is insufficient to fertilize available ovules, resulting in a reduction in fruit and/or seed production (Burd 1995; Aizen & Harder 2007). The fitness consequences of pollen limitation in variable pollination environments are diverse, and chronic pollen limitation can affect the evolution of a range of life history and reproductive traits in plant populations (Haig & Westoby 1988; Ashman et al. 2004; Morgan et al. 2005; Porcher & Lande 2005; Harder & Aizen 2010). Extended floral longevity is one trait that can increase opportunities for pollinators to visit flowers and buffer fertility in stochastic environments with few pollinators (Kerner von Marilaun 1895; Primack 1985; Ashman & Schoen 1994; Charnov 1996). Theoretical models have examined the factors influencing optimal floral longevities (Ashman & Schoen 1994; Schoen & Ashman 1995), but relatively few studies have experimentally examined the direct consequences of floral longevities on female reproductive success (but see Ashman & Schoen 1997; Rathcke 2003; Alonso 2004).

Early spring-flowering woodland herbs in temperate climates often persist in variable pollination environments characterized by low temperatures and uncertain pollinator service (Schemske et al. 1978; Motten 1986; Barrett & Helenurm 1987). *Trillium grandiflorum* (Melanthiaceae) is a self-incompatible woodland herb in eastern North America that depends on bee pollinators (mostly *Bombus* spp.) for pollen-transfer (Broyles et al. 1997; Sage et al. 2001) and populations are often pollen limited (Lubbers & Lechowicz 1989; Wright & Barrett 1999; Irwin 2000; Griffin & Barrett 2002). Flowers of *T. grandiflorum* are exceptionally long-lived (17-21 d; Ashman & Schoen 1996; Sage et al. 2001) and the goal of our study was to investigate if long floral lifespan acts as a 'sit-and-wait' strategy (e.g. Schoener 1971) to increase the chances of a pollinator visit and female reproductive success. To test this hypothesis, we experimentally manipulated the exposure time of flowers to pollinators as a proxy for floral lifespan. Our study had three objectives: (1) To characterize the pollination environment of *T. grandiflorum* by quantifying pollinator activity and ambient temperature during the flowering season, (2) To determine the extent to which fruit and seed set per fruit in *T. grandiflorum* may be pollen limited, and (3) To investigate the importance of exposure time to pollinators on these components of reproductive success.

MATERIALS AND METHODS

Study organism and site

Trillium grandiflorum is a perennial, non-clonal herb that occurs in the deciduous forest understory of eastern North America (Case & Case 1997). Populations flower for approximately three weeks in early spring and are pollinated largely by pollen-foraging bees (Case & Case 1997; Irwin 2000; Sage et al. 2001). Flowers produce no nectar and are weakly protandrous (Irwin 2000; S.C.H. Barrett unpubl.

*Corresponding author; email: edarling@sfu.ca

observations). Each plant produces a single flower with on average 25.8 ± SEM 0.9 ovules, $n = 142$, at our study site. We performed our experiments between May and June 2005 at a large population of several thousand plants covering approximately 1 ha at the Koffler Scientific Reserve at Joker's Hill in southern Ontario, Canada (44.03°N, 79.53°W; http://www.zoo.utoronto.ca/jokershill/jh.html). At this location, *T. grandiflorum* is self-incompatible and predominantly outcrossing ($t = 0.68 \pm 0.09$, Sage et al. 2001), as has been reported in other populations of the species (Broyles et al. 1997; Kalisz et al 1999).

Pollination environment

To estimate pollinator activity, we recorded the number of bees that were active during flowering within the study population. Queens of *Bombus* spp. are the primary pollinator of *Trillium grandiflorum* at our study site (Griffin & Barrett 2002) although *Andrena carlini* Cockerell was also observed visiting flowers. We recorded the number of bee flights on six transects (~100 m each) in the population for a total of 44.5 h over 18 d throughout the flowering season, an average of 2.5 hours per day. We conducted pollinator observations between mid-morning and mid-afternoon when bee activity is highest. A bee flight was considered to be any time we observed an individual flying within the population, not necessarily visiting flowers. We were careful not to recount individual pollinators; this was possible due to the low pollinator activity at the site (0.08 visits/h, see also Irwin 2000). We therefore consider our observations to reflect the potential for pollination and we assume that the amount of effective pollination is likely to increase with the number of bees active in the community.

We obtained average daily temperatures (°C) from the nearest Environment Canada weather station located at Buttonville, Ontario (43.52°N, 79.22°W), ~20 km east of the study site. We evaluated the relationships between temperature, bee abundance and the number of days since the beginning of flowering (anthesis). We used Pearson partial correlations (ρ) of temperature and pollinator abundance while controlling for days since anthesis to account for the potential non-independence of daily temperature estimates (temporal autocorrelation).

Measurement of pollen limitation

We performed a pollen supplementation experiment on *T. grandiflorum* to investigate if female fertility at our site was pollen limited. We selected 40 plants of similar size and randomly assigned each one to either an open- or hand-pollination treatment. For the hand-pollinated treatment, we applied fresh pollen from three different donor plants that were located more than 1 m but less than 10 m away from the focal plant to limit effects of biparental inbreeding. Pollen was applied in a single application to stigmas and flowers began to senesce and turn pale pink several days after supplemental pollination (E. S. Darling and S.C.H. Barrett, pers. obs.). Open-pollinated plants were left unmanipulated for the entire duration of the flowering season, which in 2005 was 25 days.

In mid-July, we determined whether each flower had set fruit and collected all fruits. We counted the number of mature seeds, shrunken seeds and apparently unfertilized ovules under a dissecting microscope (see Griffin & Barrett 2002). We calculated fruit set as the proportion of individual plants producing a fruit with mature seeds. Seed set (the proportion of mature seeds out of the total number of ovules) was calculated per fruit and per flower. Seed set per fruit considered only the subset of flowers that had set fruit with at least one mature seed, while seed set per flower considered all flowers in the sample, regardless of whether they produced seed. We used Student *t*-tests to compare female fertility between open-pollinated and hand-pollinated treatments.

Determinants of female reproductive success

We tagged 120 individuals when flowers were in bud and recorded the approximate start date of anthesis (the beginning of flowering) for each plant, judged by when flowers were fully open and pollen was visible on anthers. At anthesis, we measured the height of the flower from the ground (hereafter plant height) and the density of reproductive individuals in a 1m² area around the focal plant (neighborhood flower density). We used plant height as an indicator of investment in floral attractive structures since this measure has been shown to correlate with both flower biomass ($r = 0.51$, $P < 0.0001$) and pollen number ($r = 0.93$, $P < 0.0001$) in this population (Wright & Barrett 1999).

To evaluate the extent to which floral longevity contributes towards reproductive success, we manipulated the amount of time that *T. grandiflorum* flowers were exposed to pollinators. Prior to flowering, we randomly assigned the 120 tagged individuals to nine exposure times ranging between two and 25 days. At the end of each exposure time, flowers were emasculated and covered with bridal veil bags to prevent further pollination. We treated exposure time as a proxy for flower lifespan and we assume stigmas are receptive during the entire period that flowers are exposed to pollinators. Previous studies in the same population support this assumption and found no effect of pollination timing or flower age on fruit and seed set (Griffin & Barrett 2002). We calculated pollinator availability for each flower as the mean number of observed pollinators day^{-1} during the days the flower was exposed to pollinators. We collected mature fruits in mid-July 2005 and calculated fruit set and seed set per fruit (seed set for the subset of flowers with at least one mature seed) and per flower (seed set for all flowers) as described above.

We evaluated the relationship between exposure time and three measures of female reproductive success, fruit set, seed set per fruit and seed set per flower using linear relationships; exposure time was log-transformed prior to analysis. We evaluated the effect of plant height, neighbourhood flower density, pollinator availability and exposure time on seed set per flower using a general linear model (GLM) with a quasibinomial distribution to account for overdispersion (Zuur et al. 2009). We checked for collinearity between predictor variables using variance inflation factors (VIFs); the four predictor variables had

VIFs < 2 and were all included in the full model (Zuur et al. 2010). Model diagnostics were performed by visually evaluating the distribution of the residuals. All analyses were performed in R (version 2.11.1; www.r-project.org).

Results

Pollination environment

Flowering of *Trillium grandiflorum* in 2005 in our study population began on 10 May and continued until 4 June. Bees were observed on ten days during the flowering season with none observed during the first week of flowering (10 - 17 May). The average daily number of bees observed was 2.22 ± 2.85 (mean ± SEM) and a total of 40 separate observations of *Bombus* queens and *Andrena carlini* were made in the population during the flowering season. The vast majority of bee flights did not include flower visits and we recorded only four visits to flowers in which pollination is likely to have occurred over 44.5 hours of observation.

There was an increasing trend in mean daily temperature and bee observations throughout the flowering season (Fig. 1). Pollinator abundance was correlated with daily temperature ($r = 0.70$, $P < 0.01$) and the number of days since the beginning of flowering ($r = 0.65$, $P < 0.01$). After accounting for temporal autocorrelation across the flowering

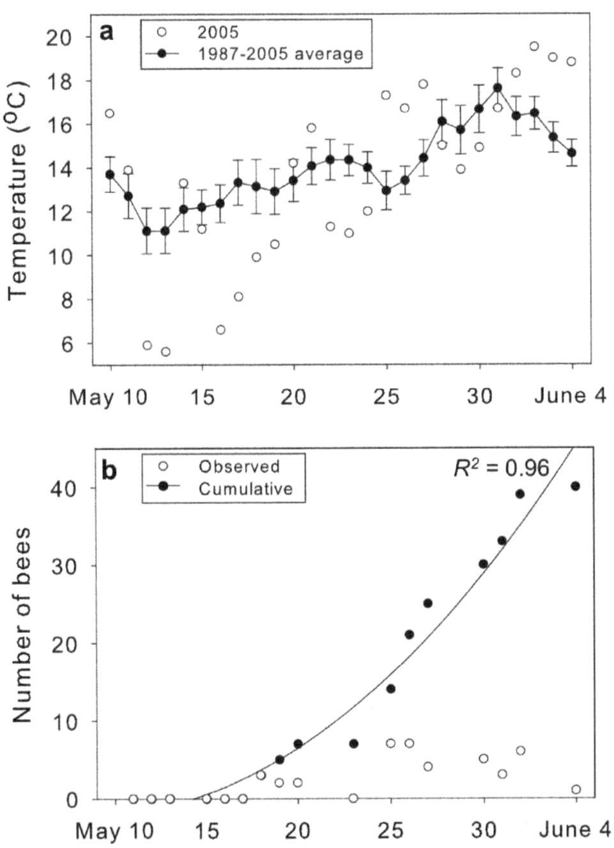

FIGURE 1. Conditions that favour pollination of *Trillium grandiflorum* increase throughout the flowering season. (a) Daily temperature (°C) in 2005 (open circles) and historical daily temperatures (mean ± SEM) between 1987 and 2005 (filled circles). (b) The number of observed (open circles) and cumulative (filled circles) bee pollinators.

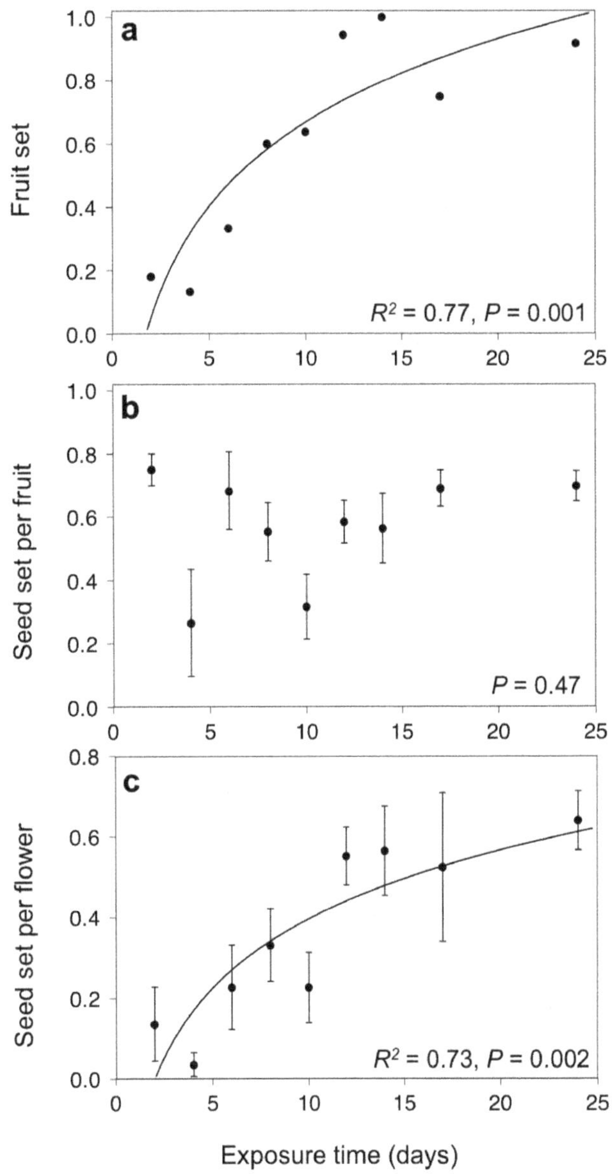

FIGURE 2. Reproductive success increases with exposure time in *Trillium grandiflorum*. There was a significant positive relationship between exposure time and (a) fruit set, and (c) mean seed set per flower, but not (b) mean seed set per fruit. Seed set values are based on seed:ovule ratios and are averages of individual plants within each exposure group treatment ± SEM. Sample sizes at exposure times of 2, 4, 6, 8, 10, 12, 14, 17, 24 d were 2, 2, 4, 9, 9, 17, 6, 5, 11 ($n = 65$) for fruits, and 11, 15, 12, 15, 13, 18, 6, 6, 12 ($n = 108$) for flowers, respectively.

season (number of days since anthesis), pollinator abundance remained significantly correlated with temperature ($\rho = 0.54$, $P = 0.03$). There was no correlation between pollinator abundance and days from the start of the flowering season after the effect of temperature was removed ($\rho = 0.39$, $P = 0.14$).

Pollen limitation

Despite low pollinator activity, supplemental hand pollinations provided no evidence that plants of *T. grandiflorum* were pollen limited in 2005. There were no significance differences between the open-pollinated (OP) and hand-pollinated (HP) treatments in fruit set (OP = 0.79, HP = 1.0; $t_{df=24}$ = 1.08, P = 0.86), seed set per fruit (mean ± SEM: OP = 0.70 ± 0.05, HP = 0.62 ± 0.06; $t_{df=24}$ = -0.571, P = 0.29), or seed set per flower (OP = 0.55 ± 0.09, HP = 0.62 ± 0.06; $t_{df=24}$ = 0.202, P = 0.58).

Relation between exposure time and reproductive success

Of the 120 plants that were tagged at the beginning of the experiment, 65 plants set fruit with at least one mature seed; 12 plants could not be recovered at the end of the flowering season. Fruit set was reduced by 27-89% in flowers with experimentally shortened exposure times. Both fruit set (Fig. 2a) and seed set per flower (seed:ovule ratio, Fig. 2c) increased significantly with exposure time to pollinators (Fig. 2c), however seed set per fruit was unrelated to exposure time (Fig. 2b). There was a significant effect of exposure time (quasibinomial GLM: t = 2.80, P = 0.006) and the number of available pollinators (t = 2.87, P = 0.005) on seed set per flower, but no significant effect of plant height (t = 0.60, P = 0.55) or flower density (t = 0.79, P = 0.43).

Discussion

Extended floral longevity in *Trillium grandiflorum* may be an adaptation to uncertain pollination environments characterized by daily variation in temperature and pollinator availability. Flowers with longer exposure times had a greater opportunity to be visited by pollinators and higher reproductive success than flowers with experimentally shortened exposure times. Individuals with flower life spans shorter than 10 days suffered significantly lower fertility, likely because of the reduced probability of visitation by pollinators. Extended floral longevity in *T. grandiflorum* thus increases the chances of a plant obtaining a pollinator visit, especially when there are few pollinators available in large populations of many flowering individuals. Under these circumstances, pollination is likely to be a largely stochastic process with many plants receiving no visitation in some flowering seasons. We characterize this process as 'sit-and-wait pollination', similar to the sit-and-wait model of animal foraging that occurs in environments characterized by uncertain rewards (Schoener 1971; Perry & Pianka 1994).

A patchy and uncertain pollinator environment can affect two different aspects of female reproductive success – percent fruit set and the number of seeds produced per fruit. Although these aspects of female fertility are often assumed to be roughly correlated (e.g. populations with high fruit set produce more seeds per fruit), the processes involved in fruit and seed set can be quite different. For example, in *T. grandiflorum* plants with longer exposure times to pollinators had higher fruit set than those with experimentally-shortened exposure times. However, there was no effect of exposure time on seed set per fruit. Thus, reproductive success in this species appears to be determined not by the total number of visits a flower receives, but rather whether individuals are visited by a single pollinator or not. This can occur when there is a low abundance of pollinators but a large number of flowers available for pollination. Furthermore, the large body size of *Bombus* queens, the primary pollinator at our site, likely facilitates the deposition of substantial numbers of pollen grains to the exposed stigmas of *T. grandiflorum* during a single visit resulting in maximal seed set.

Experimental evidence supports the idea that pollination in *T. grandiflorum* is an "all or nothing" phenomenon. Studies using genetic markers have demonstrated that most flowers of *T. grandiflorum* are pollinated by a single male parent (Broyles et al. 1997; Kalisz et al. 1999), a result consistent with very low visitation rates. Moreover, controlled pollination studies using different number of pollen donors indicate that there is no fertility benefit to receiving multiple pollinator visits (Griffin & Barrett 2002). Theoretical models of optimal floral longevity assume that female reproductive success is gradually accrued throughout the flowering season as a result of multiple visits to individual flowers (Ashman & Schoen 1994, 1996; Schoen & Ashman 1995). However, our results instead suggest that for systems with low densities of pollinators, stochastic pollination is more likely to characterize the pollination process and species may have evolved traits that maximize fitness from single pollinator visits (Harder & Wilson 1992; Burd 1995).

Despite the adaptive advantages to increased flower longevity in a stochastic pollination environment, why do flowers of *T. grandiflorum* open early during unfavourable pollinator conditions and not instead open later in the spring during more favourable conditions when pollinators are more likely to be abundant? Local weather conditions and the temporal dynamics of pollinator densities can vary from year to year (e.g. Schemske et al. 1978; Barrett & Helenurm 1987; Goodwillie 2001) and long floral life spans may serve as a bet-hedging strategy to reduce the risk of reproductive failure. Daily temperatures in 2005 at our study population were generally lower than the 19-year (1987-2005) average. This variation in the spring environment may explain why the perennial *T. grandiflorum* continues to flower early. While some years have poor early spring conditions for pollination, warm temperatures and early pollinator emergence in the other years can provide high reproductive success (e.g. bet-hedging, Schemske et al. 1978). Studies across multiple seasons are required to investigate the hypothesis that variation in the early spring environment affects reproductive success. Such annual environmental variation in pollination conditions may explain why studies of pollen limitation at our site have given contrasting results; we found no evidence for pollen limitation in 2005, however pollen limitation has been previously reported in this population in both 1998 (Wright & Barrett 1999) and 1999 (Griffin & Barrett 2002). Sit-and-wait pollination through extended floral longevity may be an adaptation to stochastic pollination environments that are common to early spring-blooming species.

Acknowledgements

The authors thank Jannice Friedman for field assistance and Lawrence Harder, Jannice Friedman, Mark Johnson and several anonymous reviewers for their comments on the manuscript and Laurence Packer for identifying *Andrena carlini*. This research was funded by a Natural Sciences and Engineering Research Council (NSERC) Undergraduate Student Research Award to ESD and an NSERC Discovery Grant to SCHB.

REFERENCES

Aizen MA, Harder LD (2007). Expanding the limits of the pollen-limitation concept: effects of pollen quantity and quality. Ecology 88:271-281

Alonso C (2004) Early blooming's challenges: extended flowering season, diverse pollinator assemblage and the reproductive success of gynodioecious *Daphne laureola*. Annals of Botany 93:61-66

Ashman, T-L, Schoen DJ (1994) How long should flowers live? Nature 371:788-791

Ashman, T-L, Schoen DJ (1996) Floral longevity: Fitness consequences and resource costs. In: Lloyd DG, Barrett SCH (eds) Floral Biology: Studies of Floral Evolution in Animal-pollinated Plants. Chapman and Hall, New York, pp 112-139

Ashman, T-L, Schoen DJ (1997) The cost of floral longevity in *Clarkia tembloriensis*: an experimental investigation. Evolutionary Ecology 11:289-300

Ashman, T-L, Knight TM, Steets JA, Amarasekare P, Burd M, Campbell DR, Dudash MR, Johnston MO, Mazer SJ, Mitchell RJ, Morgan MT, Wilson WG (2004) Pollen limitation of plant reproduction: ecological and evolutionary causes and consequences. Ecology 85:2408-2421

Barrett SCH, Helenurm K (1987) The reproductive biology of boreal forest herbs: I. Breeding systems and pollination. Canadian Journal of Botany 65:2036-2046

Broyles SB, Sherman-Broyles SL, Rogati P (1997) Evidence of outcrossing in *Trillium erectum* and *Trillium grandiflorum* (Liliaceae). Journal of Heredity 88:325-329

Burd M (1994) Bateman's principle and plant reproduction: the role of pollen limitation in fruit and seed set. The Botanical Review 60:83-139

Burd M (1995) Ovule packaging in stochastic pollination and fertilization environments. Evolution 49:100-109

Case FW, Case RB (1997) Trilliums. Timber Press Inc., Portland, Oregon

Charnov EL (1996) Optimal floral longevities. Evolutionary Ecology 10:245-248

Goodwillie C (2001) Pollen limitation and the evolution of self-compatibility in *Linanthus* (Polemoniaceae). International Journal of Plant Sciences 162:1283-1292

Griffin SR, Barrett SCH (2002) Factors affecting low seed:ovule ratios in a spring woodland herb, *Trillium grandiflorum* (Melanthiaceae). International Journal of Plant Sciences 163:581-590

Harder LD, Wilson WG (1992) Floral evolution and male reproductive success – optimal dispensing schedules for pollen dispersal by animal-pollinated plants. Evolutionary Ecology 8:542-559

Harder LD, Aizen MA (2010) Floral adaptation and diversification under pollen limitation. Philosophical Transactions of the Royal Society of London Series B. 365:529-543

Haig D, Westoby M (1988) On limits to seed production. American Naturalist 131:757-759

Irwin RE (2000) Morphological variation and female reproductive success in two sympatric *Trillium* species: evidence for phenotypic selection in *Trillium erectum* and *Trillium grandiflorum* (Liliaceae). American Journal of Botany 87:205-214

Kalisz S, Hanzawa FM, Tonsor SJ, Thiede DA, Voigt S (1999) Ant-mediated seed dispersal alters patterns of relatedness in a population of *Trillium grandiflorum*. Ecology 80:2620-2634

Kerner von Marilaun A (1895) The natural history of plants, their forms, growth, reproduction, and distribution. Henry Holt, New York

Knight TM, Steets JA, Vamosi JC, Mazer SJ, Burd M, Campbell DR, Dudash MR, Johnston, MO, Mitchell RJ, Ashman T-L (2005) Pollen limitation of plant reproduction: Pattern and process. Annual Review of Ecology, Evolution and Systematics 36:467-497

Larson BMH, Barrett SCH (2000) A comparative analysis of pollen limitation in flowering plants. Biological Journal of the Linnean Society 69:503-520

Lubbers AE, Lechowicz MJ (1989) Effects of leaf removal on reproduction vs. belowground storage in *Trillium grandiflorum*. Ecology 70:85-96

Morgan MT, Wilson WG, Knight TM (2005) Pollen limitation, population dynamics, and the persistence of plant populations: roles for pollinator preference and reproductive assurance. American Naturalist 166:169-183

Motten AF (1986) Pollination ecology of the spring wildflower community of a temperate deciduous forest. Ecological Monographs 56:21–42

Perry G, Pianka ER (1994) Animal foraging: past, present and future. Trends in Ecology and Evolution 12:360-364

Porcher E, Lande R (2005) The evolution of self-fertilization and inbreeding depression under pollen discounting and pollen limitation. Journal of Evolutionary Biology 18:497-508

Primack RB (1985) Longevity of individual flowers. Annual Review of Ecology and Systematics 16:15-37

Rathcke BJ (2003) Floral longevity and reproductive assurance: seasonal patterns and an experimental test with *Kalmia latifolia* (Ericaceae). American Journal of Botany 90:1328-1332

Sage TL, Griffin SR, Pontieri V, Drobac P, Cole WW, Barrett SCH (2001) Stigmatic self-incompatibility and mating patterns in *Trillium grandiflorum* and *Trillium erectum*. Annals of Botany 88:829-841

Schemske DW, Wilson MF, Melampy MB, Miller LJ, Verner L, Schemske KM, Best LB (1978) Flowering ecology of some spring woodland herbs. Ecology 59:351-366

Schoen DJ, Ashman T-L (1995) The evolution of floral longevity: resource allocation to maintenance versus construction of repeated parts in modular organisms. Evolution 49:131-139

Schoener TW (1971) Theory of feeding strategies. Annual Review of Ecology and Systematics 2:369-404

Wright SI, Barrett SCH (1999) Size-dependent gender modification in a hermaphroditic perennial herb. Proceedings of the Royal Society of London Series B: Biological Sciences 266:225-232

Zuur AF, Ieno EN, Walker NJ, Saveliev AA (2009) Mixed Effects Models and Extensions in Ecology with R. Springer, New York

Zuur AF, Ieno EN, Elphick CS (2010) A protocol for data exploration to avoid common statistical problems. Methods in Ecology and Evolution 1:3-14

12

STIMULATION OF FLOWER NECTAR REPLENISHMENT BY REMOVAL: A SURVEY OF ELEVEN ANIMAL-POLLINATED PLANT SPECIES

Elaine Y. Luo[1], Jane E. Ogilvie*[1,2], James D. Thomson[1,2]

[1]*Department of Ecology and Evolutionary Biology, University of Toronto, 25 Harbord Street, Toronto, ON, M5S 3G5, Canada*
[2]*Rocky Mountain Biological Laboratory, Post Office Box 519, Crested Butte, CO 81224-0519, USA*

Abstract—Understanding the interaction between reward-seeking flower feeding animals and plants requires consideration of the dynamic nature of nectar secretion. Studies on several plants suggest that nectar secretion may increase in response to its removal, but it is not clear whether the phenomenon is widespread. We determined whether 11 species of Colorado mountain wildflowers showed removal-enhanced nectar replenishment (RENR). We measured floral phenology, nectar volumes, rate of replenishment, and compared the cumulative nectar produced following five hourly removals with that accumulated after five hours. Nectar replenishment occurred rapidly, within minutes; statistically significant RENR was observed in 9 of our 11 study species, with the strongest effects in bee-pollinated species. We discuss the implications of RENR in plant species on the measurement of nectar, the adaptive advantage of RENR, and the energetic costs of RENR.

Keywords: Bumble bee, hummingbird, nectar production, pollination, Rocky Mountain Biological Laboratory

INTRODUCTION

The study of nectar secretion schedules is central to pollination ecology. From a plant's viewpoint, flower nectar schedules have the potential to influence the type, schedule, and behaviour of pollinator visits, ultimately affecting pollen transfer and fitness. From a flower-feeding animal's viewpoint, flower nectar schedules have the potential to influence foraging strategies in time and space, and ultimately energetic returns and fitness. For the study of plant-pollinator energetics in particular, some early treatments extolled the relative simplicity of measuring floral nectar and its sugar content. For example, Heinrich (1983, p. 275) considered the economics of bumble bees to be especially tractable because, "they forage for essentially pure energy resources that are *easy to quantify in the lab and in the field*" (emphasis ours). Indeed, by extracting a flower's nectar with a capillary tube at one time point and estimating its sugar content by refractometry, one can establish both volume and approximate energy value in less than a minute. This is considerably easier than determining the dietary value of a leaf or a seed. However, actually calculating the energetic return that a forager can expect while nectar-feeding at a particular flower species is complicated by the variable and dynamic nature of nectar secretion. A flower's nectar characteristics can vary with many intrinsic and extrinsic factors including, but not limited to, flower age and sexual phase, position on a plant, surrounding microenvironment, and time of day or season (reviewed in Pacini & Nepi 2007; Willmer 2011). Furthermore, flowers can dynamically control nectar production: both volume and sugar content can be responsively secreted or reabsorbed by nectaries over the lifetime of a flower (e.g., Nepi et al. 2011).

Nectar secretion can be responsive, in particular, to nectar removal by flower visitors (Castellanos et al. 2002; Ordano & Ornelas 2004). In this paper, we are concerned with removal-enhanced nectar replenishment (RENR), in which a flower that is visited repeatedly over a time period produces more total nectar than a flower visited only once at the end of the period. Particularly when multiple flower-feeders visit flowers at short intervals, and when pollen transfer increases with multiple visits to a flower, RENR seems likely and advantageous for plant fitness, at least to some point (Castellanos et al. 2002). If RENR occurs, a flower's nectar "reward" cannot be considered a simple phenotypic entity like a seed or a leaf; because it interacts dynamically with the behaviour of flower-feeders, it must be treated much as a behavioural variable itself. The plant's constitution and physiology may establish characteristic potentials and limits, but the actual expression of the trait depends on circumstances. The desirability of considering this dynamism has been pointed out for a long time (e.g., Cruden & Hermann 1983).

There is a small but growing literature on the effect of nectar extractions on the total nectar production of flowers. Multiple nectar removals can increase (e.g., Nicolson & Nepi 2005; Ornelas et al. 2007; Munguía-Rosas et al. 2009; Amorim et al. 2013; Bobrowiec & Oliveira 2012), decrease (e.g., Bernardello et al. 1994; Galetto et al. 1997; McDade & Weeks 2004; Carlson 2007), or have no effect (e.g., Nepi et al. 2011; Veiga Blanco et al. 2013) on the amount of nectar produced by flowers, relative to flowers in which nectar has accumulated over the same time period. In a meta-analysis of published studies, Ordano & Ornelas (2004) found that while the responses of different plant species varied widely, overall, nectar removals enhanced the volume produced by

*Corresponding author; email: jane.ogilvie@utoronto.ca

flowers (dataset of 15 species), but nectar removal caused an overall slight decline in the amount of sugar produced (dataset of 31 species). Most of these studies focussed on a single species or genus (e.g., Gill 1988; Rivera et al. 1996; Navarro 1999; Castellanos et al. 2002; Ornelas & Lara 2009), or within a pollinator guild of plants (e.g., McDade & Weeks 2004; Bobrowiec & Oliveira 2012), and the sampling protocols differ widely. Thus, whether RENR is common over a broad range of species and flower traits remains unclear.

To begin broadening and systematizing the study of enhanced nectar renewal, we conducted a survey of RENR in 11 species of animal-pollinated plants in the vicinity of the Rocky Mountain Biological Laboratory (RMBL) in Gothic, Colorado, USA (Fig. 1). Chosen for convenience, the plant species were abundant, with large flowers visited mostly by bumble bees or hummingbirds. Many of these plants have received much attention from pollination researchers at the RMBL (e.g., Pyke 1978; Waser 1982; Price et al. 2005). For all species, we first studied the duration and timing of events during floral anthesis, and also nectar volume and concentration, to establish the most suitable flower age to examine RENR. We then adopted a standard protocol for assessing RENR, by comparing multiple removals to single removals over five hour periods. In addition, we also measured how quickly flowers replenished their nectar volumes after being drained. To minimize damage to nectaries from repeated sampling, we blotted up nectar with absorbent paper wicks. This precluded us from obtaining concentration data, so our conclusions about RENR concern nectar volumes only.

MATERIALS AND METHODS

Study system and general protocols

We carried out observations and sampling of the 11 plant species at sites in and around the RMBL in June to August, 2011, in dry rocky subalpine meadows, more mesic meadows, and in the understorey of aspen (*Populus tremuloides*) forests at around 2900 m above sea level. Floral-ecological traits of the plant species are presented in Tab. 1, including their main floral visitors near the RMBL and the time of the season that they were sampled. All species are native to the area, except for the introduced *Linaria vulgaris*. We sampled plants at a particular site for each species, to minimise the variation in nectar due to environmental conditions. To follow known flowers, we marked flowers using permanent markers applied to the calyx or with labelled pieces of coloured tape applied to the pedicel. Because nectar secretion, reabsorption, and concentration may vary throughout the day and with weather conditions (Pleasants 1983; Gill 1988; Herrera 1990; Burquez & Corbet 1991; Rivera et al. 1996), nectar measurements were performed at set times in the early afternoon for all species, and only on warm and sunny days. Although observations at various times of day would have been informative, time constraints associated with doing a broad survey of plant species forced us to choose one generally manageable time. In these cool habitats, early afternoon often represents peak activity of bumble bees (see Fig. 1, Williams & Thomson 1998); hummingbirds become active at first light and continue until dusk. We performed all nectar measurements on plants that were bagged prior to floral anthesis to exclude flower visitors, using either sand-bottom mesh bags (Thomson et al. 2011) or polyolefin Tyvek specimen bags. A separate experiment showed that there was no significant difference in nectar volume and concentration measurements of *Linaria vulgaris* flowers bagged with the two types (E. Y. Luo, unpubl. res.).

Floral scheduling

We first established the duration and timing of events during anthesis in flowers left open to visitors, by following at least 20 individual flowers from each species from bud opening until corolla detachment. We observed the plants one to three times per day and recorded the sexual stage of the flowers. From this, we determined the number of days of floral anthesis, and the order and duration of male and female phases.

Nectar production over the floral lifetime

To determine the most suitable flower age to examine RENR, we measured nectar secretion over the lifetime of flowers for each species. At daily intervals, we measured the volume of nectar that had accumulated in flowers that spanned the range of the flower lifetime. A particular flower was sampled only once. We estimated nectar volumes by introducing one end of a thin paper wick to the nectaries and measuring the moistened length of wick with calipers. We used stiff Whatman paper sold for use as wicks in starch-gel electrophoresis. To cut strips of consistent width, we equipped an Olfa circular cutter with two blades separated by a plastic spacer of 0.71 mm thickness (nominal 0.030 inch). We had previously conducted trials with known volumes of sugar solutions; these indicated that the moistened length varied linearly with volume, although the precise relationship between volume and moistened length depended on nectar concentration. For most plant species in this study, we collected nectar from several flowers with glass microcapillary tubes and determined nectar concentration using a hand-held refractometer suitable for small volumes. We used the mean concentration value to choose the most appropriate species-specific calibration equation for converting the wick "moistened length" to volume, shown in Tab. 2. For *Mertensia fusiformis* and *Delphinium nuttallianum*, we based the "moistened length" calibration conversion on mean nectar concentrations taken from near the RMBL (M. Stang, unpubl. res.). For those species with multiple nectar reservoirs (*Aconitum columbianum*, *Aquilegia* spp., *Delphinium* spp., and *Castilleja linariifolia*), we inserted a paper wick in each and summed the moistened lengths for total nectar volume per flower.

Removal-enhanced nectar replenishment

Following Castellanos et al. (2002), we tested for RENR by comparing flowers sampled only once to flowers sampled repeatedly. For each species, we selected at least 20 pairs of flowers that were matched for location on the plant and for developmental stage. We chose fresh flowers at ages when nectar volumes were near their maximum (see above,

FIGURE 1. Photos of the flowers of the 11 animal-pollinated plant species surveyed for removal-enhanced nectar replenishment (RENR) near the Rocky Mountain Biological Laboratory, Gothic, Colorado. Top row, left to right: *Mertensia ciliata, Mertensia fusiformis* (Boraginaceace), *Distegia involucrata* (Caprifoliaceae); second row, left to right: *Aconitum columbianum, Aquilegia coerulea, Aquilegia elegantula* (Helleboraceae (Ranunculaceae)); third row, left to right: *Delphinium barbeyi, Delphinium nuttallianum* (Helleboraceae), *Ipomopsis aggregata* (Polemoniaceae); fourth row, left to right: *Castilleja linariifolia, Linaria vulgaris* (Scrophulariaceae). Photo credits: *D. involucrata* by D. W. Inouye; *M. fusiformis*, and *L. vulgaris* by E. Y. Luo; remaining by J. E. Ogilvie.

TABLE 1. Floral-ecological characteristics of the study plant species. Nomenclature follows Weber & Wittman (2012), with previous names from Barrell (1969) in parentheses to clarify relationships to older literature. Flowers were considered in male phase if anthers were dehisced, and in female phase if styles were exposed. Common flower visitors are from personal observations by J.D.T. "Bees" typically include bumble bees and a variety of smaller native species (not *Apis mellifera*), with the bumble bees likely being the most effectual pollinators of these generally large flowers (except for *Mertensia fusiformis*, which is visited commonly and effectively pollinated by *Osmia* spp.; Forrest et al. 2011). "Birds" include the common hummingbirds *Selasphorus platycercus* and *S. rufus*. Flower visitors in parentheses are seen less commonly at the flowers.

Plant family Species	Duration of anthesis (d)			Duration of sex phases (range, d)		Common visitors	Time period studied (2011)
	N	Mean	SE	Male	Female		
Boraginaceae							
Mertensia ciliata	20	3.7	0.1	0 to 4	0 to 4	bees	July 15 - 22
Mertensia fusiformis	23	2.4	0.2	0 to 3	0 to 3	bees	June 13 - 16
Caprifoliaceae							
Distegia (Lonicera) involucrata	30	4.4	0.1	0 to 4	0 to 4	birds, (bees)	July 11 - 17
Helleboraceae (Ranunculaceae)							
Aconitum columbianum	20	6.4	0.2	2 to 6	4 to 6	bees	July 28 - August 9
Aquilegia coerulea (caerulea)	20	4.5	0.1	1 to 4	2 to 4	moths for nectar, bees for pollen	July 8 - 30
Aquilegia elegantula	20	4.0	0.2	1 to 4	2 to 4	birds	June 28 - July 3
Delphinium barbeyi	20	3.6	0.2	1 to 4	3 to 4	bees, (birds)	July 24 - August 3
Delphinium nuttallianum (nelsonii)	20	9.5	0.2	2 to 9	5 to 9	bees, (birds)	June 12 - 23
Polemoniaceae							
Ipomopsis aggregata	21	4.1	0.1	1 to 4	2 to 4	birds	June 28 - July 13
Scrophulariaceae							
Castilleja linariifolia	22	8.6	0.3	1 to 8	3 to 8	birds	July 1 - 24
Linaria vulgaris	20	4.4	0.1	0 to 4	0 to 4	bees	August 4 - 18

Fig. 2, and Tab. 2 for the flower ages tested for each species). In one flower of each pair, we extracted nectar repeatedly by blotting with filter paper wicks as above, at hourly intervals for five hours; in the other we extracted only once, at the end of five hours. Using wicks eliminates the possibility of estimating sugar content refractometrically (although wicks can be dried and stored for subsequent analysis of sugars if desired, see McKenna & Thomson 1988). However, we preferred wicks to glass capillaries for studying RENR because capillaries can easily damage delicate nectary tissues (see Willmer 1980, 2011). We drained the flowers before the experiment to exclude variation in standing nectar volume from the results. To ensure that our sampling method did not damage flowers, on the day after sampling, we checked to see if the flowers were still producing nectar. In most cases, the flowers had measurable nectar. We removed those occasional flowers that did not have nectar from our data set. We summed the five hourly nectar volume measurements from the removal flowers and compared them to the single volume measurement of the paired control flowers using paired *t*-tests. To test for biases in assigning flowers to the control and removal groups, we recorded the initial amounts drained and compared the means of two sample groups using paired *t*-tests. We carried out all analyses in R version 2.15.2 (R Core Team 2012).

Nectar replenishment rate

We examined how quickly nectar replenishment occurred in flowers. We used wicks to initially empty flowers and then re-sampled them at intervals, ranging from 5 to 140 minutes, and recorded the nectar volumes. We sampled these flowers at the same age as flowers tested for RENR, and we used the same nectar concentration to convert the moistened length of wick to volume as in the RENR sampling.

RESULTS

Floral scheduling

The duration of floral anthesis and the sex phases of each species are presented in Tab. 1. Overall, mean anthesis duration varied from 2.4 days in *Mertensia fusiformis* to 9.5 days in *Delphinium nuttallianum*, and in seven species the flowers were protandrous while the remaining four species had hermaphroditic flowers with no dichogamy (Tab. 1).

Nectar production over the floral lifetime

Accumulated nectar volumes typically showed marked variation among flowers at any particular flower age; occasional empty or nearly empty flowers were frequent in this group of species (Fig. 2). Inspection of the points in Fig. 2 shows that nectar volume per flower varied continuously from low levels near zero to highs that were characteristic for that species. There is no suggestion of discontinuous variation between discrete "full" and "empty" categories in any species. The most common pattern was for nectar volumes to increase with flower age (*Mertensia ciliata, M. fusiformis, Aconitum columbianum, Aquilegia coerulea,*

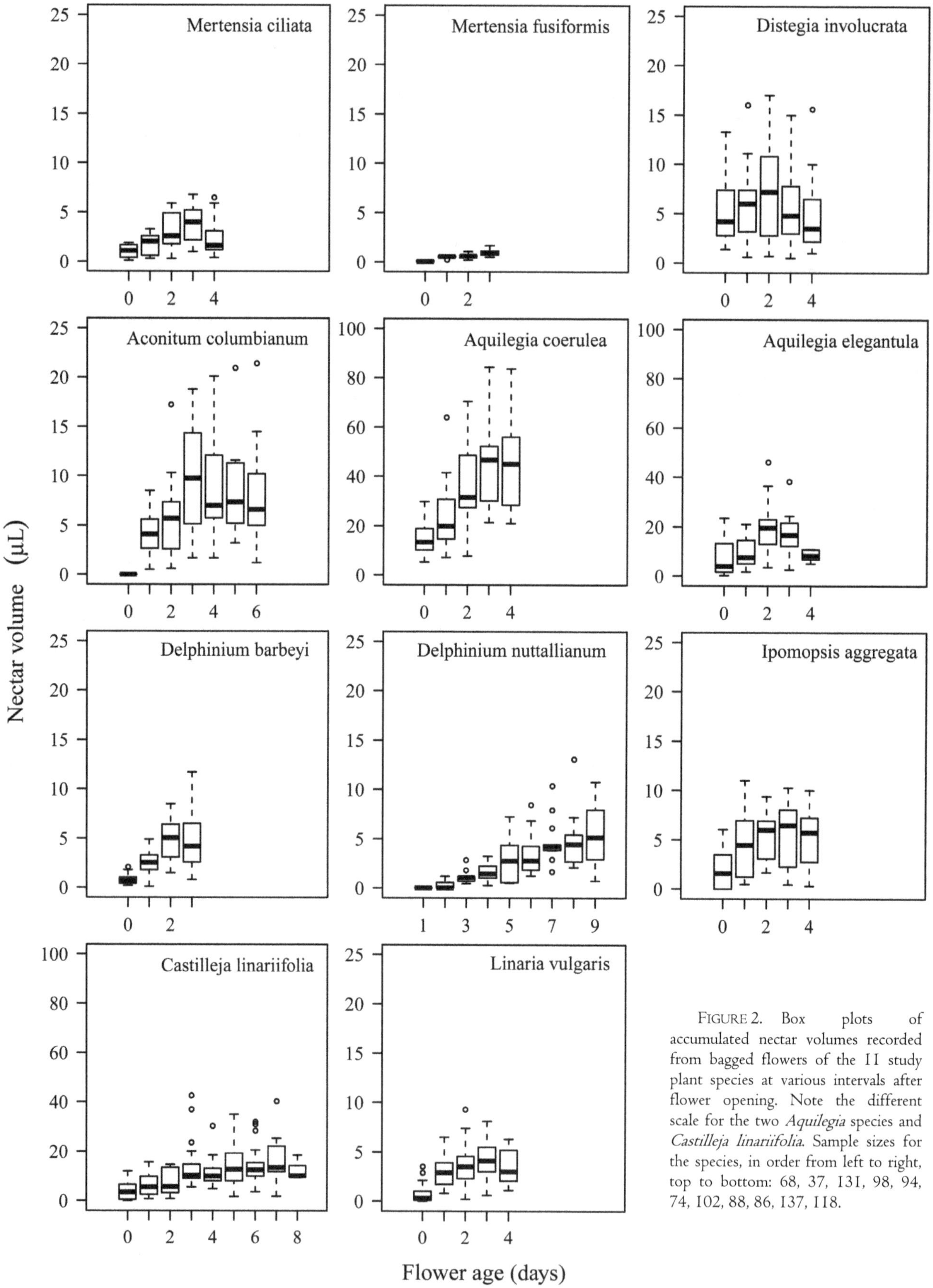

FIGURE 2. Box plots of accumulated nectar volumes recorded from bagged flowers of the 11 study plant species at various intervals after flower opening. Note the different scale for the two *Aquilegia* species and *Castilleja linariifolia*. Sample sizes for the species, in order from left to right, top to bottom: 68, 37, 131, 98, 94, 74, 102, 88, 86, 137, 118.

TABLE 2. Details of the nectar concentration and removal-enhanced nectar replenishment (RENR) sampling of the 11 study plant species. Nectar concentration was measured using microcapillary tubes while RENR was measured using filter paper wicks. The "calibration concentration" values were the concentrations used to calculate the species-specific calibration equation for converting the "moistened length" of wicks to nectar volume, informed by the mean nectar concentrations.

Plant species	Nectar concentration study		Removal-enhanced nectar replenishment study			
	N	Mean concentration (SE) (%w/w)	Calibration concentration (%w/w)	N pairs	Flower age sampled (d)	RENR ratio, 5 samples/single
Mertensia ciliata	22	18.5 (1.5)	20	29	2 – 3	2.88
Mertensia fusiformis	-	-	40	24	2	2.15
Distegia involucrata	53	18.9 (0.7)	20	25	2 – 3	1.52
Aconitum columbianum	21	51.5 (3.2)	50	20	3 – 5	1.62
Aquilegia coerulea	27	18.9 (0.7)	20	20	2 – 3	2.26
Aquilegia elegantula	24	31.3 (1.1)	30	20	2 – 3	1.29
Delphinium barbeyi	22	49.8 (2.0)	50	25	2	1.37
Delphinium nuttallianum	-	-	40	22	5 – 7	2.10
Ipomopsis aggregata	25	16.1 (2.2)	20	25	2 – 3	1.19
Castilleja linariifolia	53	18.7 (1.0)	20	24	3 – 7	2.55
Linaria vulgaris	24	40.9 (4.5)	40	25	2 – 3	1.57

Delphinium barbeyi, D. nuttallianum), producing roughly triangular spreads of nectar volumes. This pattern suggests continuing secretion of new nectar as a flower ages. In *A. columbianum, D. barbeyi* and *D. nuttallianum,* flower openings are sequential and highly structured within inflorescences, so that older flowers occupy lower positions, producing vertical gradients of nectar volumes (see Pyke 1978). In *Ipomopsis aggregata* and *Linaria vulgaris,* nectar volumes tended to be uniform across flower age, as if nectar secretion ceased after flower opening. *Castilleja linariifolia* showed an intermediate pattern in which nectar volume initially increased with flower age but subsequently levelled off. This species was sampled for more days than most others, probably because the colourful bracts persist; the observed levelling off may well have occurred after true anthesis had finished. In two species, *Aquilegia elegantula* and *Distegia involucrata,* nectar volumes peaked at intermediate flower ages, producing unimodal distributions suggestive of initial secretion followed by evaporation or reabsorption.

Removal-enhanced nectar replenishment

In all species except *Aquilegia elegantula* and *Ipomopsis aggregata,* flowers produced significantly more nectar when they were repeatedly drained compared to those sampled only once over the same time period (paired *t*-tests, $P < 0.05$; Fig. 3). For a quantitative index of RENR, we calculated a ratio of the total volume measured from the repeatedly sampled flowers to the total volume measured from the once-sampled control flowers (Tab. 2). Values range from 1.19 in *Ipomopsis aggregata* to 2.88 in *Mertensia ciliata* (Tab. 2). Thus, repeatedly draining a single flower tended to extract about twice the amount of nectar as a flower would accumulate over the same period if left unvisited. The nectar volumes removed from all flowers at the start of the 5-hour sampling period did not differ between removal and control groups for all species (paired *t*-tests, $P > 0.05$), indicating no bias in assigning the flowers to the two groups. Across the 11 study species, there was no correlation between mean or median nectar volumes at the focal flower ages and the RENR ratios (Spearman's rank correlation, $P > 0.05$).

Nectar replenishment rate

In Fig. 4, measures of nectar replenished during various accrual periods after initial emptying indicate the rate of nectar replenishment. Among species, the rate of replenishment generally increased with greater standing nectar volumes. Most flowers exhibited the ability to replenish measurable nectar in minutes rather than hours, suggesting that nectar replenishment occurs rapidly after emptying.

DISCUSSION

In most of our study species, flowers drained repeatedly produced 1.5 to 2.9 times more nectar than flowers drained only once, despite variation in species-specific maximum nectar volumes. Although we have expressed this property in terms of "replenishment," the same pattern has elsewhere been attributed to "reabsorption" (Nicolson 1995). The "reabsorption" viewpoint supposes that nectar is secreted continuously, but that some of it is reabsorbed unless it is removed at frequent intervals. Ascertaining the exact mechanism through which repeated withdrawals harvest more nectar than a single withdrawal would require careful physiological study, but from the standpoint of plant-pollinator relations, the mechanism is less important than the phenomenon. Our measurements suggest that researchers working on flowers adapted for large insect visitors like bumble bees should expect to find removal-enhanced replenishment of nectar; this phenomenon appears to be more the rule than the exception.

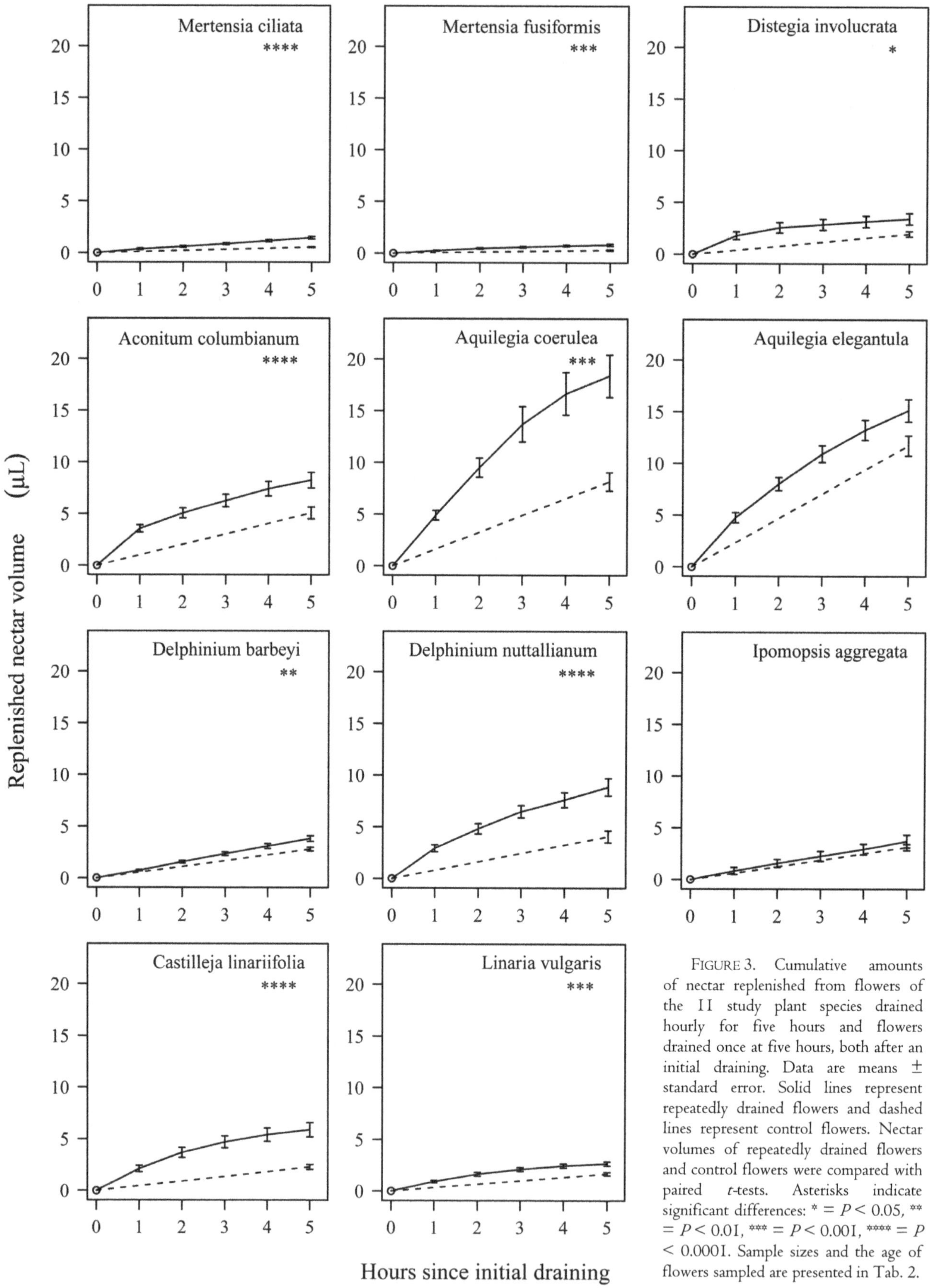

FIGURE 3. Cumulative amounts of nectar replenished from flowers of the 11 study plant species drained hourly for five hours and flowers drained once at five hours, both after an initial draining. Data are means ± standard error. Solid lines represent repeatedly drained flowers and dashed lines represent control flowers. Nectar volumes of repeatedly drained flowers and control flowers were compared with paired *t*-tests. Asterisks indicate significant differences: * = $P < 0.05$, ** = $P < 0.01$, *** = $P < 0.001$, **** = $P < 0.0001$. Sample sizes and the age of flowers sampled are presented in Tab. 2.

RENR as a complication

Our findings serve to reinforce the strengthening consensus that measuring nectar is not as easy as some earlier treatments assumed (Zimmerman 1988; Willmer 2011). In most of the species we examined, the volume of nectar that flower visitors can extract depends on how many extractions are made. Furthermore, we found that one-time nectar volume measurements were not related to the extent of replenishing that was characteristic of a plant species. Removal-responsive replenishing by flowers must affect investigations of how much energy gain visitors can achieve by visiting a particular species, and how much metabolic cost a plant will incur by providing nectar. Nectar replenishment will be especially important in considering the economics of trapline foraging, in which flower visitors return to plants at frequent intervals and compete with each other to control the nectar flow. Ohashi and others (Ohashi & Thomson 2005; Ohashi et al. 2008) explored the consequences of such competition, theoretically and empirically, but those models did not consider RENR, and the laboratory apparatus used in those experiments produced fixed nectar secretion. Ohashi et al.'s qualitative conclusions regarding the general advantages of traplining would probably be robust, but the quantitative results would surely be modified in plants that display RENR (indeed, plants that are actively visited by traplining animals may be likely to show RENR).

An important caveat of our studies is that we examined how flowers responded to nectar removals by measuring nectar volume and not sugar content. Most plant species that show a significant increase in nectar volume in response to nectar removal also show an increase in total sugar production (e.g., Navarro 1999; Nicolson & Nepi 2005; Ornelas & Lara 2009; Bobrowiec & Oliveira 2012; Amorim et al. 2013, though see Galetto et al. 1994). If sugar concentration remains approximately the same in extracted samples, total sugar content produced by flowers will increase in response to removal (e.g. Navarro 1999). However, we cannot assume that flowers of all species and under varied environmental conditions will replenish their nectar at a constant sugar concentration. Even ignoring environmental effects such as evaporative concentration or hygroscopic dilution, it is easy to imagine that a flower's replenished nectar might be more dilute than the first nectar secreted by that flower. Indeed, in flowers of *Penstemon speciosus*, water seemed to replenish more quickly than sugar (Castellanos et al. 2002). Nectar concentration may be controlled by passive diffusion, while nectar volume may be controlled by hydrostatic pressure differences (Pacini & Nepi 2007). If sugar diffusion is slower than replenishment of water, nectar concentrations may be lower than normal shortly after removal. In addition, continual replenishment of nectar incurs a metabolic cost that can lower female reproduction (Pyke 1991; Ordano & Ornelas 2005; Ornelas & Lara 2009); reducing nectar sugar concentration in situations with continuous visitation may compensate for this cost.

If newly secreted nectar is more dilute, the RENR rate measured for nectar volume (as in our study) would overestimate the rate for sugar content. For questions that hinge on energy content, it would be critical to supplement the sort of measurements we made with additional measurements of sugar concentrations. There is a further complication, too: if replenished nectar is considerably more dilute, it will change the relationship between the moistened length of a paper wick and the volume of solution taken up. Instead of a single regression equation, one would need to devise a more complicated function to account for trends in concentration.

Residual nectar and RENR

Foraging bumble bees frequently leave behind residual nectar in visited flowers (Hodges & Wolf 1981; Zimmerman 1983; Cresswell 1999). Since nectar production may be mediated by a hydrostatic pressure difference (Pacini & Nepi 2007), residual nectar may lower the rate of nectar replenishment in nature compared to that experienced from experimental removal, which might deplete nectar more completely (e.g., our study). This question hinges on how well blotting with filter-paper wicks compares to extraction by flower visitors. If wicks remove more nectar, the rate and metabolic costs of replenishment (Ordano & Ornelas 2005; Ornelas & Lara 2009) may be overestimated in experimental studies relative to natural conditions.

RENR as an object of study

Aside from the negative message that RENR complicates the quantification of nectar energetics, the variation that we observed across species hints that RENR might be an informative trait for comparative studies of flower function. In general, we might expect that the extent and rate of RENR will covary with pollination system. In a parallel example, Castellanos et al. (2006) argued that species of *Penstemon* that were adapted for pollination by hummingbirds differed in their schedules of pollen presentation from those adapted for pollination by hymenopterans. The adaptive function of more gradual pollen presentation by bee-adapted flowers was hypothesized to arise from the grooming behaviour of the bees, which imposed diminishing returns on the relationship between pollen presentation and subsequent delivery. Diminishing returns mean that bee-adapted flowers will achieve the most pollen export if they can present small doses of pollen to many visitors (Harder & Thomson 1989; Thomson & Thomson 1992). In accordance with this prediction, bee-adapted species have anthers that open less widely than those of related bird-adapted species (Castellanos et al. 2006). However, the potential advantages of gradual pollen presentation cannot be realized without a high visitation rate. Therefore, we would predict that bee-adapted flowers would show higher RENR than bird-adapted relatives, because greater nectar replenishment should encourage heavier visitation.

Ideally, such comparisons should be made among near-sister taxa within a strict phylogenetic framework, as in the *Penstemon* anther study. Our opportunistic survey does not allow a rigorous test of the prediction, but there are suggestive indications that the hypothesis is worth pursuing. Consider the eleven species in Fig. 3. Of the three species

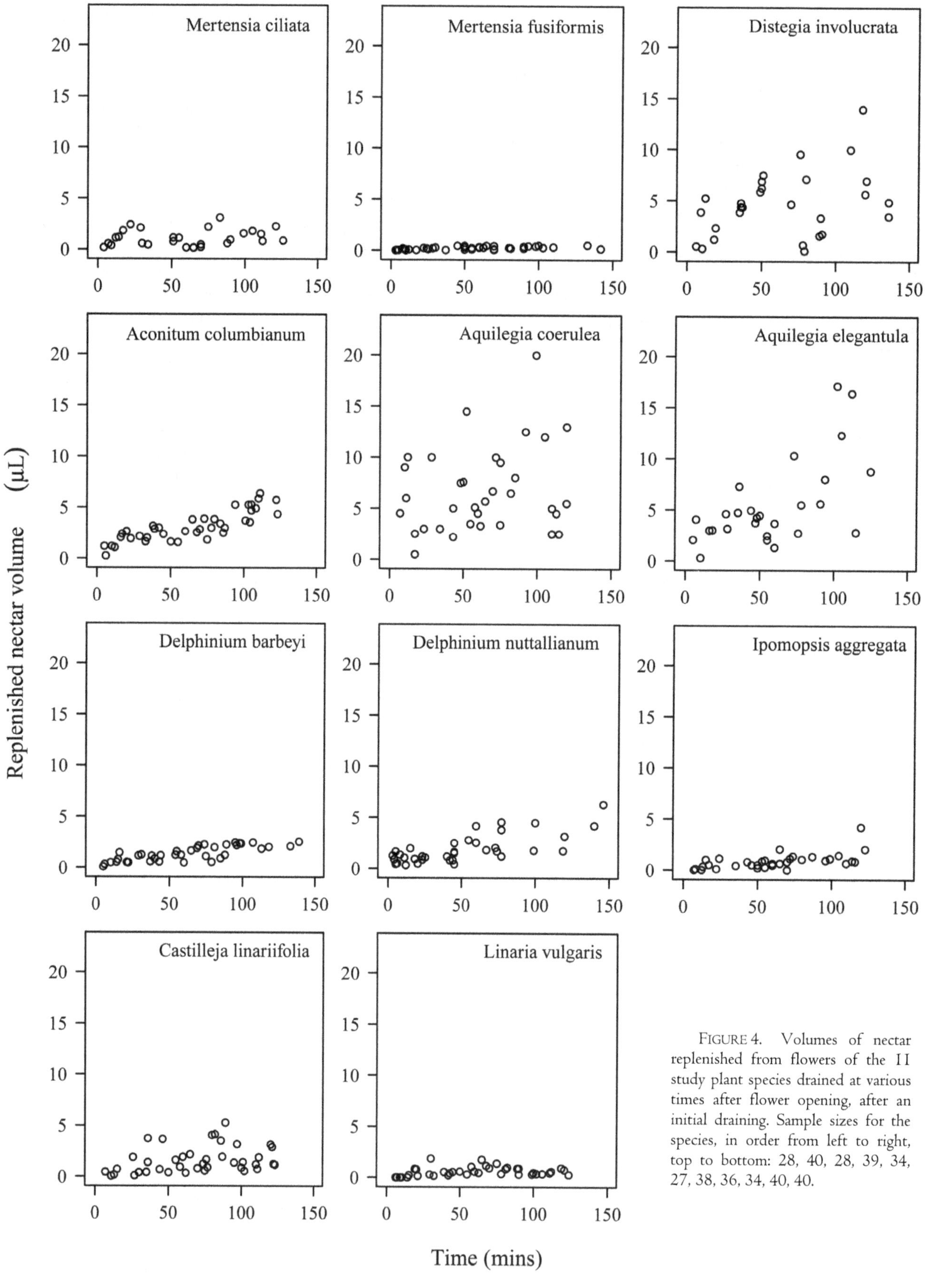

FIGURE 4. Volumes of nectar replenished from flowers of the 11 study plant species drained at various times after flower opening, after an initial draining. Sample sizes for the species, in order from left to right, top to bottom: 28, 40, 28, 39, 34, 27, 38, 36, 34, 40, 40.

most exclusively visited by hummingbirds, two (*Ipomopsis aggregata* and *Aquilegia elegantula*) have the lowest RENR values in the study. Pleasants (1983) also found that nectar production in *I. aggregata* flowers was not stimulated by its removal. The next two lowest values are *Distegia involucrata* and *Delphinium barbeyi*, both of which receive many bird visitors in addition to bees. The only primarily bird-visited species to show high RENR is an Indian paintbrush, *Castilleja linariifolia*. Unfortunately, we lack data from any of the local bee-visited paintbrushes (*C. sulphurea* or *C. occidentalis*), so we cannot tell whether their RENR might be even higher. Nevertheless, there is variation in the RENR ability of bird-visited plants in different genera: some show decreased nectar production in response to nectar removals (e.g., Gill 1988; Bernardello et al. 1994; Rivera et al., 1996; McDade & Weeks 2004; Carlson 2007), others increased production (e.g., Nicolson & Nepi 2005; Ornelas & Lara 2009), and others have similar nectar production to controls (e.g., McDade & Weeks 2004). Pursuing this hypothesis would seem to be more worthwhile within a single genus like *Penstemon* that shows numerous bee-to-bird transitions that can be phylogenetically localised.

Conclusion

Our study confirms that replenishment of nectar volume is commonly a rapid, dynamic process that may be mediated by the frequency of pollinator visits. It appears that most of our study species possessed some ability to regulate replenishment, demonstrating RENR. A full understanding of secretion dynamics would require coordinated investigation of sugar concentration changes in addition to replenishment of volume, but such measurements were beyond the scope of this study. It would be interesting to see if our hypothesized bee-bird difference holds up in more comparisons of related flower species adapted to different pollinators. Potentially common RENR by plants complicates key questions in plant-pollinator interactions, though further investigation will help illuminate these questions.

Acknowledgements

We thank the Rocky Mountain Biological Laboratory for field support, Martina Stang for sharing her knowledge of Gothic flowers, and Takashi Makino for advice. This work was supported by a Natural Sciences and Engineering Research Council of Canada Discovery Grant to J.D.T.

References

Amorim FW, Galetto L, Sazima M (2013) Beyond the pollination syndrome: nectar ecology and the role of diurnal and nocturnal pollinators in the reproductive success of *Inga sessilis* (Fabaceae). Plant Biology 15:317-327.

Barrell J (1969) Flora of the Gunnison Basin: Gunnison, Saguache, and Hinsdale counties, Colorado. Natural Land Institute, Rockford.

Bernardello L, Galetto L, Rodriguez IG (1994) Reproductive biology, variability of nectar features and pollination of *Combretum fruticosum* (Combretaceae) in Argentina. Botanical Journal of the Linnean Society 114:293-308.

Bobrowiec PED, Oliveira PE (2012) Removal effects on nectar production in bat-pollinated flowers of the Brazillian Cerrado. Biotropica 44:1-5.

Burquez A, Corbet SA (1991) Do flowers reabsorb nectar? Functional Ecology 5:369-379.

Carlson JE (2007) Male-biased nectar production in a protandrous herb matches predictions of sexual selection theory in plants. American Journal of Botany 94:674-682.

Castellanos MC, Wilson P, Thomson JD (2002) Dynamic nectar replenishment in flowers of Penstemon (Scrophulariaceae). American Journal of Botany 89:111-118.

Castellanos MC, Wilson P, Wolfe A, Keller S, Thomson JD (2006) Anther evolution: pollen presentation strategies when pollinators differ in efficiency. American Naturalist 167:288-296.

Cresswell JE (1999) The influence of nectar and pollen availability on pollen transfer by individual flowers of oil-seed rape (*Brassica napus*) when pollinated by bumblebees (*Bombus lapidarius*). Journal of Ecology 87:670-677.

Cruden RW, Hermann SM (1983) Studying nectar? Some observations on the art. In: Bentley B, Elias T (eds) The biology of nectaries. Columbia University Press, New York, pp 223–241.

Forrest JRK, Ogilvie JE, Gorischek AM, Thomson JD (2011) Seasonal change in a pollinator community and the maintenance of style length variation in *Mertensia fusiformis* (Boraginaceae). Annals of Botany 108:1-12.

Galetto L, Bernardello G, Rivera GL (1997) Nectar, nectaries, flower visitors, and breeding system in five terrestrial Orchidaceae from Central Argentina. Journal of Plant Research 110:393-403.

Galetto L, Bernardello LM, Juliani HR (1994) Characteristics of secretion of nectar in *Pyrostegia venusta* (Ker-Gawl.) Miers (Bignoniaceae). New Phytologist 127:465-471.

Gill FB (1988) Effects of nectar removal on nectar accumulation in flowers of *Heliconia imbricate* (Heliconiaceae). Biotropica 20:169-171.

Harder L, Thomson JD (1989) Evolutionary options for maximizing pollen dispersal of animal-pollinated plants. American Naturalist 133:325-334.

Heinrich B (1983) Insect foraging energetics. In: Jones CE, Little RJ. Handbook of experimental pollination biology. Scientific and Academic Editions, New York, pp 187-214.

Herrera CM (1990) Daily patterns of pollinator activity, differential pollinating effectiveness, and floral resource availability, in a summer-flowering Mediterranean shrub. Oikos 58:277-288.

Hodges CM, Wolf LL (1981) Optimal foraging in bumblebees: why is nectar left behind in flowers? Behavioural Ecology and Sociobiology 9:41-44.

McDade LA, Weeks JA (2004) Nectar in hummingbird-pollinated neotropical plants II: Interactions with flower visitors. Biotropica 36:216-230.

McKenna M, Thomson JD (1988) A technique for sampling and measuring small amounts of floral nectar. Ecology 69:1036-1037.

Munguía-Rosas MA, Sosa VJ, Jácome-Flores ME (2009) Pollination system of the *Pilosocereus leucocephalus* columnar cactus (tribe Cereeae) in Eastern Mexico. Plant Biology 12:578-586.

Navarro L (1999) Pollination ecology and effect of nectar removal in *Macleania bullata* (Ericaceae). Biotropica 31:618-625.

Nepi M, Cresti L, Guarnieri M, Pacini E (2011) Dynamics of nectar production and nectar homeostasis in male flowers of *Cucurbita pepo* L. International Journal of Plant Sciences 172:183-190.

Nicolson SW (1995) Direct demonstration of nectar reabsorption in the flowers of *Grevillea robusta* (Proteaceae). Functional Ecology 9:584-588.

Nicolson SW, Nepi M (2005) Dilute nectar in dry atmospheres: nectar secretion patterns in *Aloe castanea* (Asphodelaceae). International Journal of Plant Sciences 166:227-233.

Ohashi K, Leslie A, Thomson JD (2008) Trapline foraging by bumble bees: V. Effects of experience and priority on competitive performance. Behavioural Ecology 19:936-948.

Ohashi K, Thomson JD (2005) Efficient harvesting of renewing resources. Behavioural Ecology 16:592-605.

Ordano M, Ornelas JF (2004) Generous-like flowers: nectar production in two epiphytic bromeliads and a meta-analysis of removal effects. Oecologia 140:495-505.

Ordano M, Ornelas JF (2005) The cost of nectar replenishment in two epiphytic bromelialds. Journal of Tropical Ecology 21:541-547.

Ornelas JF, Lara C (2009) Nectar replenishment and pollen receipt interact in their effects on seed production of *Penstemon roseus*. Oecologia 160:675-685.

Ornelas JF, Ordano M, Lara C (2007) Nectar removal effects on seed production in *Moussonia deppeana* (Gesneriaceae), a hummingbird-pollinated shrub. Ecoscience 14:117-123.

Pacini E, Nepi M (2007) Nectar production and presentation. In: Nicolson SW, Nepi M, Pacini E (eds) Nectaries and nectar. Springer, Netherlands, pp 167-214.

Pleasants JM (1983) Nectar production patterns in *Ipomopsis aggregata* (Polemoniaceae). American Journal of Botany 70:1468-1475.

Price MV, Waser NM, Irwin RE, Campbell DR, Brody AK (2005) Temporal and spatial variation in pollination of a montane herb: A seven-year study. Ecology 886:2106-2116.

Pyke GH (1978) Optimal foraging in bumblebees and their coevolution with their plants. Oecologia 36:281-293.

Pyke GH (1991) What does it cost a plant to produce floral nectar? Nature 350:58-59.

R Core Team (2012) R: A language and environment for statistical computing. R Foundation for Statistical Computing, Vienna, Austria. http://www.R-project.org/.

Rivera GL, Galetto L, Bernardello L (1996) Nectar secretion pattern, removal effects, and breeding system of *Ligaria cuneifolia* (Loranthaceae). Canadian Journal of Botany 74:1996-2001.

Thomson JD, Forrest JRK, Ogilvie JE (2011) Pollinator exclusion devices permitting easy access to flowers. Journal of Pollination Ecology 4:24-25.

Thomson JD, Thomson BA (1992) Pollen presentation and viability schedules in animal-pollinated plants: consequences for reproductive success. In: Wyatt R (ed) Ecology and evolution of plant reproduction: New approaches. Chapman and Hall, New York, pp 1-24.

Veiga Blanco T, Galetto L, Machado IC (2013) Nectar regulation in *Euphorbia tithymaloides* L., a hummingbird-pollinated Euphorbiaceae. Plant Biology 15:910-918.

Waser NM (1982) A comparison of distances flown by different visitors to flowers of the same species. Oecologia 55:251-257.

Weber WA, Wittmann RC (2012) Colorado flora: Western slope. University Press of Colorado, Boulder.

Williams NM, Thomson JD (1998) Trapline foraging by bumble bees: III. Temporal patterning of visits. Behavioral Ecology 9:612-621.

Willmer PG (1980) The effects of insect visitors on nectar constituents in temperate plants. Oecologia 47:270-277.

Willmer P (2011) Pollination and floral ecology. Princeton University Press, Princeton.

Zimmerman M (1983) Calculating nectar production rates: residual nectar and optimal foraging. Oecologia 58:258-259.

Zimmerman M (1988) Nectar production, flowering phenology, and strategies for pollination. In: Lovett Doust J, Lovett Doust L (eds) Plant reproductive ecology: patterns and strategies. Oxford University Press, Oxford, pp 157-178.

13

REPRODUCTIVE BIOLOGY OF POINTLEAF MANZANITA (*ARCTOSTAPHYLOS PUNGENS*) AND THE POLLINATOR-NECTAR ROBBER SPECTRUM

Leif L. Richardson*[1] and Judith L. Bronstein[2]

[1]*Department of Biology, Dartmouth College, Hanover, New Hampshire 03755, USA*
[2]*Department of Ecology and Evolutionary Biology, University of Arizona, Tucson 85721, USA*

Abstract—Floral visitor species are often assumed to act either mutualistically towards plants (as pollinators) or to exploit them (as nectar-robbers or as nectar or pollen thieves). We investigated the reproductive biology of pointleaf manzanita (*Arctostaphylos pungens* K. Kunth), a regionally abundant North American shrub, in relation to the wide spectrum of behaviours displayed by its flower-visiting insects. We recorded *A. pungens* population-level flowering phenology and nectar standing crop, and conducted experiments documenting its breeding system, in an Arizona upland habitat in 1998 and 1999. Floral visitors were observed over 38 hr. We recorded frequencies of six foraging behaviours within and among individuals of each visitor species.

Arctostaphylos pungens flowers in late winter. During this period it is the only abundant floral resource for a diverse array of generalist insects in its habitat. We observed 1206 floral visits by 46 taxa. Most floral visitors pursued mixed behaviours: at the species and/or individual level, they foraged both legitimately and as nectar-robbers or thieves. The most commonly mixed behaviours were legitimate pollen collection (which likely resulted in pollen transfer) and secondary nectar-robbing (which was highly unlikely to do so). As *A. pungens* was found to be largely self-incompatible, robbing and thieving visits should be detrimental to reproductive success.

Although theoretical analyses often assume that exploiters must be punished or excluded for mutualism to persist evolutionarily, exploitation is in fact ubiquitous within cooperative interactions in nature. In manzanita, very few floral visitor species could be classified as exclusively beneficial or detrimental to plants: rather, they exhibited multiple foraging strategies, with no evidence of plant control. Such pollinator-nectar robber spectra appear to be common, and constitute an important challenge to current understanding of how mutualism can persist.

Keywords: Arctostaphylos, cheating, exploitation, foraging, manzanita, mutualism, nectar-robbing, pollination.

INTRODUCTION

Mutualisms involve a mutually profitable exchange of commodities between two species. One phenomenon common to virtually all mutualisms, however, is the presence of exploiters (often termed cheaters or parasites). These are organisms that obtain the commodities mutualists offer, but that deliver none in return (Yu 2001; Bronstein 2001, 2003). The exploitation of mutualism has recently attracted extensive attention because of its potential evolutionary significance. Individuals that obtain the benefits mutualists offer while avoiding investment into mutualistic commodities should experience a fitness advantage. Unless exploitative strategies are countered, they are therefore expected to replace mutualism over evolutionary time. Paradoxically, however, mutualisms are ubiquitous in nature, and exploitation seems ubiquitous within them (Bronstein 2001, 2003). These observations underscore the need for further empirical studies of exploitation, particularly those devoted to understanding how mutualism can persist in its presence.

Plant/pollinator interactions have become model systems for studying exploitation. In particular, a great deal is known about the natural history of nectar-robbing (Irwin et al. 2010). Nectar-robbers are floral visitors that collect nectar from punctures in the corolla, made either by themselves or by other individuals. In so doing, they usually bypass direct contact with the stamens and stigmas. Nectar-robbing exhibits two features typical of exploitation. First, exploitation varies widely in the costs it inflicts upon the partner. Many robbers neither pick up nor deposit pollen during a visit; pollinators may subsequently avoid robbed flowers, with consequent reductions in seed set and/or pollen donation (e.g., Irwin & Brody 1999, 2000). Other robbers, however, do move pollen, or else cause changes in pollinator behaviour (e.g., inducing pollinators to leave flowers earlier than they otherwise would) that lead to greater pollen transfer. In those cases, exploitation may actually benefit plants (e.g., Navarro 2000; Maloof 2001; Richardson 2004). Second, what appears to be one form of exploitation is actually a suite of related phenomena. The most thoroughly studied nectar-robbers never visit the

*Corresponding author; email: leif.l.richardson.gr@dartmouth.edu

species that they rob as legitimate pollinators. However, in some floral visitor species, certain individuals exclusively rob and others exclusively pollinate the same plant species; in other visitor species, each individual both robs and pollinates the same plant (Irwin et al., 2010). Although mixed pollinating-robbing strategies have been reported previously (e.g., Free 1968; Rust 1979; Morris 1996; Villalobos & Sherry 1996; Olesen 1996; Dedej & Delaplane 2005; Newman & Thompson 2005), little is known of their causes, consequences, or ecological distribution.

Here, we document the spectrum of potentially mutualistic and antagonistic behaviours exhibited by floral visitors to a regionally abundant North American shrub, pointleaf manzanita (*Arctostaphylos pungens* HBK, Ericaceae, Vaccinoideae). We first present data on the phenology, floral biology, and breeding system of *A. pungens*. These features have not previously been described, and are necessary for examining implications of its floral visitors' foraging behaviours. We then describe those behaviours, based on over 38 hours of observation of 46 insect species over two years. We document (1) the range of foraging behaviours that floral visitors use, (2) the frequency with which single visitor species and single individuals mix behaviours, and (3) the behaviours that are most commonly mixed at the species and individual level. We then combine data on plant breeding system and visitor behaviour to discuss the implications of the foraging spectrum from the perspective of the insects themselves, the plants they visit, and the mutualisms they sometimes form.

MATERIALS AND METHODS

Study system

Arctostaphylos pungens (pointleaf manzanita) occurs over much of the southwestern United States and as far south as Oaxaca, Mexico (Kearney & Peebles 1960; Diggs 2009), often forming a significant component of Madrean evergreen woodland and chaparral vegetation types (Brown 1982). It is a sprawling evergreen shrub with simple ovate leaves and distinctive, smooth red stems. Flowering occurs from Jan-Apr at most sites; however, sporadic flowering can take place at almost any time of year. Flowering is minimal to absent in drought years (L. Richardson and J. Bronstein, pers. obs.). The small, pendulous flowers are borne in dense inflorescences. A single plant can bear several hundred open flowers at one time during peak flowering. Flowers are bell-shaped, with fused corollas. Flowers range from white to deep pink and average 6 + 1 mm in length and 3 + 1 mm at their widest point; they are constricted at their apical ends to about 2 mm in diameter when fully open (n = 485 flowers). Nectar is produced in depressions in a basal part of the corolla slightly occluded by the ovary. The 10 stamens bear distinctive burgundy-coloured anthers. Pollen is produced in tetrads and is dehisced toward the style via apical pores on the anthers. The superior ovary contains 5-10 ovules. The fruit is a fleshy berry with 1-5 stony segments, each containing 1-3 seeds.

Study Sites

This study was conducted at two locations in the Santa Catalina Mountains, Arizona, USA (32.3374, -110.6918) where *A. pungens* was the dominant shrub species: Molino Basin (1280 m elevation) and the Gordon Hirabayashi Recreation Site (1500 m elevation). Both sites receive ~35-40 cm of precipitation per year (Sabino Canyon, the nearest weather station, receives a yearly average of 33.1 cm of precipitation at 840 m elevation; Western Regional Climate Center, 2012). Most precipitation falls during summer monsoons and a mid-winter rainy season. Summer temperatures typically exceed 35°C, and winter temperatures below 0°C have been recorded annually since record keeping began (Western Regional Climate Center, 2012).

Floral Biology

Population-level floral phenology was documented at Molino Basin during Jan- Mar 1998 and 1999. Plants (4 in 1998; 12 in 1999) were haphazardly chosen at 10 meter intervals along a roughly linear transect. On each plant, stems were arbitrarily chosen (2 in 1998; 1 in 1999) and marked. Every 2-10 d, the following information was recorded for individually marked inflorescences on these stems: number of flowers open, number of flowers with corolla perforations, number of flowers missing (flower and/or pedicel abscised due to abortion or herbivory), and number of fruits initiated. Similar numbers of flowers were censused each year. In 1999, a severe drought cut short the flowering season. Censusing ended in early March, as flowers were no longer opening or abscising from the plant.

Nectar standing crop was measured from open flowers between 0930 and 1045 on 1-2 February 2001 at Hirabayashi, using a 1 µl micropipette. On 17 plants, 1-3 pairs of adjacent flowers, in which one flower of the pair was undamaged while the other had been perforated by a nectar-robber earlier the same morning, were sampled.

Breeding System

In 1998, the mating system of *A. pungens* was studied at Molino Basin. Before flowers opened, a group of inflorescences (~75 flowers) were bagged on 4 stems of each of 4 plants to exclude floral visitors. A fifth stem was randomly chosen to serve as an unbagged control. Flower buds on all inflorescences of the 5 stems were counted before anthesis. Four treatments were performed on the bagged branches; on each plant, each bag received a different treatment. In the bagged control treatment, bags were removed and then immediately replaced in order to test for autogamous reproduction. The outcrossing treatment involved collecting pollen from three or more donor plants located 5-50 m away by sonicating anthers with a tuning fork, then using a pinhead to apply this pollen to stigmas. The geitonogamous self-pollination treatment involved collecting pollen by sonicating flowers within a bag, then distributing that pollen onto stigmas of other flowers within the same bag. Finally, in the buzz self-pollination treatment, all open flowers within a bag were sonicated with a tuning fork. Each treatment was repeated periodically as new flowers opened within the bags. Fruits were collected approximately 6 wk after flowering ceased. The number of fruits produced was divided by the total number of flowers to calculate fruit set for that treatment. Counting seeds to measure reproductive success was not feasible: in *A. pungens*, the persistent, woody endocarp surrounding seeds often causes them to adhere so tightly to one another that individual seeds cannot be distinguished. Instead, fresh fruits were weighed individually to the nearest hundredth of a gram with an electronic balance. Fruit weights were compared among treatments with analysis of variance, followed by Tukey-Kramer post-hoc tests.

Floral Visitors

Floral visitors were observed at both sites. In 1998, visits were recorded only anecdotally, so most of the data reported here are from 1999. Focal plants were chosen haphazardly and a different plant was observed each day. Observers stood 0.5-2.0 m from the focal plant and watched as many flowers as possible (usually several hundred) for 30-180 min. Most observations were made during two daily peaks of insect activity, between 0900 and 1200 and between 1500 and 1700. When an insect arrived at the plant, it was continuously observed until it disappeared from sight or left the plant. A single visitor was observed each time, even if other insects were arriving, departing, and foraging simultaneously. Each observed trip to a flower is subsequently termed a "visit," and a series of visits during which an individual visitor was monitored continually until it left a plant is termed a "bout." In the descriptions that follow, all visits are included, regardless of their sequence within a bout, and each visit is treated as a separate event. We consider this to be appropriate because (a) proportions of each visit type (described below) recorded for the most common visitor species during their first visit were indistinguishable from proportions for the second, third, and fourth visit, and (b) proportions of behaviour types in visits 1-4 were not significantly different from proportions of those behaviours in the entire data set (data not shown).

For each visit, one of six possible behaviours was recorded. We briefly describe them here; further details are provided in the Results. Three types of visits were likely to lead to pollen transfer between flowers. During a *legitimate nectar-collecting* visit, an insect collected nectar by inserting its mouthparts through the corolla opening. In a *legitimate pollen-collecting* visit, an insect reached inside the corolla to glean fallen pollen from the tuft of hairs situated underneath the anthers. In a *buzz pollen-collecting* visit, the insect hung from the corolla and sonicated the flower to release pollen from the poricidal anthers. Three additional types of visits were recorded that were likely not to result in pollen transfer; our terminology follows Inouye (1980). During a *primary nectar-robbing* visit, the insect used its mouthparts to perforate the corolla near the base of the flower, then collected nectar without contacting the stigma or anthers. A *secondary nectar-robbing* visit involved the collection of nectar from a pre-existing perforation. Finally, *nectar thievery* was recorded when the insect forced its mouthparts through the closed corolla of a bud without damaging it, collecting nectar before the flower opened.

Floral visitors were regularly collected during observation periods. The majority of bees were identified to genus; most other insects were identified at least to family with available reference texts (McAlpine 1981-1989; Borror et al., 1989; Michener et al., 1994). Determinations were checked against specimens in the University of Arizona Entomology collection. Although it was not possible to obtain complete identifications for every visitor we observed, all but the Lepidoptera were easily identifiable to operational taxonomic unit on the wing. The > 12 Lepidoptera species have been combined in this analysis, as they displayed similar foraging strategies, and individually, were insignificant as visitors. Voucher specimens have been deposited in the University of Arizona Entomology collection.

RESULTS

Floral Biology

Arctostaphylos pungens individuals typically produced several hundred inflorescences each year. An average of 10.2 (+ 0.2 S.E.) flowers were produced per inflorescence in 1998 (n = 16 inflorescences). Individual flowers remained open an average of 3.3 d (n = 30 flowers). Flowers within an inflorescence usually matured sequentially, although in some cases most were open on the same day. Open, unrobbed flowers sampled at mid-morning contained 0.42 + 0.78 μl of nectar (n = 38 flowers on 17 plants).

In 1998, an exceptionally wet year, flowering began on 1 February, peaked on 4 March, and ended on 25 March. Approximately 36% of marked flowers set fruit. In contrast, 1999 was extremely dry. Flowering began about 2 weeks earlier than in 1998, but there was never a strong population-level bloom. By 1 February most inflorescences were static, neither initiating new flowers nor dropping old, dry corollas. Desiccated flowers remained on the plants for approximately 4 weeks and then abscised. No fruits matured at Molino Basin in 1999.

Mating System

Arctostaphylos pungens appears to be largely self-incompatible. Although selfed flowers produced as many fruits as did hand-outcrossed flowers, an ANOVA revealed that fruit weights differed significantly among pollination treatments ($F_4 = 114.54$, $P < 0.0001$; Table 1), and many fruits derived from selfed flowers apparently did not contain viable seeds. Tukey-Kramer post-hoc tests showed that hand-outcrossed flowers and unbagged control flowers produced significantly heavier fruits than those produced by the three self-pollination treatments (bagged control, geitonogamous self-pollination, and buzz self-pollination). Fruits from two of the self-pollination treatments (bagged control and geitonogamous self-pollination) were of very low weight and not significantly different from each other. Those from the buzz self-pollination treatment were significantly heavier than those of the other self-pollination treatments.

Floral Visitors

Floral visitors were observed for 38 hr. Data from 1206 visits by 46 insect taxa during 565 foraging bouts are summarized in Table 2. Appendix 1 lists data by taxon. Bouts ranged from 1-24 visits in length; mean bout length was 2.1 (+ 0.1) visits. Visits by 12 insect species (10 bees and 2 flies) accounted for 93% of visits and 88% of bouts

TABLE 1. Arctostaphylos pungens produced significantly heavier fruits ($F_{4,387}$ = 114.54, $P < 0.0001$) and produced fruits from a greater proportion of flowers ($F_{4,143}$ = 3.32, $P < 0.0124$) when experimentally outcrossed than when selfed. Fruit set was calculated as the number of fruits/number of flowers initiated. Data are presented ± 1 S.E. Values with the same lower case postscript are not significantly different.

Experimental Treatment	Sample Size (no. fruits)	Mean Fruit Weight (g)		Sample Size (no. flowers)	Mean Fruit Set	
Hand outcrossed	157	0.35 ± 0.01	a	42	0.51 ± 0.04	d
Hand selfed	97	0.09 ± 0.01	b	24	0.48 ± 0.05	d
Buzz selfed	25	0.17 ± 0.02	c	24	0.33 ± 0.05	e
Bagged control	35	0.08 ± 0.02	b	24	0.36 ± 0.05	e
Unbagged control	78	0.33 ± 0.01	a	34	0.35 ± 0.04	e

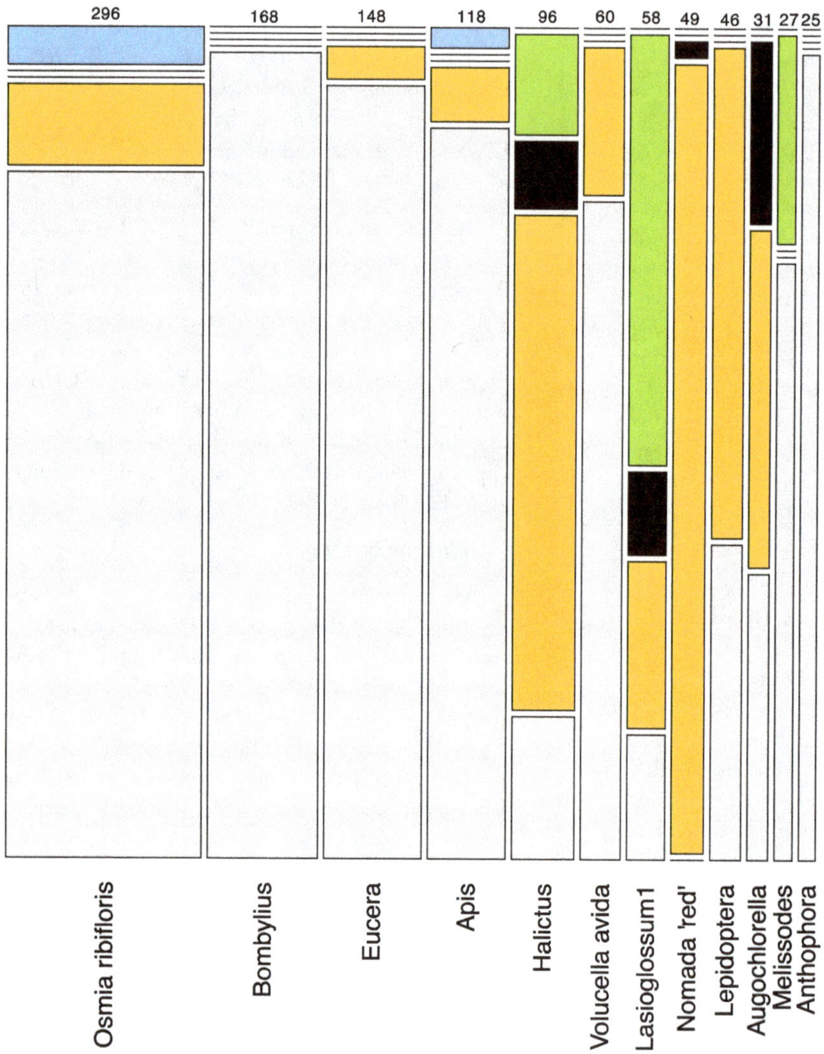

FIGURE 1. Mosaic plot of foraging behaviours of the 12 most common visitors to *Arctostaphylos pungens*, indicating that most flower visitors employed more than one foraging strategy. Visitors are arranged from left to right according to abundance, and sample sizes are given above each bar. From bottom to top, behaviors are: legitimate nectar collection (grey), secondary nectar-robbing (orange), primary nectar-robbing (red), buzz pollination (green), and legitimate pollen collection (blue). A horizontal line indicates that the taxon did not perform a behavior. Observations of nectar thievery have been omitted due to small sample size. All but two of these exhibited mixed foraging behaviours. Details on individual taxa are provided in the Appendix.

observed. The most common visitors were bees of the genera *Osmia, Augochlorella, Lasioglossum, Halictus, Nomada,* and *Eucera*; also common were bee flies (*Bombylius*) and flower flies (*Volucella*). Multiple individuals of all but two of these 12 species exhibited mixed foraging behaviours (Fig. 1).

Legitimate visits for pollen and nectar. -- The most common behaviour (70% of all observed visits) was legitimate nectar collection, i.e., a visit in which an insect inserted its mouthparts through the opening of the corolla tube and probed the nectaries at the base of the ovary. Ten

TABLE 2. Foraging behaviours of insects visiting *Arctostaphylos pungens* flowers. Based on 1202 visits by 46 insect species observed during 565 foraging bouts. Behaviours are defined in the text.

Foraging behaviour	Number of observed visits	As % of all observations	% of insect species employing behaviour
Legitimate nectar collection	846	70,4	65,7
Secondary nectar robbing	263	21,9	65,7
Primary nectar robbing	24	2	14,3
Buzz pollen collection	52	4,3	11,4
Legitimate pollen collection	17	1,4	5,7
Nectar thievery	4	0,3	2,1

bee species and 10 fly species, as well as small butterflies (65.7% of all visitor species), were observed to forage this way (Appendix 1). Five bee species sonicated ("buzzed") the flowers for pollen, accounting for 4.3% of all insect visits (Table 2). When these visitors collected pollen, they usually clung to the flowers and vibrated their flight muscles, causing pollen to rain out of the poricidal anthers. Two other bee species collected pollen by inserting the forelegs into the flower and scraping dehisced pollen from the trichomes on the internal surface of the corolla (1.4% of all visits, Table 2).

Illegitimate visits. -- Six bee species acted as primary nectar robbers. This behaviour was recorded in only 2% of all visits observed in 1998 (Table 2), even though ~ 70-90% of corollas were perforated that year. Primary nectar robbers were observed to cut slits or oval punctures through the corolla, then inserted their mouthparts without contacting the stigma or anthers. They occasionally made a second perforation in the same flower immediately after making the first, possibly because they could not access all of the nectar from the first hole. We assumed that all perforations were observed (which were highly consistent in size, shape, and placement) were due to the actions of primary nectar robbers.

Primary robbing significantly lowered nectar standing crop in 2001, even though nectar-robbing rates were extremely low that year (6% of flowers robbed; n = 302 flowers) relative to the same time period in 1998. By mid-morning, robbed flowers contained 82% less nectar on average than did adjacent unperforated ones (paired t-test, $t = 2.8$, $P = 0.008$, n = 38 flower pairs).

Secondary nectar robbers foraged at the holes made previously by primary robbers. Secondary robbing was the most common foraging behaviour after legitimate nectar collection, representing 22% of observed visits (Table 2). Eleven bee species, 2 wasp species, and 11 fly species, as well as diverse small butterflies (66% of all visitor species), fed this way (Appendix 1).

Nectar thieves fed by forcing their mouthparts through the petal lobes of nearly open, nectar-containing buds. Nectar thievery represented < 1% of visits (Table 2), all of them by the bee *Osmia ribifloris* (Megachilidae), which was overall the most common floral visitor.

Mixed foraging behaviours. -- About half of all floral visitor species exhibited mixed foraging behaviours at either the species or individual level, using two, three, or four of the six foraging strategies we recorded (Appendix 1). Among visitor species exhibiting a single behaviour, six were only seen foraging legitimately for nectar, two only foraged for pollen by sonicating flowers, one acted only as a primary robber, and 10 acted only as secondary robbers. It is likely that many of these species too had mixed foraging strategies, but that they went undetected because these species were relatively rare. Indeed, of the 13 visitor species observed 25 or more times, all but two exhibited mixed behaviours (Fig. 1).

Overall, 14 of the 15 insect species that foraged legitimately for pollen or nectar also included at least one illegitimate behaviour in their foraging repertoire. The most common behavioural combinations employed by foraging insects were legitimate nectar collection and secondary robbing (13 species); legitimate collection of nectar and sonication to obtain pollen (three species); legitimate nectar collection and primary robbing (three species); and primary and secondary robbing (three species).

Some species that mixed foraging behaviours were doing so at the within-individual level. These individuals switched behaviours within a foraging bout on an individual plant, and occasionally even while visiting a single flower. We observed switches in behaviour in 7.4% of 565 recorded foraging bouts (Table 3); the actual frequency is certainly higher, since many visitors disappeared from view while still foraging. Of these behavioural switches, more than 90% involved shifts between illegitimate and legitimate visits. The large majority of switches were between legitimate nectar collection and secondary nectar-robbing (Table 3).

In the large majority of foraging bouts (92.6%), however, individuals were observed to use a single foraging behaviour. For instance, some *Halictus* species bees were seen to bypass open, unperforated flowers while searching for perforated flowers to secondarily rob; at the same time, other individuals of the same species were foraging legitimately at undamaged flowers on the same plant.

TABLE 3. Switches in foraging behaviours during a bout by the most common floral visitors observed to switch. An "X" indicates that a species was observed to switch between that pair of behaviours during at least 1 bout. The total number of bouts during which a species switched behaviours is given, followed by the percentage of all bouts in which behavioural shifts were observed. Abbreviations: LN = legitimate nectar collection; LP = legitimate pollen collection; BZ = buzz pollination; 1NR = primary nectar robbing; 2NR = secondary nectar robbing; NT = nectar thievery.

Taxon	Behavioural Switches Observed					No. switching Bouts	Switching as % of all bouts
	LN/2NR	LN/1NR	BZ/2NR	BZ/1NR	1NR/2NR		
Augochlorella sp.	1	1			1	3	15
Halictus sp.	2	1	1	1	1	6	13,6
Lasioglossum sp. 1	3	2	1		1	7	23,3
Nomada sp. 'red'					1	1	4,3
Osmia ribifloris	10					10	6,7
Apis mellifera	1					1	2
Bombylius sp.	3					3	4,2
Bombyliadae sp.	1					1	100
Volucella avida	5					5	20
Lepidoptera spp.	3					3	8,8

Other floral visitors. -- Two other groups of visitors were commonly recorded on *A. pungens* flowers. Three species of thrips were frequently observed feeding within flowers. The most abundant of these, *Orothrips kelloggii* (Thysanoptera: Aelothripidae), was found in 68% of all flowers and on every plant. It damaged flowers by laying eggs in the pedicels and by feeding on corolla sap, nectar, and pollen. When males dispersed from their natal flowers, they passively transported pollen between flowers (Richardson 1999). Anna's hummingbirds (*Calypte anna*) defended breeding territories at Molino Basin and occasionally foraged at *A. pungens* flowers. Bill insertions usually resulted in severe damage to floral tissue. Because nectar standing crop was so low, we speculate that these birds were primarily feeding on thrips, a food source reported for other hummingbird species (Bené 1945).

DISCUSSION

Arctostaphylos pungens blooms exceptionally early in the year in the southwestern United States. Even though flower abundance and phenology appear to be strongly influenced by late winter and early spring weather and nectar volumes per flower are rather low, it is apparently a critical resource for a large and diverse community of flower-visiting insects, probably because it is usually the only plant in abundant flower in its habitat at these times of year. As shown here, most of these visitors do not consistently visit in ways likely to result in pollination (i.e., 'legitimate' nectar foraging or buzz pollen collecting). Rather, a significant portion of the time they function as nectar robbers and thieves unlikely to benefit the plant. In many of these insect species, individuals pursue a single behaviour while foraging on an individual plant, even though other individuals (and possibly the same individual at other times) exhibit a different foraging strategy. Individuals of other visitor species shift among as many as four different behaviours within a foraging bout on a single plant. These observations raise a number of issues with regard to the ecological and evolutionary consequences of these behaviours to the plants they visit, to the insects themselves, and to the mutualisms that at times exist between them.

The Plant Perspective

Arctostaphylos pungens produces significantly heavier fruits when it has the potential to outcross (Table 1), underscoring the importance of floral visitors. However, a variety of foraging behaviors were employed by these visitors, and, as in other pollination systems (e.g. Herrera 1987; Irwin et al. 2001), it is likely that some foragers are more effective pollen vectors than others. Although we did not measure the individual contributions of foragers to pollination of this plant, the specifics of the different foraging behaviours permit some reasonable inferences.

It is likely that the two common foraging behaviours we refer to as "legitimate" – buzz pollen collection and nectar collection via the corolla opening – resulted in pollen transfer and were thus mutualistic in nature, although not necessarily to the same degree. Buzz-pollinating species forage by clinging to the pendulous flowers and vibrating their flight muscles, causing pollen to rain out of the poricidal anthers. These behaviours appear to bring them into contact with the stigma, effecting pollination. Furthermore, female bees that vibrated the flowers for pollen were densely hairy and, when captured, were typically carrying large loads of pollen both in their scopae and dispersed over their bodies (L.L. Richardson, *personal observation*). It was more difficult to observe if and when pollen transfer took place during legitimate nectar-collecting visits, due to the flowers' small size and tightly constricted openings. However, most nectar-collecting insects probably made contact with the prominent stigma, as well as with the recurved anther appendages. We did note that the two most common legitimately nectar-foraging bees, *Osmia ribifloris* and *Eucera* sp., commonly had pollen adhering to their facial hairs. The morphology and foraging behaviours of the two flies that made many legitimate nectar visits (*Bombylius* sp. and *Volucella avida*) likely rendered them poor pollinators.

In contrast to its legitimate visitors, the nectar thieves and robbers of *A. pungens* are unlikely to transfer pollen. Nectar thieves force their way into *A. pungens* buds to feed upon nectar. As the anthers do not dehisce until buds have opened (L.L. Richardson, pers. obs.), it is unlikely that these visitors collect pollen, although it is possible that they deposit some if they have recently made a legitimate visit to another flower. Nectar-robbers do pick up and deposit pollen in some plants (Higashi et al. 1988; Scott et al. 1993; Navarro 2000). However, these species generally have stigmas and anthers that are either highly exserted or else that can be contacted via the corolla perforation. The structure of the *A. pungens* flower makes these phenomena unlikely.

It is possible, however, that nectar-robbing visits could lead to changes in pollinator behaviour that would benefit the plant. Nectar-robbing significantly reduced nectar volumes. In some plant species, pollinators make briefer visits to heavily robbed plants (Irwin 2003), increasing outcrossing rates and hence reproductive success (Maloof 2001; Richardson 2004). In other species, however, reduced nectar volumes lead pollinators to abandon heavily robbed plants or even to avoid them entirely, reducing plant reproductive success (Irwin 2000). Reduced nectar levels could even have induced pollinators to shift to robbing strategies (e.g., if robbing is a more effective method of obtaining the last dregs of nectar from nearly drained flowers), but we are unaware of any report of this phenomenon in the literature. Further experimental studies would be necessary to provide evidence for these effects in *A. pungens*.

The Insect Perspective

Research on nectar-robbing has typically focused on its costs and benefits to plants, rather than on its nature as a foraging strategy (Irwin et al. 2010). Perhaps for this reason, it has commonly been overlooked that robbing and pollinating often are not in fact distinct behaviours performed by different visitor species. Rather, many robbers also act as pollinators of the same or other plants, whereas at least some pollinators also rob. Honey bees have repeatedly been noted to function simultaneously as legitimate pollen-collectors, legitimate nectar-collectors and secondary nectar-robbers on a single plant species (Weaver 1956; Free 1968, 1993; Joubert et al. 1977; Villalobos & Shelly, 1996; Dedej & Delaplane 2005), much as they did on *A. pungens* (Appendix 1). Honey bees and other insects may also function as non-mutualist pollen thieves (*sensu* Inouye 1980), even if they sometimes pollinate the same plants when they forage for nectar (Hargreaves et al. 2009, 2012). Similarly, some bumblebee species both pollinate (often via sonication) and act as either robbers or thieves of the same plant (Kwak 1977; Free 1993; Morris 1996; Oleson 1996; Newman & Thomson 2005). Mixed pollinating-robbing behaviours have also been noted in a few other Hymenoptera (Hurd & Linsley 1963; Rust 1979). The present study indicates that complex mixed behaviours may be common in a much broader spectrum of flower-feeding insects, including Diptera and Lepidoptera.

Factors that might favor mixed foraging behaviours within and among floral visitors and the rules that might govern behavioural switches have not yet been investigated in this system, and little relevant information is available from any other system (but see Dedej & Delaplane 2005; Newman & Thomson 2005). Data from other studies generate two observations, however, that may shed light on the nature of these phenomena.

First, certain generalist pollinators unable to obtain nectar of a certain plant species by legitimate means shift to nectar-robbing after a few visits, rather than abandoning that plant to forage legitimately at other species (Joubert et al. 1977; Rust 1979). When *A. pungens* flowers, small generalist insects have few other floral resources available to them, and they may be forced to forage upon it in any way that they can. Such forced switches may also explain the strategy-switching we commonly observed within individual foraging bouts of certain visitor species. In a second North American *Arctostaphylos* species, *A. pringlei*, flowers vary greatly in corolla width; we observed bees (*Nomada* sp.) foraging legitimately for nectar on flowers with relatively wide corollas and acting as primary robbers on flowers with more restricted entrances (Richardson 1999).

Secondly, among insects unable to chew holes through corollas, shifts from legitimate visitation to secondary nectar-robbing have in some cases been shown to be contingent upon the actions of primary nectar-robbers (Hawkins 1961; Free 1968; Barrows 1980; Villalobos & Sherry 1996). In one detailed study, Free (1968) showed that the abundance of honey bees on runner beans (*Phaseolus multiflorus*) increased greatly in the presence of primary nectar-robbing bumblebees. When bumblebees were present, most honey bees shifted from legitimate nectar collection visits to secondary nectar-robbing, suggesting that secondary robbing was a more efficient strategy, but one constrained to certain community contexts. In *Vaccinium ashei*, Dedej and Delaplane (2005) have demonstrated that honey bees in fact receive higher energetic returns from secondary robbing than from legitimate visits. Secondary nectar robbing, as well as facultative switching from legitimate nectar foraging to secondary robbing, may thus be more common at plants frequented by diverse guilds of generalist foragers that include primary nectar robbers., such as in *Impatiens capensis* (Rust 1979) and *A. pungens*.

Implications for Mutualism

That floral visitors mix beneficial and detrimental foraging behaviours is ecologically interesting, but does it matter to the evolution of pollination mutualisms? We would argue that it does. First, selection for increased specificity within mutualisms is expected to be weak or absent in the face of spatial and temporal variation in mutualist quality (Howe 1984; Schemske & Horvitz 1984). In pollination systems, both across-individual variation in foraging behaviours within visitor species and within-individual, context-dependent behavioural shifts undoubtedly generate low predictability as to which visitors will be the most effective pollinators at a given time and place. Thus, variable foraging strategies may be one more factor that keeps most plant/pollinator interactions highly generalized (Waser et al. 1996).

Second, it has become of great interest to understand how mutualisms can persist at all in the face of exploitation. Theoreticians have argued that without mechanisms to retaliate against or to exclude exploiters, mutualisms cannot be evolutionarily stable (Axelrod & Hamilton 1981; Bull & Rice 1991; Yu 2001). A variety of floral traits have been interpreted as defences against nectar robbing, including thickened calyces, toxic nectar and glandular trichomes (Prys-Jones & Willmer 1992; Agrawal et al. 2000; Irwin et al. 2010), yet many plants have no apparent defences and are heavily robbed. The behavioural spectrum documented here demonstrates that mutualism and exploitation can be intertwined in surprisingly complex ways, perhaps obviating the (theoretical) requirement for such defences. Plants able to deter visits by some exploiters might well experience a net increase in mutualist visits. In the present case, if *A. pungens* floral morphology prevented nectar robbing, behaviourally flexible nectarivores like *O. ribifloris* would presumably forage in a more cooperative fashion (i.e., as legitimate foragers), thereby increasing plant fitness. However, it is equally possible that if *A.* pungens could prevent primary nectar robbing, some facultative secondary robbers would abandon the plant altogether, leading to a loss of legitimate visits that result in pollination. These competing hypotheses should be tested in a plant system where nectar robbing can be effectively manipulated at the population level.

One of the most striking features of the behavioural spectrum exhibited by floral visitors is that species with foraging choices visit flowers legitimately as frequently as they do, even when robbing might be more efficient and plants do not defend against robbing. In this study, 13 of the 14 nectar-robbing species that exhibited foraging flexibility regularly, and indeed more commonly, visited flowers in ways likely to have led to pollination. Theoretical studies of exploitation generally assume that because cheating is often more efficient than foraging legitimately, organisms with choices should cheat whenever possible. Indeed, the only study to date of the relative economics of legitimate vs. robbing visits has shown that the latter are more energetically advantageous to honey bees (Dedej & Delaplane 2005). Further work on the exploitation of pollination mutualisms is needed to explain how these strategies can coexist.

Appendix

Additional supporting information may be found in the online version of this article:

Appendix I. Insects observed to visit flowers of *Arctostaphylos pungens* during 1998 and 1999.

Acknowledgements

For field assistance, we thank Melinda McElveen, Jennifer Skinder, and Margaret McIntosh, as well as Lynn Adler, who collected the 2001 nectar data. Rebecca Irwin, Lucinda McDade, Molly Hunter, Sarah Richardson, Nicole Rafferty, and Goggy Davidowitz provided helpful comments on the manuscript. This work was supported by grants from the Department of Ecology and Evolutionary Biology and the Vice President's Office for Research and Graduate Studies of the University of Arizona. It represents part of a master's thesis submitted by LLR to the Graduate College of the University of Arizona.

References

Agrawal AA, Rudgers JA, Botsford LW, Cutler D, Gorin, JB Lundquist CJ, Spitzer BW, Swann AL (2000) Benefits and constraints on plant defense against herbivores: spines influence the legitimate and illegitimate flower visitors of yellow star thistle, *Centaurea solstitialis* L. (Asteraceae). Southwestern Naturalist 45: 1-5.

Axelrod R, Hamilton WD (1981) The evolution of cooperation. Science 211: 1390-1396.

Barrows EM (1980) Robbing of exotic plants by introduced carpenter and honey bees in Hawaii, with comparative notes. Biotropica 12: 23-29.

Bené, F (1945) The role of learning in the feeding behavior of black-chinned hummingbirds. The Condor 47(1): 3-22.

Borror DJ, Triplehorn CA, Johnson NF (1989) An Introduction to the Study of Insects. Saunders College Publishing, Philadelphia.

Bronstein JL (2001) The exploitation of mutualisms. Ecology Letters 4: 277-287.

Bronstein JL (2003) The scope for exploitation within mutualistic interactions. In: Hammerstein P (ed) Genetics and Cultural Evolution of Cooperation. MIT Press, Cambridge, pp 185-202.

Brown DE (1982) Biotic communities of the American Southwest - United States and Mexico. Desert Plants 4: 1-342.

Bull JJ, Rice WR (1991) Distinguishing mechanisms for the evolution of cooperation. Journal of Theoretical Biology 149: 63-74.

Dedej S, Delaplane KS (2005) Net energetic advantage drives honey bees (*Apis mellifera* L) to nectar larceny in *Vaccinium ashei* Reade. Behavioral Ecology and Sociobiology 57: 398-403.

Diggs GM, Jr (2009) *Arctostaphylos*. In: Flora of North America Editorial Committee, (eds) Flora of North America North of Mexico, volume 3. Oxford University Press, New York and Oxford, pp 356-357.

Free JB (1968) The behaviour of bees visiting runner beans (*Phaseolus multiflorus*). Journal of Applied Ecology 5: 631-638.

Free JB (1993) Insect pollination of crops. Academic Press, London.

Hargreaves, AL, Harder LD, Johnson SD (2009) Consumptive emasculation: the ecological and evolutionary consequences of pollen theft. Biological Reviews 84(2): 259-276.

Hargreaves, AL, Harder LD, Johnson SD (2012) Native pollen thieves reduce the reproductive success of a hermaphroditic plant, *Aloe maculata*. Ecology 91(6): 1693-1703.

Hawkins RP (1961) Observations on the pollination of red clover by bees. I. The yield of seed in relation to the numbers and kinds of pollinators. Annals of Applied Biology 49: 55-65.

Herrera CM (1987) Components of pollinator "quality": comparative analysis of a diverse insect assemblage. Oikos 50: 79-90.

Higashi S, Ohara M, Arai, H, Matsuo K (1988) Robber-like pollinators: overwintered queen bumblebees foraging on *Corydalis ambigua*. Ecological Entomology 13: 411-418.

Howe HF (1984) Constraints of the evolution of mutualism. American Naturalist 123: 764-777.

Hurd PD Jr, Linsley EG (1963) Pollination of the unicorn plant (Martyniaceae) by an oligolectic, corolla-cutting bee (Hymenoptera: Apoidea). Journal of the Kansas Entomological Society 36: 248-252.

Inouye DW (1980) The terminology of floral larceny. Ecology 61: 1251-1253.

Irwin RE (2000) Hummingbird avoidance of nectar-robbed plants: spatial location or visual cues. Oikos 91: 499-506.

Irwin RE (2003) Impact of nectar robbing on estimates of pollen flow: conceptual predictions and empirical outcomes. Ecology 84: 485-495.

Irwin RE, Brody AK (1999) Nectar-robbing bumble bees reduce the fitness of *Ipomopsis aggregata* (Polemoniaceae). Ecology 80: 1703-1712.

Irwin RE, Brody AK (2000) Consequences of nectar robbing for realized male function in a hummingbird-pollinated plant. Ecology 81: 2637-2643.

Irwin RE, Brody AK, Waser NM (2001) The impact of floral larceny on individuals, populations, and communities. Oecologia 129: 161-168.

Irwin RE, Bronstein JL, Manson JS, Richardson L (2010) Nectar-robbing: ecological and evolutionary perspectives. Annual Review of Ecology, Evolution and Systematics 41: 271-292.

Joubert G, Tasei JN, Delaude A (1977) Pollinisation du trèfle violet diploide (*Trifolium pratense* L.) par les apoides et particulièrement par les abeilles domestiques hybride (*Apis mellifica causasica* x *A. mellifica ligustica*). Apidologie 8: 259-280.

Kearney T, Peebles RH (1960) Arizona flora. University of California Press, Berkeley.

Kwak M (1977) Pollination ecology of five hemiparasitic, large-flowered Rhinanthoideae with special reference to the pollination behaviour of nectar thieving, short-tongued bumblebees. Acta Botanica Neerlandica, 26: 97-107.

Maloof JE (2001) The effects of a bumble bee nectar robber on plant reproductive success and pollinator behavior. American Journal of Botany 88: 1960-1965.

McAlpine JF (ed) 1981-1989. Manual of nearctic Diptera. Ottawa: Research Branch, Agriculture Canada.

Michener CD, McGinley RJ, Danforth BN (1994) The bee genera of North and Central America (Hymenoptera: Apoidea). Smithsonian Institution Press, Washington.

Morris WF (1996) Mutualism denied? Nectar-robbing bumble bees do not reduce female or male success of bluebells. Ecology 77: 1451-1462.

Navarro L (2000) Pollination ecology of *Anthyllis vulneraria* subsp. *vulgaris* (Fabaceae): Nectar robbers as pollinators. American Journal of Botany 87: 980-985.

Newman DA, Thomson JD (2005) Effects of nectar robbing on nectar dynamics and bumblebee foraging strategies in *Linaria vulgaris* (Scrophulariaceae). Oikos 110: 309-320.

Olesen JM (1996) From naiveté to experience: Bumblebee queens (*Bombus terrestris*) foraging on *Corydalis cava* (Fumariaceae). Journal of the Kansas Entomological Society 69: 274-286.

Prys-Jones OE, Willmer PG (1992) The biology of alkaline nectar in the purple toothwort (*Lathraea clandestina*): ground level defenses. Biological Journal of the Linnean Society 45: 373-388.

Richardson L (1999) Mutualist and antagonist flower visitors: a pollination study of two species of manzanita (*Arctostaphylos:* Ericaceae). Masters Thesis, University of Arizona, Tucson.

Richardson SC (2004) Are nectar-robbers mutualists or antagonists? Oecologia 139: 246-254.

Rust RW (1979) Pollination of *Impatiens capensis*: pollinators and nectar robbers. Journal of the Kansas Entomological Society 52: 297-308.

Schemske DW, Horvitz CC (1984) Variation among floral visitors in pollination ability: a precondition for mutualism specialization. Science 225: 519-521.

Scott PE, Buchmann SL, O'Rourke MK (1993) Evidence for mutualism between a flower-piercing carpenter bee (*Xylocopa californica*) and ocotillo (*Fouquieria splendens*): use of pollen and nectar by nesting bees. Ecological Entomology 18: 234-240.

Villalobos EM, Shelly TE (1996) Temporal and spatial variation in the foraging behavior of honey bees (Hymenoptera: Apidae) at Chinese violets. Florida Entomologist 79: 398-407.

Waser NM, Chittka L, Price MV, Williams NM, Ollerton J (1996) Generalization in pollination systems, and why it matters. Ecology 77: 1043-1060.

Weaver N (1956) The foraging behaviour of honeybees on hairy vetch. I. Foraging methods and learning to forage. Insectes Sociaux 3: 537-549.

Western Regional Climate Center (2012) Sabino Canyon, Arizona. www.wrcc.dri.edu/cgi-bin/cliMAIN.pl?azsabi

Yu DW (2001) Parasites of mutualisms. Biological Journal of the Linnean Society 72: 529-546.

14

How well do we understand landscape effects on pollinators and pollination services?

Blandina F. Viana[1]*, Danilo Boscolo[2], Eduardo Mariano Neto[1], Luciano E. Lopes[3], Ariadna V. Lopes[4], Patrícia A. Ferreira[1], Camila M. Pigozzo[5] and Luis M. Primo[1]

[1]*Instituto de Biologia, Universidade Federal da Bahia, Salvador, Bahia, 40170-210, Brazil*

[2]*Universidade Federal de São Paulo – UNIFESP, Campus Diadema, Zip 09972-270, Diadema, São Paulo, Brazil*

[3]*Centro de Ciências Biológicas e da Saúde, Universidade Federal de São Carlos, São Carlos, 13565-905, Brazil*

[4]*Centro de Ciências Biológicas, Universidade Federal de Pernambuco, Recife, Pernambuco, 50372-970, Brazil*

[5]*Centro Universitário Jorge Amado, Salvador, Bahia, 41745-130, Brazil*

Abstract— Many studies in the past decade, mostly in temperate countries, have documented the effects of habitat loss and fragmentation on species richness, composition, and abundance and the behaviour of pollinators. Changes in landscape structure are considered to be the primary causes of the limitation of pollination services in agricultural systems. Here, we review evidence of general patterns as well as gaps in knowledge that could be used to support the development of policies for pollinator conservation and the restoration of degraded landscapes. Our results indicate a recent increase in the number of studies on the relationships between pollination processes and landscape patterns, with some key trends already being established. Many authors indicate, for example, that the spatial organization of a landscape has a great influence on the survival and dispersal capacity of many pollinators, as spatial organization affects resource availability and determines the functional connectivity of the landscape. Additionally, the shape, size and spatial arrangement of the patches of each type of natural environment, as well as the occurrence of different types of land use, can create sites with different degrees of connectivity or even barriers to movement between patches, which can deeply modify pollinator flows through the landscape and consequently the success of cross-pollination. However, there are still some gaps, such as in the knowledge of which critical values of habitat loss can lead to drastic increases in pollinator extinction rates, information that is needed to evaluate at what point plant-pollinator interactions may collapse. We also need to concentrate research effort on improving a landscape's capacity to facilitate pollinator flow (connectivity) between crops and nesting/foraging areas.

Keywords: Pollinator-friendly landscapes, conservation, land management, matrix, land use, agricultural systems, pollinator crisis.

Introduction

The past decade has seen a worldwide concern over pollinator decline (see COP 5 CBD section II decision v/5). This concern has sparked a remarkable increase in studies that identify threats to pollinators and quantify the impact of pollinator decline on pollination services in natural and agricultural systems. Most studies point to landscape changes resulting from intensive land use and leading to habitat loss and fragmentation as one of the primary threats to pollination services (Kremen et al. 2002; Steffan-Dewenter & Westphal 2008; Winfree et al. 2009). There is evidence that several crops are directly affected by changes in landscape structure, resulting in productivity loss that endangers both biodiversity and the stability of food production in the world (Steffan-Dewenter et al. 2005; Tscharntke et al. 2005; Chacoff & Aizen 2006, Carvalheiro et al. 2010; Isaacs & Kirk 2010).

Recent reviews indicate that pollinator loss in agroecosystems is faster in the tropics than in temperate regions. According to Aizen et al. (2009), agriculture has become more pollinator-dependent over time, and this trend is more pronounced in the developing world, which comprises almost all tropical regions, than the developed world. They propose that a shortage of pollinators will intensify the demand for agricultural land, a trend that will also be more pronounced in the developing world. Thus, the increases in total cultivated area needed to compensate for pollination deficits would be smaller in developed countries and larger in other parts of the world. This difference suggests a future increase in land conflicts in the tropics as well as the acceleration of deforestation processes and intensification of human pressure on natural tropical vegetation remnants, which has important practical consequences such as increased species loss and the subsequent deterioration of plant-pollinator networks, thus further weakening pollination services (Carvalheiro et al. 2011, Garibaldi et al. 2011). In this context, there is an acute need to quickly identify and overcome knowledge gaps regarding the interplay between landscape patterns and pollination processes and to directly apply new knowledge to the management of productive and sustainable landscapes.

Previous reviews and meta-analyses of pollination deficit focused on particularly relevant questions: i) the importance of changes in the abundance of foraging plants to bee

*Corresponding author; email: blandefv@ufba.br

conservation (Carvell et al. 2006, based on 14 datasets); ii) the effect of habitat fragmentation on plant reproduction (Aguillar et al. 2006, based on 54 studies); and iii) the relationship between the distance from natural or semi-natural habitats and pollination service (Ricketts et al. 2008, based on 23 studies).

Our present review is based on 219 studies, 166 of which specifically address effects on pollinators, including a broad range of questions about plant-pollinator interactions from an ecological landscape perspective. This survey was then summarized into a scientometric analysis that enabled the quantitative and qualitative evaluation of temporal trends in the knowledge produced in this research area beginning when this subject first appeared in the literature. Thus, our study presents the results of a broad literature review of the effects of landscape changes on pollinators and pollination services. Through this review, we sought to identify underlying patterns and gaps in knowledge as well as to help with the development of guidelines for research and conservation that could support new policies at the landscape level to minimize pollination deficits in areas degraded by intensive land use.

First, we present a brief description of the methods used for the literature review and data analysis. Next, we discuss the conceptual problems identified in the publications reviewed, which were related to landscape ecology, and the terminology standardization used in our study. Then, we present the most important general patterns observed and our primary findings. Finally, we note gaps in the existing knowledge that must be addressed by research programs aimed at meeting the demands for conservation and sustainable management of pollinators at the landscape level.

MATERIAL AND METHODS

Database

The survey was carried out in late July 2011 in the Web of Science - Science Citation Index Expanded (http://portal.isiknowledge.com/), using the combination of the keywords 'Landscape AND Pollinat*'. The search was made with the filter 'topic' that searches for words defined in the title, keywords and in the body of the text. At a preliminary stage, we selected all articles that dealt directly with the effects of landscape changes on pollinators and pollination services. For analyzing the articles, a database was created that included a standardized list of the articles, which enabled compiling and quantifying the characteristics of those studies. This database is composed of 25 fields (Tab. I) selected for the purpose of extracting information from each analyzed publication within a scientometric perspective, enabling us to measure and quantify the scientific and technological progress within this research topic.

From this database, chronological changes and other possible interrelationships between any fields could easily be extracted to perform exploratory analyses with the aim of identifying general patterns and the temporal evolution of worldwide trends in the scientific literature on the relationship between landscape structure and the availability of pollinators and pollination services. These trends were summarized into graphs and percentage analyses to allow a critical understanding of the knowledge accrued over the years. To facilitate the visual comparison of database fields with numerous categories, some graphs of chronological change are presented with the data clustered into quinquennial groups of surveyed works.

Conceptual problems and terminology standardization

The definition of concepts, and in some cases their standardization, is an important step in surveying scientific knowledge for environmental management purposes. The greatest problem with the inadequate conceptualization and standardization of scientific terms is the risk of attributing the effect of a certain entity to a similar but essentially different factor, which can make it difficult to understand the text or lead to mistaken conclusions.

Some of the analyzed articles clearly confounded the processes of habitat loss and habitat fragmentation. According to Fahrig (2003), this kind of confusion should be avoided because the consequences of those processes for biological conservation are essentially different, although they usually occur together in nature. This confusion is less frequent, but not absent, in papers from 2004 on.

Another controversial concept was the definition of a 'landscape matrix', a term that has been used in several ways in the landscape ecology literature. The most-used concepts define the matrix as a 'dominant unit of the landscape' (spatially or functionally) or as a 'set of non-habitat units' (Metzger 2001); both characterize the matrix as a landscape feature. However, in many of the studies analyzed in the present review, the word 'matrix' was frequently used without an explicit definition, for example, to mean the areas adjacent to the studied fragment, which could be better defined as context or type of fringing environments, depending on the situation. Many times, those definitions were unspecific, thus making direct interpretations difficult. The same problem was also identified for the concepts of patch, fragment, corridor and even landscape.

Hence, to avoid making mistaken conclusions derived from conceptual problems, we decided to standardize all of the technical terminology related to landscape ecology in accordance with the review by Metzger (2001) and also with the formal concepts of fragmentation and habitat loss proposed by Fahrig (2003). Therefore, in the present study, landscape is defined as 'a heterogeneous mosaic formed by interactive units, given that this heterogeneity exists for at least one factor, according to one observer and at a given scale of observation' (Metzger 2001). This concept is relatively broad, as it enables a wide variety of units/habitats of different sizes to be considered landscapes. However, it establishes minimum criteria for us to use to separate the landscape from its constitutive elements such as patches, corridors, edges and different types of environments.

This standardization was used to organize the structure of our database to eliminate a priori the conceptual confusion from some articles and to standardize our analysis. Hence, information from all articles was revised and

TABLE I. Information selected for the analysis and its respective categories. For all items we included also a category: 'information not mentioned by the author'. The items "Landscape context" and "Landscape approach level" were described following Metzger (2001) and Fahrig (2003).

Items for analysis	Categories
Publication year	–
Surname of the first author	–
Country of the first author	–
Journal title	–
Geographic location of the study area (country and geographic coordinates)	–
Ecoregion/climatic zone where the study was carried out	1. tropical (including subtropical), 2. temperate (including boreal)
Ecosystem	1. agriculture, 2. forest, 3. grasslands, 4. savannah, 5. desert, 6. urban, 7. agroforestry system (AFS)
The kind of landscape matrix was explicitly declared by the	1. yes, 2. No
Kind of matrix	1. natural, 2. silviculture, 3. agriculture, 4. urban
Landscape context	1. natural, 2. agriculture, 3. urban, 4. mixed
Nature of the study	1. empiric, 2. review, 3. meta-analysis, 4. modelling, 5. conceptual, 6. opinion, 7. editorial
Nature of the method	1. descriptive, 2. bibliographic survey, 3. observational (sampling), 4. experimental, 5. modelling, 6. meta-analysis
Nature of the objectives of the study	1. descriptive, 2. establishing relationships, 3. modelling, 4. review
Landscape approach level	1. landscape, 2. buffers, 3. patches, 4. intra-patches
Size of the study area (total sampling range)	1. up to 1 ha, 2. between 1 and 10 ha, 3. between 10 and 100 ha, 4. between 100 and 1 000 ha, 5. > 1 000 ha, 6. global, 7. not specified
Level of biological organization analyzed (unit mentioned by the author)	1. individuals; 2. populations, 3. communities
Study object	1. pollinator, 2. plant, 3. plant/pollinator interactions
Response variables described by the author	–
Independent explanatory variables described by the author	–
Type of relationship between the explanatory and response	1. directly proportional, 2. inversely proportional, 3. no relationship
Functional or taxonomic group studied	–
Sampling method	1. pan-trap, 2. trap-nest, 3. entomological net, 4. Focal observation, 5. counting of the frequency of visitors, 6. translocation, 7. baits, 8. others
Pollinator specialization as described by the author	1. generalist, 2. specialist, 3. not specified
Pollinator sociality	1. social, 2. solitary
Number of citations until July 2011	–

standardized according to the concepts used in Metzger (2001) and Fahrig (2003) and corrected in our data matrices by including standardized columns, which are presented in Tab. 1.

Another important aspect for properly assessing the studies was the minimum relevant information that should be reported in the publications, but that, in most cases, was omitted by the authors. Some of the missing information was, for example, the extent of the study area and the types of land uses surrounding the studied patches (Tab. 1). When that type of information was implicit in the article we made an effort to recover and explicitly include it in our analysis.

RESULTS AND DISCUSSION

General patterns

In the present review, we found 219 studies focusing on ecological landscapes and pollination processes. Those studies had primarily been published since 2001 (Fig. 1) in 60 journals. Thirty-five of those journals represented only one article, whereas the journal Biological Conservation published 25 of the analyzed studies (11.4%; Appendix II).

In general, we observed a dominance of purely scientific journals specialized in ecology and conservation with barely any journals focusing on agricultural sciences, applied land management or food production technologies. This indicates that the relationship between landscape patterns and pollinator availability is not yet perceived as economically

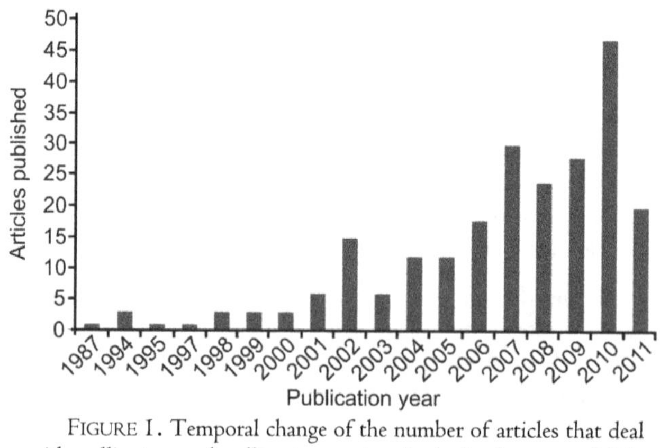

FIGURE 1. Temporal change of the number of articles that deal with pollinators and pollination services in the landscape until July 2011 (N = 219).

relevant outside of the biological and/or ecological sciences. Most studies were empirical/observational (166 studies; 75.8%), but some were reviews (29; 13.24%) that aimed to consolidate information, with a trend toward an increasing annual production of reviews (Fig. 2). We also identified nine modelling studies (4.1%), all from 2007 on.

The most frequently assessed biological organization levels were community (118 studies, 53.9%) and population (63 studies, 28.8%). Only 8 studies (3.6%) encompassed both communities and populations, and 7 studies (3.19%) focused on the behavioural responses of individual pollinators. From all 219 of the published studies initially selected by searching the online database, 166 (75.8%) directly addressed pollinators' response to landscape changes. Out of these 166 studies, 145 (87.3%) aimed to establish causal relationships between landscape patterns and pollinators availability, and 4 studies (2.41%) had merely descriptive approaches.

The first scientific study that explicitly analyzed the effects of changes in the spatial distribution of habitat on the activity of pollinator species was carried out in Brazil in the central Amazon (Powell & Powell 1987). However, between its publication and the year 2000, there was no meaningful increase in the number of studies on the relationship between changes in landscape structure and pollination processes, the literature being limited to only 15 (6.8%) papers in 13 years. It is worthwhile to highlight that among those publications are the empirical studies carried out by Aizen & Feinsinger (1994a; b) in Argentina, which used a community approach and reported a decrease in the richness and abundance of native pollinators in small and isolated fragments compared to continuous environments, with strong consequences for pollination. Those early studies can be considered the pioneers in establishing relationships between landscape structure, pollinator diversity and pollination processes.

It was only after 2000 that we could identify an increase in the number of papers on the effects of landscape structure on pollinators (Fig. 1). We found only 10 publications (6.2% of the 161 studies) between 1996 and 2000, whereas in the two subsequent quinquennia 51 (23.3%) and 151 (68.9%) papers were published, respectively. This increase

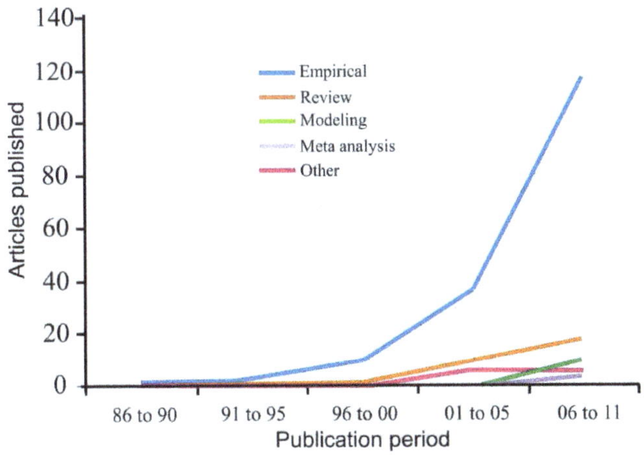

FIGURE 2. Number of studies of each type in 4-year time intervals from January, 1986 to July, 2011. (N = 219)

was due primarily to the effort of work-groups from the United States, Germany and England (Fig. 3). This increase was most likely stimulated by the worldwide pollinator crisis identified in the previous decade (Buchmann & Nabhan 1996; Kearns et al. 1998), which created a scientific demand for an understanding of the causes of pollinator decline, particularly the decline of bee populations in the northern hemisphere.

This temporal pattern of increasing publications was uneven between the climatic zones where the studies were carried out (Fig. 4). Our overall results indicated a clear trend toward an increase in the number of studies developed in temperate zones compared to tropical zones. Between 2001 and 2005, both climatic zones were similarly represented in terms of the number of publications with slightly more works carried out in tropical environments (16 studies in temperate regions and 19 in tropical regions). However, in the five following years (2006 to 2011), there was a great increase in the number of studies produced in temperate areas (79) that was not accompanied by an increased number of studies produced in the tropics in the same period (43). Among all 219 studies, most compared patches of the same landscape and few compared different

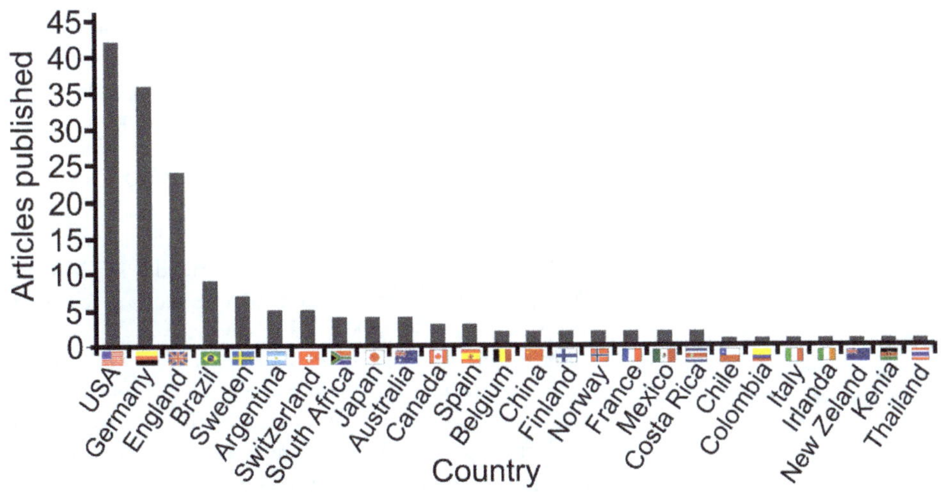

FIGURE 3. Country of the first authors of studies which are directly related to pollinators responses to landscape changes, revisions were excluded from this analysis (N = 146).

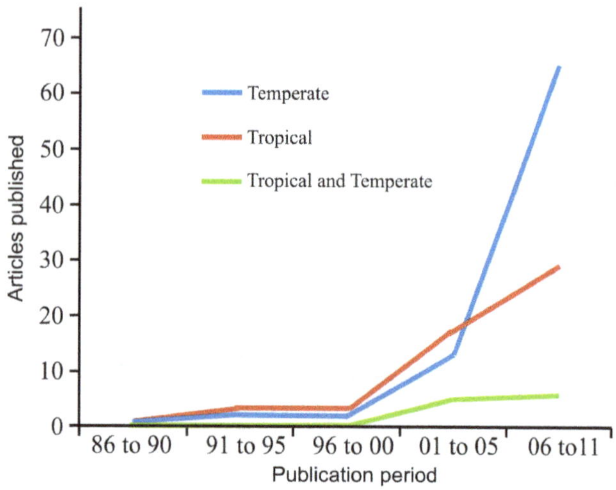

FIGURE 4. Overall number of studies produced in each one of the Climate zones by period (N= 219)

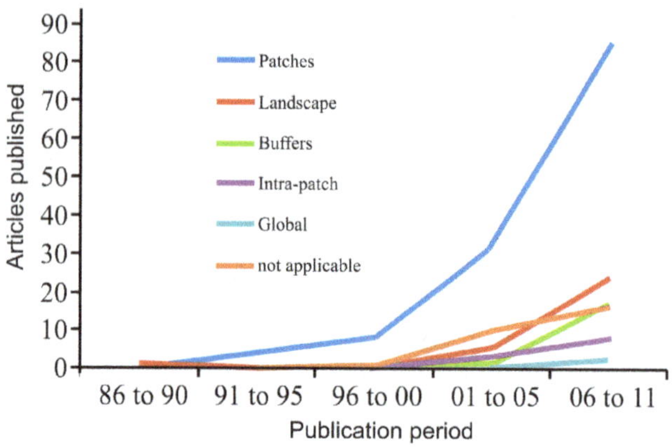

FIGURE 5. Temporal changes (number of studies/period) of the approach Level of the study (N= 219).

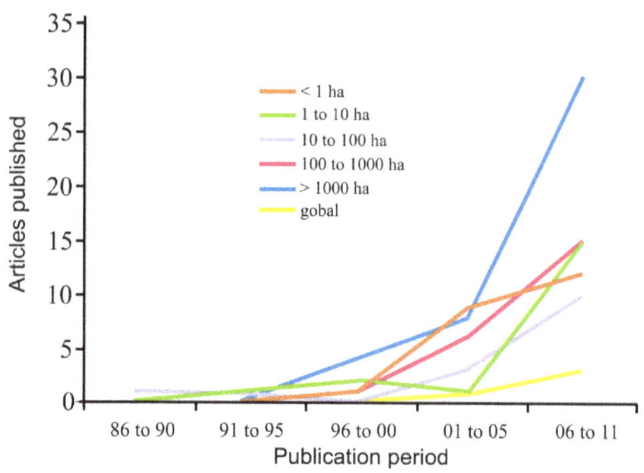

FIGURE 6. Temporal changes (number of studies/period) of landscape or patch size where the studies were carried out. Only those studies that specified patch or landscape size were considered. (N= 124).

landscapes, indicating that most of the available information is at the patch rather than the landscape level; this continued to occur in the last six years (Fig. 5). Only a few studies approached the problem at hand using whole landscapes as sampling units; about half of those studies used fixed distances around sampling sites (buffers) to assess how landscape structure affected pollinators. There was a noticeable increase in the number of publications using this type of approach after 2005 (Fig. 5). Additionally, we observed a large variation in the absolute meaning of patch size categories among studies, with no standardization of categories. The same patch size could be categorized as "large" in one study but "small" in another, even when the two studies addressed closely related organisms or processes. In addition, 77 studies did not provide any information on the size of the landscapes/patches studied. Out of the studies that provide information on habitat size (124), most (42) assessed patches or landscapes larger than 1 000 ha (Fig. 6).

We could also identify in our review a temporal change in the importance assigned to the inter-habitat matrix. References to matrix characteristics were scarce until 2005 (less than a third of articles) (Appendix III A). However, after 2006 the matrix began to be mentioned more frequently. In this period, the number of studies that did not cite matrix characteristics or left them implicit was very similar to the number of articles that commented explicitly about this matter. The proportion of studies that mention the matrix is also increasing among the full set of analyzed studies in the last decade (Appendix III A). This indicates a significant increase in interest in this aspect of the landscape, whose importance had already been noted in some existing publications (Clergeau & Burel 1997; Develey & Stouffer 2001; Rejinfo 2001). Among the 166 studies that focused on the relationship between the landscape and pollinators, most reported matrices composed of mixed, agricultural or natural environments (Appendix III B). The matrices composed of pastures, urban environments and silvicultures were few and represented similar numbers. After 2006, only a few empirical studies provided no information on the spatial context where the work was carried out.

Primary qualitative findings

In spite of the need for advancing the work of existing studies, some relationships could be clearly established in the surveyed literature. Many authors demonstrated that the spatial organization of the landscape has a great influence on the survival and dispersal capacity of many pollinator species, as spatial organization affects resource availability (Andersson et al. 2007, Jha & Vandermeer 2010; Cruz-Neto et al. 2011; Roulston & Goodell 2011) and determines functional connectivity, i.e., 'the capacity of the landscape (or landscape units) to facilitate biological flows' (Metzger 2001) in a given region (Steffan-Dewenter & Tscharntke 1999; Brosi et al. 2007). The shape, size and spatial arrangement of the patches of each type of natural environment as well as the occurrence of different types of land use can create sites with different degrees of connectivity or even barriers that impede animal movement through the landscape (Kreyer et al. 2004; Ekroos et al.

2008; Ricketts et al. 2008), which can strongly modify pollinator flows and consequently the success of cross-pollination (Gathmann & Tscharntke 2002; Goverde et al. 2002).

It is also widely known that higher proportions of natural or semi-natural environments in the landscape enable the maintenance of pollinator species that would otherwise go locally extinct with the suppression of the native vegetation (Laurance et al. 2002; Lennartson 2002; Taki & Kevan 2007; Hadley & Betts 2009). Although many types of crops are able to provide food to many pollinator species (Westphal et al. 2003; Albrecht et al. 2007; Holzschuh et al. 2008; Klein et al. 2008), food is not the only resource needed for pollinators' survival. In addition to food, animals need adequate sites for nesting and reproduction (Steffan-Dewenter & Tschanrtke, 1999; Knight et al., 2009). For many pollinator species, those sites are found more frequently in natural environments (Gathmann & Tscharntke. 2002, Westphal et al., 2003, Dixon, 2009) composed of primary vegetation or as the result of natural regeneration or restoration initiatives (Goverde et al. 2002; Kremen et al. 2007).

Many studies demonstrated that reductions in the size and number of natural remnants in the landscape can have deleterious effects on many species (e.g., Beier et al. 2002; Kremen et al. 2004; 2007, Cortes-Delgado & Perz-Torres 2011; Cruz-Neto et al. 2011). As large patches tend to have higher environmental diversity, they usually maintain more diverse communities of pollinators than smaller patches (Aizen & Feinsinger 1994b; Tscharntke & Brandl 2004). In fact, Steffan-Dewenter et al. (2001) postulate that the number of visitors to a flower in sites with intensive agricultural management in central Germany increases significantly in landscapes with more types of semi-natural environments, which characterizes them as more complex in comparison to landscapes dominated by monospecific crops. Those studies reinforce the importance of preserving native vegetation patches with enough area to maintain several pollinator species in the landscape, particularly in the case of specialist species restricted to only a few habitat types (Tscharntke & Brandl 2004; Tscharntke et al. 2005). However, the minimum size needed for native vegetation patches to ensure the survival of those pollinator species and the conservation of their pollination services is uncertain and most likely differs for different ecosystems and species (Aizen & Feinsinger 1994a; Kremen et al. 2004; Ricketts et al. 2004; Tscharntke et al. 2008).

In addition to the extent of natural environments, several authors indicate that the distance between patches has been one of the primary factors to affect the long-term maintenance of many species (Wiens 1995; Lennartson 2002; Smith & Hellmann 2002, Lander et al. 2010). This occurs because variations in the distances between habitat remnants change the landscape's connectivity, which can restrict the movement of individuals through the landscape and the establishment of new populations. Large inter-patch distances may also directly affect the accessibility of floral resources for individuals, threatening the floral visitors' populations. Studies indicate that the abundance and richness of native pollinators may increase significantly as the distance to natural environments decreases, which also affects agricultural production (Steffan-Dewenter & Tschanrtke 1999; Greenleaf & Kremen 2006, Klein 2009; Tscharntke et al. 2011). In the state of Minas Gerais in southeastern Brazil, for example, De Marco & Coelho (2004) reported that the proximity of crops to native vegetation resulted in a 14.6% higher coffee production compared to farms further from native vegetation. Similar results were observed in Indonesia (Klein et al. 2003) and in Costa Rica (Ricketts et al. 2004). In some cases, these variations in the landscape structure can even lead to behavioural changes in native pollinators (Osborne et al. 1999; Goverde et al. 2002), affecting their interactions with other species. In North American landscapes, for example, it was observed that the proximity of natural vegetation patches to sunflower crops increases the number of native bees, which compete with exotic bees (*Apis mellifera*). This surplus of pollinators leads to a more efficient cross-pollination process, incurring higher seed production (Greenleaf & Kremen 2006).

In addition to proximity to natural habitats, the matrix surrounding native vegetation patches, which is usually composed of different types of crops and agricultural management regimes, also exerts a strong influence on the behaviour and local maintenance of pollinators (Osborne et al. 1999). What has been established in the literature, primarily from the early 1990s on, is that the isolation level of native habitats and crops depends on the interaction between the pollinators' biological characteristics and the hostility of the matrix, resulting in both negative and positive effects depending on the species (Jules & Shahani 2003; Tscharntke & Brandl 2004; Ricketts et al. 2008; Brittain et al. 2010; Carvalheiro et al. 2010; 2011). Indeed, some studies have indicated that the characteristics of the matrix are an essential factor in the maintenance of a landscape's functional connectivity, with some pollinator species even benefiting from some agricultural activities (Westphal et al. 2003; Klein et al. 2007). Based on a compilation of 22 years of research in fragmented landscapes in the Amazon, Laurance et al. (2002) concluded that, due to the existence of a wide variety of responses to fragmentation, the inter-forest matrix can maintain high species richness, even leading to an increase in beta diversity in fragmented landscapes. However, because species that avoid the matrix tend to be the first to go extinct after fragmentation (Jules & Shahani 2003; Westphal et al. 2003; Tscharntke & Brandl 2004, Cussans et al. 2010; Kamm et al. 2010), we believe that the maintenance of a high diversity of native pollinators in the landscape requires that both the matrix and the surrounding natural habitats be sufficiently diversified.

The services provided by pollinators include the pollination of native plant species in patches of natural vegetation as well as the pollination of crops in agricultural areas. The magnitude of the effect of habitat fragmentation on plant reproduction in natural vegetation remnants was analyzed by Aguilar et al. (2006) based on 54 studies. These authors observed that reductions in patch size and increases in the isolation of natural habitat fragments have generally negative effects on pollination and on fruit and seed production in the native species studied. Among the effects that we compiled from our survey, the most evident are related to the size of the native vegetation remnants where pollinator populations live. In small native vegetation fragments, plants tend to receive limited pollination services and exhibit lower fruit and seed production (Donaldson et al. 2002; Brys et al. 2004; Kolb 2008; González-Varo et al. 2009; Taki et al. 2010). Moreover, we found studies demonstrating that fragmentation and habitat loss can lead to a dramatic simplification of pollinator interaction networks due to a decrease in the availability of specialized and rare pollinators (e.g., Donaldson et al. 2002; Brosi et al. 2007) For instance, the loss of some pollinator groups such as birds, flies and non-flying mammals impaired pollination and consequently the reproduction of several plant species in some fragments of the Brazilian Atlantic forest (Lopes et al. 2009). As a result, second growth forest patches developed plant assemblages with a higher frequency of species and individuals that are pollinated by generalist vectors with approximately 30% lower functional diversity of pollination interactions in comparison to continuous mature forest areas (see Girão et al. 2007; Lopes et al. 2009; Cruz-Neto et al. 2011).

Furthermore, in our review, we also observed that self-incompatible species, which depend completely on pollinators for sexual reproduction, are the most susceptible to habitat fragmentation (Aguilar et al. 2006; Girão et al. 2007; Lopes et al. 2009; Tabarelli et al. 2010; Cruz-Neto et al. 2011). With these results, we suggest that fragmentation promotes a remarkable change in the relative abundance of certain reproductive attributes of Atlantic forest trees and that it largely reduces the reproductive functional diversity of tree assemblages. Therefore, in fragmented landscapes, it is likely that small fragments and narrow forest corridors, both dominated by edge effects (Murcia 1995), are not sufficient on their own to conserve the complete diversity of life histories of trees and their mutualists (Lopes et al. 2009).

Regarding agricultural landscapes, Aizen et al. (2009) estimated that in the absence of pollinating services rendered by animals, world agricultural production might decrease by up to 8%, with clear social and economic impacts. According to these authors, worldwide agricultural production has historically increased at a rate of approximately 1.5% per year, and there is no evidence of global variation in the productivity of species that are dependent on vs. independent of pollinators (Aizen et al. 2009). However, we found several studies showing evidence of the effects of the composition (i.e., which ecosystems are there and in what proportion) and disposition (i.e., how those ecosystems are distributed in space) of landscape elements on agricultural production (Klein et al. 2003; De Marco & Coelho 2004; Ricketts et al. 2004, Gemmill-Herren & Ochieng 2008, Phalan et al. 2011), which indicates that unplanned changes in landscape pattern can reduce the productivity of pollinator-dependent crops.

Knowledge gaps to be explored

All of the information discussed above highlights the importance of areas of natural vegetation for the maintenance of pollinator species and their associated flora. Based on our results, we suggest that, in general, agricultural landscapes must be interspersed with natural and semi-natural vegetation patches for the maintenance of proper pollination services in native and human-made environments. Obviously, variations among different ecosystems and species and their interactions with their surroundings must be considered. The relationship between landscape structure and pollinator survival and behavioural responses is a very complex subject that has only recently begun to be revealed (Bélisle 2005). Therefore, to advance our knowledge of those relationships and understand how the composition and configuration of the landscape affects plant-pollinator interactions, more studies are needed that address the diversity of pollinators and population attributes (such as density fluctuations and survival) and can explain changes in diversity and behavioural attributes (such as mobility and foraging patterns) that could modify the efficiency of individuals as pollinators.

That type of research is particularly needed for tropical ecosystems, where the recent increase in the number of studies has been lower than in temperate regions and where the higher diversity of plants and pollinators impedes a more thorough knowledge of these systems. Due to the high worldwide importance of those regions for the production of food and primary agricultural goods, more attention should be given to the development of knowledge of pollinators and pollination processes in complex tropical landscapes. High-diversity tropical regions are usually located in developing countries, which commonly have limited funds and specialized personnel with which to conduct high quality environmental research (e.g., Stocks et al. 2008). International scientific cooperation could reinforce national research programs.

Starting from a broader perspective, the complexity of plant-pollinator systems and the broad spatial scale necessary for studies of landscape ecology indicate that larger research teams and greater cooperation are needed to compare landscapes rather than patches within the same landscape, the more frequent approach. One aspect that is strongly highlighted in the literature is the risk that the processes of fragmentation and reduction of habitat area pose to the conservation of pollinators and the maintenance of pollination services worldwide. Lennartson (2002) states that the processes of habitat loss and fragmentation can lead to abrupt qualitative changes in landscape structure, limiting the survival and movement of pollinators. Those processes may result in species extinctions due to threshold dynamics, deeply affecting the viability of biological populations and communities beyond a certain degree of habitat loss (Andrén 1994; Metzger & Décamps 1997). Local extinctions could lead to the disruption of plant-pollinator interactions with unpredictable consequences for the maintenance of biodiversity and environmental services (e.g., Steffan-Dewenter & Tscharntke 1999; Girão et al. 2007; Brosi et al. 2008; Lopes et al. 2009; Tabarelli et al. 2010). The identification of those thresholds and their implications is of extreme importance for the conservation of natural environments. To properly conserve biological diversity and its associated processes, habitat loss should never reach such extinction thresholds (Radford et al. 2005). Public policies and legal enforcement should be substantially based on scientific knowledge and precautionary principles to guarantee that the status of the landscape remains well above these modification-inducing levels. To do this, it is necessary to determine which critical values of habitat loss can lead to drastic increases in pollinator extinction rates so that we can evaluate at what point plant-pollinator interactions may collapse. However, determining such values for each region is an extremely difficult task, as this most likely varies among systems. Ideally, conservation strategies should be grounded in research conducted in the same region or in areas with similar ecological characteristics.

Similarly, we should concentrate efforts on acquiring the necessary knowledge to improve the landscape's capacity to facilitate pollinator flow (connectivity) between crop areas and nesting and foraging habitats. However, due to huge interspecific variations in pollinators' capacity to use and move between landscape components such as natural vegetation corridors or large extents of anthropogenic environments, our search for new techniques for integrated landscape management must aim at the maintenance of not only structural but functional connectivity between environments. In this way, it would be possible to ensure effective pollen flow and, consequently, fruit and seed production. Therefore, it is essential that we understand how pollen is dispersed, and to do so we must investigate the factors that affect pollinator mobility. However, to complete this task, methodological and technological obstacles must be overcome. Because most pollinators are tiny animals such as bees, most telemetry equipment is still too large or heavy to be carried by individuals. This hinders the direct assessment of movement behaviour in the landscape and consequently most of the presently available information on this subject comes from indirect observations and/or deductive conclusions based on basic biology. Nevertheless, new technologies have been developed in the last decade, such as harmonic radars for honeybee-sized animals (Osborne et al. 1999) or radio-tracking for bumble bees (Hagen et al. 2011), which will enable important developments in our understanding of pollinator spacing behaviours in coming years. The development of better individual tracking technologies will inevitably lead to more detailed studies on pollinator movement through the landscape, which together with the knowledge already available in the literature will lead to the development of better tools and guidelines for the management and design of landscapes with highly efficient ecosystem services, also ensuring the long-term conservation of pollination services in agro-natural systems.

The integrated management of landscapes based on scientific knowledge can compensate for global habitat loss (Ricketts et al. 2004; Tscharntke & Brandl 2004), as structurally complex landscapes with good habitat connectivity have proven to be more efficient in the maintenance of species diversity (Tscharntke et al. 2007). Additionally, simple changes in agricultural management such as the restricted use of pesticides and the sowing of species used by pollinators along crop edges can result in important improvements in landscape quality (Kremen et al. 2004, Gabriel et al. 2010; Carvalheiro et al. 2011; Krauss et al. 2011; Kovács-Hostyánszki et al. 2011).

Furthermore, for the integrated management of landscapes to become a reality for land use planning, in addition to incentives for excellent scientific research in areas related to the topics discussed herein (which will provide a technical basis for decisions on land use), public policies are needed to stimulate the following: 1) incentives for the creation of land use categories that enable pollinator flow and enhance pollen flow and the sexual cross-reproduction of native and cultivated plants in agrosystems and their surroundings; 2) the natural regeneration or restoration of pollinator-friendly habitats in sites where those environments have been degraded; 3) the monitoring of plant species that provide feeding and nesting resources for pollinators; 4) the conservation of key elements such as habitat patches and corridors that support the structural and functional complexity of the landscape; 5) population increases in managed native pollinator species in the vicinity of crops, thus ensuring their access to floral resources; and 6) planned reforestation at a regional scale, using multiple native species that will serve as resources for pollinators and that must be positioned in such a way as to improve pollen flow in the landscape.

Acknowledgements

We thank professors Dr. Vera L. Imperatriz Fonseca, Dr. Antônio Mauro Saraiva from USP and Dr. Dora A.L. Canhos from the Reference Center on Environmental Information - CRIA, the coordinators of the Brazilian Research Council - CNPq project 'Avaliação do uso sustentável e conservação dos serviços ambientais realizados pelos polinizadores no Brasil', for the support given to our study. We also thank the anonymous referees whose comments improved the final version of the manuscript. BFV and AVL received productivity fellowships from CNPq.

Appendices

Additional supporting information may be found in the online version of this article:

Appendix I. List of articles selected for analysis

Appendix II. Graphical representation of the number of analyzed articles published per journal

Appendix III. Graphical representation of the temporal changes of studies that mention the matrix (A) and matrix type (B).

References

Aguilar R, Ashworth L, Galetto L, Aizen MA (2006) Plant reproductive susceptibility to habitat fragmentation: review and synthesis through a meta-analysis. Ecology Letters 9:968-980.

Aizen MA, Feinsinger P (1994a) Forest Fragmentation, Pollination, and Plant Reproduction in a Chaco Dry Forest, Argentina. Ecology 75:330-351.

Aizen MA, Feinsinger P (1994b) Habitat fragmentation, native insect pollinators, and feral honey bees in Argentine Chaco Serrano. Ecological Applications 4:378-392.

Aizen MA, Garibaldi LA, Cunningham SA, Klein AM (2009) How much does agriculture depend on pollinators? Lessons from long-term trends in crop production. Annals of Botany 103:2005-2010.

Albrecht M, Duelli P, Müller C, Kleijn D, Schmid B (2007) The Swiss agri-environment scheme enhances pollinator diversity and plant reproductive success in nearby intensively managed farmland. Journal of Applied Ecology 44:813-822, 2007.

Andersson E, Barthel S, Ahrne K (2007) Measuring social–ecological dynamics behind the generation of ecosystem services. Ecological Applications 17:1267-1278.

Andrén H (1994) Effects of habitat fragmentation on birds and mammals in landscapes with different proportions of suitable habitat: a review. Oikos 71:355-366.

Beier P, Van Drielen M, Kankam BO (2002) Avifaunal collapse in West African forest fragments. Conservation Biology 16:1097-1111.

Bélisle, M. (2005). Measuring landscape connectivity: The challenge of behavioural landscape ecology. Ecology 86: 1988–1995.

Brittain C, Bommarco R, Vighi M, Settele J, Potts SG (2010) Organic farming in isolated landscapes does not benefit flower-visiting insects and pollination. Biological Conservation 143:1860-1867.

Brosi BJ, Daily GC, Ehrlich PR (2007) Bee community shifts with landscape context in a tropical countryside. Ecological Applications 17:418-430.

Brosi BJ, Daily GC, Shih TM, Oviedo F, Durán G (2008) The effects of forest fragmentation on bee communities in tropical countryside. Journal of Applied Ecology 45:773–783.

Brys R, Jacquemyn H, Endels P, Rossum FV, Hermy M, Triest L, Bruyn L, Blust GDE (2004) Reduced reproductive success in small populations of the self-incompatible Primula vulgaris. Journal of Ecology 92:5-14.

Buchmann SL, Nabhan GP (1996) The pollination crisis - The plight of the honey bee and the decline of other pollinators imperils future harvests. Science 36:22-27.

Carvalheiro LG, Seymour CL, Veldtman R, Nicolson SW (2010) Pollination services decline with distance from natural habitat even in biodiversity-rich areas. Journal of Applied Ecology 47:810-820.

Carvalheiro LG, Veldtman R, Shenkute AG, Tesfay GB, Pirk CWW, Donaldson JS, Nicolson SW (2011) Natural and within-farmland biodiversity enhances crop productivity. Ecology letters 14:251-259.

Carvell C, Roy DB, Smart SM, Pywell RF, Preston CD, Goulson D (2006) Declines in forage availability for bumblebees at a national scale. Biological Conservation 132:481–489.

Chacoff NP, Aizen MA (2006) Edge effects on flower-visiting insects in grapefruit plantations bordering premontane subtropical forest. Journal of Applied Ecology 43:18-27.

Clergeau P, Burel F (1997) The role of spatio-temporal patch connectivity at the landscape level: an example in a bird distribution. Landscape and Urban Planning 38:37-43.

Cortes-Delgado N, Perez-Torres J (2011) Habitat edge context and the distribution of phyllostomid bats in the Andean forest and anthropogenic matrix in the Central Andes of Colombia. Biodiversity Conservation 20:987–999.

Cruz-Neto O, Machado IC, Duarte JA, Lopes AV (2011) Synchronous phenology of hawkmoths (Sphingidae) and Inga species (Fabaceae–Mimosoideae): implications for the restoration of the Atlantic forest of northeastern Brazil. Biodiversity and Conservation 20:751-765.

Cussans J, Goulson D, Sanderson R, Goffe L, Darvill B, Osborne JL (2010) Two bee-pollinated plant species show higher seed production when grown in gardens compared to arable farmland. PloS ONE 5:e11753.

De Marco P, Coelho FM (2004) Services performed by the ecosystem: forest remnants influence agricultural cultures pollination and production. Biodiversity and Conservation 13:1245-1255.

Develey PF, Stouffer PC (2001) Effects of roads on movements by understory birds in mixed-species flocks in central Amazonian Brazil. Conservation Biology 15:1416-1422.

Dixon KW (2009) Pollination and Restoration. Science 325:571-573.

Donaldson J, Nanni I, Zachariades C, Kemper J (2002) Effects of habitat fragmentation on pollinator diversity and plant reproductive success in renosterveld shrublands of South Africa. Conservation Biology 16:1267-1276.

Ekroos J, Piha M, Tiainen J (2008) Role of organic and conventional field boundaries on boreal bumblebees and butterflies. Agriculture, Ecosystems and Environment 124:155-159.

Fahrig L (2003) Effects of habitat fragmentation on Biodiversity. Annual Review of Ecology, Evolution, and Systematics 34:487–515.

Gabriel D, Sait SM, Hodgson JA, Schmutz U, Kunin WE, Benton TG (2010) Scale matters: the impact of organic farming on biodiversity at different spatial scales. Ecology letters 13:858-869.

Garibaldi LA, Steffan-Dewenter I, Kremen C, Morales JM, Bommarco R, Cunningham SA, Carvalheiro LG, Chacoff NP, Dudenhöffer JH, Greenleaf SS, Holzschuh A, Isaacs R, Krewenka K, Mandelik Y, Mayfield MM, Morandin LA, Potts SG, Ricketts TH, Szentgyörgyi H, Viana BF, Westphal C, Winfree R, Klein AM (2011) Stability of pollination services decreases with isolation from natural areas despite honey bee visits. Ecology Letters 14:1062–1072.

Gathmann A, Tscharntke T (2002) Foraging ranges of solitary bees. Journal of animal ecology 71:757-764.

Gemmill-Herren B, Ochieng AO (2008) Role of native bees and natural habitats in eggplant (Solanum melongena) pollination in Kenya. Agriculture, Ecosystems and Environment 127:31–36.

Girão LC, Lopes AV, Tabarelli M, Bruna EM (2007) Changes in Tree Reproductive Traits Reduce Functional Diversity in a Fragmented Atlantic Forest Landscape. PLoS One 2:e908.

Goverde M, Schweizer K, Baur B, Erhardt A (2002) Small-scale habitat fragmentation effects on pollinator behaviour: experimental evidence from the bumblebee Bombus veteranus on calcareous grasslands. Biological Conservation 104:293-299.

González-Varo JP, Arroyo J, Aparicio A (2009) Effects of fragmentation on pollinator assemblage, pollen limitation and seed production of Mediterranean myrtle (Myrtus communis). Biological Conservation 142:1058-1065.

Greenleaf SS, Kremen C (2006) Wild bees enhance honey bees pollination of hybrid sunflower. Proceedings of the National Academy of Sciences 103:13890-13895.

Hadley AS, Betts MG (2009) Tropical deforestation alters hummingbird movement patterns. Biology Letters 5:207-210.

Hagen M, Wikelski M, Kissling WD (2011) Space Use of Bumblebees (Bombus spp.) Revealed by Radio-Tracking. PLoS ONE 6(5): e19997.Holzschuh A, Steffan-Dewenter I, Tscharntke T (2008) Agricultural landscapes with organic crops support higher pollinator diversity. Oikos 117:354-361.

Holzschuh A, Steffan-Dewenter I, Tscharntke T (2008) Agricultural landscapes with organic crops support higher pollinator diversity. Oikos 117:354–361.

Isaacs R, Kirk AK (2010) Pollination services provided to small and large highbush blueberry fields by wild and managed bees. Journal of Applied Ecology 47:841-849.

Jha S, Vandermeer JH (2010) Impacts of coffee agroforestry management on tropical bee communities. Biological Conservation 143:1423-1431.

Jules ES, Shahani PA (2003) broader ecological context to habitat fragmentation: Why matrix habitat is more important than we thought. Journal of Vegetation Science 14:459-464.

Kamm U, Gugerli F, Rotach P, Edwards P, Holderegger R (2010) Open areas in a landscape enhance pollen-mediated gene flow of a tree species: evidence from northern Switzerland. Landscape Ecology 25:903-911.

Kearns CA, Inouye DW, Waser NM (1998) Endangered mutualisms: The conservation of plant-pollinator interactions. Annual Review of Ecology and Systematics 29:83-112.

Klein AM, Steffan-Dewenter I, Tscharntke T (2003) Fruit set of highland coffee increases with the diversity of pollinating bees. Proceedings of the Royal Society B-Biological Sciences 270:955-961.

Klein AM, Vaissière BE, Cane JH, Steffan-Dewenter I, Cunningham SA, Kremen C, Tscharntke T (2007) Importance of pollinators in changing landscapes for world crops. Proceedings of the Royal Society B-Biological Sciences 274:303-313.

Klein AM, Cunningham SA, Bos M, Steffan-Dewenter I (2008) Advances in pollination ecology from tropical plantation crops. Ecology 89:935-943.

Klein AM (2009) Nearby rainforest promotes coffee pollination by increasing spatio-temporal stability in bee species richness. Forest Ecology and Management 258:1838-1845.

Knight ME, Osborne JL, Sanderson RA, Hale RJ, Martin AP, Goulson D (2009) Bumblebee nest density and the scale of available forage. Insect Conservation and Diversity 2:116-124.

Kremen C, Williams NM, Thorp RW (2002) Crop pollination from native bees at risk from agricultural intensification. Proceedings of the National Academy of Sciences of the United States of America 99:16812-16816.

Kremen C, Williams NM, Bugg RL, Fay JP, Thorp RW (2004) The area requirements of an ecosystem service: crop pollination by native bee communities in California. Ecology Letters 7:1109-1119.

Kremen C, Williams NM, Aizen MA, Gemmill-Herren B, LeBuhn G, Minckley R, Packer L, Potts SG, Roulston T, Steffan-Dewenter I, Vazquez DP, Winfree R, Adams L, Crone EE, Greenleaf SS, Keitt TH, Klein AM, Regetz J, Ricketts TH (2007) Pollination and other ecosystem services produced by mobile organisms: a conceptual framework for the effects of land-use change. Ecology Letters 10:299-314.

Kreyer D, Oed A, Walther-Hellwig K, Frankl R (2004) Are forests potential landscape barriers for foraging bumblebees? Landscape scale experiments with Bombus terrestris agg. and Bombus pascuorum (Hymenoptera, Apidae). Biological Conservation 116:111-118.

Kolb A (2008) Habitat fragmentation reduces plant fitness by disturbing pollination and modifying response to herbivory. Biological Conservation 141:2540-2549.

Kovács-Hostyánszki A, Batáry P, Báldi A (2011) Local and landscape effects on bee communities of Hungarian winter cereal fields. Agricultural and Forest Entomology 13:59-66.

Krauss J, Gallenberger I, Steffan-Dewenter I (2011) Decreased functional diversity and biological pest control in conventional compared to organic crop fields. PloS One 6:e19502.

Lander TA, Boshier DH, Harris SA (2010) Fragmented but not isolated: Contribution of single trees small patches and long-distance pollen flow to genetic connectivity for Gomortega keule an endangered Chilean tree. Biological Conservation 143:2583-2590.

Laurance WF, Lovejoy TE, Vasconcelos HL, Bruna EM, Didham RK, Stouffer PC, Gascon C, Bierregaard RO, Laurance SG, Sampaio E (2002) Ecosystem Decay of Amazonian Forest Fragments: a 22-Year Investigation. Conservation Biology 16:605-618.

Lennartson T (2002) Extinction thresholds and disrupted plant – pollinator interactions in fragmented plant populations. Ecology 83:3060-3072.

Lopes AV, Girão LC, Santos BA, Peres CA, Tabarelli M (2009) Long-term erosion of tree reproductive trait diversity in edge-dominated Atlantic forest fragments. Biological Conservation 142:1154-1165.

Metzger JP (2001) O que é ecologia de paisagens? Biota Neotropica 1:1-9.

Metzger JP, Décamps H (1997) The structural connectivity threshold: an hypothesis in conservation biology at the landscape scale. Acta Ecologica 18:1-12.

Murcia C (1995) Edge effects in fragmented forests: implications for conservation. Tree 10:58-62.

Osborne JL, Clark SJ, Morris RJ, Williams IH, Riley JR, Smith AD, Reynolds DR, Edwards AS (1999) A landscape-scale study of bumble bee foraging range and constancy, using harmonic radar. Journal of Applied Ecology 36:519-533.

Phalan B, Onial M, Balmford A, Green, RE (2011) Reconciling food production and biodiversity conservation: Land sharing and land sparing compared. Science 333: 1289-1291.

Powell AH, Powell GVN (1987) Population dynamics of male euglossine bees in Amazonian forest fragments. Biotropica 19:176-179.

Radford, JQ. Bennett, AF, Cheers, GJ (2005). Landscape-level thresholds of habitat cover for woodland-dependent birds. Biol Conserv 124, 317-337.

Rejinfo LM (2001) Effect of natural and anthropogenic landscape matrices on the abundance of subandean bird species. Ecological Applications 11:14-31.

Ricketts TH (2004) Tropical forest fragments enhance pollinator activity in nearby coffe crops. Coservation Biology 18:1262-1271.

Ricketts TH, Daily GC, Ehrlich PR, Michener CD (2004) Economic value of tropical forest to coffee production. Proceedings of the National Academy of Sciences 101:12579-12582.

Ricketts TH, Regetz J, Steffan-Dewenter I, Cunningham SA, Kremen C, Bogdanski A, Gemmill-Herren B, Greenleaf SS, Klein AM, Mayfield MM, Morandin LA, Ochieng A, Viana BF (2008) Landscape effects on crop pollination services: are there general patterns? Ecology Letters 11:499-515.

Roulston TH, Goodell K (2011) The role of resources and risks in regulating wild bee populations. Annual review of entomology 56:293-312.

Smith JNM, Hellmann JJ (2002) Population persistence in fragmented landscapes. Trends in Ecology and Evolution 17:397-399.

Steffan-Dewenter I, Tscharntke T (1999) Effects of habitat isolation on pollinator communities and seed set. Oecologia 121:432-440.

Steffan-Dewenter I, Westphal C (2008) The interplay of pollinator diversity, pollination services and landscape change. Journal of Applied Ecology 45:737-741.

Steffan-Dewenter I, Münzenberg U, Tscharntke T (2001) Pollination, seed set and seed predation on a landscape scale. Proceedings of the Royal Society B 286:1685-1690.

Steffan-Dewenter, I, Potts SG, Packer L (2005) Pollinator diversity and crop pollination services are at risk. Trends in Ecology and Evolution 20:651-652.

Stocks G, Seales L, Paniagua F, Maehr E, Bruna E. (2008). The Geographical and Institutional Distribution of Ecological Research in the Tropics. Biotropica 40(4): 397–404

Tabarelli M, Aguiar AV, Peres CA, Girão LC, Lopes AV (2010) Effects of pioneer tree species hyperabundance on forest fragments in northeastern Brazil. Conservation Biology 24:1654-1663.

Taki H, Kevan PG (2007) Does habitat loss affect the communities of plants and insects equally in plant – pollinator interactions? Preliminary findings. Biodiversity and Conservation 16:3147–3161.

Taki H, Kevan PG, Yamaura Y (2010) Effects of Forest Cover on Fruit Set in the Woodland Herb, Maianthemum canadense (Liliaceae). The Canadian Field-Naturalist 122:234-238.

Tscharntke T, Brandl R (2004) Plant-insect interactions in fragmented landscapes. Annual Review of Entomology 49:405-430.

Tscharntke T, Klein AM, Kruess A, Steffan-Dewenter I, Thies C (2005) Landscape perspectives on agricultural intensification and biodiversity – ecosystem service management. Ecology Letters 8:857-874.

Tscharntke T, Bommarco R, Clough Y, Crist TO, Kleijn D, Rand TA, Tylianakis JM, Nouhuys SV, Vidal S (2007) Conservation biological control and enemy diversity on a landscape scale. Biological Control 43:294-309.

Tscharntke T, Sekercioglu CH, Dietsch TV, Sodhi NS, Hoehn P, Tylianakis JM (2008) Landscape constraints on functional diversity of birds and insects in tropical agroecosystems. Ecology 89:944-951.

Tscharntke T, Clough Y, Bhagwat SA, Buchori D, Fau H, Hertel D, Holscher D, Juhrbandt J, Kessler M, Perfecto I, Scherber C, Schroth G, Veldkamp E, Wanger TC (2011) Multifunctional shade-tree management in tropical agroforestry landscapes – A review. Journal of Applied Ecology 48:619–629.

Westphal C, Steffan-Dewenter I, Tscharntke T (2003) Mass flowering crops enhance pollinator densities at a landscape scale. Ecology Letters 6:961-965.

Wiens JA (1995) Habitat fragmentation: island v landscape perspectives on bird conservation. Ibis 137:97–104.

Winfree R, Aguilar R, Vazquez DP, LeBuhn G, Aizen MA (2009) A meta-analysis of bees responses to anthropogenic disturbance. Ecology 90:2068-2076.

15

Implications of Enemy Escape on Chemically Mediated Interactions with Mutualists: Wild Parsnip Pollination in Two Hemispheres

Tania Jogesh[1], Arthur Zangerl[1], Margaret C. Stanley[2], and May R. Berenbaum[1]

[1]*Department of Entomology, University of Illinois at Urbana-Champaign, Urbana, IL*
[2]*Centre for Biodiversity and Biosecurity, School of Biological Sciences, Tamaki Campus, University of Auckland, Private Bag 92019, Auckland, New Zealand*

Abstract—When plant species invade new areas, they can escape from specialist enemies and thereby reduce investment in chemical defense. Enemy release may have other impacts on plant chemistry; in the absence of specialists, plants may be able to increase production of volatiles that enhance attractiveness to pollinators. In the United States (US), the introduced Eurasian wild parsnip, *Pastinaca sativa,* has long been subject to attack by an introduced coevolved florivore, *Depressaria pastinacella,* the parsnip webworm. In 2004, webworms were found for the first time attacking parsnips in New Zealand (NZ). Relative to US *P. sativa,* NZ *P. sativa* produces lower levels of defenses but higher levels of certain floral volatiles, suggesting that escape from its specialist florivore may have resulted in changes in volatile profiles to increase pollinator attraction.

In this study, we examined the influence of wild parsnip floral volatiles on pollinator attraction and seed production in NZ and the US. While many insect groups were observed on parsnip umbels, the percentage of flowers that set seed could significantly be predicted by the occurrence of large calyptrate flies and small syrphids in both localities. In the US, β-pinene, γ-terpinene, hexyl butyrate, octyl butyrate, germacrene D and an unknown monoterpene were all positively correlated with visitation by large calyptrates and small syrphids. In NZ, trans-ocimene, carene and octyl butyrate were positively correlated with visitation. Remarkably, most compounds positively associated with visitation are produced in significantly higher proportions in NZ flowers, suggesting that NZ flowers, in the absence of specialized florivores, may be chemically better constituted to attract pollinators.

Keywords: Coevolution, floral volatiles, florivory, plant-pollinator interactions, invasive species, pollination

Introduction

When plant species invade new areas, they often escape from coevolved specialist enemies; escape in certain circumstances can be accompanied with reduced investment in chemical defense. This relationship in fact forms the basis of multiple hypotheses aimed at predicting the success of invasive plants (Maron & Vila 2001; Keane & Crawley 2002; Joshi & Vrieling 2005; Liu & Stiling 2006). Escape from coevolved specialists, however, may have impacts on the chemistry of plants other than reducing investment in defense. Some plants avoid herbivores by reducing the production of attractants used by adapted specialists to find host plants (Feeny 1977). Colonizing new habitats where coevolved enemies are absent may result in altered selection on volatile attractants, particularly if these chemicals mediate interactions with mutualists such as pollinators. While floral volatiles render plants attractive to beneficial pollinators, they may also make them conspicuous to detrimental specialist herbivores that have evolved the ability to recognize host-specific volatiles (Theis 2006, Theis et al. 2007, Raguso 2008, Kessler & Baldwin 2011, Kessler et al. 2012, Theis & Adler 2012).

In the presence of coevolved specialist herbivores, plant species may experience considerable selection pressure to alter their floral profiles to reduce their apparency (Baldwin et al. 1997). A potential cost of this alteration is a reduction in pollinator attraction, especially if the same components of the floral blend are attractive to both the herbivores and pollinators. For example, although benzyl acetone is a pollinator attractant for the flowers of *Nicotiana attenuata* Torr., it is emitted at low levels because plants with elevated production are subject to greater herbivory (Kessler et al. 2008). As well, benzaldehyde and phenylacetaldehyde emitted by flowers of *Cirsium arvense* (L.) Scop. are attractive to both florivores and pollinators (Theis, 2006). Increased scent production in *Curcurbita pepo* results in greater herbivore damage and reduced fitness (Theis and Adler 2012). Understanding how volatiles mediate both antagonistic and mutualistic interactions is thus a prerequisite for predicting the phytochemical and ecological consequences of enemy release.

A system in which the chemical consequences of herbivore escape for pollinator attractiveness may be

*Corresponding author; email: tjogesh@life.illinois.edu

examined is the Eurasian weed *Pastinaca sativa* L., wild parsnip, and its European florivorous specialist, *Depressaria pastinacella* Duponchel, the parsnip webworm. The defensive chemistry of this interaction has been well characterized both in its native Europe and in the introduced range in North America, where the wild parsnip has been part of the flora for over 300 years and where it has been reassociated with the parsnip webworm for more than 150 years (Bethune 1869; Riley 1888).

In 2004, webworms were discovered on wild parsnips in New Zealand (NZ), where the plants had grown free from significant herbivory since the introduction and establishment of the weed in the mid-nineteenth century (Zangerl et al. 2008). This accidental introduction provided a unique opportunity for a real-time analysis of evolution in this interaction. Extracts of parsnip buds and flowers in NZ contained lower amounts of certain chemicals, including linear furanocoumarins, octyl butyrate, and myristicin, which function in defense against webworms in the US (Berenbaum & Zangerl 1996; Carroll & Berenbaum 2002, Mao et al. 2008). NZ flower and fruit parsnip extracts also contained higher amounts of octyl acetate, an olfactory attractant for larval webworms (Carroll & Berenbaum 2002). Reduced levels of costly defensive compounds, such as the furanocoumarins, are a predictable consequence of enemy escape, but increased production of other compounds is consistent with a response to other agents of selection, such as pollinators.

Throughout its native and invasive range, *P. sativa* is thought to be "promiscuously pollinated" by a wide range of insects (Bell 1971), including, in central Illinois, dipterans in nine families (Lohman et al. 1996). Nectar is secreted by a stylopodium that is easily accessed by generalist visitors, particularly flies. Many of the volatile constituents of parsnip flowers, including octyl esters and sesquiterpenes (Borg-Karlson et al. 1993), are attractive to saprophilic pollinating flies (Jürgens et al. 2006). Visitation by a wide variety of generalist pollinators has been suggested to contribute to the ability of parsnips and other apiaceous species to establish a broad geographic distribution and to succeed in new habitats after introduction (Bell 1971). Yet, despite an abundance of potential pollinator species, pollination success is rarely 100% in US populations. In many populations, a substantial proportion of ovules go unfertilized, as evidenced by a high frequency of parthenocarpy (normal-sized seeds lacking endosperm or embryo) (Zangerl et al. 1991). Parsnips prevent selfing by protandry, so seed set is heavily limited by visitation. Thus, pollinator visitation can impose strong selection on reproductive fitness; release from its specialist enemy in a newly invaded area may free *P. sativa* from constraints on production of volatile attractants and enhance its ability to attract pollinators.

The invasion of webworms in NZ allows us to compare floral chemical traits of wild parsnips in habitats with and without a coevolved specialist. Tradeoffs between conspicuousness to antagonists and to mutualists should be particularly striking in interactions between florivores and their hostplants. Although the interaction between parsnips and webworms has been studied in detail for over 30 years, little information is available on the interaction between parsnips and their pollinators and how this interaction might shape floral traits. We hypothesized that volatiles are important in mediating interactions between parsnips and their pollinators and that changes in volatile chemistry accompanying range expansions may reflect differences in the composition of the pollinator community. Thus, the primary goals of this study were to: 1) evaluate how variation in floral volatiles influences pollinator visitation and seed production; and 2) determine if differences in parsnip floral chemistry associated with introduction into NZ are consistent with a role in increasing pollinator attraction. In this study, we examined the floral bouquet of wild parsnips in NZ and US and its influence on pollinator attraction and pollination success. Volatiles and floral visitors were collected, identified, and quantified from parsnips in the US and NZ, and we utilized the natural variation in the volatile profiles to determine which chemical differences among individual plants explained differences in pollination success and consequently in fitness gained through seed production.

MATERIALS AND METHODS

Experiments were conducted in June 2009 in the "Overpass" parsnip population (40.121173, -88.143400) in Champaign County, Illinois, USA, which has long hosted webworm populations (Zangerl & Berenbaum 2009) and in December 2009 in the "Cottage" population in Seacliff, Otago, New Zealand (45.678541, 170.625247), known to be infested by webworms only since 2004 (Zangerl et al. 2008). Identical methods were used in both localities to measure insect visitation, volatile production, and pollination success.

Insect visitation and volatile collection

Flowering in *P. sativa* commences with the maturation of the primary apical umbel followed by flowering in the secondary umbels, which branch out laterally from the main stalk. Insect visitation and volatile production were monitored on separate secondary umbels on the same plant. *P. sativa* is protandrous and all measurements were made when the flowers were receptive to pollen (i.e., when most of the inflorescence had matured into the female stage). For all plants in this study, we removed primary umbels to ensure that sufficient resources were allocated to the full development of secondary umbels and that fruit development was not resource-limited.

P. sativa flowers transition from male to female within the span of a few days. We secured fine nylon mesh bags over secondary umbels with male flowers that were beginning to lose their anthers and develop stigmal structures but were still unreceptive to pollen. We identified 4 to 5 plants per day that were intermediate in size and uninfested by webworms and bagged two secondary umbels per plant to prevent insect visitation. Bagged plants were checked daily for the appearance of mature female flowers (denoted by the presence of bifurcated stigmas). On maturation, bags were removed from both umbels. A camera was positioned above one of the umbels to photograph visitors and volatiles were collected from the other umbel. Photographs and volatiles were collected from a total of 52 plants in Illinois and 32 plants in NZ.

The camera and the volatile collection apparatus were affixed to a customized aluminum pole that was inserted into the ground next to the plant. Four adjustable arms that extended from the pole held the camera, the volatile collection apparatus and two umbels in place securely. The camera (Canon Powershot S31S, 6 megapixels) was placed directly over one of the umbels and programmed to take high-resolution photographs every 5 minutes for 5 hours (100 pictures per plant). All experiments were conducted over a standardized time period within a day, 9 am to 2 pm for the US component and 10 am to 3 pm for the NZ component. A small picture-card containing printed text was placed on the surface of the umbel and used to focus manually on the flowers to obtain sufficient image resolution of floral visitors for identification. The card was removed prior to the experiment. The height of the camera over the umbel was adjusted to obtain the best resolution of the picture-card.

Volatiles were collected from the second umbel on the same plant over the same time period by dynamic headspace sampling. An inverted Erlenmeyer flask was secured around the umbel. Air passing through the flask entered via a small tube filled with activated charcoal for filtration. Air from the flask was drawn out at the rate of 0.5L/min through an adsorbent Porapak Q cartridge with the aid of a vacuum pump (Airchek XR5000, Eighty four, PA). The cartridge consisted of 150 mg of Porapak Q packed inside stainless steel tubing, 5 cm in length and 2.5 mm in diameter. Both the charcoal filter and the Porapak Q were held in place by insertion into a foam stopper that plugged the opening of the Erlenmeyer flask. The flask was covered with an aluminum sun-shield (Zangerl & Berenbaum, 2009) to keep temperature inside the flask close to ambient temperature. At the end of the 5-hour period, the umbels were re-bagged to prevent further pollination and remained enclosed for 2-3 weeks until fruits completed development.

All photographs were scored for the presence and identity of visitors. The number of visits per morphospecies was generally very low so we combined visitor groups based on taxonomy and size (Table 1). The detail in the photographs allowed us to identify certain insects to species (*Apis mellifera, Chauliognathus pennsylvanicus*), whereas others were identifiable to family (Diptera: Syrphidae, Bibionidae, Stratiomyidae; Hymenoptera: Vespidae, Formicidae; Coleoptera: Coccinellidae, Lampyridae, Cerambycidae), groups of families (calyptrate Diptera, acalyptrate Diptera), or order (Thysanoptera). In addition, we used the photographs to count number of flowers as an estimate of umbel size and the amount of nectar available per umbel. A glossy sheen, generally clearly visible in the photographs, indicates the presence of nectar on the flower; the last photograph for each plant (of the 100 taken) was scored for the proportion of flowers bearing nectar. *P. sativa* flowers do not age synchronously within an umbel but age centripetally. Owing to the non-synchronous maturation of flowers in an umbel, we counted the number of receptive stigmas from the photograph to measure pollination success, the percentage of receptive flowers that had set seed. The number of receptive stigmas was obtained from the final photograph for each umbel following the protocols in Zangerl & Berenbaum (2009).

Porapak Q cartridges from the volatile apparatus were rinsed with 0.5 ml of hexane containing a 0.0001 % tridecane internal standard and the contents were eluted into 2 ml GC-vials. They were then cleaned with pentane and dried with nitrogen gas. Eluents were analyzed by gas chromatography with flame ionization detection and subjected to gas chromatography-mass spectrometry (GC-MS) to identify the volatile compounds. US samples were analyzed by GC-MS on the date of collection and NZ samples were stored in a – 20° C freezer and shipped to the US for analysis. A small number of the NZ samples unavoidably volatilized in transit, leaving a total of 19 NZ samples to be analyzed for floral volatiles.

One μL per sample was injected in the GC-MS (Shimadzu QP2010 Plus, SHRXI-5MS capillary column, 30m × 2.5mm × 0.25u) in splitless mode for 1.5 minutes with an inlet temperature of 250° C and helium as the carrier. The initial oven temperature was set at 60° C, held for 30 seconds, increased at a rate of 10° C/min until 80° C, at a rate of 30 C/min until 120° C, followed by 3° C/min until 150° C and finally at a rate of 30° C/min for 1 minute. The spectra were recorded in Single Ion Mode (SIM) and the ions detected were m/z 93, 56, 71, 161, 192. Volatile peaks from *P. sativa* flowers were identified in this study from a small sample of flowers after completing a full-scan of total ions; these have also been identified in a prior study (Zangerl and Berenbaum, 2009). We used SIM to increase the sensitively of peak detection. We selected five ions because of their relative abundance. Peaks were identified by comparison to the NIST08 library, to Kovat indexes obtained from an alkane ladder, and to standards when available. Amounts were quantified based on peak area integrations of total ion chromatograms. Areas were adjusted with the peak area of the internal standard (tridecane). The absolute areas of each compound were converted to a proportion of the total emission. Using proportions, we were able to standardize floral emissions and compare the composition of floral volatiles between US and NZ populations.

Bagged umbels with fully developed fruit were clipped and brought to the laboratory, where all fruits were evaluated for the presence of endosperm. The parsnip fruit is a schizocarp; each of the two mericarps of an individual fruit was placed over a light source to determine whether it was "empty" or filled with endosperm (Zangerl et al. 1991). The presence of an endosperm on each side of the mericarp fruit indicates a successful pollination of the corresponding stigma on the flower.

Pollen analysis

On days when experiments were in progress, insect visitors were collected from female secondary umbels that were not photographed in this study. Insect were collected with an aspirator and placed in a 2-ml tube with 70% ethanol. Pollen counts for all captured insects were estimated following the protocols in Zangerl & Berenbaum (2009). Vials with insects were sonicated for two minutes, insects were removed, and the remaining ethanol was centrifuged at 15,000X for 30 seconds. Five □L of fuschin dye solution (in glycerol) were mixed with the pellet and pipetted onto a glass slide, which was then covered with a cover slip. Parsnip pollen grains were counted in three microscope fields (109 objective, field width = 1.9 mm) of an Olympus CH-2 compound microscope. Pollen counts determined if an insect visitor was carrying parsnip pollen when captured and accordingly had the potential to be a successful pollinator.

Statistical analysis

Statistical analyses were conducted in R version 2.13.0 (R Development Core Team 2011) and PASW Statistics version 18. Data sets from the US and NZ study were analyzed separately using Generalized Linear Models. Pollination success was analyzed as a binomial variable with successes measured as the number of fertilized fruit and failures as the number of receptive flowers that were not fertilized (total number of receptive flowers minus the number of fertilized fruit). Pollination success was fitted using a General Linear Model with a quasi-binomial error structure, with visitor pollinator potential (frequency of visitation*average pollen load) as the explanatory variable. We included only frequency of visitation for NZ thrips and acalyptrate flies as we did not have pollen load data for these visitor groups. Visitors categorized as "very small insects" were frequently observed but their very small size precluded both taxonomic identification and quantification of pollen load.

We used the "drop1" function in R to determine the significance of each explanatory variable in the model. The drop1 function drops each explanatory variable in turn and each time compares the difference between the partial model deviance and the full model to a Chi-square distribution (Zuur *et al.*, 2009). Based on the results, we could identify visitor groups that are most likely to be important in predicting pollination success. Insect groups with very rare visits (less than an average of 5 visits per umbel per 5 hours) were considered unlikely pollinators and thus were not included in the analysis. After calculating correlations and VIF values for all explanatory variables, we found that two groups, large flies and small syrphids, were highly correlated in the US data set (VIF: Large flies = 6.26, small syrphids = 4.68, r = 0.83). Because of the difficulty in separating out the individual effects of each group, we ran the model once with large flies and once with small syrphids.

Absolute peak areas for each volatile were converted to a relative proportion of the total and proportions were compared between countries using multivariate analysis of variance (MANOVA). We used a multivariate distance-

TABLE 1: Frequency of insect visitation on *Pastinaca sativa* umbels. Observations were made from 100 photographs (5 hours) for each umbel and the frequencies were averaged over all umbels Visitation frequencies were calculated as the average number of times the insect was observed in the span of 5 hours (in 100 photographs). Pollen counts were divided by the weight of the insect to obtain a pollen count per mg measure, which was averaged over all specimens collected.

	Pollinators	Frequency of visitation/5 hours		Average pollen count per mg (US/NZ)
		US	NZ	
COLEOPTERA	*Chauliognathus pennsylvanicus*	8.94		107.0
	Cerambycidae		0.03	
	Coccinellidae	0.96	0.07	39.3/50.3
	Lampyridae	2.86		51.4
	Other Coleoptera		4.87	
DIPTERA	Acalyptrate flies		36.47	
	Anthomyiidae	7.98		
	Bibionidae		465.97	96.1
	Large calyptrate Diptera (Calliphoridae, Muscidae, Tachinidae, Sarcophagidae)	14.39	58.27	64.7/69.8
	Large Syrphidae	1.14	2.80	679.3/20.2
	Small Syrphidae	44.67	9.83	64.6/136.1
	Stratiomyidae		0.97	206.1
HYMENOPTERA	*Apis mellifera*	0.08		75.4
	Large Formicidae	13.69		55.2
	Small Formicidae	10.63		84.5
	Small Hymenoptera	0.41	2.03	
	Vespidae	0.18		2.9
LEPIDOPTERA	Lepidoptera		0.10	
THYSANOPTERA	Thysanoptera		671.47	
< 1mm, unidentified		63.84		

based (square root transformed, Bray-Curtis distances) approach DISTLM in Permanova+ for Primer 6.1 to identify the relative contribution of individual volatiles to pollinator visitation. Absolute values, adjusted with the internal standard, were used in the Permanova+ analysis. A stepwise selection procedure based on AIC measures of fit was used to determine which volatiles in the bouquet could best predict large fly and small syrphid visitation. In US plants, cis-ocimene and trans- ocimene were highly correlated and we excluded trans-ocimene from the stepwise selection procedure. In the NZ data, isomers of alpha-farnesene were higher correlated and only (E,E) alpha-farnesene was included in the model. Individual volatiles deemed important from the DISTLM analysis were plotted against total visitation to discern the overall direction of the relationship.

We also fitted a GLM model with a negative binomial error structure to determine if visitation counts of insect groups were dependent on nectar (proportion of flowers bearing nectar) and umbel size (total number of flowers per umbel).

RESULTS

Visitation and pollination success

Pollen loads were correlated with insect size (by weight) in both the U.S. and New Zealand (US $r^2 = 0.45$, NZ $r^2 = 0.22$), but certain groups of large insects, such as honey bees and large syrphid flies, were observed too infrequently to have contributed to pollination success in this study irrespective of pollen load size (Table 1). In both the US and NZ, wild parsnip flowers were pollinated primarily by large calyptrate and syrphid flies, but the identity of dipteran visitors varied by country (Table 1). In the US, although small syrphids were the most frequent identifiable visitors, they carried relatively small pollen loads; large calyptrate flies (Calliphoridae, Muscidae, Sarcophagidae, Tachinidae) were likely the most important pollinators as they carried relatively large pollen loads and were also frequent visitors (Table 1). In contrast with the US, wild parsnips received proportionately fewer visits from syrphids in NZ, where large calyptrate flies carried large amounts of pollen and were also frequent visitors. Despite striking geographic differences in visitation frequency, in both the US and NZ the percentage of flowers that set seed could significantly be predicted by the occurrence of large calyptrate flies and small syrphids (Table 2).

Insects with high pollen loads did not necessarily predict pollination success. In the US, soldier beetles (*C. pennsylvanicus*) carried large amounts of pollen and were also frequent visitors but their visitation was not significantly associated with pollination success (Table 2). Soldier beetles appeared to utilize the large, flat umbels of *P. sativa* as a mating site and movements between umbels were not observed. Similarly, NZ umbels were visited primarily by bibionid flies, which carried relatively large pollen loads but were not predictive of pollination success (Table 1, 2). Like

TABLE 2. Deviance (Chi-sq) tests for GLMs (quasi-binomial model) predicting pollination success by insect visitation frequency x average pollen load in US and NZ parsnips. Significant differences at $P<0.001$ are indicated by *** and $P<0.05$ by *

	Df	Deviance	Scaled deviance	p-value	
US: Pollination success: with large calyptrate Diptera					
Full model		4,765.60			
Anthomyiidae	1	4,786.40	0.20	0.65	
Large calyptrate Diptera	1	7,442.40	26.13	0.00	***
Small Formicidae	1	4,811.70	0.45	0.50	
Large Formicidae	1	5,071.40	2.99	0.08	.
Chauliognathus pennsylvanicus	1	4,852.20	0.85	0.36	
US: Pollination success: with small Syrphidae					
Full model		4,798.40			
Anthomyiidae	1	4,993.60	1.84	0.17	
Small Syrphidae	1	7,442.40	24.93	0.00	***
Small Formicidae	1	4,817.60	0.18	0.67	
Large Formicidae	1	4,982.20	1.73	0.19	
Chauliognathus pennsylvanicus	1	4,817.50	0.18	0.67	
NZ: Pollination success					
Full model		1561.30			
Large calyptrate Diptera	1	1957.00	5.10	0.02	*
Small Syrphidae	1	2260.30	9.02	0.00	**
Bibionidae	1	1763.90	2.61	0.10	
Thysanoptera (frequency only)	1	1664.90	1.34	0.24	
Acalyptrate flies (frequency only)	1	1632.50	0.91	0.33	

C. pennsylvanicus in the US, they aggregate and mate in large numbers on the umbels without moving often from flower to flower.

We did not directly compare pollinator visitation rates between the US and NZ because of local differences in insect diversity and because visitation rate is not necessarily correlated with effective pollination. Although pollination success in NZ appeared to be higher (32.1% of plants experienced >50% fertilization) than in the U.S. (20% of plants experienced >50% fertilization), this difference was only marginally significant ($P = 0.056$, as determined by a binomial GLM model).

Volatile chemistry

The floral bouquet of US parsnips comprised 16 distinct compounds, 14 of which could be identified; two monoterpenes could not be definitively characterized. The most abundant constituents were cis- (30. 81%) and trans-ocimenes (23.69%) (Figure 1). As in the US, the floral bouquet of *P. sativa* in NZ was dominated by cis- (32.27%) and trans- (49.12%) ocimenes and octyl butyrate. NZ umbels emitted low quantities (<1%) of five compounds absent in the US floral bouquet: linalool, carene, allo-ocimene and two isomers of α-farnesene (Figure 1). We found a total of 20 identifiable constituents but we were unable to identify two compounds present in low concentrations. A comparison of US and NZ floral volatiles revealed significant differences in the proportional representation of seven compounds (Wilks λ $F_{13,57}$ = 680.64, $P<0.01$); NZ flowers emit a greater proportion of beta-pinene, trans-ocimene, gamma-terpinene, octyl butyrate and germacrene whereas US flowers emit a greater proportion of octanol and octyl acetate.

Association between pollination and volatile chemistry

In the US, volatiles that best explain the variation in the pollinator matrix were β-pinene, γ-terpinene, hexyl butyrate, octyl butyrate, germacrene D and an unknown monoterpene. This was the model with the lowest AICc score (AICc = 336.23, R^2 = 0.34). These compounds were all positively associated with pollinator visitation (large calyptrate flies and small syrphids, Figure 2). In NZ, volatiles that best explain

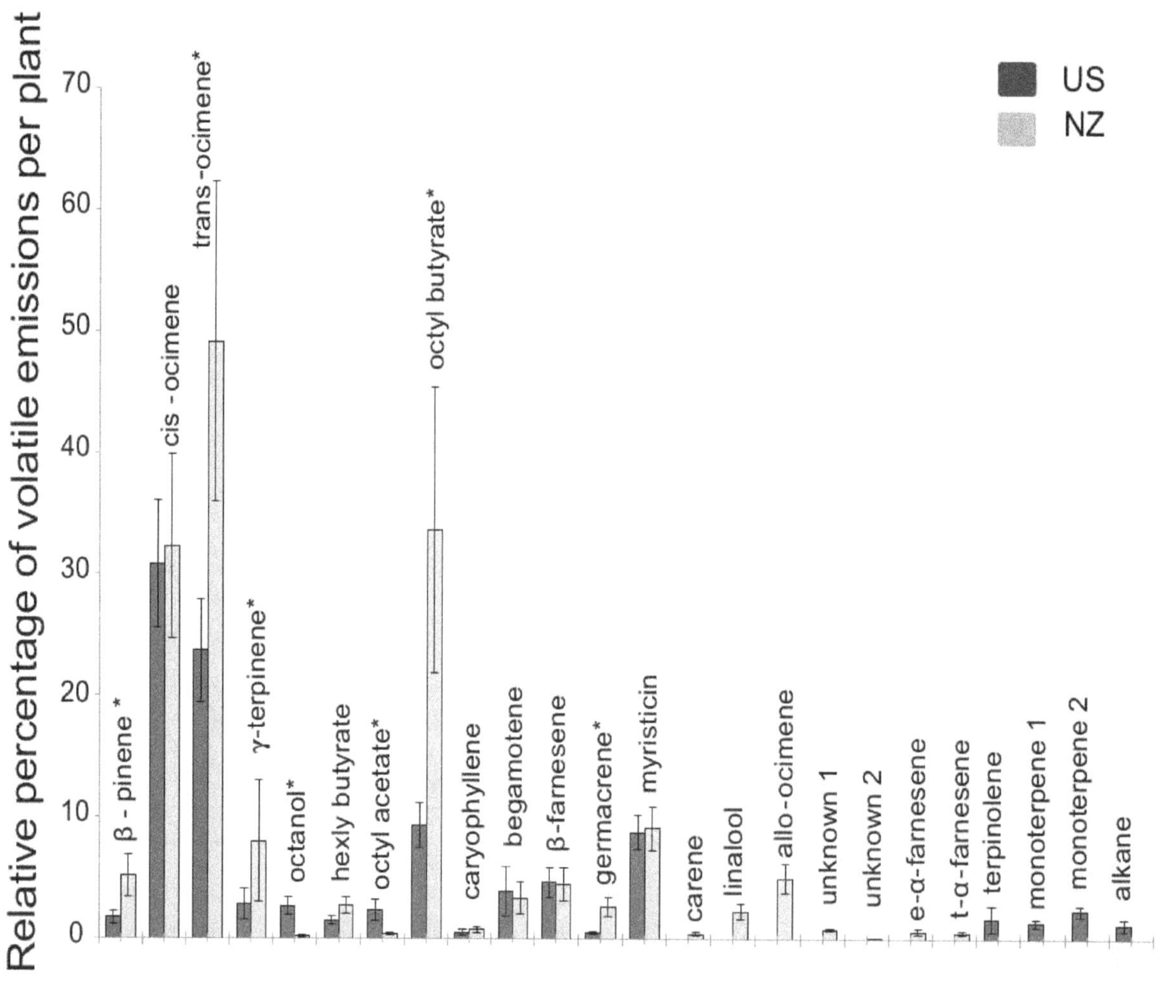

FIGURE 1. A comparison of the composition of volatile emitted from US and NZ flowers. Error bars represent standard error. Significant differences (MANOVA) indicated by *

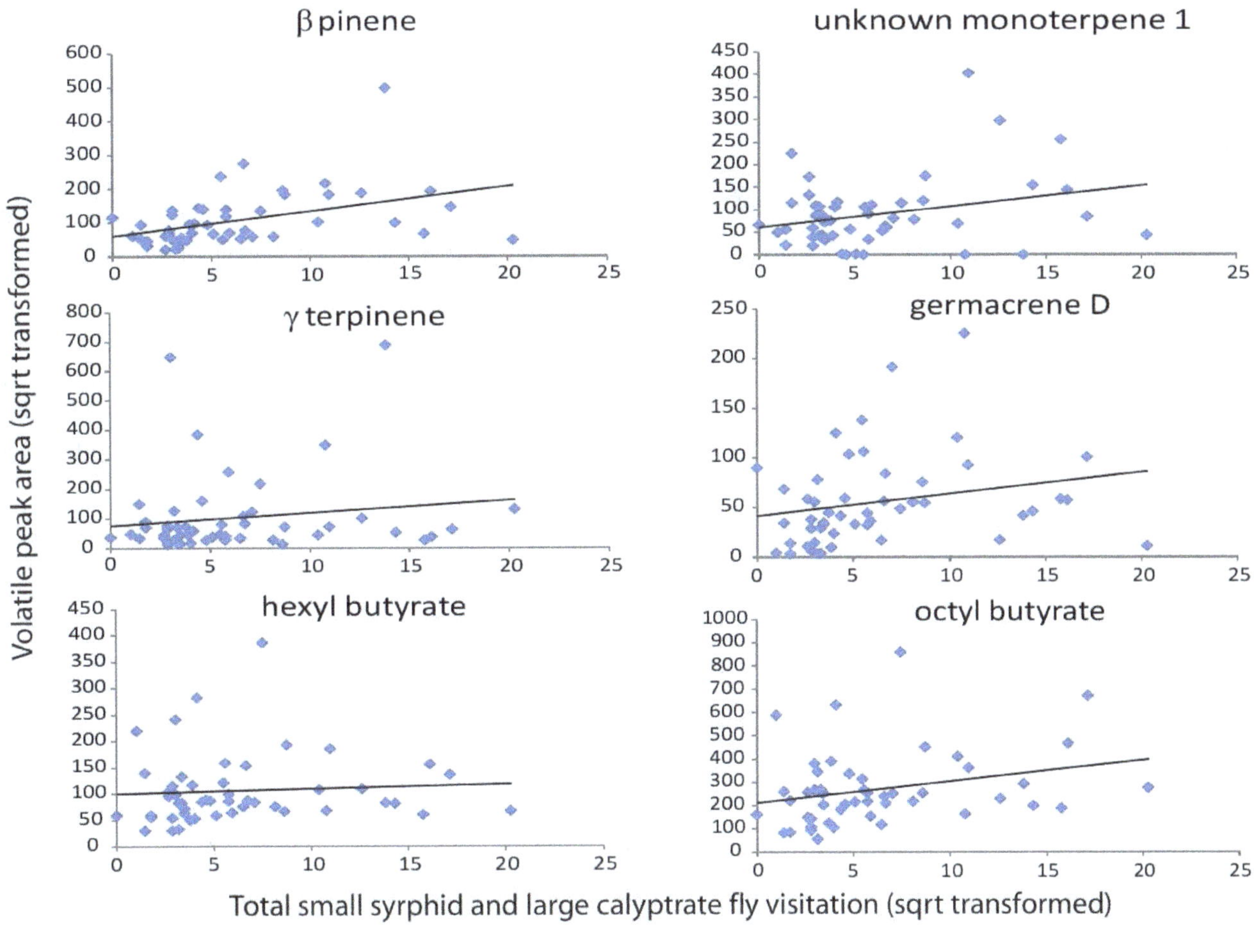

FIGURE 2. Compounds that best explain variation in the US pollinator matrix (DISTLM analysis: AICc = 336.23, R^2 = 0.34). Graphs represent the direction of relationship between pollinator visitation (counts of large calyptrate flies and small syrphids) and individual volatile compounds in US parsnip flowers. The Kovat index for the unknown monoterpene (monoterpene 1) calculated from C9 and C10 alkanes at 80°C is 916 and m/z ion fragments in order of relative abundance are 93, 56, 71, 192, 161

the variation in the pollinator matrix were trans- ocimene, carene, octyl butyrate and octyl acetate. This model had the lowest AICc score (AICc = 103.95, R^2 = 0.69). Trans-ocimene, carene and octyl butyrate were positively associated with pollinator visitation while octyl acetate was negatively associated with visitation (large calyptrate flies and small syrphids) (Figure 3). Most compounds positively associated with visitation (β-pinene, γ-terpinene, octyl butyrate, germacrene D, trans-ocimene and carene) were emitted in higher proportions in the NZ plants, whereas octyl acetate, which is negatively associated with visitation, is emitted in lower proportions in NZ plants.

Association between pollination and nectar availability

Large calyptrate and small syrphid visitation was significantly associated with nectar availability in the US but not in NZ (Tables 3a and b, Figures 4a and b). The number of visitors was positively associated with umbel size for large calyptrate and anthomyiid flies in the US and large calyptrate, small syrphid and bibionid flies in NZ (Tables 3a and 3b).

DISCUSSION

Wild parsnips depend heavily on insect visitation for reproduction, indicating that seed production in this monocarpic biennial is highly contingent on the plant's ability to attract effective pollinators. Easily accessible nectar and large conspicuous inflorescences render wild parsnips a target for visitation by a wide range of insects. Many (possibly most) frequent visitors, however, do not appear to be effective pollinators. Frequency of visitation and pollen loads did not prove to be a reliable indicator of pollinator efficacy in this study; *C. pennsylvanicus* in the US and bibionid flies in NZ, for example, were extremely abundant and carried substantial amounts of pollen but were poor predictors of pollination success, possibly because of their tendency to aggregate and remain on the inflorescence for long periods (personal observations). Increased duration of visits per flower likely reduces the frequency with which other flowers are visited, an important determinant of pollinator effectiveness (Madjidian et al. 2008). Similarly, large ants, although common on flowers, do not appear to contribute proportionately to pollination success.

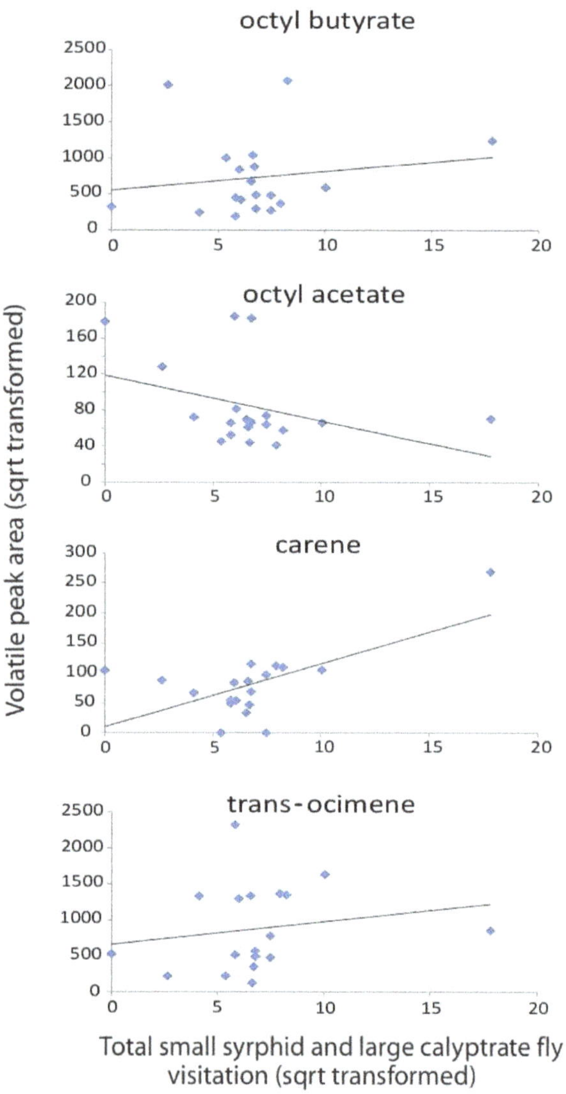

FIGURE 3. Compounds that best explain variation in the NZ pollinator matrix (DISTLM analysis: AICc = 103.95, $R^2 = 0.69$). Graphs represent the direction of relationship between pollinator visitation (counts of large calyptrate flies and small syrphids) and individual volatile compounds in NZ parsnip flowers.

Despite an apparent "promiscuous" strategy (*sensu* Bell 1971), in both the US and NZ the insects chiefly responsible for parsnip pollination in this study were species in the order Diptera. In fact, this so-called "promiscuously pollinated" plant may be somewhat specialized for pollination by flies. Visits by large calyptrate flies and small syrphids were significantly predictive of seed set in both the US and NZ. Moreover, the composition of floral volatiles of *P. sativa* is suggestive of some degree of specialization for dipteran pollinators. Although containing representatives from phytochemical classes widespread among angiosperms, the floral volatiles of *P. sativa* and closely related species in the genus *Heracleum* are essentially unique in containing octyl esters (Borg-Karlson et al. 1993, Knudsen et al. 2006). Evidence suggests that some dipterans may rely on aliphatic esters as cues for locating floral hosts (Knudsen & Stahl, 1994). Similarly, attractiveness to flies among some plant species has been linked to production of trans-ocimene (e.g., *Prestoea schultzeana,* Knudsen et al. 2001; *Silene otitis,* Dotterl et al. 2012). Indeed, we found that many of these floral volatiles have a significant role in attracting dipteran pollinators to *P. sativa.* Pollinators appear to respond mainly to monoterpenes, sesquiterpenes and esters in the floral bouquet. Trans-ocimene and octyl butyrate dominate the floral bouquets of both NZ and US flowers (Figure 1) and are positively associated with visitation. In addition to floral volatiles, pollinator visitation is also significantly driven by nectar rewards (Figure 4a) and the conspicuousness of the umbels (umbel size, Tables 3a and 3b).

Remarkably, most compounds positively associated with visitation are produced in significantly higher proportions in NZ flowers. In the absence of herbivory, NZ plants might be under selection to increase fitness by producing more pollinator attractants. However, in this study, the tradeoff between compounds that mediate pollinator and herbivore attraction/deterrence may be constrained or otherwise influenced by biosynthetic considerations. Octyl butyrate is a known webworm deterrent but also a pollinator attractant, whereas octyl acetate is a webworm attractant and a pollinator deterrent, suggesting that selection should favor plants with high octyl butyrate and low octyl acetate. However, octyl acetate and octyl butyrate are extremely polymorphic in *P. sativa* and their production might be genetically constrained via pleiotropic associations with other chemical pathways. In a heritability study conducted on North American wild parsnip mature fruit, Carroll et al. (2000) determined that amounts of each ester are independently phenotypically correlated with linear furanocoumarins; if these phenotypic correlations reflect underlying genetic correlations, each ester may thus respond differently to selection on furanocoumarins or on other defense compounds. Despite absence of direct evidence of conflicting selection on floral chemistry in wild parsnips, the floral bouquet of wild parsnips in NZ reflects overall a profile of volatiles that is more attractive to pollinators.

A previous examination of wild parsnip floral chemistry in the US and NZ compared the composition of ethyl acetate extracts of buds and flowers (Zangerl et al. 2008). However, our focus in this study was attraction of pollinators and we compared headspace volatiles (Knudsen et al. 1993) as more ecologically relevant. Accordingly, the content and composition of floral volatiles differ between studies, with greater proportional representation of monoterpenes and lower representation, or absence, of compounds of higher molecular weight (e.g., palmitolactone) in this study.

The presence of herbivores can also have a direct impact on pollination. In *P. sativa,* florivory by webworms can greatly reduce the number of floral visitors to *P. sativa* (Lohman et al. 1996). Even a single webworm can reduce visitation; webworms not only web together and consume flowers, reducing the size of the floral display and restricting access of pollinators to flowers, they also deposit frass in their webs, altering the composition of the floral bouquet. Zangerl et al. (2009) demonstrated that webworm frass contributes octanol to the floral bouquet, through webworm

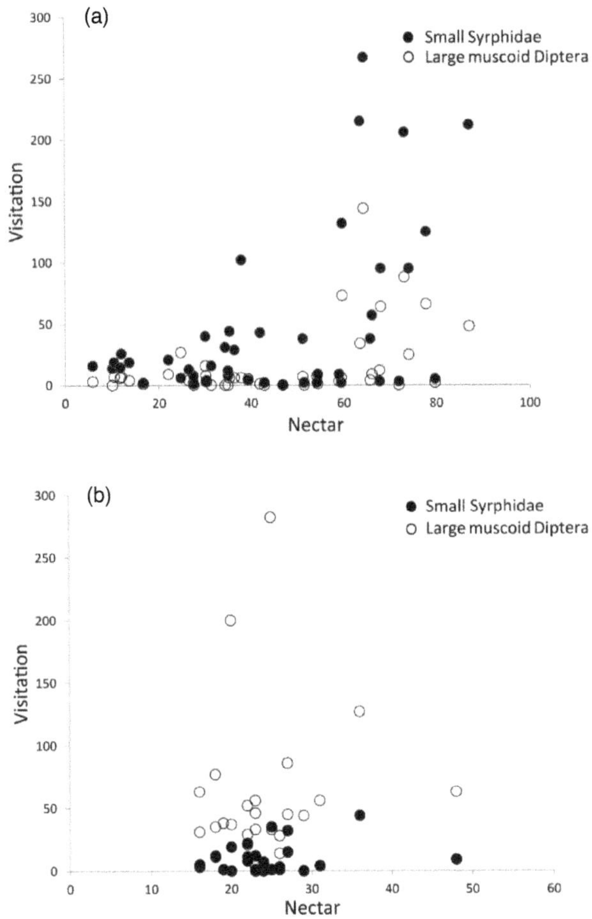

FIGURE 4. Relationship between the number of visits and nectar availability (percentage of flowers per umbel bearing nectar on the day's last photograph) on a) US parsnips and b) NZ parsnips for large calyptrate flies (US: $P<0.001^*$, NZ: $P=0.75$, GLM negative binomial) and small syrphids (US: $P=0.01^*$, NZ: $P=0.93$ GLM negative binomial)

detoxification of the esters octyl acetate and octyl butyrate. Conversion of octyl esters to octanol may thus contribute to reductions in pollinator visitation via reducing the abundance of a potential pollinator attractant and increasing the abundance of an otherwise rare volatile component.

In undertaking this study, we found evidence consistent with the hypothesis that floral chemical traits in wild parsnips free from webworm florivory will reflect a volatile profile that maximizes fitness via the attraction of pollinators. Our findings suggest that future studies of phytochemical impacts of enemy release should address interactions with mutualists as well as interactions with enemies, in order to obtain a more comprehensive understanding of both the ecology and evolution of plant-insect interactions in non-indigenous environments.

ACKNOWLEDGEMENTS

We thank Allen Lawrance, Alan Yanahan, Andrew Lovdahl, Daniel Pearlstein, and Terry Harrison in Illinois and Kelly Booth, Richard Toft, and John Dugdale in New Zealand for assistance with fieldwork, pollen counts, analysis of photographs, and insect identification. This work was funded by National Science Foundation grant NSF0816616.

REFERENCES

Andersson S, Dobson H (2003) Behavioral foraging responses by the butterfly *Heliconius melpomene* to *Lantana camara* floral scent. Journal of Chemical Ecology 29: 2303-2318

Ashman TL, Bradburn M, Cole DH, Blaney BH, Raguso RA (2005) The scent of a male: The role of floral volatiles in pollination of a gender dimorphic plant. Ecology 86: 2099-2105

Baldwin IT, Preston C, Euler M, Gorham D (1997) Patterns and consequences of benzyl acetone floral emissions from *Nicotiana attenuata* plants. Journal of Chemical Ecology 23: 2327-2343

Bell CR (1971) Breeding systems and floral biology of the Umbelliferae or evidence for specialization in unspecialized flowers. In: Heywood V H (ed) The Biology and Chemistry of the Umbelliferae Supplement I to Botanical Journal of the Linnean Society 64: 93-108

Berenbaum MR (1981) Patterns of furanocoumarin production and insect herbivory in a population of wild parsnip (*Pastinaca sativa* L). Oecologia 49: 236-244

Borg-Karlson A, Valterová I, Nilsson LA (1993) Volatile compounds from flowers of six species in the family Apiaceae: Bouquets for different pollinators? Phytochemistry 35: 111-119

Carroll MJ, Berenbaum MR (2002) Behavioral responses of the parsnip webworm to host plant volatiles. Journal of Chemical Ecology 28: 2191-2201

Carroll MJ, Zangerl AR, Berenbaum MR (2000). Brief communication. Heritability estimates for octyl acetate and octyl butyrate in the mature fruit of the wild parsnip. Journal of Heredity 91: 68-71.

Dobson HEM (2006) Relationship between floral fragrance composition and type of pollinator In Pichersky NA, Dudareva E (eds) The Biology of Floral Scents ,CRC Press, Boca Raton, Florida pp 147 – 199

Dotterl S, Jurgens A, Seifert K et al. (2006) Nursery pollination by a moth in *Silene latifolia*: The role of odours in eliciting antennal and behavioural responses. New Phytologist 169: 707-718

Dotterl S, Jahreiss K, Umma Salma J et al. (2012) Temporal variation of flower scent in *Silene otites* (Caryophyllaceae): a species with a mixed pollination system. Botanical Journal of the Linnean Society 169: 447-460.

Feeny P (1977) Defensive ecology of the Cruciferae. Annals of the Missouri Botanical Garden 64: 221-234

Galen C, Kaczorowski R, Todd SL, Geib J, Raguso RA (2011) Dosage-dependent impacts of a floral volatile compound on pollinators, larcenists, and the potential for floral evolution. American Naturalist 177: 258-272

Janzen DH (1975) Behavior of *Hymenaea courbaril* when its predispersal agent is absent. Science 189: 145-147

Joshi J, Vrieling K (2005) The enemy release and EICA hypothesis revisited: Incorporating the fundamental difference between specialist and generalist herbivores. Ecology Letters 8: 704-714

Jürgens A, Dötterl S, Meve U (2006) The chemical nature of fetid floral odours in stapeliads (Apocynaceae-Asclepiadoideae-Ceropegieae). New Phytologist 172: 452-468

Jürgens A, Dötterl S, Liede-Schumann S et al (2008) Chemical diversity of floral volatiles in Asclepiadoideae-Asclepiadeae (Apocynaceae). Biochemical Systematics and Ecology 36: 842-852

Keane RM, Crawley MJ (2002) Exotic plant invasions and the enemy release hypothesis. Trends in Ecology and Evolution 17: 164-170

TABLE 3. Log-likelihood ratio (Chi-sq) tests for GLMs (negative binomial model) predicting large fly and small syrphid visitation by nectar availability in US and NZ parsnips. Significant differences at $P<0.001$ are indicated by *** and $P<0.05$ by *

a) US

	Df	Deviance	AIC	LRT	*P*-value	
Large calyptrate Diptera						
Full Model		52.43	294.42			
Nectar	1	63.29	303.22	10.802	0.00	***
Umbel size	1	69.40	309.33	16.911	0.00	***
Small Syrphidae						
Full Model		51.95	409.18			
Nectar	1	62.89	418.12	10.943	0.00	***
Umbel size	1	55.48	410.71	3.5304	0.06	
Anthomyiidae						
Full Model		49.80	266.16			
Nectar	1	56.03	269.84	5.678	0.01	*
Umbel size	1	55.48	270.38	6.226	0.01	*
Chauliognathus pennsylvanicus						
Full Model		50.07	240.93			
Nectar	1	50.91	270.49	0.84	0.35	
Umbel size	1	50.74	270.31	0.67	0.41	
Small Formicidae						
Full Model		43.90	240.93			
Nectar	1	44.93	239.96	1.02	0.311	
Umbel size	1	46.37	241.40	2.46	0.11	
Large Formicidae						
Full Model		43.49	228.94			
Nectar	1	46.10	262.06	2.60	0.10	
Umbel size	1	43.52	259.48	0.03	0.86	

b) NZ

	Df	Deviance	AIC	LRT	*P*-value	
Large calyptrate Diptera						
Full Model		33.09	270.80			
Nectar	1.00	33.19	268.89	0.09	0.75	
Umbel size	1.00	39.23	274.94	6.14	0.01	*
Small Syrphidae						
Full Model		31.48	176.88			
Nectar	1.00	31.49	174.89	0.06	0.93	
Umbel size	1.00	39.76	183.18	8.27	0.00	*
Bibionidae						
Full Model		31.01	385.77			
Nectar	1.00	31.66	384.41	0.64	0.42	
Umbel size	1.00	35.36	388.12	4.34	0.03	*
Acalyptrate flies						
Full Model		31.60	252.96			
Nectar						
Umbel size	1.00	32.57	251.94	0.97	0.32	
Thysanoptera						
Full Model		32.06	415.06			
Nectar	1.00	31.65	251.01	0.04	0.82	
Umbel size	1.00	32.78	413.78	0.72	0.39	

Kessler D, Baldwin IT (2011) Back to the past for pollination biology. Current Opinions in Plant Biology 14: 429-434

Kessler D, Gase K, Baldwin IT (2008) Field experiments with transformed plants reveal the sense of floral scents. Science 321: 1200-1202

I. Kessler D, Diezel C, Clark DG, Colquhoun TA, Baldwin IT (2012) *Petunia* flowers solve the defence/apparency dilemma of pollinator attraction by deploying complex floral blends. Ecology Letters doi: 10.1111/ele.12038

Knudsen JT, Stahl B (1994) Floral odors in the Theophrastaceae. Biochemical Systematics and Ecology 22: 259-268

Knudsen JT, Eriksson R, Gershenzon J, Stahl B (2006) Diversity and distribution of floral scent. Botanical Review 72: 1-120

Knudsen JT, Tollsten L, Ervik F (2001) Flower scent and pollination in selected Neotropical palms. Plant Biology 3: 642-653

Liu H, Stiling P (2006) Testing the enemy release hypothesis: A review and meta-analysis. Biological Invasions 8: 1535-1545

Lohman D, Zangerl AR, Berenbaum MR (1996) Impact of floral herbivory by parsnip webworm (Oecophoridae: *Depressaria pastinacella* Duponchel) on pollination and fitness of wild parsnip (Apiaceae: *Pastinaca sativa* L). American Midland Naturalist 136: 407-412

Mao W, Zangerl AR, Berenbaum MR, Schuler MA (2008) Metabolism of myristicin by *Depressaria pastinacella* CYP6AB3v2 and inhibition by its metabolite. Insect Biochemistry and Molecular Biology 38: 645-651

Maron JL, Vilà M (2001) When do herbivores affect plant invasion? evidence for the natural enemies and biotic resistance hypotheses. Oikos 95: 361-373

R Development Core Team (2008) R: A language and environment for statistical computing R Foundation for Statistical Computing, Vienna, Austria ISBN 3-900051-07-0, URL http://wwwR-projectorg

Raguso RA (2008) Wake up and smell the roses: The ecology and evolution of floral scent. Annual Review of Ecology and Evolution S 39: 549-569

Shuttleworth A, Johnson SD (2010) The missing stink: Sulphur compounds can mediate a shift between fly and wasp pollination systems. Proceedings of the Royal Society B-Biological Sciences 277: 2811-2819

Theis N (2006) Fragrance of Canada thistle (*Cirsium arvense*) attracts both floral herbivores and pollinators. Journal of Chemical Ecology 32: 917-927

Theis N and Adler LS (2012) Advertising to the enemy: enhanced floral fragrance increases beetle attraction and reduces plant reproduction. Ecology 93:430–435

Wootton JT (1994) The nature and consequences of indirect effects in ecological communities. Annual Review of Ecology Evolution and Systematics 25: 443-466

Zangerl AR, Berenbaum MR (2009) Effects of florivory on floral volatile emissions and pollination success in the wild parsnip. Arthropod-Plant Interactions 3: 181-191

Zangerl AR, Stanley MC, Berenbaum MR (2008) Selection for chemical trait remixing in an invasive weed after reassociation with a coevolved specialist. Proceedings of the National Academy of Sciences 105: 4547-4552

Zangerl AR, Berenbaum MR (2005) Increase in toxicity of an invasive weed after reassociation with its coevolved herbivore. Proceedings of the National Academy of Sciences 102: 15529-15532

Zangerl AR, Berenbaum MR (1993) Plant chemistry, insect adaptations to plant chemistry, and host plant utilization patterns. Ecology 74: 47-54

Zangerl AR, Berenbaum MR (1990) Furanocoumarin induction in wild parsnip - genetics and populational variation. Ecology 71: 1933-1940

Zangerl AR, Arntz A, Berenbaum MR (1997) Physiological price of an induced chemical defense: Photosynthesis, respiration, biosynthesis, and growth. Oecologia 109: 433-441

Zangerl AR, Berenbaum MR, Nitao JK (1991) Parthenocarpic fruits in wild parsnip - decoy defense against a specialist herbivore. Evolutionary Ecology, 5, 136-145

Zuur AF, Ieno EN, Walker NJ, et al (2009) Mixed Effects Models and Extensions in Ecology with R. Springer, New York, NY, USA

16

HUMMINGBIRDS AT ARTIFICIAL FLOWERS MADE TO RESEMBLE ORNITHOPHILES VERSUS MELITTOPHILES

Wyndee A. Guzman, Paul Wilson*

Department of Biology, California State University, Northridge, CA 91330-8303

Abstract—Certain floral characteristics are associated with specific pollinators. Hummingbird-pollinated flowers are usually red, lack a landing platform, lack nectar guides, and contain a high amount of dilute sucrose-rich nectar. Here we test hypotheses concerning the reasons for these characters to the extent that they involve hummingbird responses. An array was set up of 16 artificial plants, each with five artificial flowers. (1) Flowers made to differ only in colour elicited a slight preference for red. (2) When colour was associated with nectar offerings, and birds generally learned to visit flowers that provided much more nectar but did not associatively learn differences as little as 2 μL. (3) Birds were offered 8 μL of 12% sucrose versus 2 μL of 48% hexose, and they did *not* prefer the dilute nectar; they showed no evidence of discerning sucrose from hexose; however, they preferred 48% over 12% sucrose when both were offered in the same quantity. (4) Birds preferred flowers that lacked landing platforms over those with landing platforms. (5) Birds were offered flowers with nectar guides, associated with differing nectar volumes, and they did not associate the higher nectar reward with either flower type. In summary, the feedback from hummingbirds reflects some of the differences between bird- and bee-adapted flowers, but nectar seemed less predictive than expected. Factors other than the behavioural proclivities of hummingbirds, such as adaptation to discourage bees, are discussed as additional causes for the differences between the syndromes. We also discuss significance testing for field experiments involving one unreplicated array.

Keywords: *hummingbird, landing platform, nectar, nectar guide, pollination syndrome*

INTRODUCTION

Flowers adapted to birds are ornithophiles. Flowers adapted to bees are melittophiles. A suite of characters is associated with ornithophily as compared to melittophily. Most are red, often without much colour patterning, have high quantities of diluted sucrose-rich nectar, lack landing platforms, have long narrow floral tubes, and have outwardly exserted anthers and stigmas (Faegri & van der Pijl 1979; Mitchell & Paton 1990; Thomson et al. 2000). A number of factors might be responsible for the syndrome differences, namely, pollinator efficiency, floral costs, the deterrence of less beneficial visitors, and the behavioural proclivities of pollinators. Here we address hypotheses that involve the behavioural proclivities of hummingbirds visiting ornithophilous versus melittophilous artificial flowers.

In western North America there is a guild of ornithophiles that have reddish flowers (Grant & Grant 1968; Brown & Kodric-Brown 1979; Bleiweiss 2001). Using floral colour and colour patterns, pollinators are cued into nectar rewards, and colouration may also help pollinators extract nectar more quickly (Baker 1961). Though ornithophiles are usually red, hummingbirds visit not only red flowers. They can be conditioned to visit flowers of other colours, and when other colours are associated with nectar more to their liking, they show little long-term preference for red (Bene 1941; Collias & Collias 1968; Stiles 1976). An alternative explanation to the view that hummingbirds simply prefer red has been well articulated by Rodríguez-Gironés and Santamaría (2004) who suggest that bees are less likely to visit red flowers because they lack red colour receptors that would allow them to easily distinguish between red flowers and green foliage. If birds are more efficient pollinators than bees, then flowers adapted to birds should have characters that deter pollen-wasteful bees from emptying the nectaries and scattering the pollen. Under this theory, birds are constantly finding that there is more nectar in red flowers, which have been under-visited by bees, so they come to expect the rule to hold (Raven 1972).

In ornithophiles nectar is the only floral reward, and it is often hidden deep in the flower. Hummingbirds rely on cues, such as colour, to locate which flowers are rewarding and which are not (Hurly & Healy 1996). If ornithophiles evolved partly or solely as a result of hummingbird preferences, then we would expect that hummingbirds should prefer to visit artificial flowers that mimic the nectar properties of ornithophiles as opposed to melittophiles. For example, the ornithophile *Ipomopsis aggregata* produces between 1 and 5 μL of 20-25% sucrose-rich nectar (Pleasants 1983; Irwin & Brody 2000). Ornithophilous penstemons, on average, offer 7.76 μL of 26% sucrose-rich nectar (Wilson et al. 2006). Aside from secreting copious nectar, the other properties of ornithophilous nectar have been explained as reducing viscosity (Baker 1975, cf. Roberts 1995, 1996). The thinking is that birds prefer less

*Corresponding author; email: paul.wilson@csun.edu

viscous nectar because they can extract more energy per unit time. Although we expect birds to prefer ornithophilous nectar over melittophilous nectar, ornithophiles may not secrete what birds prefer. Nectar production can be costly for flowers. *Blandfordia nobilis* plants have been found to expend as much as 37% of their energy on nectar production (Pyke 1991), and nectar can be a substantial water expense (Baker 1975; Pyke & Waser 1981).

Ornithophiles tend to lack landing platforms compared with melittophiles (Faegri & van der Pijl 1979). It has been suggested that landing platforms would obstruct hummingbirds from extracting nectar quickly, and would make pollen placement less precise (Castellanos et al. 2004). In spite of the rule, some ornithophiles retain a lower lip, and Smith et al. (1996) present experimental evidence that it reduces the number of errors hummingbirds make in attempting to probe flowers. In other words, loss of what was a landing platform in a melittophilous ancestor should not be considered uniformly adaptive for ornithophiles.

The last feature we consider is spotting on a flower that is presumed to function as a "nectar guide" in melittophiles. Such patterns are often lost when a lineage shifts to ornithophily. For example, there is no patterning in the ornithophilous *Delphinium cardinalis*, in contrast to the many species of melittophilous *Delphinium*. Studying melittophilous *D. nuttallianum*, Waser and Price (1985) showed that patterning represents a nectar guide that speeds up handling time by bees. Medel et al. (2003) found disruptive selection on nectar guides in *Mimulus luteus* in which bees selected for larger nectar guides that point toward the mouth of the floral tube while hummingbirds selected for smaller heart-shaped nectar guides. Perhaps birds are less sensitive than bees to details that are seen only after the animals are very close to the flower (Gould & Gould 1988; Giurfa & Lehrer 2001).

MATERIALS AND METHODS

General experimental set-up

An array of artificial flowers was set up during the years 2008 and 2009 at a site in the southern Sierra Nevada of California, near the unincorporated town of Kennedy Meadows (36° N, 118° W). Visiting the array were hummingbirds of several species, sexes, and age classes: male and female *Archilochus alexandri*, male and female *Calypte anna*, and female *Selasphorus rufus*. Species and sex were recorded, but analyses were done pooling together all birds. As detailed in Haley (2010), the array consisted of 16 artificial plants spaced 1 m from one to the next. Each plant was a stalk with five artificial flowers spaced at intervals of 7.62 cm. Petals were made of brightly coloured ribbon fabric, and artificial nectar was pipetted into plastic nectar spurs made of 200 μL PCR tubes. The 16 plants typically differed by treatment. For example, on certain experimental days half of them were red with 2 μL of nectar and the other half were purple with 4 μL of nectar, while on other experimental days half were purple with 2 μL and the other half red with 4 μL. At the start of each hour of an experimental day, we replaced artificial plants with fresh models that had just been filled with nectar. Within experimental hours, each observation period was 40 minutes long. During that period hummingbird visits were followed and video taped. Generally, there were six 40-minute observation periods during a day (though on one stormy day, we only observed for five periods).

We scored four dependent variables (Wilson & Jordan 2009), which are reported on fully in Haley (2010): the kind of plant upon which a bird *initiated* a bout of foraging at the array; the number of *plants visited* of the two kinds of plants offered; the *number of probes* at each plant of the two types; and the *time per probe* at each plant. (Time per probe, or "handling time," is not expected to be minimized since birds with longer handling times were often extracting more calories, i.e., it is not "handling efficiency." In general, we are not trying to measure the physiological costs and benefits to the bird on a per visit basis, merely the feedback that the pollination environment gives in response to floral offerings.) Here we condense our treatment to just *plants visited* and *time per probe*. In general, *bout initiations* gave similar results to *plants visited*, and in general *number of probes* gave similar results to *time per probe* (Haley 2010). That *bout initiations* were similar to *plants visited* indicates that birds are able to adjust their behaviour not only within a bout but also between bouts. That *number of probes* was similar to *time per probe* was less expected since *number of probes* is an inverse part of *time per probe*, so the similarity in the responses of these two variables must mean that the duration of visits counteracted the inverse part-whole relationship.

Experiment 1: colour preference

Red is associated with ornithophily, whereas melittophilous flowers are often purple (or yellow: Haley 2010). To get started, we set up the 4 × 4 array with half of the plants having red flowers, and the other half having purple flowers. Equal amounts of nectar (2.5 μL of 23% sucrose solution) were pipetted into each of the artificial flowers at the beginning of each hour of study. Flower type was alternated within the array, with the first row being red-purple-red-purple, the second row being purple-red-purple-red, etc. On the next experimental day, the colour of alternating plants was switched. This experiments was run for six days.

Experiment 2: nectar volume

A second set of experiments was done in order to test preference between different amounts of 23% sucrose solution associated with different colours. Nectar of different amounts (0 versus 2 μL; 0 versus 4; 0 versus 6; 2 versus 4; 2 versus 6; 4 versus 6; 2 versus 8) was placed into differently coloured flowers. Birds were expected to use floral colour as a cue to find nectar when the difference in amount was large enough. On half of the days, red flowers contained one nectar volume and purple flowers contained the other nectar volume, and on the other half of the days the colour of the cue was reversed. Each nectar volume combination was set up for a span of four experimental days. Positions of the different treatments were rotated on different days. The various sub-experiments were done consecutively with different combinations of nectar amounts contrasted.

Experiment 3: effects of nectar composition

Three sub-experiments were done in order to test bird preferences for or against nectar concentrations and the size of sugar molecules, both of which affect nectar viscosity. Three solutions were made: 12% sucrose weight to volume, 48% sucrose, and 48% hexose comprised of half glucose and half fructose. Each sub-experiment consisted of a difference in nectar. In Experiment 3A, we used nectars differing in volume *and* concentration *and* sugar molecule size to see if hummingbirds show a preference for a higher volume of lower concentrated sucrose nectar. We used 8 μL of 12% sucrose versus 2 μL of 48% hexose. The goal was to have two solutions offering equal amounts of energy but different amounts of water. This sub-experiment confounds the volume difference with the concentration difference and with the chemical differences in the same way that natural syndromes confound these three variables, and the multifaceted difference was deemed a priori the most likely to have an effect. In Experiment 3B, birds were offered equal amounts of the same concentration of differing sugar types: 5 μL of 48% sucrose versus 48% hexose. In Experiment 3C, birds were given the choice between two solutions that only differed by concentration, and had equal volumes of the same sugar: 8 μL of 12% sucrose versus 48% sucrose. Each of the three sub-experiments was replicated on four days, switching flower colours and positions in all combinations.

Experiment 4: preference against landing platforms

Landing platforms are generally present in melittophilous flowers and often absent in ornithophilous flowers. An array was set up so that half of the plants consisted of artificial flowers that had landing platforms (i.e., an elongated tube that was cut to resemble a flower with five petals), whereas the other half of the stakes consisted of artificial flowers without landing platforms (i.e., elongated tubes with the bottom two petals cut out: Haley 2010). Each artificial flower in the array contained 5 μL of 25% sucrose solution. The artificial petals were always red. This experiment was run for four days in 2008 and two days in 2009. As usual, inflorescences were set up in an alternating spatial pattern, and positions were rotated each day.

Experiment 5: patterning on the petals

The purpose of the final experiment was to test if birds could use a flower cue other than the colour of the perianth as a whole to associate with great nectar volume. An array was set up where half of the plants had a pattern on their petals – the pattern was two black spots on the perianth – whereas the other half of the stakes had unmarked flowers (Haley 2010). This was an associative-learning experiment in which varying nectar reward was associated with presence or absence of pattern. All perianths were red. This experiment was replicated during four days in 2008 and four more days in 2009. Stakes were set up in an alternating pattern, where flowers contained 8 or 2 μL. The flower types and nectar amounts were rotated to ensure that flower types were in every position, solutions were in every position, and solutions were combined with each flower type in every position. Nectar in 2008 was 21% sucrose solution, and in 2009 was 23% sucrose solution (the difference being unintentional).

Statistics, with worries about replication

Our null hypothesis H_O was generally that manipulations in flower characters would not elicit preferences by hummingbirds. There would be a 1:1 ratio for plant visits, and birds would spend the same amount of time at a plant per probe regardless of experimental treatment. Our expectations were as follows:

1. H_A based on the systematic differences between ornithophiles and melittophiles, along with past experimental results, we expected hummingbirds to visit red more than purple, and we did not have expectations for time per probe.

2. H_A we expected birds to learn to prefer any colour associated with a higher volume of nectar and to take more time to extract more nectar.

3. H_A when presented with high versus low concentrations of nectar, both containing the same amounts of sugar, we predicted birds would prefer flowers that contain higher volume (less concentrated) nectar. Flowers that have a higher volume of more dilute nectar were predicted to have a decreased time per probe because their nectar is less viscous (though perhaps this prediction is naïve since a higher volume would take more time to consume if concentration had been equal).

4. H_A birds were expected to prefer flowers that lack landing platforms and to have increased time per probe when visiting flowers with landing platforms.

5. H_O for the final experiment on nectar guides, we predicted that the null hypothesis would be accepted. Because ornithophilous flowers often lack nectar guides compared to close relatives that are melittophilous, hummingbirds ought *not* to distinguish patterns on the perianth.

We summarize visits to plants as the percentage of visits that were to red inflorescences. This is a categorical variable that we analysed with G^2 goodness-of-fit tests to 1:1. Heterogeneity in goodness-of-fit was tested to see if days were similar enough to be pooled ($P > 0.25$), or if they had to be separated. When pooling was permitted, we report on the simplified analysis. Time per probe is a continuous variable, so we used ANOVA, and scrutiny of residuals compelled us to do analyses on log(time per probe). Before doing the tests, we searched for outliers, and as noted in the Results, a few were winsorized to the next highest number. The ANOVAs often started out split-plot mixed-models because day was a random variable whereas treatments such as colourcue and nectartype were fixed variables. Days were nested within cues, and plants were replicates within nectartype. We calculated F-values as in Quinn and Keough (2002, p. 314). For these split-plot designs, the effect of nectartype, which is the within-plot treatment, is the one of primary interest, and in the full model it would be tested over nectartype × day ⊂ cue. When $P > 0.25$, terms were sequentially pooled. For example, if $P > 0.25$ for nectartype × day ⊂ cue, then the effect of nectartype could be tested over the pooled residual with far more degrees of freedom. Cue, the among-plots treatment, was not of primary interest, and in the full model it would be tested over day ⊂ cue, which would be a weak test because there were few days. Pooling higher order terms when $P > 0.25$ often allowed for a much simpler and more powerful final model than the original split-plot design.

In using these significance tests, we are pretending as though each visit to an artificial plant composed of five flowers were a replicate representing a larger universe of such visits. However, by the very nature of these types of experiments (in which animals are acquiring information as they forage) a visit to a plant by a bird is not a unitary indicator-of-behaviour that is independent of the other visits to that array by that bird or by other birds that are visiting the array. Unfortunately, there was only one array. A dozen arrays each with different birds surrounded by naturally varying flowers would have allowed arrays to be used as replicates. Then a variate would have been the percent of visits to, say, sucrose-filled plants in an array or the average time per probe at sucrose-filled plants (Rodríguez et al. 2004). Such an effort was far beyond our means.

Many other pollination biologists have done array experiments using a level of replication below whole arrays. Schemske and Ågren (1995), Meléndez-Akerman et al. (1997), Wesselingh and Arnold (2000), Gómez et al. (2008) and Owen and Bradshaw (2011) all chose to do significance tests such as we did using the visit as a replicate. Carlson (2008), likewise, used visits as the unit of replication, although her experiments have higher level blocking factors (like our day effect), but she does not take the average response in a block as a variate. Another option would be to use an observation period as a replicate. Majetic et al. (2009) used 10-minute observation periods as replicates, and Hersch and Roy (2007) used 45-minute observation periods. The analog in our study would have been to use our 40-minute observation periods. This would not have eliminated pseudoreplication. Also presumably number of visits in 40 minutes would not have followed a normal distribution being a meristic variable with a low mean, which would have compelled using generalized (rather than merely general) linear modelling. Observation period is arbitrary in the sense that it is fixed by the scientist, so pollination biologists sometimes choose to use foraging bout as a replicate (Aldridge & Campbell 2007, Campbell et al. 2010), but foraging bouts are still pseudoreplicates of the same individual pollinator who is accumulating experience with the treatments. If bout were used, the outcome would then be reported as the mean of a sample of numbers like 1/1, 1/5, 2/3, 10/10 each based on counts with continuity problems and lack of normality. Also, a great deal of data would be gathered on a very few individuals while many individuals would be represented by only one or two bouts, so using bout would be a poor approximation of using individual bird, which itself was nearly as far beyond our means as replicating experimental arrays.

The most sensible minimization of pseudoreplication that is feasible within the confines of using one array for only two years is to use each "plant" as a replicate (Schemske & Bradshaw 1999, Fenster et al. 2006). In addition to the analyses we report using visits as replicates, we also elaborate on selected results using the positions in the array during the course of a day as a replicate. This is akin to using plant except that we replaced the artificial plants each hour. Using positions as replicates, *visits to plants* becomes *number of visits to a position*, and *log(time per probe)* becomes *mean log(time per probe)* averaged across all visits observed in a day at that position in the array. There were $n = 8$ such replicates for each of two treatments during any one day. Position is less pseudoreplicated than visit, but it is still not equivalent to replicating the experiment. We cling to using visits as replicates in our primary analyses for several reasons. First, it seems the most natural unit of response by which birds interact with plants. A bird is more or less making decisions to visit one plant or another and to take time during its probes at a plant, whereas the number of visits during six 40-minute observations on a day or the average time per probe during those visits is a mere artificial aggregation; it's not fitness because the flowers are plastic, and it's not about how a representative bird treats a plant behaviourally. Second, we have a number of non-significant results for which it seems best to report the most powerful test, which is the test that uses visits as replicates, so only when there is a significant and interesting result at the lower level of replication do we feel compelled to check the significance at the level of positions within a day. Third, one would not expect the various "plants" that occupied a position during the course of a day to be particularly similar to one another. In fact, looking at the data there was generally no evidence of such an effect; for example, in the experiment with the most replication (Experiment 5), a nested ANOVA of log(time/probe) yielded $P = 0.991$ for position nested within treatment-day combinations. Values of $P > 0.25$ are widely accepted as permission to pseudoreplicate (Quinn & Keough 2002).

To reiterate, then: tests at neither level allow for inferential statistics beyond the circumstances of our one array visited by a small number of birds with relatively scant access to naturally occurring flowers at our arid field site. Significance tests such as ours for experiments of an array merely provide a conventional cut-off for credibility that a pattern is non-random in that one context.

RESULTS

Experiment 1: colour preference

Birds had a slight preference for red. Plant visits: On days 1 and 2, birds significantly preferred red inflorescences ($day_1 = 69.6\%$ red; $day_2 = 69.0\%$; $P_{day1} < 0.001$; $P_{day2} = 0.012$), but on days 3 through 6, birds showed no significant preference for either flower colour ($day_3 = 61.9\%$; $day_4 = 52.3\%$; $day_5 = 55.6\%$; $day_6 = 59.3\%$:

TABLE 1. Nectar volume experiments.

		% visits to inflorescences (n)	nectar quantity[a]	log(handling time, s/probe)	nectar quantity[b]
2 vs 0 μL	red with 2 vs purple with 0	70.4 vs 29.6 (206)	***	-0.135 vs -0.140	NS
	purple with 2 vs red with 0	58.2 vs 41.9 (184)	*	-0.168 vs -0.150	
4 vs 0 μL	red with 4 vs purple with 0	63.8 vs 36.3 (323)	***	-0.079 vs -0.147	***
	purple with 4 vs red with 0	56.6 vs 43.4 (258)	*	-0.104 vs -0.149	
6 vs 0 μL	red with 6 vs purple with 0	69.0 vs 31.1 (168)	***	-0.038 vs -0.185	***
	purple with 6 vs red with 0	64.7 vs 35.3 (269)	***	-0.091 vs -0.144	
4 vs 2 μL	red with 4 vs purple with 2	50.4 vs 49.6 (500)	NS	-0.055 vs -0.102	***
	purple with 4 vs red with 2	51.7 vs 48.3 (208)	NS	-0.063 vs -0.101	
6 vs 2 μL	red with 6 vs purple with 2	58.0 vs 42.0 (226)	*	-0.081 vs -0.150	NS
	purple with 6 vs red with 2	55.9 vs 44.1 (195)	†	-0.076 vs -0.105	
6 vs 4 μL	red with 6 vs purple with 4	50.9 vs 49.1 (340)	NS	-0.037 vs -0.081	*
	purple with 6 vs red with 4	50.2 vs 49.8 (251)	NS	-0.057 vs -0.061	
8 vs 2 μL	red with 8 vs purple with 2	63.1 vs 36.9 (425)	***	-0.078 vs -0.114	NS
	purple with 8 vs red with 2	55.5 vs 44.5 (290)	†	-0.111 vs -0.169	

NS $P > 0.1$; †$P < 0.1$; *$P < 0.05$; **$P < 0.01$; ***$P < 0.001$

[a]goodness-of-fit test to 1:1; on different days for each experiment, nectar quantity preference may have been complicated by colourcue preference; days were heterogeneous for colourcue and often could not be pooled

[b]split-plot ANOVA; for 6 vs 4 μL nectar quantity × colourcue yielded $P < 0.05$; otherwise the interaction and colourcue yielded $P > 0$.

$G^2_{heterogeneity} = 6.89$, $df = 5$, $P = 0.229$). Pooling all days, red was preferred ($G^2_{pooled} = 19.29$, $df = 1$, $P < 0.001$). Replicating with position, colour was still significant ($P_{colour} = 0.013$). Time per probe: The interaction of colour × day yielded $P = 0.161$ so could not be pooled. Generally, birds spent significantly more time at red than at purple flowers ($P_{colour} = 0.009$), and days varied in the amount of time birds spent per flower ($P_{day} < 0.001$). Replicating with position, colour × day yielded $P = 0.010$, which meant that the colour effect had to be tested over the interaction, and it signalled no significant effect beyond the interaction ($P_{colour} = 0.073$). Further results with colour combinations involving yellow are given in Haley (2010).

Experiment 2: nectar volume

Zero versus some nectar: Table 1 shows that birds significantly preferred flowers with nectar over empty flowers. Plant visits: They visited inflorescences that contained nectar significantly more for each sub-experiment. Time per probe: Birds spent significantly more time at flowers that had 4 or 6 μL versus 0 μL, but there was no significant difference when the filled flowers had only 2 μL.

Some versus more nectar: When nectar differed by a larger amount (≥4 μL), birds chose the higher nectar volume (Table 1). Plant visits: They showed a significant preference for higher nectar volumes when the volumes differed by at least 4 μL. In other cases, the difference was not significant. Time per probe: Birds spent significantly more time at the higher flower volume when the two volumes differed by only 2 μL. There was no significant difference when flowers differed by ≥4 μL, which was an unexpected result.

Experiment 3: effects of nectar composition

2 μL 48% hexose versus 8 μL 12% sucrose: A difference in viscosity elicited very little difference in bird behaviour (Fig. 1). Plant visits: There was not a consistent preference for plants with one of the nectar types over the other. If anything, there is simply a suggestion that birds prefer the flowers that are red regardless of nectar type. For the analysis with positions as replicates, day within colorcue interacted with nectar type ($P_{nectartype \times day \subset colourcue} = 0.003$), but simpler terms involving nectar type were not significant ($P < 0.05$). Time per probe: Using plant visits as replicates, birds seemed to take significantly more time at 2 μL 48% hexose than at 8 μL 12% sucrose ($P_{nectartype} = 0.002$). However, replicating with position, no term involving nectartype was significant ($P < 0.05$).

Sucrose versus hexose: Birds seemed unaffected by sugar type (Fig. 2). Plant visits: They showed no significant preference for sucrose versus hexose ($P > 0.05$ for each day). Time per probe: One datum was winsorized from 0.4293 log(seconds/flower) to 0.2975 log(seconds/flower). The interaction of nectartype × day ⊂ colourcue yielded $P = 0.197$, so this term could not be pooled. Nectartype × colourcue yielded $P = 0.685$. There was added variance in how much time was spent at sugar types on different days ($P_{day \subset colourcue} < 0.001$), but birds did not significantly spend more or less time depending on sugar type ($P_{nectartype} = 0.544$) or the colour of the cue ($P_{colourcue} = 0.984$). The analysis of position as the unit of replication did not change the outcomes of these tests.

Day	G^2	P
1. purple@8	10.113	0.001
2. purple@8	0.718	0.397
3. red@8	0.964	0.326
4. red@8	1.871	0.171

	G^2	P
purple@8		
heterogeneity	2.709	0.100
pooled	8.122	0.004
red@8		
heterogeneity	0.030	0.863
pooled	2.804	0.094

	df	MS	*P*
Colorcue	1	0.090	0.286
Nectartype	1	0.783	0.002
Error	342	0.079	

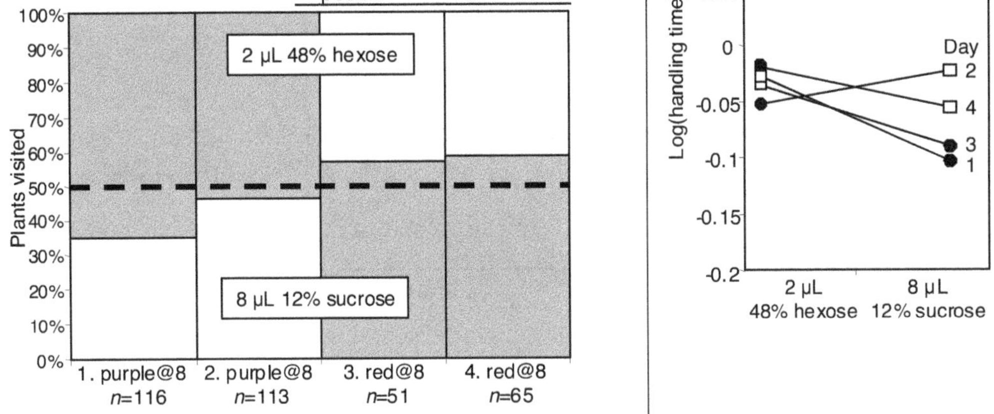

FIGURE 1. 2 µL 48% hexose versus 8 µL 12% sucrose. This experiment tested preferences for bee- versus bird-type nectar. (grey bars: red flowers; white bars: purple flowers; ● red flowers; □ purple flowers)

12% versus 48% sucrose: Birds generally preferred 48% over 12% sucrose (Fig. 3). Plant visits: Overall, birds significantly preferred 48% sucrose on two days, and showed marginal significance on one day. Replicating by position, there was a highly significant preference for the higher concentration ($P_{nectartype} < 0.001$). Time per probe: Birds spent significantly more time visiting more concentrated nectar ($P_{nectartype} < 0.001$). They also spent different amounts of time depending on the colour of the cue ($P_{colourcue} = 0.017$). Replicating with position, they still took significantly more time per probe ($P_{nectartype} < 0.001$).

Experiment 4: preference against landing platforms

As expected, birds preferred flowers lacking landing platforms (Fig. 4). Plant visits: Birds significantly visited inflorescences without landing platforms over those with landing platforms on five of the six days ($P < 0.05$), and on the non-significant day, the deviation was in the same direction. There was significant heterogeneity among days ($P = 0.020$), but the total G^2 for all days was highly significant ($P < 0.001$). This effect was significant when positions were used as replicates ($P_{flowertype} = 0.002$, based on dividing $MS_{flowertype}/MS_{interaction}$ in mixed model).

Day	G^2	P
1. purpleS	0.591	0.442
2. purpleS	1.569	0.210
3. redS	0.087	0.768
4. redS	2.599	0.107

	G^2	P
purpleS		
heterogeneity	2.099	0.147
pooled	0.061	0.805
redS		
heterogeneity	1.377	0.241
pooled	1.309	0.253

	df	MS	*P*
Colorcue	1	<0.001	0.984
Nectartype	1	0.015	0.544
Nectartype x colorcue	1	0.165	0.685
Day ⊂ colorcue	2	0.284	<0.001
Nectartype x day ⊂ colorcue	2	1.635	0.197
Error	292	0.017	

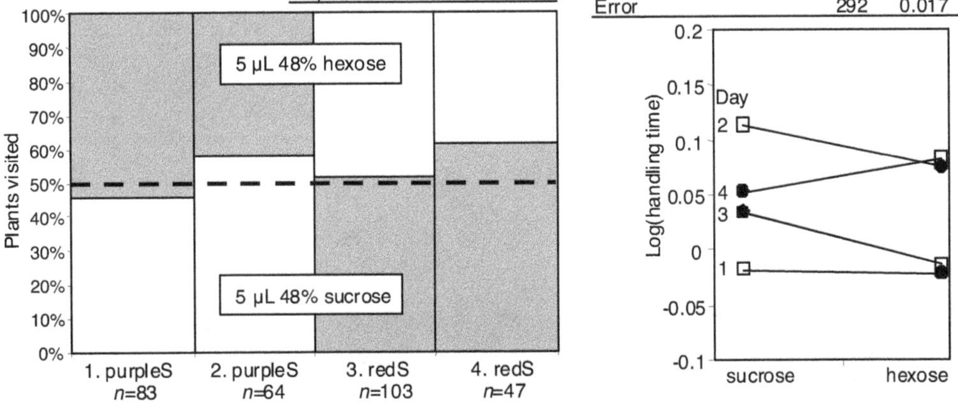

FIGURE. 2. Sucrose versus hexose. This experiment tested preferences for sugar type. (grey bars: red flowers; white bars: purple flowers; ● red flowers; □ purple flowers)

Day	G^2	P
1. purple@48	3.221	0.073
2. purple@48	5.437	0.020
3. red@48	1.950	0.163
4. red@48	9.942	0.002

	G^2	P
purple@48		
heterogeneity	0.213	0.644
pooled	8.445	0.004
red@48		
heterogeneity	1.912	0.167
pooled	9.979	0.002

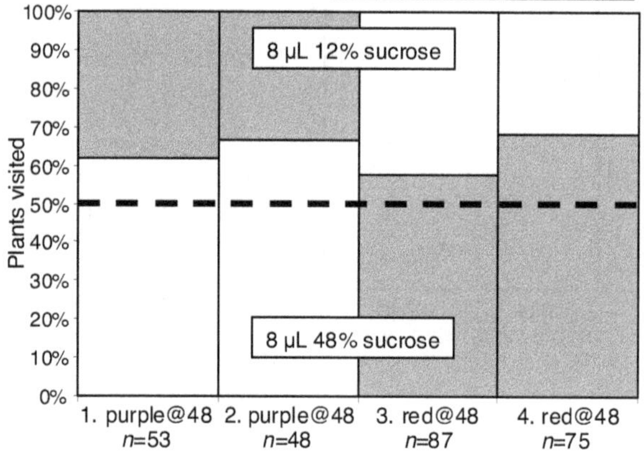

	df	MS	P
Colorcue	1	0.111	0.017
Nectartype	1	1.647	<0.001
Error	264	0.019	

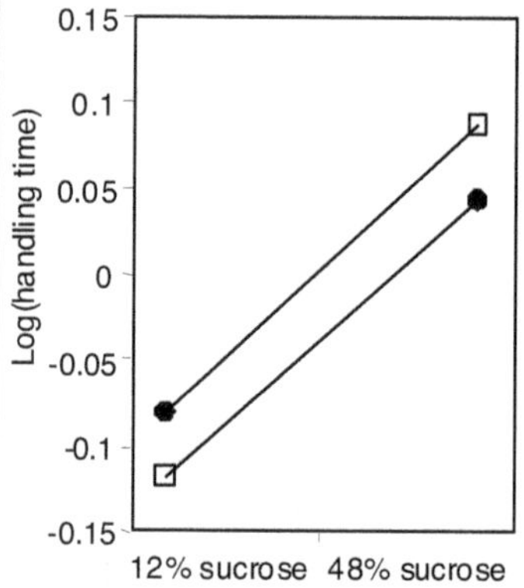

FIGURE 3 12% versus 48%. This experiment tested preferences for nectar concentrations. (grey bars: red flowers; white bars: purple flowers; ● red flowers; □ purple flowers)

Time per probe: Birds took significantly less time per probe visiting flowers lacking landing platforms than flowers with landing platforms ($P_{flowertype} < 0.001$). This was also highly significant when using position as a replicate ($P_{flowertype} < 0.001$).

Experiment 5: patterning on the petals

Birds did not use patterns as a cue to find the higher nectar source, although they definitely paid attention to higher nectar volumes once they were at an inflorescence (Fig. 5). Plant visits: They visited the two flower types comparably, regardless of which volume was associated with patterning ($P_{all\ days\ pooled} = 0.933$). Using positions as replicates, nectartype did not interact with the pattern of the cue ($P_{nectartype \times patterncue} = 0.0765$), nor did nectartype have an effect ($P_{nectartype} = 0.977$ after pooling), and the effect of the pattern of the cue was not significant when testing over the highly significant day effect ($P_{patterncue} = 0.363$ after all other terms had been pooled). Time per probe: One datum was winsorized from 0.9841 to 0.3971 log(seconds/flower). Birds spent significantly more time at 8 µL than 2 µL, but the amount of time spent varied by day ($P_{nectartype \times day \subset patterncue}$

	G2	P
2008 day 1	21.482	<0.001
2008 day 2	10.743	0.001
2008 day 3	6.180	0.013
2008 day 4	0.978	0.323
2009 day 5	18.536	<0.001
2009 day 6	12.172	<0.001
total	70.091	<0.001
pooled	56.760	<0.001
heterogeneity	13.331	0.020

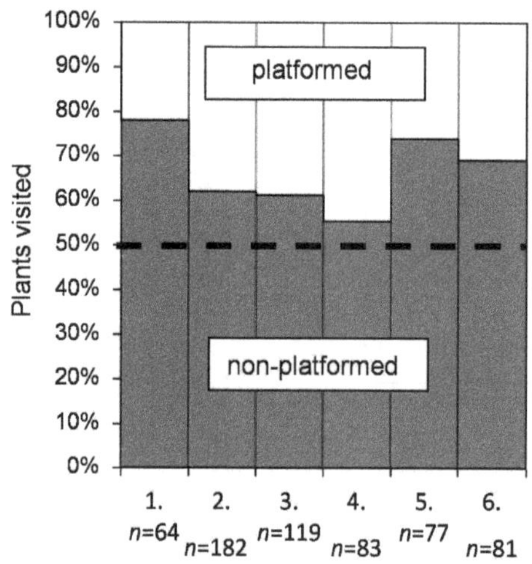

	df	MS	P
Flowertype	1	2.501	<0.001
Error	604	0.021	

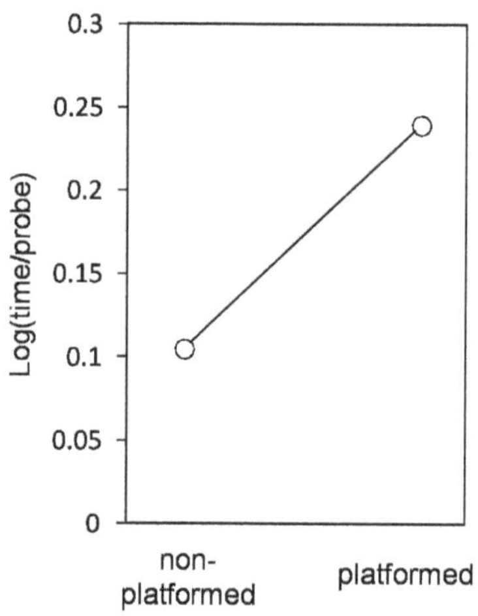

FIGURE 4. Platformed versus non-platformed. This experiment tested preferences against landing platforms. (○ pooled days)

Day	G^2	P
1. pattern@8	1.099	0.295
2. pattern@8	0.288	0.591
3. nopattern@8	0.157	0.692
4. nopattern@8	0.825	0.364
5. pattern@8	0.065	0.799
6. pattern@8	0.006	0.939
7. nopattern@8	0.134	0.714
8. nopattern@8	0.143	0.705

	G^2	P	df
pattern@8			
heterogeneity	1.348	0.718	3
pooled	1.110	0.741	1
nopattern@8			
heterogeneity	1.191	0.755	3
pooled	0.068	0.794	1
all days			
heterogeneity	0.170	0.680	2
pooled	0.007	0.933	1

	df	MS	P
Patterncue	1	0.057	0.479
Nectartype	1	1.68	0.007
Nectartype x pattercue	1	0.002	0.740
Day ⊂ patterncue	6	0.099	<0.001
Nectartype x day ⊂ pattercue	6	0.103	<0.001
Error	1253	0.020	

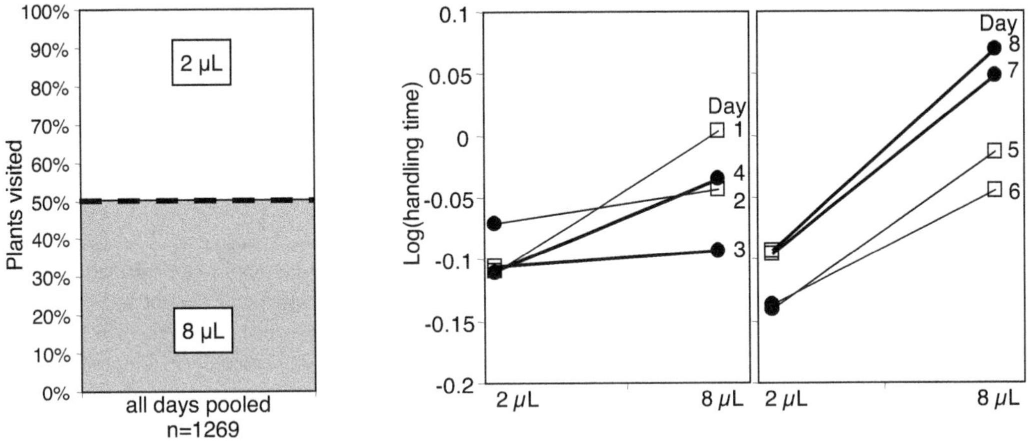

FIGURE 5. Patterned versus non-patterned flowers. This experiment tested whether hummingbirds would use pattern cues to find more rewarding nectar. (grey bar: non-patterned flowers; white bar: patterned flowers; ● non-patterned flowers; □ patterned flowers; — days that were patterned@ higher volume; — days that were non-patterned@ higher volume)

< 0.001). Furthermore, there was no interaction between nectartype and the pattern of the cue ($P_{nectartype \times patterncue} = 0.740$), though replicate days varied greatly by the amount of time spent at a pattern type ($P_{day \subset patterncue} < 0.001$). Birds spent more time at flowers with 8 µL over those with 2 µL ($P_{nectartype} = 0.007$), and they spent similar amounts of time regardless of the pattern of the cue ($P_{patterncue} = 0.479$). Using position days as replicates, significance tests were the same. Most notably, there was no significant interaction between nectar amount and patterncue ($P_{nectartype \times patterncue} = 0.848$) and there was a significant effect of nectar amount ($P_{nectartype} = 0.004$).

DISCUSSION

In our experiments hummingbirds sometimes showed a slight preference for red over purple. They definitely preferred flowers with nectar over nectarless flowers. When nectar volumes were 2 µL or more, birds were not consistently discerning between nectar volumes in terms of visits to inflorescences. The effects seemed more consistent for time spent per probe. When nectar was energetically equivalent, birds generally visited the two types of flowers similarly with a slight bias for red cues. Birds visited flowers with sucrose and hexose in equal frequencies when the two sugar types were offered in equal amounts and concentrations. They generally preferred 48% sugar over 12% sugar. Birds preferred flowers without landing platforms, and took less time probing flowers without landing platforms. Regardless of the presence or absence of spots on the flowers, birds visited both types of flowers equally, evidently failing to associate the spots with the difference in nectar volume. The constitution of these specific hummingbird-flower characteristics (flower colour, a higher nectar volume, and the lack of landing platforms) seems to be, in part but only in part, due to the behavioural proclivities of hummingbirds for the flowers with ornithophilous features. The main inconsistency with simple predictions stemming from the recognition of pollination syndromes is that birds visiting our array did not always prefer flowers with nectar at the extreme of hummingbird-adapted flowers.

How did the traits associated with ornithophily evolve? Floral characters might be adaptations to principle pollinators, either because their pollinators have specific behavioural preferences such as we studied or because of the mechanics of pollen transfer (Wilson 1995). Alternatively, ornithophiles might be adapted to discourage visits by bees that take rewards while transferring less pollen (Castellanos et al. 2004). And there are many environmental factors other than visitor preferences that affect floral evolution, such as herbivory (Kang et al. 2008; Gomez et al. 2009), temperature stress (Tyrrell & Tyrrell 1985), and length of reproductive season (Elliott & Irwin 2009).

Comments on ornithophilous features one at a time

Colour. Floral colour is a cue, or advertisement, that pollinators use to find rewards (Waser 1983). The behavioural propensity for red was evident, but after a short period of time birds were able to recognize purple flowers as rewarding and visit them in nearly equal frequencies. Using Stiles' (1976) terminology, there appears to be a weak holdover period for colour preference. The remaining curiosity is that birds spent more time per probe at red flowers than equally rewarding purple flowers of similar construction. This was also true of red compared to yellow flowers (Haley 2010).

Nectar quantity. Hummingbirds will under-visit flowers that in the past they have found to be empty (Hurly & Healy 1996). Our birds preferred flowers with nectar over those without nectar, but any threshold below which birds disdain flowers of a certain appearance would seem to be less than 2 μL. Gass and Roberts (1992) pointed out that some ornithophiles produce large volumes of nectar, whereas others produced very small volumes. In our experiments, birds showed no preferences for higher nectar volumes when flowers differed by 2 μL. Carlson (2008) reports similar results. The choice of volumes in our experiments may have clouded our efforts to test other aspects of bird preferences: 1 versus 4 μL might have yielded different results than 2 versus 8 μL. One might expect amount of reward to be the most important property affecting hummingbird preferences during pollination (Blem et al. 2000) and for there to be some volume at which the flower and the pollinator come to an evolutionary compromise (Zimmerman 1988). Ornithophiles should produce a high enough volume to entice birds into becoming frequent visitors, some amount above the birds' minimum threshold, but it should not be vastly above that number because doing so would waste energy and water (Southwick 1984; Pyke 1991) and would fail to encourage the birds to move on to the next plant. Our study suggests that a nectar volume of 2 μL may be more than enough to keep birds coming back to flowers. We could speculate, however, that in less arid environments the minimum reward might well be higher than at our field site.

Nectar viscosity and its components. Melittophiles tend to have a much higher sugar concentration than ornithophiles: on average 35% but in some species as high as 80% sugar (Baker 1975; Pyke & Waser 1981), and melittophilous nectar tends to be more hexose-rich (Baker & Baker 1983a; Wilson et al. 2006). Possibly bees under-visit flowers with dilute nectar (Harder 1986; Cnaani et al. 2006). The extra water in more dilute nectar takes longer for bees to get rid of. In contrast, hummingbirds have extremely efficient kidneys that excrete water quickly (Tyrrell & Tyrrell 1985). Since hovering is expensive, hummingbirds are expected to minimize the amount of time expended while foraging, hence to prefer flowers with less viscous nectar (Baker & Baker 1983b; Powers and Nagy 1988; Wilson et al. 2006). As nectar viscosity increases, the amount of nectar extracted per lick decreases and licking takes longer. Kingsolver and Daniel (1983) suggested that if a bird were able to extract all floral nectar in a single lick (1.9 μL for a Rufus Hummingbird), then it would prefer lower concentrations. However, Roberts (1995, 1996) casts doubt on whether energy intake rate calculated over the duration of a visit to a flower would actually be higher at lower concentrations for flowers with 4-8 μL. Our findings were not particularly consistent with the Kingsolver-Daniel theory. In Experiment 3, birds showed little response to different nectar types except when the amount of sugar contained within them was different. Stiles (1976) showed that hummingbirds had a preference for sucrose-containing nectar, rather than solutions that contained glucose or fructose. Contrary to Stiles, our birds had no preference for sugar type, but similar to Stiles, they preferred more concentrated nectar. It is possible that the dilute nectar of ornithophiles is to discourage bees, because bees tend to have a lower net energy gain when extracting dilute nectar (Harder 1986), so the discrepancies we see between bird behaviour and the syndrome difference might be due to bee preference, not bird preference. An explanation for the syndrome difference not based on viscosity has been put forward by Schondube and Martínez del Rio (2003). They suggest that hummingbirds prefer sucrose only at high concentrations and that they do so because sucrose has a lower osmotic potential than does hexose at the same concentration. Their birds preferred sucrose when artificial nectars were presented at 40%. Our birds failed to confirm this preference when presented nectars at 48% sugar.

Landing platforms. The lack of landing platforms in ornithophiles could be explained in a number of ways. First, landing platforms might obstruct hummingbirds from extracting nectar (Sutherland & Vickery 1993; Castellanos et al. 2004), thereby making them prefer flowers without landing platforms, as they did in Experiment 4. Consistent with this, birds took more time extracting nectar when there were landing platforms than when landing platforms were absent. Second, hummingbirds may be sensitive when flowers obscure their view of potential predators and competing hummingbirds (Lima 1991). Third, the absence of landing platforms might not be a response to bird preference but instead to deter inferior visitors. In other words, the absence of landing platforms may be to decrease competition between hummingbirds and bees for nectar (Valido et al. 2002; Rodríguez-Gironés & Santamaría 2004) thereby reducing visits by bees that tend to waste more pollen (Castellanos et al. 2003). Fourth, landing platforms might hinder the loading of pollen onto and the pickup of pollen from the bird's head. Landing platforms make the contact against anthers and stigmas imprecise (Castellanos et al. 2004; Fenster et al. 2004; Armbruster et al. 2009), so selection for increased pollen transfer efficiency would eliminate the landing platform. Studying the melittophilous *Mimulus lewisii*, Owen and Bradshaw (2011) found that mutants without landing platforms were under-visited by bumblebees, and even when the bees did visit, they did not crawl into the mutant flowers in the usual way, so were not transferring pollen as effectively. The theory that landing platforms are not valuable in ornithophiles was not, however, supported by Temeles and Rankin (2000) who compared ornithophilous *Monarda didyma* flowers with the lower lip removed to normal-lipped flowers. They found that removing the lip nearly halved the amount of pollen picked up by hummingbirds. Thus, lips are maintained in *M. didyma*, unlike in many ornithophiles. Questions remain, but we can say that, when given the choice, birds avoided flowers with landing platforms, plausibly because it takes longer to extract nectar.

Nectar guides. In shifts from melittophily to ornithophily, flowers sometimes lose colour patterning. Waser (1982) suggested that the presence of patterns on perianths in melittophiles might be used as nectar guides to lead bees to rewards, thereby improving the bees' handling time and making the flowers more attractive. Waser and

Price (1985) supported this proposition with the finding that foragers have increased handling times when visiting albino flowers that lack nectar guides. In *Mimulus lewisii*, mutants without nectar guides are under-visited compared to wildtype flowers (Owen & Bradshaw 2011). Bees seem to easily discern rewards based on associated patterning (Gould 1985). Perhaps birds are not guided the way bees are. In Experiment 5, hummingbirds did not use black dots to find rewarding nectar, although when they visited flowers with 8 μL, they not only probed them more often (Haley 2010), they spent more time per probe. It would be good to confirm hummingbird inattention to nectar guides using other markings (such as white stripes), and using reward differentials to which hummingbirds are more consistently sensitive (not 2 versus 8 μL). However, our results taken at face value are consistent with the syndrome difference. The lack of patterning in ornithophiles could be a result of hummingbird negligence – not recognizing patterns that might help them find rewarding nectar. Nectar guides may be at a visual scale that birds are insensitive to. Our results do not suggest that hummingbirds would select *against* nectar guides. In contrast, the results of Medel et al. (2003) suggest that hummingbirds do notice the red against yellow nectar guides of *Mimulus luteus*, and select them to be smaller and of a different shape than would be optimal for attracting bees.

Ending

Can the results from the various experiments be integrated? Bird preferences for some characters seemed stronger than others. For example, bird preference for higher volumes in Experiment 2 seemed not as definitive as preference for higher concentrations in Experiment 3C. This result follows a concentration-before-volume criterion hierarchy in bees (Cnaani et al. 2006). Stiles (1976) likewise suggested that hummingbirds have a preference hierarchy whereby birds take into account factors most important to them first and less important factors later. Smith et al. (2008) supported this hypothesis in showing that hummingbirds consider nectar reward and number of flowers in an inflorescence as more important than floral colour. Our study further supports this idea, suggesting that amount of sugar might make birds change behaviour, whereas nectar with almost any amount of water is treated similarly in terms of teaching birds to visit flowers. Hexose versus sucrose would be the least important aspect of nectar. In contrast, the landing platform difference was responded to quickly, strongly and stably, perhaps because it didn't require an assessment of hidden rewards. Comparing our several results is provocative when speculating on bird behaviours, but it does little to address the topic of multi-trait integration in pollination syndromes (Fenster et al. 2004). For example, reddish colouration might be considered the sin qua non of ornithophily in the western U.S., but our results do not indicate that red colouration is necessary for hummingbird pollination. Hummingbird behavioural proclivities *add* some characters to the syndrome, but to the extent that there is a functional *interaction* between traits at their optimum it would seem to be based on a harmony of several morphological characters working mechanically well together in transferring pollen onto and off of particular pollinators (Reynolds et al. 2010).

ACKNOWLEDGEMENTS

Thanks to Sigma Xi for funding, to M. Fernandez for help with sugars, to C. Ryan, M. Steele, F. Hertel and various anonymous reviewers for comments on experimental design and various other topics, and to the following family members for support and assistance: T., A., L. and E. Haley, A., S., J. and E. Baldwin, and N. Guzman.

REFERENCES

Aldridge G, Campbell DR (2007) Variation in pollinator preference between two *Ipomopsis* contact sites that differ in hybridization rate. Evolution 61: 99-110.

Armbruster WS, Hansen TF, Pelabon C, Perez-Barrales R (2009) The adaptation of flowers: measurement and microevolutionary patterns. Annals of Botany 103: 1529-1545.

Baker HG (1961) The adaptation of flowering plants to nocturnal and crepuscular pollinators. Quarterly Review of Biology 36: 64-73.

Baker HG (1975) Sugar concentrations in nectar from hummingbird flowers. Biotropica 7: 37-41.

Baker HG, Baker I (1983) Floral nectar sugar constituents in relation to pollinator type. In: Jones CE, Little RJ (eds) Handbook of experimental pollination biology. Van Nostrand Reinhold, New York, NY, pp 117-141.

Baker HG, Baker I (1983b) A brief historical review of the chemistry of floral nectar. In: Bently B, Elias T (eds) The biology of nectaries. Columbia University Press, New York, NY, pp 126-152.

Bene F (1941) Experiments on the color preference of black-chinned hummingbirds. Condor 43: 237-242.

Bleiweiss, R. 2001. Mimicry on the QT(L): Genetics of speciation in *Mimulus*. Evolution 55: 1706-1709.

Blem, CR, Blem LB, Felix J, van Gelder J (2000) Rufous hummingbird sucrose preference: precision of selection varies with concentration. Condor 102: 235-238.

Brown, JH, Kodric-Brown A (1979) Convergence, competition, and mimicry in a temperate community of hummingbird-pollinated flowers. Ecology 60: 1022-1035.

Campbell DR, Bischoff M, Lord JM, Robertson AW (2010) Flower color influences insect visitation in alpine New Zealand. Ecology 91: 2638-2649.

Carlson JE (2008) Hummingbird responses to gender-biased nectar production: are nectar biases maintained by natural or sexual selection? Proceedings of the Royal Society London B 275: 1717-1726.

Castellanos MC, Wilson P, Thomson JD (2003) Pollen transfer by hummingbirds and bumblebees, and the divergence of pollination modes in *Penstemon*. Evolution 57: 2742-2752.

Castellanos MC, Wilson P, Thomson JD (2004) 'Anti-bee' and 'pro-bird' changes during the evolution of hummingbird pollination in *Penstemon* flowers. Journal of Evolutionary Biology 17: 876-885.

Cnaani J, Thomson JD, Papaj DR (2006) Flower choice and learning in foraging bumblebees: effects of variation in nectar volume and concentration. Ethology 112: 278-285.

Chittka L, Waser N (1997) Why red flowers are not invisible to bees. Israel Journal of Plant Sciences 45: 169-183.

Collias NE, Collias EC (1968) Anna's hummingbirds trained to select different colors in feeding. Condor 70: 273-274.

Elliott S, Irwin RI (2009) Effects of flowering plant density on pollinator visitation, pollen receipt, and seed production in *Delphinium barbeyi* (Ranunculaceae). American Journal of Botany 96: 912-919.

Faegri K, van der Pijl L (1979) The Principles of Pollination Ecology, 2nd ed. Pergamon Press, New York.

Fenster CB, Armbruster WS, Wilson P, Dudash MR, Thomson JD (2004) Pollination syndromes and floral specialization. Annual Review of Ecology and Systematics 35: 375-403.

Fenster CB, Cheely G, Dudash MR, Reynolds RJ (2006) Nectar reward and advertisement in hummingbird-pollinated *Silene virginica* (Caryophyllaceae). American Journal of Botany 93: 1800-1807.

Gass CL, Roberts WM (1992) The problem of temporal scale in optimization: three contrasting views of hummingbird visits to flowers. American Naturalist 5: 829-853.

Giurfa M, Lehrer M (2001) Honeybee vision and floral displays: From detection to close-up recognition. In Chittka L, Thomson JD (eds.), Cognitive ecology of pollination: Animal behavior and floral evolution, Cambridge University Press, Cambridge, pp. 61-82.

Giurfa M, Vorobyev M, Kevan P, Menzel R (1996) Detection of colored stimuli by honeybees: minimum visual angles and receptor specific contrasts. Journal of Comparative Physiology A 178: 699-709.

Gómez JM, Bosch J, Perfectti F, Fernández JD, Abdelaziz M, Camacho JPM (2008) Spatial variation in selection on corolla shape in a generalist plant is promoted by the preference patterns of its local pollinators. Proceedings of the Royal Society of London B 275: 2241-2249.

Gómez JM, Perfectti F, Bosch J, Camacho JPM (2009) A geographic selection mosaic in a generalized plant-pollinator-herbivore system. Ecological Monographs 79: 245-263.

Gould JL (1985) How bees remember flower shapes. Science 227: 1492-1494.

Grant, KA, Grant V. 1968. Hummingbirds and their flowers. Columbia University Press, New York.

Haley WA (2010) Hummingbird choices at artificial flowers made to resemble ornithophiles versus melittophiles. M.S. thesis, California State University, Northridge. http://hdl.handle.net/10211.2/685

Harder LD (1986) Effects of nectar concentration and flower depth on flower handling efficiency of bumble bees. Oecologia (Berlin) 69: 309-315.

Hersch EI, Roy BA (2007) Context-dependent pollinator behavior: an explanation for patterns of hybridization among three species of indian paintbrush. Evolution 61: 111-124.

Hurly TA, Healy SD (1996) Memory for flowers in rufous hummingbirds: location or local visual cues? Animal Behaviour 51: 1149-1157.

Irwin RE, Brody AK (2000) Consequences of nectar robbing for realized male function in a hummingbird-pollinated plant. Ecology 81: 2637-2643.

Kang Y, Armbruster D, Kuang Y (2008) Dynamics of a plant-herbivore model. Journal of Biological Dynamics 2: 89-101.

Kingsolver JG, Daniel TL (1983) Mechanical determinants of nectar feeding strategy in hummingbirds: energetics, tongue morphology, and licking behavior. Oecologia (Berlin) 60: 214-226.

Kodric-Brown A, Brown JH (1978) Influence of economics, interspecific competition, and sexual dimorphism on territoriality of migrant rufous hummingbirds. Ecology 59: 285-296.

Lima SL (1991) Energy, predators and the behaviour of feeding hummingbirds. Evolutionary Ecology 5: 220-230.

Majetic CJ, Raguso RA, Ashman T-L (2009) The sweet smell of success: floral scent affects pollinator attraction and seed fitness in *Hesperis matronalis*. Functional Ecology 23: 480-487.

Medel R, Botto-Mahan C, Kalin-Arroyo M. (2003) Pollinator-mediated selection on the nectar guide phenotype in the Andean monkey flower, *Mimulus luteus*. Ecology 84: 1721-1732.

Melendez-Ackerman E, Campbell DR, Waser NM (1997) Hummingbird behavior and mechanisms of selection on flower color in *Ipomopsis*. Ecology 78: 2532-2541.

Mitchell RJ, Paton DC (1990) Effects of nectar volume and concentration on sugar intake rates of Australian honeyeaters (Meliphagidae). Oecologia (Berlin) 83: 238-246.

Owen CR, Bradshaw HD. (2011) Induced mutations affecting pollinator choice in *Mimulus lewisii* (Phrymaceae). Arthropod-Plant Interactions 5: 235-244.

Pleasants JM (1983) Nectar production in *Ipomopsis aggregata* (Polemoniaceae). American Journal of Botany 70: 1468-1475.

Powers DR, Nagy KA (1988) Field metabolic rate and food consumption by free-living Anna's Hummingbirds (*Calypte anna*). Physiological Zoology 6: 500-506

Pyke GH (1991) What does it cost a plant to produce floral nectar? Nature 350: 58-59.

Pyke GH, Waser NM (1981) The production of dilute nectar by hummingbird and honeyeater flowers. Biotropica 13: 260-270.

Quinn G, Keough M (2002). Experimental design and data analysis for biologists. Cambridge University Press, Cambridge, United Kingdom.

Raven PH. 1972. Why are bird-visited flowers predominantly red? Evolution 26: 674,

Reynolds RJ, Dudash MR, Fenster CB (2010) Multiyear study of multivariate linear and nonlinear phenotypic selection on floral traits of hummingbird-pollinated *Silene virginica*. Evolution 64: 358-369.

Roberts WM (1995) Hummingbird licking behavior and the energetics of nectar feeding. Auk 112: 456-463.

Roberts WM (1996) Hummingbirds' nectar concentration preferences at low volume: the importance of time scale. Animal Behaviour 52: 361-370.

Rodríguez-Gironés MA, Santamaría L (2004) Unsolved mystery: why are so many bird flowers red? Public Library of Science Biology 2: 1515-1519.

Rodríguez I, Gumbert A, Hempel de Ibarra N, Kunze J, Giurfa M. (2004) Symmetry is in the eye of the 'beeholder': innate preference for bilateral symmetry in flower-naïve bumblebees. Naturwissenschaften 91: 374-377.

Schemske DW, Ågren J (1995) Deceit pollination and selection on female flower size in *Begonia involucrate*: an experimental approach. Evolution 49: 207-214.

Schemske DW, Bradshaw HD (1999) Pollinator preference and the evolution of floral traits in monkeyflowers (*Mimulus*). Proceedings of the National Academy of Sciences USA 96: 11910-11915.

Schondube JE, Martínez del Rio C (2003) Concentration-dependent sugar preferences in nectar feeding birds: mechanisms and consequences. Functional Ecology 17: 445-453.

Smith CE, Stevens JT, Temeles EJ, Ewald PW, Hebert RJ, Bonkovsky RL (1996) Effect of floral oriface width and shape on hummingbird-flower interactions. Oecologia (Berlin) 106: 482-492.

Smith SD, Ané C, Baum DA (2008) The role of pollinator shifts in the floral diversification of *Iochroma* (Solanaceae). Evolution 62: 793-806.

Southwick EE (1984) Photosynthate allocation to floral nectar: a neglected energy investment. Ecology 65: 1775-1779

Stiles FG (1976) Taste preferences, color preferences, and flower choice in hummingbirds. Condor 78: 10-26.

Sutherland S, Vickery Jr. RK (1993) On the relative importance of floral color, shape, and nectar rewards in attracting pollinators to *Mimulus*. Great Basin Naturalist 53: 107-117.

Temeles EJ, Rankin AG (2000) Effect of the lower lip of *Monarda didyma* on pollen removal by hummingbirds. Canadian Journal of Botany 78: 1164-1168.

Thomson JD, Wilson P, Valenzuela M, Malzone M (2000) Pollen presentation and pollination syndromes, with special reference to *Penstemon*. Plant Species Biology 15: 11-29.

Tyrrell EQ, Tyrrell RA (1985) Hummingbirds, their life and behavior a photographic study of the North American species. Crown, New York, NY

Valido A, Dupont YL, Hansen DM (2002) Native birds and insects, and introduced honeybees visiting *Echium wildpretii* (Boraginaceae) in the Canary Islands. Acta Oecologia 6: 413-419.

Waser NM (1982) A comparison of distances flown by different visitors to flowers of the same species. Oecologia (Berlin) 55: 251-257.

Waser NM (1983) The adaptive nature of floral traits: ideas and evidence. In Real L (ed) Pollination biology. Academic Press, Orlando, FL, pp 241-285.

Waser NM, Price MV (1985) The effect of nectar guides on pollinator preference: experimental studies with a montane herb. Oecologia (Berlin) 67: 121-126.

Wesselingh RA, Arnold ML (2000) Pollinator behaviour and the evolution of Louisiana iris hybrid zones. Journal of Evolutionary Biology 13: 171-180.

Wilson P (1995) Selection for pollination success and the mechanical fit of Impatiens flowers around bumble bee bodies. Biological Journal of the Linnean Society 55: 355-383.

Wilson P, Castellanos MC, Wolfe A, Thomson JD (2006) Shifts between bee- and bird-pollination among penstemons. In: Waser NM, Ollerton J (eds) Plant-pollinator interactions: from specialization to generalization. University of Chicago Press, Chicago, IL, pp 47-68.

Wilson P, Jordan EA (2009) Hybrid intermediacy between pollination syndromes in *Penstemon*, and the role of nectar in affecting hummingbird visitation. Botany 87: 272-282.

Zimmerman M (1988) Nectar production, flowering phenology and strategies for pollination. In: Doust JL, Doust LL (eds) Plant reproductive ecology. Oxford University Press, Oxford, England, pp 157-178.

17

TARAXACUM OFFICINALE POLLEN DEPRESSES SEED SET OF MONTANE WILDFLOWERS THROUGH POLLEN ALLELOPATHY

Deirdre Loughnan[1], James D. Thomson[1,2*], Jane E. Ogilvie[1,2] and Benjamin Gilbert[1]

[1] *Department of Ecology and Evolutionary Biology, University of Toronto, 25 Harbord Street, Toronto, ON, M5S 3G5, Canada*
[2] *Rocky Mountain Biological Laboratory, Post Office Box 519, Crested Butte, CO 81224-0519, USA*

Abstract—Plant species that share pollinators can suffer from interspecific pollen deposition. Male reproductive success is inevitably reduced by the loss of pollen to flowers of another species. Female reproductive success can be affected by reduced stigmatic area or, more strongly, through allelopathic effects by which the admixture of some foreign pollen reduces seed or fruit set. We tested for allelopathic effects of *Taraxacum officinale* (Asteraceae) pollen on the seed set of montane wildflowers *Erythronium grandiflorum* (Liliaceae) and *Erysimum capitatum* (Brassicaceae), by hand-pollinating plants with pollen mixtures. *Taraxacum* is a common invasive species, which produces allelopathic chemicals in its root and vegetative tissue, making it a likely candidate for pollen allelopathy. Flowers of both species produced fewer well-developed seeds when pollinated with pollen mixtures containing *Taraxacum* pollen. The pollen-allelopathic potential of weedy dandelion may add to its ability to disrupt communities that it invades.

Keywords: Pollen allelopathy, Taraxacum officinale, Erythronium grandiflorum, Erysimum capitatum, *interspecific pollen transfer, phenological disruption*

INTRODUCTION

Animal-pollinated plant species often occur in plant communities in which they share flowering times and pollinators with other plant species. A plant species can alter the pollination success of a co-flowering plant if its presence changes the flower-feeding behaviour of shared pollinators (Rathcke 1983; Mitchell et al. 2009). In particular, when pollinators move between the flowers of multiple plant species, they can cause interspecific pollen deposition (IPD) on plant stigmas. IPD can directly or indirectly reduce seed production and influence competitive dynamics among species (Levin & Anderson 1970; Waser 1978; Thomson et al. 1981). Phenological overlap in flowering is a necessary condition for IPD, raising the possibility that changes in phenological timing with climate change or changes in interacting species following invasion may lead to an increased incidence of IPD.

In the Rocky Mountains of the western United States, warmer spring temperatures and changing snowpack levels are causing earlier snowmelt and earlier onset of plant growth (Inouye 2008; Forrest et al. 2010). As a result, species that were once unlikely to co-occur, either within a season or habitat, are increasingly interacting; species that flower early in the spring are most affected, with more species co-flowering at this time (Forrest et al. 2010). This increased overlap among species may be exacerbated by the potential for increased colonization by exotic species in the area (Walther et al. 2009), with some exotic species such as *Taraxacum officinale* also flowering in the early spring.

An adaptation believed to play a role in the superior competitive abilities and invasive nature of some exotic species is allelopathy, the release of chemicals that decrease the fitness of neighboring plants (Inderjit et al. 2008). These chemicals can be found in different plant tissues, including pollen grains. Through IPD on stigmas, pollen allelopathy has been inferred from reductions in pollen germination (Kanchan & Jayachandra 1980; Murphy & Aarssen 1995), pollen tube formation (Kanchan & Jayachandra 1980), and seed development (Thomson et al. 1981). The degree of pollen allelopathy depends first on the amount and frequency of interspecific pollen that is transferred to stigmas, which is governed by the flower constancy and visitation rate of pollinators when foraging, and the locations of pollen on a pollinator's body (Morales & Traveset 2008). Once pollen grains are deposited on stigmas, the degree of pollen allelopathy depends on the concentration of allelopathic chemicals in the donor pollen and the sensitivity of the recipient species to the chemicals (Inderjit et al. 2008).

We tested for pollen allelopathy in the common dandelion *Taraxacum officinale* (hereafter *Taraxacum*), a common, globally distributed invasive species capable of establishing under diverse environmental conditions (Holm et al. 1997). We hypothesized that this species would be likely to produce allelopathic pollen, as it is known to possess allelopathic chemicals in both its vegetative and root tissues (Jankowska et al. 2009). Anecdotal support for this hypothesis is mixed, however. Decreased seed set of *Erythronium grandiflorum* has been observed in the presence of *Taraxacum* pollen (J. D. Thomson, unpublished data), but the presence of neighboring *Taraxacum* plants had no

*Corresponding author; email: james.thomson@utoronto.ca

detectable effect on seed set of *Delphinum nuttallianum* (Jones 2004).

To test whether allelopathic chemicals in *Taraxacum* affect the seed set of species native to the Colorado Rocky Mountains, we performed a hand pollination study on two species of animal-pollinated flowering plants that show some IPD with *Taraxacum*: *Erythronium grandiflorum* and *Erysimum capitatum* (hereafter *Erythronium* and *Erysimum*, respectively). We predicted that pollen mixtures including *Taraxacum* pollen would result in decreased seed set in the two native plant species, with greater proportions of *Taraxacum* pollen causing greater declines.

MATERIALS AND METHODS

Study site and plant species selection

We worked in subalpine meadows in the vicinity of the Rocky Mountain Biological Laboratory (RMBL) in Gothic, Colorado, USA, in May-July 2012. Around the RMBL, *Taraxacum officinale* (Asteraceae) grows along trails and roads and in disturbed meadows, and flowered from May to July in 2012. We wished to study the effect of *Taraxacum* pollen on the seed set of focal plant species that receive *Taraxacum* pollen in the field. To search for candidate species, we sampled stigmas of seven plant species that flowered with and grew near *Taraxacum*. We collected stigmas from 10 plants of each species across multiple sites. The stigmas were stained and mounted using basic fuchsin jelly on microscope slides, to distinguish the pollen of different species present on a stigma (Kearns & Inouye 1993). *Taraxacum* pollen was not found on most of the species sampled, but was found at low amounts on the stigmas of three species, from which we selected our two focal species. We also carried out pollinator observations in areas where *Taraxacum* and the candidate species co-flowered, to see if the plants shared pollinators. These observations were carried out for ten-minute periods between 12:00 and 14:00 on multiple days. We selected *Erythronium* as a target species in part because of these observations, but also based on the promise of previous observations (J. D. Thomson, unpublished data). The presence of *Taraxacum* pollen on stigmas and observed shared visitors were taken to indicate the potential for pollen transfer to naturally occur between *Taraxacum* and *Erythronium* and *Erysimum*. *Erythronium grandiflorum* ('glacier lily'; Liliaceae) is a small perennial herb that flowers May to June and has yellow pendent flowers that are visited in the area principally by bumble bee queens (*Bombus* spp.) but also by hummingbirds and smaller solitary bees (Thomson 2010). *Erysimum capitatum* ('wallflower'; Brassicaceae) is a perennial herb that flowers June to July and typically has pale yellow upright flowers that are likely visited by a variety of insects including bumble bees. We observed visits by *Bombus bifarius*, *Bombus appositus* and halictid bees visiting flowers of *Taraxacum* and our two focal species.

Hand-pollination trials

Erythronium trials were conducted at four sites, using two-budded plants that grew in direct sunlight. We removed the second bud from each plant to ensure ample resource allocation to the first flower. Shortly before the first bud opened, we separated the tepals and covered the pistil completely with a 2.5 cm length of plastic drinking straw. Straws with different diameters were used to ensure a snug friction fit around the ovary, without causing damage. We also placed sand-base mesh pollinator exclusion bags (Thomson et al. 2011) over the plant for the following two days, to ensure the plant was not pollinated in the event that the straw was dislodged. Due to the below freezing temperatures at night, plants were covered using inverted Styrofoam drinking cups. These cups were secured by piercing them with pin flags. The cups were removed early in the mornings and replaced over the flowers close to dusk.

Each plant used in the experiment was randomly assigned to one of the three groups as they matured within a site. We performed all pollinations during a twenty day period, from May 13 to June 1, 2012. Bagged flowers were pollinated using one of three *Taraxacum* pollen concentrations: a pure conspecific control including no *Taraxacum* pollen, and two mixtures that included 5-10% or 50% *Taraxacum* pollen, on a grain-count basis. The pollen of *Erythronium* came from buds collected the morning prior to pollination and left overnight for all anthers to dehisce. The buds of *Taraxacum* were collected the day of the pollination since they were unlikely to reopen if left overnight. Pollen mixtures were made by removing pollen from ten or more plants of *Erythronium* and *Taraxacum* and tapping pollen from the entire flower onto a pane of glass. We made fresh pollen mixtures daily and verified the proportions of pollen grains in each mixture by staining the pollen with fuchsin and counting the number of pollen grains of each species within two fields of view under a compound microscope. We made adjustments when necessary to produce the desired proportions. All mixtures for the lower concentration treatment contained between 5-10% *Taraxacum* pollen, while the 50% treatment was deemed suitable with a deviation of less than 2.5%, allowing for 5% error.

Each flower was hand pollinated using a Microbrush® disposable applicator dipped into the pollen mixture. Pollen was applied until a pollen layer was clearly visible on all three lobes of an *Erythronium* stigma. After a pollinated plant's ovaries began to swell, we removed the straws and allowed the fruits to develop and dry. At this time, fruits were collected and the well-developed seeds counted. We were able to quantify seed development for 25 of our control plants, 22 plants of the 5-10% treatment group, and 21 plants of the 50% treatment group.

We used a similar procedure for trials involving *Erysimum*. Plants at four sites were pollinated from June 14 to June 28. Because *Erysimum* produces inflorescences with numerous flowers, we removed 25% of the flowers in order to increase the resource allocation to those remaining. Two replicates of the three treatments were applied to individual flowers on a given plant. Depending on the height of the plant, we used either a swatch of bridal veil or an exclusion bag to exclude pollinators in these trials. Pollen mixtures were prepared as for *Erythronium* and applied the day after the buds first opened. We collected the fruits once they developed, and counted the number of well-developed seeds. Some fruits were excluded from our analysis as insect larvae within the fruits had caused damage to the developing seeds. We were able to quantify the number of well-developed seeds for 29 of our control replicates, 28 replicates of the 5-10% treatment group, and 25 replicates of the 50% treatment group.

Data analysis

Seed counts for *Erythronium* treatments produced a zero-inflated dataset. Since we were unable to transform the data to create a normal distribution, we applied both a non-parametric test and a zero-inflated count data regression model. As both analyses produced similar results, we are only reporting the results of the Kruskal-Wallis test, a non-parametric test that is similar to an ANOVA. For *Erythronium,* we found a statistically significant depression of seed set for only the lowest treatment level of *Taraxacum* pollen; we therefore combined the data from the 5-10% and 50% treatments to increase the sample size and repeated the analysis to test whether the presence of *Taraxacum* pollen, at any level, had an effect. We performed a posteriori tests using the *kruskalmc* function in the pgirmess package of R version 3.0.0 (R Core Team 2013) to make pairwise comparisons between treatments.

The residuals produced by the seed counts for *Erysimum* were not normally distributed and were therefore square root transformed before analysis. We tested for effects of the pollen treatments using a linear mixed-effects model, using the *lme* function in the nlme package of R version 3.0.0 (R Core Team 2013). The model included the pollen treatment as a fixed effect. Individual plant identities were included as a random effect to account for the use of multiple treatments on single plants.

RESULTS

Taraxacum pollen significantly reduced seed set, but *Erythronium* and *Erysimum* differed in the patterns of their responses. *Taraxacum* had a significant effect on the development of seeds for *Erythronium* (χ^2 = 6.163, P = 0.04, 2 df). By performing a posteriori comparisons, we found a significant effect from the application of the 5-10% *Taraxacum* pollen mixture on the development of *Erythronium* seeds (P = 0.046; Fig. 1), with a reduction of 50% in the number of well-developed seeds produced. Surprisingly, pollinating with 50% *Taraxacum* pollen produced a smaller mean reduction (31%) that did not differ from either the control group or the 5-10% treatment group (Fig. 1). Grouping the results from the two treatments, and testing the effect of the presence of *Taraxacum* pollen confirmed that there was a significant reduction in the seed set with the application of *Taraxacum* pollen (χ^2 = 5.126, P = 0.02, 1 df).

The application of *Taraxacum* pollen had a significant effect on the seed set of *Erysimum* ($F_{2,55}$ = 7.7, P = 0.001). Pair-wise contrasts of the least square means showed no significant difference in the seed set of the 0% *Taraxacum* and 5-10% *Taraxacum* pollen mixture treatments; however, both the control and 5-10% treatment groups had a

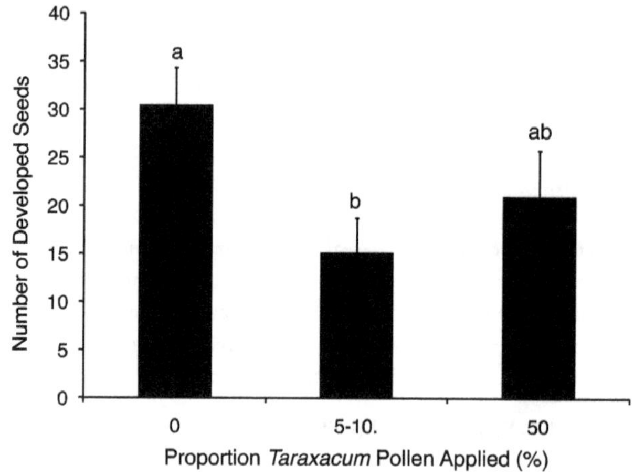

Figure 1. Seed set in *Erythronium grandiflorum* flowers when hand-pollinated with mixtures of conspecific and varying amounts of *Taraxacum officinale* pollen. Error bars represent standard errors. Different letters indicate significant differences between groups. n=25 for the 0% treatment group, n=22 for the 5-10% treatment group, and n=21 for the 50% treatment group.

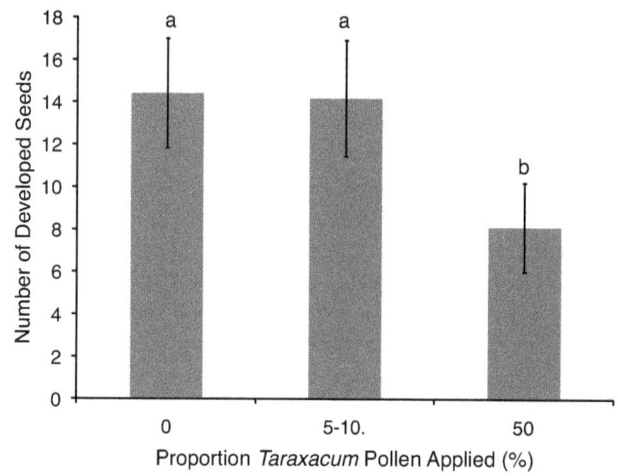

Figure 2. Seed set in *Erysimum capitatum* flowers when hand-pollinated with mixtures of conspecific and varying amounts of *Taraxacum officinale* pollen. Error bars represent the standard errors. Different letters indicate significant differences between groups. n=29 for the 0% treatment group, n=28 for the 5-10% treatment group, and n=25 for the 50% treatment group.

significantly higher seed set than that of the 50% *Taraxacum* pollen treatment (P = 0.016 and P = 0.008, respectively). The application of pollen containing 5-10% *Taraxacum* pollen resulted in a 12.2% decline in the number of well-developed seeds produced by this species; however, our analysis showed this to be a non significant decline. The deleterious effects of the *Taraxacum* pollen increased greatly when the applied pollen consisted of 50% *Taraxacum* grains, with seed production declining by 67.5% (Fig. 2).

DISCUSSION

In our two recipient species, seed set was depressed when stigmas were pollinated with mixtures containing *Taraxacum officinale* pollen in addition to sufficient conspecific pollen to fertilize all ovules. Presumably, this reduction arose from chemicals associated with the *Taraxacum* pollen itself, although the mode of action also remains to be demonstrated. *Taraxacum* thus joins a few other genera known to produce potent pollen grains (Kanchan & Jayachandra 1980; Murphy & Aarssen 1995; Sukhada & Jayachandran 1980; Thomson et al. 1981). Interestingly, three of these genera (*Parthenium, Hieracium,* and *Taraxacum*) are Asteraceae that are also noted for apomixis (Powers & Rollins 1945; Richards 1973; Koltunow et al. 2011). Much broader taxonomic sampling would be necessary to determine whether this suggestive pattern is robust enough to need further investigation.

The more specific questions raised by our data concern dosage responses and possible ecological importance. First, the pattern of effects on *Erythronium* is perplexing: pollen mixtures with higher doses of *Taraxacum* pollen appeared to have weaker effects than milder doses. Because it is hard to imagine a mechanism for this pattern, and because the 50% treatment did not differ significantly from either of the others, we expect that this essentially represents a statistical anomaly. Uncontrolled environmental variation contributes to substantial variation in seed counts within treatments. More experiments in more controlled environments should be able to reduce the error variance and clarify the dosage response.

Second, the strength of inhibition shown by *Taraxacum* pollen on seed production in both of our recipient species is markedly less than in our earlier experiments with *Hieracium* pollen on *Diervilla lonicera* (Thomson et al. 1981). For example, the highest levels of seed reduction observed in this study were 67% (*Erysimum*) and 50% (*Erythronium*), whereas Thomson et al. (1981) showed that pollen mixtures with more than 20% *Hieracium* pollen often reduced seed set of *Diervilla* by more than 80%. This tentatively suggests that dandelion pollen may be more innocuous than hawkweed pollen, or alternatively, that some recipient species may be more sensitive to allelopathic pollen than others. Parallel trials of the two noxious pollens across a range of common recipients would be necessary to distinguish between these possibilities.

Does allelopathy from dandelion pollen play a role in this weedy plant's invasion of native communities? Murphy (2000) provided a rigorous experimental and observational framework necessary to demonstrate meaningful ecological effects of pollen allelopathy in plant-plant interactions in the field. Our results fall far short of such a demonstration, but our aim was simply to establish whether dandelion pollen might exert an allelopathic effect on other species, and our two recipient species were chosen partly for convenience. Our preliminary observations did detect *Taraxacum* pollen on stigmas of *Erythronium* and *Erysimum,* but we did not quantify the amount of IPD in this community and therefore cannot gauge the importance of the effect. Because of a tendency for flower-visiting animals to show flower constancy, two plant species can share pollinating species without much sharing of pollinating individuals. This may well contribute to Jones' (2004) finding that *Delphinium nuttallianum* flowers were unaffected by the presence of *Taraxacum officinale.* The flowers in Jones' study differ radically in both colour and form, which will tend to encourage flower constancy and thus reduce IPD (Chittka et al. 1999; Gegear & Laverty 2005).

Our results do indicate that the nearly ubiquitous dandelion has at least the potential to affect its co-flowering neighbours through interspecific pollen deposition. Although *Taraxacum* pollen may be less toxic than that of *Hieracium,* concentrations as low as 5-10% suffice to depress seed set in *Erythronium.* Having established that modest admixtures of dandelion pollen can inhibit female reproductive success in other plants, we now can turn to investigating whether such effects are important in the field. If IPD occurs often enough to produce meaningful depression of reproductive output, future studies should examine the factors that determine the temporal and spatial proximity of dandelions to various recipient species. Moreover, species-specific changes in flowering times with climate change (Forrest et al. 2010) raise new possibilities about the potential for allelopathic interactions through IDP, as the overlaps between some species pairs intensify.

Pollen allelopathy is a previously unrecognized means by which *Taraxacum officinale* interacts with its surrounding plant community. Because this species is so widespread, and can be locally abundant, it is important to investigate whether this mode of interaction might have serious effects on vulnerable native taxa.

ACKNOWLEDGEMENTS

We would like to thank the Rocky Mountain Biological Laboratory for their institutional support and for providing such a positive environment to work in. This research was supported by an NSERC Discovery Grant to JDT and an award to DL from the Centre for Global Change Science, University of Toronto.

REFERENCES

Chittka L, Thomson JD, Waser NM (1999) Flower constancy, insect psychology, and plant evolution. Naturwissenschaften 86:361-377.

Forrest J, Inouye DW, Thomson JD (2010) Flowering phenology in subalpine meadows: Does climate variation influence community co-flowering patterns? Ecology 91:431-440.

Gegear RJ, Laverty TM (2005) Flower constancy in bumblebees: a test of the trait variability hypothesis. Animal Behaviour 69:939-949.

Holm L, Doll L, Holm E, Pacheco J, Herberger J (1997) World weeds: natural histories and distributions. Wiley, New York.

Inouye DW (2008) Effects of climate change on phenology, frost damage and floral abundance of montane wildflowers. Ecology 89:353- 362.

Inderjit TR, Seastedt, RM, Callaway JL, Pollock JK (2008) Allelopathy and plant invasions: traditional, congeneric, and bio-geographical approaches. Biological Invasions 10:875-890.

Jankowska J, Ciepiela GA, Sosnowski J, Kolczarek R, Jankowski K (2009) The allelopthatic effect of *Taraxacum* officinale F.G. Wiff on the seeds germination and initial growth. Acta Agrobotanica 62:207-212.

Jones K (2004) Do dandelion flowers influence seed set of a native plant (*Delphinium nuttallianum*) in subalpine meadows? American Midland Naturalist 151:201-205.

Kanchan S, Jayachandra (1980) Pollen allelopathy--A new phenomenon. New Phytologist 84:739-746.

Kearns CA, Inouye DW (1993) Techniques for pollination biologists. University Press of Colorado, Niwot, Colorado, USA.

Koltunow AM, Johnson SD, Okada T (2011) Apomixis in hawkweed: Mendel's experimental nemesis. Journal of Experimental Botany 62:1699-1707.

Levin DA, Anderson WW (1970) Competition for pollinators between simultaneously flowering species. American Naturalist 104:455-467.

Mitchell RJ, Flanagan RJ, Brown BJ, Waser NM, Karron JD (2009) New frontiers in competition for pollination. Annals of Botany 103:1403-1414.

Morales CL, Traveset A (2008) Interspecific pollen transfer: magnitude, prevalence and consequences for plant fitness. Critical Reviews in Plant Sciences 27:221-238.

Murphy SD, Aarssen LW (1995) Allelopathic pollen extract from *Phleum pratens* L. (Poaceae) reduces germination, in vitro of pollen of sympatric species. International Journal of Plant Science 156:425-434.

Murphy SD (2000) Field testing for pollen allelopathy: a review. Journal of Chemical Ecology 26:2155-2172.

Powers L, Rollins RC (1945) Reproductive and pollination studies on guayule (*Parthenium argentatum* Gray and *P. incanum* H.B.K.) Journal of the American Society of Agronomy 37:96-112.

R Core Team (2013) R: A language and environment for statistical computing. R Foundation for Statistical Computing, Vienna, Austria. http://www.R-project.org/.

Rathcke B (1983) Competition and facilitation among plants for pollination. In: Real L (ed). Pollination Biology. Academic Press, New York, pp 305-329.

Richards AJ (1973) The origin of *Taraxacum* agamospecies. Botanical Journal of the Linnean Society 66:189-211.

Thomson JD (2010) Flowering phenology, fruiting success and progressive deterioration of pollination in an early-flowering geophyte. Philosophical Transactions of the Royal Society B 365:3187-3199.

Thomson JD, Andrews BJ, Plowright RC (1981) The effect of a foreign pollen on ovule development in *Diervilla lonicera* (Caprifoliaceae). New Phytologist 90:777-783.

Thomson JD, Forrest JRK, Ogilvie JE (2011) Pollinator exclusion devices permitting easy access to flowers. Journal of Pollination Ecology 4:24-25.

Walther GR, Roques A, Hulme PE, Sykes MT, Pysek P, Kuhn I, Zobel M, Bacher S, Botta-Dukat Z, Bugmann H, Czucz B, Dauber J, Hickler T, Jarosik V, Kenis M, Klotz S, Minchin D, Moora M, Nentwig W, Ott J, Panov VE, Reineking B, Robinet C, Semenchenko V, Solarz W, Thuiller W, Vila M, Vohland K, Settele J (2009) Alien species in a warmer world: risks and opportunities. Trends in Ecology & Evolution 24:686-693.

Waser NM (1978) Competition for hummingbird pollination and sequential flowering in two Colorado wildflowers. Ecology 59: 934-944.

18

HONEST SIGNALLING AND THE BILLBOARD EFFECT: HOW HELICONIID POLLINATORS RESPOND TO THE TRICHROMATIC COLOUR CHANGING *LANTANA CAMARA* L. (VERBENACEAE)

Gyanpriya Maharaj[1, 2, 3] and Godfrey R. Bourne[1, 3]

[1]*Department of Biology, University of Missouri-St. Louis, One University Boulevard, St. Louis, MO 63121-4400, USA*

[2]*Centre for the Study of Biological Diversity and Department of Biology, University of Guyana, Turkeyen Campus, P.O. Box 10-1110, Greater Georgetown, Guyana*

[3]*CEIBA Biological Center, Linden Highway, Madewini, Guyana*

Abstract—Plants communicate with their pollinators through an astonishing range of signals that serve as either honest or deceptive cues which draw in and inform potential visitors of possible rewards. In wild type sweet sage, *Lantana camara,* floral colour signals were associated with nectar volume and sucrose concentration, and many pollinator taxa quickly learned to associate these varying colour signals with rewards. We tested the hypothesis that if sweet sage is employing a generalist pollinator strategy based on a trichromatic changing floral presentation system of honest rewards for pollinators, then the following predictions will be realized: 1) pre-change yellow flowers will be visited more frequently by pollinators than post change orange, or red flowers; 2) pre-change yellow flowers will produce higher quality and greater quantities of sucrose rewards than post-change orange, or red flowers; 3) inflorescences with higher ratios of rewarding flowers to unrewarding flowers are more attractive at short distances; and 4) inflorescences with a combination of pre-change rewarding and post-change rewarding and unrewarding flowers will act as a multi-coloured advertising billboard and as such be most attractive at long distances. We found corroboration for all of the aforementioned predictions. Thus, sweet sage evolved a generalized pollination visitation system based on honest signalling—of reward quantity and quality tied to colour changing visual signals acting in consort to produce a billboard that was easily perceived and deciphered. These resulted in high visitation rates by many different taxa of pollinators, thus contributing to higher individual plant fitness.

Keywords: colour change, Guyana, honesty signals, billboard effect, Lantana camara, pollinator

INTRODUCTION

The evolution of the great array of floral traits seen in Angiosperms rely on the diversity of animal pollinators to visit regularly and inadvertently transfer pollen efficiently from anthers of one flower to the stigmas of conspecifics (Graham et al. 2003; Kaesar et al. 2006). About 90% of the more than 240,000 species of flowering plants are pollinated by more than 200,000 animal species (Graham et al. 2003; Holland 2011). These plants employ three broad strategies for achieving pollination: (1) deception, where "gullible" animals are tricked by mimicry of "false" rewards into providing pollen transfer among flowers (Wickler 1968; Ackerman 1986; Nilsson 1992; Graham et al. 2003); (2) imprisonment, where flowers attract insects most of which are already covered with conspecific pollen, and delayed for several hours until pollen is released (Lack & Diaz 1991; Proctor et al. 1996; Gibernau et al. 2004; Bolin et al. 2009); and (3) honesty, in which the plant produces something of value to the animal (Nilsson 1992; Graham et al. 2003). Here the plant usually invests in food rewards—nutritious nectar fortified by sugars and amino acids, modified food pollen devoid of sperm; or provides safe and food-rich oviposition sites for insects to lay eggs, or produce fragrances that enhance males' mating success through female choice (Simpson & Neff 1981; Seymour & Matthews 2006; Wright & Schiestl 2009; Goodrich 2012). In honest signalling, these rewards are positively correlated with the presence and intensity of display signals (Kaesar et al. 2006; von Arx 2012).

Plants signal honestly to a wide range of organisms using many types of signals involving both vegetative and reproductive parts (Hamilton & Brown 2001; Schaefer et al. 2004). Many plants employ sensory signals which include colour, morphology, odour, among others, which in turn acting in concert with each other to become "sensory billboards" (Weiss & Lamont 1997; Raguso 2004; Willmer et al. 2009) . These sensory signals can function "honestly" in their communication with pollinators by reliably signalling the presence and/or quality of nectar, pollen, oil, or fragrance rewards (Nilsson 1992; Proctor et al. 1996; Schaefer et al. 2004; Raguso & Willis 2005; von Arx 2013). Colour signals are of particular importance to pollinators as they are able to perceive and distinguish colours and many show innate and learned colour preferences due to reward associations (Campbell et al. 2012). As such, plant colour signals, especially floral colour, exemplify the evolution of

*Corresponding author: gyanpriya.maharaj@mail.umsl.edu

floral traits driven by ecological interactions between plants and pollinators (Weiss 1997; Oberrath & Böhning-Gaese 1999; Ida & Kudo 2003; Ida & Kudo 2010; Willmer et al. 2009; Suzuki & Ohashi 2014).

Flower colour can remain constant during the entire anthesis stage or it can experience colour change due to multiple factors including; the environment, age or receptivity status (Weiss 1991; Yoshida et al. 2009). Changes in colour which occur in fully turgid flowers differ from fading or darkening associated with floral senescence (Weiss 1995). These changes differ in the locations which they affect and may take place in any of the four floral whorls. It may affect the entire whorl, several whorls or parts of whorls in combination, or it may be completely localized to specific areas (Weiss 1995). The location of colour changes in Angiosperms are dependent on pollinator type, for example, plants pollinated by bats or moths generally have colour changes in the entire flower while those that are butterfly, bee and fly pollinated usually have localized changes to specific floral parts, whereas bird pollinated flowers can encompass both types of changes (Weiss 1995). However, regardless of area affected it provides important information for pollinators that benefit both plants (communicator) and animals (receiver)—with pre-change flowers signalling the provision of rewards and the availability of receptive stigmas (Weiss 1991; Kudo et al. 2007). While post change flowers, are often retained though unrewarding and sexually inviable as plants benefit from larger floral displays that attract pollinators over long distances and indicating, at close range, pre-change flowers that are still viable (Gori 1989; Weiss 1991; Weiss 1995; Willmer et al. 2009, Ida & Kudo 2010). Von Linne (1793) (cited in Oberrath & Böhning-Gaese 1999) noted that floral colour change is a common phenomenon among flowering plants with diverse life histories and growth forms from over 78 families and 250 genera of angiosperms, distributed worldwide, visited by approximately 15 families of insects and four of birds (Weiss 1991; Weiss 1995; Weiss & Lamont 1997, Oberrath & Böhning-Gaese 1999). Despite the wide prevalence of flower colour change (Ida & Kudo 2010) and the well–developed hypotheses offered to explain the adaptive nature of this trait, this phenomenon has been experimentally examined in only a few species (Weiss 1995; Oberrath & Böhning-Gaese 1999). In addition, many of these studies focus on non-lepidopterans (see Ida & Kudo 2003; Ida & Kudo 2010; Pereira et al. 2011, Suzuki et al. 2014) or multiple groups of pollinators (Weiss & Lamont 1997; Oberrath & Böhning-Gaese 1999). Our study is unique because we compare the feeding behaviours of two major lepidopterans in a natural setting. Thus offering a unique perspective of how colour change of one plant differentially affects two pollinators that share a similar feeding niche (G. Maharaj unpubl. data). Our goal was to examine the relationships among floral colour change, and nectar volume and sucrose concentration in wild type sweet sage, *L. camara* on pollinator visitation at CEIBA Biological Center, Madewini, Guyana. Specifically, we asked the following questions of the sweet sage pollinator system: (1) Do younger yellow flowers produce greater quantities and higher sucrose concentration nectar than older orange and red flowers? (2) Do newly opened yellow flowers attract more *L. camara* pollinators than older orange and red flowers? And (3) how does inflorescence size and ratio of rewarding to unrewarding flowers influence butterfly pollinator visitation? Thus, we tested the hypothesis that if *L. camara* is employing a generalist pollinator strategy based on a trichromatic colour changing floral presentation system of honest rewards for pollinators, then the following predictions will be realized: (P1) first stage yellow flowers will attract more pollinators because they contain higher concentrations and volumes of sucrose than later orange and red stages; (P2) inflorescences with greater proportions of rewarding to unrewarding flowers will be more attractive over short distances as this will result in multiple visits to an individual plant due to butterflies' tendencies to visit particular colours that are associated with greater sucrose rewards; and (P3) inflorescences with a combination of rewarding yellow and orange flowers and unrewarding red will be most attractive to butterflies over long distances as these large multi-coloured inflorescences will provide large landing platforms (Barrows 1976) and serve as an advertising billboard drawing in potential pollinators from greater distances (Barrows 1976; Weiss 1991; Raguso 2004; Nuttman et al. 2005; Willmer et al. 2009).

Materials and Methods

Study site

Experiments on the pollination biology of sweet sage, *L. camara* were conducted at CEIBA Biological Center (CEIBA; 06°29/57//N, 58°13/06//W), on the Soesdyke-Linden Highway, Madewini, Guyana, South America. Observations were conducted in a sustainable demonstration farm site (320m2) filled with numerous *L. camara* stands. The study plot was bordered by a seasonally flooded white podsolized sand area comprised of low seasonal forest dominated by the fast-growing *Eperua falcate* (Caesalpiniaceae), and tall primary growth flooded forests dominated by *Mora excelsa* (Fabaceae) (Hughes 1947; Bourne & Bourne 2010).

Study species

Sweet sage, *Lantana camara* is a perennial shrub of the Vervain or Teak family (Verbenaceae) (Munir 1996) native to tropical regions of Central and South America (Graham 1963; Myint 1994). It is a readily available, easily tractable, common plant of CEIBA found in open habitats, especially on human disturbed sites (Sharma & Singh 2005) that provides food to a variety of pollinators. This plant has been the focus of many studies on colour vision and colour preference (Weiss 1991; Weiss 1997). This hairy herb with very aromatic leaves sometimes assumes either climbing or woody shrub growth forms. Wild type *L. camara* usually attains heights between 1 and 2 m and has square stems armed with short coarse spines (Ghisalberti 2000). *L. camara* plants used in this study were large shrubs that were approximately 1 m in height as these smaller plants were easier to work with i.e. manipulate. Leaves are simple and opposite, emanating at right angles from each node to leaves of the nearest neighbouring node. Leaf surfaces are wrinkled and scabrous or rough textured, while leaf edges are regularly serrate. In addition, leaf shapes vary from broadly lanceolate to cordate with distinctive pointed drip tips; leaves vary in measurement from 75.0–102.4 mm long by 25.3–56.7 mm wide, and with petiole lengths of 21.2–32.8 mm (G.R. Bourne unpubl. data). When leaves or stems are damaged, a distinctive odour is released. There are many horticulture varieties of *Lantana* that have small 5-lobed flowers in a variety of colours which include white, yellow, orange, pink, red and purple that are often mixed in the same cluster (Ghisalberti 2000; Sharma & Singh 2005). Inflorescences of our studied variety (wild-type) present trichromatic succession flowers (i.e. yellow to orange to red), held in close heads of umbel form, ranging from 31.3–42.6 mm in diameter, and with 9–30 flowers with four stamens. Thus, the inflorescences of *L. camara* allow for manipulation experiments testing the effect of colour of rewarding and unrewarding flowers on short and long distance attractiveness. Regular floral visitors include ants, carpenter bees, honey bees, black and brown stingless bees (usually as nectar robbers), wasps and hummingbirds (Weiss 1991), but especially butterflies belonging to diverse families such that many Guyanese classify sweet sage as a butterfly bush (G.R. Bourne and G. Maharaj pers. obs.). Fruits are smooth, round, two–celled berries (Graham 1963) with diameters of 4.2–6.6 mm presented in ball–like clusters 21.4–31.7 mm in diameter. When immature they are a shiny lime green in colour changing to indigo blue when ripe (Sharma & Singh 2005), and whose seeds are dispersed by many bird taxa including barbets, flycatchers, and tanagers.

We focused our experiments on two common butterfly species (Nymphalidae, Heliconiinae) at CEIBA, *Heliconius melpomene* and *Dryas iulia*. The first species is characterised by black wings with a red blurred patch on forewing (fwl ~ 41mm) and a yellow line on underside of hind wing curves towards the posterior. This species is often encountered as solitary individuals along forest edges and old second growth groves (DeVries 1987), and is frequently observed feeding on *Lantana camara* (Verbenaceae) (G. Maharaj & G.R. Bourne pers. obs.). Whereas, *D. iulia* is characterised by bright orange wings with black margins and with elongate forewings (fwl ~ 85 mm); males are typically brighter than females (DeVries 1987). This species is usually found in second growth forest imbibing nectar from many flower species, it is also a noted gregarious feeder of *L. camara* (G. Maharaj & G.R. Bourne unpubl. data). We chose to work with species of Heliconiinae because they are tractable to study in the laboratory and the wild, and have been the focus of a large body of work in evolutionary biology, genetics, and animal behaviour (Hsu et al. 2001). Heliconiids also vary considerably in the way they use visual signals to find flowers (food sources), mates, and communicate (Hsu et al. 2001).

General sampling protocols

Flower colour and diurnal sucrose measurements

In order to determine flower colour and respective rewards offered we used destructive sampling to measure daily diurnal spectral reflectance change, nectar volume and sucrose concentrations. For each flower, we used type colour swatches in Smithe (1975) to measure and name flower colour (as perceived by humans). Flowers were placed directly onto swatch and colour was determined by researcher and research assistant. If both investigators were unable to agree on colour nomenclature, a third researcher was consulted. Although human colour nomenclature and just noticeable differences were used in this study, we do recognise the need to refer to colour differences in terms of insect perceptions as both study species and most insect classes possess three classes of opsin genes, ultraviolet (UVRh λmax ~350nm), blue (BRh λmax ~ 440nm) and long-wavelength (LWRh λmax ~ 530nm) (Briscoe & Chittka 2001; Sison-Mangus et al. 2006; Briscoe 2008; Yuan et al. 2010). In this study we first aimed to establish whether there is an actual difference in behaviours of butterflies to flower colour changes as seen by humans. However, we do acknowledge that butterflies typically have long-wavelength opsins, screening pigments that modify spectral sensitivities, and the possible presence of lateral filtering pigments, that filter short wavelength light thus shifting the sensitivity of the visual pigments to the longer wavelengths (such as red filtering pigments seen in *Heliconius erato*) and furthermore some have four or five photoreceptor types (Zaccardi et al. 2006, Stavenga & Arikawa 2006, Briscoe 2008). Consequently, our study species are not only capable of distinguishing changes in long wave length red markings, they are likely to have a visual system significantly different from that of human observers, in terms of both short- and long-wavelength sensitivity, and colour discrimination abilities. In a different study we investigated these floral colour changes in our respective butterflies' colour spaces (Maharaj et al. manuscript in prep.).

For this current study, we used 1μL Drummond Microcap® tubes and a digital calliper to estimate nectar volumes, and a SPER Scientific Sugar–Brix Refractometer to measure nectar concentration (Waser & Price 1981). In order to determine colour and sucrose measurements of the three major colour stages, a total of 20 flowers were used for each cohort of each colour type. These flowers were haphazardly selected from several inflorescences of ten marked plants at 09:00 h for three days during May 2010. Day 1(yellow) — was taken as the first day after buds bloomed, day 2 (orange) — was taken the morning after that and day 3 (red) — was taken on the following morning. To estimate colour and sucrose measurements of the nine sub-divisions of these three major colour stages a total of 25 flowers were selected for nine days (July 2014). The colour stages were as follows: Stg. 1 — orange yellow centre with spectrum orange edges, Stg. 2 — orange yellow centre with chrome orange edges, Stg. 3 — orange yellow with flame scarlet edges, Stg. 4— spectrum orange with flame scarlet edges, Stg. 5 — chrome orange with flame scarlet edges, Stg. 6— chrome orange, Stg. 7— flame scarlet, Stg. 8— flame scarlet with scarlet edges and Stg. 9 — scarlet (colour swatches in Smith 1975). These flowers were picked during four 3-hour time blocks (TBs); (TBI 06:00–9:00 h, TBII 09:00–12:00 h, TBIII 12:00–15:00 h and TBIV 15:00–18:00 h).

All sampled flowers, from both the three day and nine stage experiments, were fresh, turgid and picked from previously bagged inflorescences. These inflorescences were placed in light-admitting bags as buds initially and remained bagged for the duration of the study to prevent nectar consumption by pollinators. Our sampled colour change flowers did not include colour changes from bud to flower or wilted flowers.

Pollinator species & fruit set

In order to determine which visitor taxa were effective pollinators we conducted visitation watches and fruit set experiments. We counted the number of diurnal animals visiting wildtype *L. camara* inflorescences of nine selected plants that differed in floral density, number of inflorescences per 0.5 m^2, (high [20+], medium [6-19], and low [2-5]), and colour for 60 d during May–July 2010 at CEIBA. Each 0.5 m2 quadrat was sampled for 2 min only during sunny periods in time block II (TB II, 09:00–11:59 h), the peak pollinator activity period at CEIBA. Every visiting animal taxon was photographed to aid in identification using various guides (Barcant 1970; Borror & White 1970; Milne & Milne 1980; Pyle 1981; DeVries 1987; Opler 1992; Restall et al. 2007a, b; Marshall 2008; Maharaj et al. 2010), and foraging behaviours recorded. If a floral visitor had pollen on any part of its body, it was considered a pollinator of *L. camara*. A checklist was made of all pollinators, a special focus was made on butterflies due to their proclivity for visiting this plant and the two major pollinators (as characterized by frequency of visits and abundance), *Dryas iulia* (Fabricius, 1775) (Nymphalidae) and *Heliconius melpomene* (Linnaeus, 1758) (Nymphalidae) (DeVries 1987) were our focal animals for our inflorescence manipulation experiments due to ease of observations.

Fruit-set experiments were carried out by inclusion (focal taxon)/exclusion (other taxa) in 1 m^3 mesh cages (fine gauge mosquito netting 3000 holes per cm^2). Prior to initiation of fruit-set studies, 240 immature inflorescences on 10 *L. camara* bushes were bagged using see-through, home-made pollinator bags during May 2010. Pollinator bags, 133 × 99 mm were constructed from perforated (by safety-pins, 26 ± 8 holes per cm^2) white printing paper on one side and clear plastic from ZipLock® freezer bags on the other, stitched together by 17 mini-staples. As inflorescences matured some were unbagged and the mesh cages set up in the evening (18:00 h) after diurnal pollinator activity ceased. Each pollinator tested (included all captured as we did not control for pollinator sex) was introduced by hand and held for a 72 h period. Inflorescences were then rebagged to prevent cross species visitation, after which the mesh cages were removed. Hummingbird diets were supplemented by 25% sucrose solution and adult fruitflies (*Drosophila* spp.).

Flower colour preference and billboard effect

In order to demonstrate whether or not pollinators exhibited a pattern of colour preferences, and that clustering of floral displays had a billboard effect, we conducted two field experiments in which we manipulated *L. camara* inflorescence densities by removing variable numbers of individual coloured flowers to create multiple treatments. We then observed visitation to these treatments. The first experiment, called, colour preference, was a generalized study that examined colour choices of all animal visitors to *L. camara* flowers. The second experiment, entitled, billboard effect, followed two focal butterfly species, *H. melpomene* and *D. iulia*. These were major visitors to *L. camara* where they navigated different concentrations of colour combinations in their choices of inflorescences that may explain why non-rewarding red flowers persist in displays.

Experiment 1 — Colour preference

These field studies were conducted over 60 days during May–July 2010. We first measured pollinator visitation rates (counts/2 mins) of all *L. camara* visitors during sunny periods, from 09:00-14:59 h, to vases with mixed inflorescences (all three flower colours), yellow, orange and red only inflorescences matched by floral numbers. Vials of flowers were presented on wooden dowels, randomly arrayed across the study site. The number of flowers in each treatment was standardized at nine and sample size was established at 15. This experiment was repeated by removing flowers from inflorescences on randomly selected plant stands to determine whether patterns of general pollinator visitation patterns were similar for flowers detached from plants (vase presentation) and those still attached to plants (natural presentation).

Experiment 2 — Billboard effect

For Experiment 2, a 0.5-m^2 quadrat was placed on an individual *L. camara* plant to delineate the area in which inflorescences were manipulated to reflect treatments described below. After the quadrat was removed, the entire plant with the exception of the 0.5-m^2 manipulated portion was covered with fine gauge mosquito netting, 3000 holes per cm, to prevent access of pollinators to un-manipulated inflorescences. Visitation of *H. melpomene* and *D. iulia* were observed for 30 d (June–July 2014) and 10 d (December–January 2015), with the total number of visits estimated over a 2-hour observational period. A butterfly was characterized as a visitor if it perched on the inflorescence. Visits were further categorized as either, long distance—number of approaches to a single plant stand or short distance—number of successive visits to multiple inflorescences on a single *L. camara* plant (Oberrath & Böhning-Gaese 1999). All experimental manipulations were done at 08:00 h just after yellow flowers had first opened but before focal butterfly species had begun to forage (G. Maharaj unpubl. data). All visitation observations were initiated 2-hours after experimental set-up. Each treatment was replicated five times on different plants randomly chosen form 25 marked plants with each replication carried out on different days to account for variability in butterfly behaviours and weather conditions.

Treatments were as follows: — (A) same size (app. 19-24 flowers) different colour: Each day we randomly selected a total of eight plants (two per treatment). Inflorescences on these plants were manipulated in the following ways — (i) control (un-manipulated mixed i.e. comprised of all three colour morphs in natural combinations), (ii) manipulated mixed (25:25:50 - this is an inflorescence comprised of 25% yellow, 25% orange and 50% red flowers), (iii) yellow and orange (50:50, made up of 50% yellow and 50% orange flowers), and (iv) All red (100% red flowers) (Gori 1989).

(B) different size different colour:

We modified a total of six plants per day (two per treatment) to offer the following treatments — (i) large red inflorescences (20 red flowers), (ii) large mixed inflorescences (three yellow, five orange and 12 red), and (iii) small yellow (five yellow flowers) (Weiss 1991). These three treatments were offered in the following three pairs of choices(i) large red inflorescences versus large mixed inflorescences, (ii) large red versus small yellow, and (iii) large mixed versus small yellow.

Statistical analyses

A Kruskal–Wallis 1-Way Analysis of Variance (ANOVA) model with pairwise comparisons (Tukey Test) was employed to assess differences among volumes for each of the major floral colour stages (yellow, orange and red) in wild-type sweet sage. A 1-Way ANOVA with pairwise comparisons (Holm-Sidak method) was used to compare sucrose concentrations of the three major flower stages and for the nine sub-colour stages. We employed a generalized linear model (Poisson distribution) to test for significant differences among treatments, while a binomial logistic regression was employed to predict the probability of either short or long distance attraction based on single versus multiple visits (coded as a dichotomous variable) to each treatment plant. All statistical analyses were carried out using IBM SPSS Statistics Version 23 (IBM Corp. 2015) and R Version 3.2.2 (R Development Core Team 2015).

RESULTS

Flower colour and sucrose measurements

We observed a significant decrease in both sucrose volume (H^2 = 49.06, $P < 0.001$, $N = 20$, statistically significant differences between all pairwise comparisons) and sucrose concentration ($F_{2,57} = 619.84$, $P < 0.001$, $N = 20$, statistically significant differences for pairwise comparisons, Tab. 1), as flowers changed from yellow to orange to red.

A more detailed look at the wildtype *L. camara* flower colour change system revealed that it can be subdivided into nine stages characterized by variations of the three main colours, yellow, orange, and red, with earlier stages characterized by lower volumes and higher concentration and later stages having higher volumes and lower concentrations, with the exception of the final stage that offered no reward. Overall measurements of sucrose concentration and volume by colour stage showed substantial variability (Tab. 2). However, there were significant differences among stages for both nectar sucrose concentrations and volumes. The 1-Way ANOVA for volume ($F_{8,216} = 12.906$, $P < 0.05$, $N = 25$) and post-hoc analyses (Tukey Test) revealed that Stg. 1 flowers were statistically different from stages 4, 8 and 9, Stg. 2 was different from 8 and 9, Stg. 3 differed significantly from 4 and 5, while Stg. 4 differed from Stg. 9 (Tab. 2). For our concentration measurements analyses ($F_{8,216} = 117.32$, $P < 0.05$, $N = 25$) we found that Stgs. 1, 2, 3, 4 and 5 were statistically significantly different from 6, 7, 8, and 9, and Stgs. 6, 7, and 8 were different from 9 (Tab. 2).

Pollinator taxa & fruit set

When percentage fruit set is considered, butterflies were the most effective taxon of diurnal pollinators, followed by carpenter bees (Apidae; *Xylocopa* spp.), and hummingbirds (Trochilidae; Fig. 1). Controls, butterflies, carpenter bees and hummingbirds had significantly better fruit set percentages than Trigonid bees, wasps and ants (Fig. 1). The numbers of diurnal pollinator butterfly taxa observed on *L. camara* are presented in Tab. 3. We focused our study on butterflies due to their proclivity to visit *L. camara*, and their efficacy as pollinators. Of the butterflies, the most frequent visitors were *Heliconius melpomene* followed by *Heliconius sara*, *Dryas iulia* and *Heraclides thoas* (syn *Papilio thoas*) (as seen in Tab. 3), however only *H. melpomene* and *D. iulia* were used in experiments because of their high abundancies.

TABLE 1: Mean and SD values of sucrose (nectar) volumes and concentrations for the three gross colour stages indicated declining production with time.

	Yellow	Orange	Red
Volume	0.91 (± 0.144)	0.67 (± 0.128)	0.03 (± 0.028)
Concentration	27.00 (± 2.046)	21.06 (± 1.769)	8.11 (± 1.312)

TABLE 2. Comparison of means and SD values of sucrose volumes and concentration for fine temporal colour stages (Stg.1= Stage 1 etc.) showing an increase in volume after stage 3 and no reward offered in stage 9 and a decrease in concentration in later stages (5-9). Stage 1 correspond to yellow flower colour, Stage 4 to orange and Stage 7 to red as identified in Tab. 1.

	Stg. 1	Stg. 2	Stg. 3	Stg. 4	Stg. 5	Stg. 6	Stg. 7	Stg. 8	Stg. 9
Volume	0.62 (±0.23)	0.77 (±0.41)	0.58 (±0.20)	1.18 (±0.88)	1.09 (±0.96)	0.91 (±0.41)	0.85 (±0.32)	1.36 (±0.75)	0.00 (±0.00)
Concentration	21.41 (±1.25)	20.52 (±2.57)	20.40 (±2.42)	20.74 (±2.39)	19.03 (±1.96)	16.02 (±5.40)	15.99 (±2.93)	16.22 (±4.81)	0.00 (±0.00)

TABLE 3. Frequency of Lepidopteran pollinators observed foraging on *L. camara* over a 15-day period. Top four foragers include *Heliconius melpomene, H. sara, Dryas iulia* and *Papilio thoas.*

Butterfly Species	Count/Total no. visits (15d)
Agraulis vanillae	1
Aphrissa boisduvalii	9
Aphrissa statira	37
Astraptes fulferator	1
Battus polydamas	4
Dryas iulia	115
Emesis aurimna	3
Emesis fatimella	8
Heliconius bruneiye	27
Heliconius erato	1
Heliconius hecale	22
Heliconius melpomene	357
Heliconius numata	14
Heliconius sara	212
Lemonias emylius	2
Mechanitis lysimnia	1
Melinaea lilis	3
Nymphidium ascolia	1
Nymphidium oleamun	5
Papilio thoas	95
Philatria dido	3
Phoebis argante	11
Phoebis sennae	7
Phoebis statira	1
Stalachitis phlegia	9
Synargis tytia	2
Thecla sp.	7

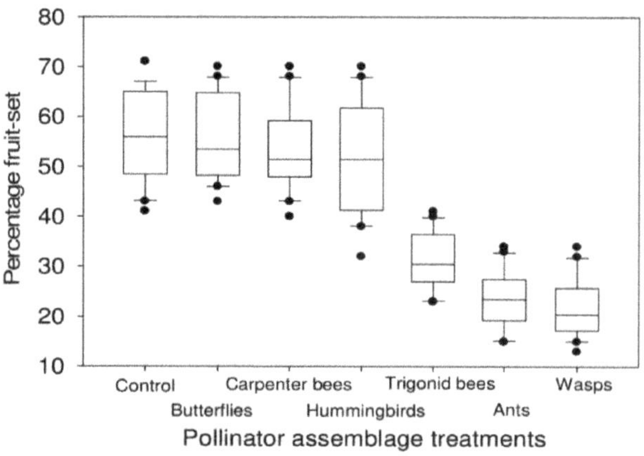

FIGURE 1. Pollinator taxa and effectiveness of visits on the percentage of fruit set in *Lantana camara.* The Control variable consists of the effects of all pollinating taxa visits on fruit set. Butterflies, Carpenter bees and Hummingbirds were found to be more effective pollinators than Trigonid bees, ants and wasps.

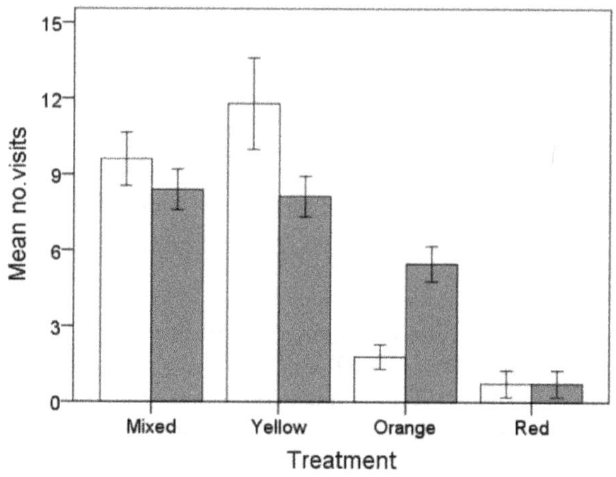

FIGURE 2. Mean (± 2SE) pollinator visitation of inflorescences presented in nature (white bars) and in vials (grey bars). Highest visitation observed at plants with mixed and all yellow inflorescences but lowest at orange and red inflorescences.

Flower colour preference & billboard effect

Experiment 1 — Colour preference

Our Poisson regression showed that there was no difference in visitation by set-up (vials vs nature), however there was a statistically significant difference in which treatments were visited by pollinators. Pollinator interest (mean pollinator visitation rates – number per 2 min) in the arrays of *L. camara* bouquets presented in vials away from the plants and in nature had highest and statistically similar visitation rates at the yellow and mixed inflorescences (Exp (B) = 0.903, 95% CI 0.766-1.065, $P = 0.224$). Pollinator visitation significantly declined when presented with orange inflorescences (Exp (B) = 0.365, 95% CI 0.293-0.454, $P < 0.001$) and even more so with red inflorescences (Exp (B) = 0.074, 95% CI 0.048-0.113, $P < 0.001$) (Fig. 2).

Experiment 2A — Billboard effects

A Poisson regression was run to predict the number of visits to a *L. camara* plant based on the butterfly species and the type of treatment carried out to the inflorescence of that *L. camara* plant. Our results show that there are statistically significant differences by species and treatment in visitation. *D. iulia* (Exp (B) = 0.300– CI 95% 0.255-0.353, $P < 0.001$) visited fewer plants than *H. melpomene*. Butterflies were most likely to visit large mixed inflorescences (Exp (B) = 1.298– CI 95% 1.095-1.539, $P = 0.003$) and large red inflorescences (Exp (B) = 1.204 – CI 95% 1.013-1.432, $P < 0.035$) in comparison to small yellow inflorescences. With *H. melpomene* visiting large mixed and large red more than small yellow, while *D. iulia* preferred large mixed and small yellow to large red (Fig. 3). A logistic regression analysis predicted the likelihood that our focal butterflies ($N_{H.\ melpomene} = 633$, $N_{D.\ iulia} = 190$) visited either single or multiple inflorescences on a single plant. For our model we used species and the three treatments (Small Yellow, Large Mixed and Large Red) as predictors. We did this to elucidate how species and treatment affects long and short distance attraction. A test of the full model against a constant only

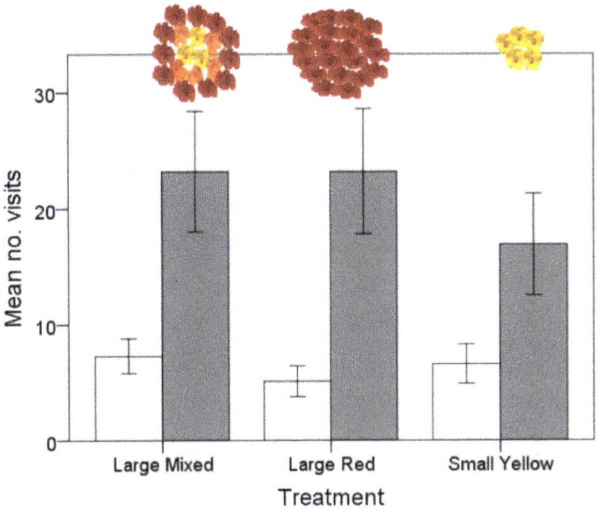

FIGURE 3. Mean (± SE) pollinator visitation of inflorescences by study species. Highest visitation by *D. iulia* (white bars) observed at Large Mixed plants while *H. melpomene* (grey bars) visited both Large Mixed and Large Red inflorescences more than Small Yellow.

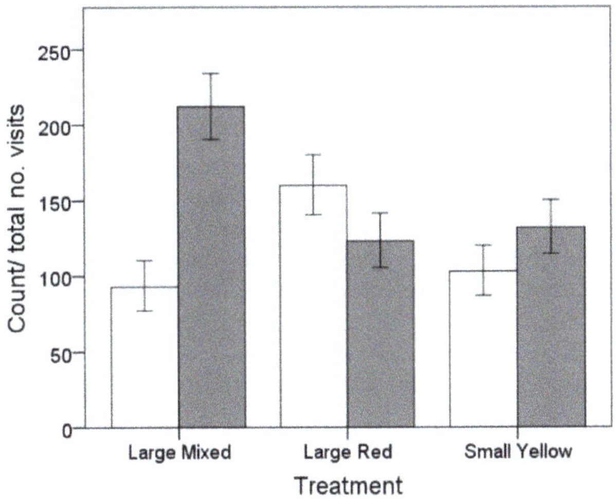

FIGURE 4. Total number (95% CI) of single (white bars)/multiple (grey bars) visits to inflorescences by study species. Highest number of multiple visits by butterflies occurred at Large Mixed inflorescence whereas, highest number of single visits occurred at Large Red inflorescences.

model was statistically significant indicating that the predictors as a set reliably distinguished between single versus multiple flower visitation, $\chi^2_3 = 41.23$, $P < 0.001$. The Wald criterion demonstrated that of the two predictor variables, only treatment, was statistically significant, i.e. butterflies were more likely to visit one inflorescence (long distance attraction) or multiple inflorescence (short distance attraction) on a plant based on type of inflorescence treatment only. Butterflies visiting Large Mixed (Exp (B) = 1.783, – CI 95% 1.250-2.544, $P = 0.001$) and small yellow inflorescences were more likely to visit multiple inflorescences on a plant, whereas butterflies visiting Large Red inflorescences were less likely to visit multiple flowers (Exp (B) = 0.603, – CI 95% 0.425-0.857, $P = 0.005$), i.e. they were more likely to visit a single inflorescence only (Fig. 4).

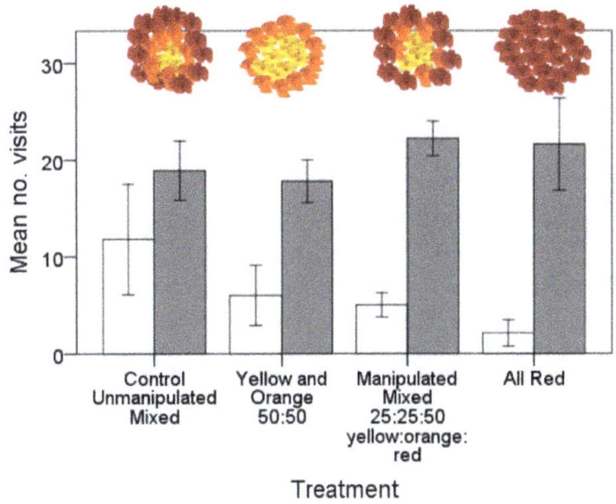

FIGURE 5. Mean (± SE) pollinator visitation of inflorescences by study species. Highest visitation by *D. iulia* (white bars) was observed at Control plants while *H. melpomene* (grey bars) visited Manipulated Mixed inflorescences and All Red inflorescences the most.

Experiment 2B — Billboard effects

Our Poisson regression predicted the number of visits to a *L. camara* plant based on the butterfly species and the type of treatment on that *L. camara* plant. Our results show that there is a significant difference in visitation by species and treatment. Similar to before, *D. iulia* butterflies (Exp (B) = 0.309 – CI 95%0.268-0.357, $P < 0.001$) visited fewer plants in comparison to *H. melpomene*. Overall, butterflies were most likely to visit control (un-manipulated mixed) inflorescences (Exp (B) = 1.295 – CI 95% 1.093-1.535, P = 0.003), followed by manipulated mixed (25:25:50-yellow: orange: red) (Exp (B) = 1.148 – CI 95% 0.965-1.366, $P = 0.121$), yellow and orange (50:50) (Exp (B) = 1.004 – CI 95% 0.839-1.202, $P = 0.963$), and least likely to visit all red inflorescences. However, visits to mixed manipulated, yellow and orange and all red inflorescences were not statistically significantly different from each other. We found that *H. melpomene* visited all red and manipulated mixed more in comparison to the other treatments while *D. iulia* preferred control and yellow and orange (Fig. 5). A logistic regression analysis was employed to predict the likelihood that our focal butterflies ($N_{H.\ melpomene} = 805$, $N_{D.\ iulia} = 249$) visited either single or multiple inflorescences on a single plant—in this model we used species and the four treatments (Control (un-manipulated mixed), manipulated mixed (25:25:50 - yellow:orange:red), yellow and orange (50:50), and All Red) as predictors. We did this to determine how species and treatment affects long and short distance attraction. A test of the full model against a constant only model was statistically significant indicating that the predictors as a set reliably distinguished between single versus multiple flower visitation, $\chi^2_4 = 60.954$, $P < 0.001$. The Wald criterion demonstrated that both predictor variables were statistically significant. We found that *H. melpomene* (Exp (B) = 1.430, 95% CI 1.057-1.935, $P = 0.02$) was more likely to visit multiple inflorescences than *D. iulia*. Overall, butterflies were more likely to visit multiple inflorescences on the following plants, i.e. yellow and orange

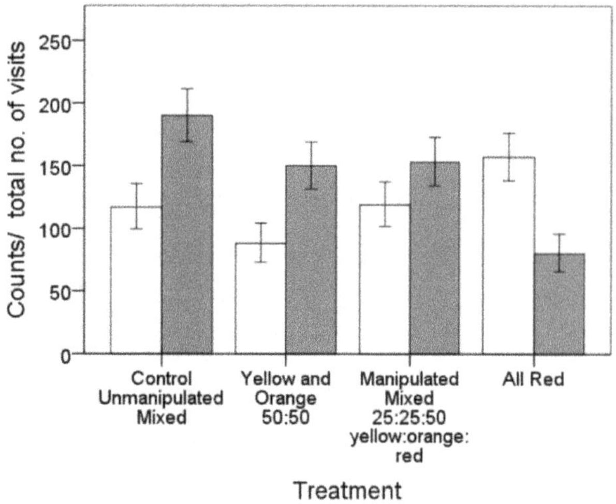

FIGURE 6. Total number (95% CI) of single (white bars) /multiple (grey bars) visits to inflorescences by study species. Highest number of multiple visits by butterflies occurred at Control inflorescence whereas, highest number of single visits occurred at All Red inflorescences.

(50:50) (Exp (B) = 3.563, 95% CI 2.433-5.219, *P* < 0.001), control (un-manipulated mixed) (Exp (B) = 3.562, 95% CI 2.464-5.148, P < 0.001), and manipulated mixed (25:25:50 yellow: orange: red) (Exp (B) = 2.618, 95% CI 1.822-3.761, *P* < 0.001), in comparison to all red inflorescences (Fig. 6).

DISCUSSION

Flower colour and sucrose measurements

When we examined flowers for three consecutive days our results, pre-change yellow flowers having higher sucrose concentration and volume in comparison to day 2, orange, and day 3, red, post-change flowers, mirrored that of Fritz Müller who reported to Charles Darwin (1877) that *Lantana camara* flowers in Brazil are viable for three days, changing from yellow on day—one, to orange on day— two, and red on day—three with these floral colour signals correlating with nectar volume and sucrose concentration in many varieties (Darwin 1877). Thus *L. camara* flowers signal honestly to their pollinator as each colour stage reliably conveys information about an associated reward. However, when we examined nine colour stages (Tab. 2) we noticed as time progressed i.e., as the flowers aged, there was not a significant change in sucrose volume, although we noted earlier stages 1-3 having lower volumes than the later 5 stages with the exception of stage 9 that did not offer sucrose. This was also evident for concentration with stages 1-5 having higher mean concentration than the later 4 stages including the final scarlet stage when no reward was offered. Additionally, although we were able to distinguish a colour change using the colour swatches in the first three stages they offered statistically indistinguishable rewards. However, when we only examined three colour stages (Tab. 1) we noted a decrease in both sucrose volume and concentration. The lower nectar volumes noted for initial stages of these the nine stages could be caused by environmental differences in temperature, relative humidity and soil moisture (Wolff 2006) between the two field seasons as we were unable to control for these factors at our study site. Carrión-Tacuri et al. (2012), showed that the nectar volumes of bagged *L. camara* flowers did not change significantly throughout the day but nectar volumes oscillated between 0.9 and 1.1 μl. These variations were probably reflected in the measurements of the 9-stage *L. camara* readings but not in the 3-stage because these readings were only taken once per day.

Pollinator species & fruit set

Colour change in *L. camara* occurs for several reasons, these include attraction of pollinators such as hummingbirds, bees, wasps, ants, but especially butterflies (G.R. Bourne and G. Maharaj unpubl. data). The pollination syndrome hypothesis posits that different pollinators prefer different floral cues, with butterflies and bees preferring colours ranging from ultraviolet to yellow or red coloured flowers, and birds, orange, deep-pink and red flowers (Proctor et al. 1996; Weiss 1997; Johnson & Steiner 2000; Graham et al. 2003). Therefore, the presence of different colours on individual inflorescences seem to serve the purpose of attracting the high taxon diversity of pollinators observed (Ostler & Harper 1978; Kampny 1995; Campbell & Hanula 2007; Suzuki & Oashi 2014). We do acknowledge that in order to test the direct effect of colour on diversity we would have to manipulate inflorescences to reflect individual colour morphs and observe changes in visit diversity. However, from our study we did observe that inflorescences with all colour morphs were visited more often by all pollinators, although butterflies, carpenter bees and hummingbirds were the main visitors and most effective pollinators. The fact that *L. camara* attracts these three ubiquitous pollinator groups, birds as seed dispersers and humans who commercialized many colourful forms may account for its spread globally (Ghisalberti 2000).

Flower colour preference and billboard effect

Our findings suggested that *L. camara* signals honestly as their colour cues correlated with nectar rewards, with early more receptive yellow stages offering better rewards (higher concentration of sucrose, although volume was variable). While sexually inviable stages such as final stage scarlet flowers offered no reward (Oberrath & Böhning-Gaese 1999; Keasar et al. 2006). Therefore, this floral colour change is an adaptive trait that benefits both the plant and its insect pollinators by cuing the insects to visit the flowers at the optimal reproductive stage and thus minimizing the probability of illegitimate visits to non-reproductive flowers by changing colour and reward value, as we have seen with yellow, orange and red flowers in our experiments (Willmer et al. 2009). Our evidence clearly supported prediction one (P1) that first stage yellow flowers attract more pollinators as they contain greater quality of rewards (greatest concentration of sucrose) than later orange and red stages (lower quality reward). Thus the pollinators of *L. camara* displayed a greater preference for these pre-change yellow flowers than orange or red flowers. We do acknowledge that our results may represent a combination of innate and learned preferences since many pollinators are able to associate colour with reward (Menzel 1967, 1985; Waser & Price 1985; Weiss 1991; Waser et al. 1996; Weiss 1997; Campbell et al. 2012). In order to determine whether our pollinators have innate colour biases for these colours we would have to test naïve pollinator or carry out experiments in which post-change flowers offer greater rewards than pre-change flowers (Lanau & Maier 1995; Weiss 1997).

When we tested for the billboard effect we noted that although pollinators were more attracted to yellow flowers there were other factors that also affected visitations rates. While foraging, pollinators increase foraging efficiency by making two decisions based on distance—at long distances pollinators decide: 1) which plants should be approached, and at shorter distances, i.e., when they are on the plant, and 2) which flower(s) should be visited. Both of these decisions are based on visual attractiveness of plants and flowers, respectively (Oberrath & Böhning-Gaese 1999). Work by Gori (1989), Weiss (1995) and Willmer et al. (2009) also demonstrate that plants benefit from larger floral displays that attract pollinators over long distances. Plants offering both rewarding pre-change flowers and provision less post-change flowers served as a superior attractant to pollinators at greater distances—a strategy that results in increased pollinator visitation (Barrows 1976; Weiss 1991; Nuttman et al. 2005). These results corroborated our findings and supported our second and third predictions. We observed inflorescences with greater proportions of yellow and orange flowers i.e., small yellow, manipulated mixed (25:25:50 - yellow:orange:red), yellow and orange (50:50), control (un-manipulated mixed) and large mixed were more attractive over short distances (P2) as this resulted in multiple visits to individual flowers on each because butterflies learned to associate colour with reward, thus pre-change yellow flowers were favoured at close range (Gori 1989; Weiss 1995; Willmer et al. 2009). Inflorescences with unrewarding red flowers were found to be most attractive to pollinators over long distances, as inflorescences on these plants were only visited once. While overall, most visits to plants were made to large mixed and control (un-manipulated mixed) due to the billboard effect that results from larger multi-coloured displays (P3) (Barrows 1976; Gori 1989; Weiss 1991; Weiss1995; Nuttman et al. 2005; Willmer et al. 2009). The retention of provision less scarlet flowers or red flowers that produce little reward function to increase the inflorescence size, and advertisement attractiveness so making a bigger landing platform for large butterflies (Barrows 1976), thereby making these inflorescences more attractive than just small yellow all rewarding inflorescences of our focal plant system. Although we found that retention of red flowers benefitted our study plants Ida and Kudo (2003) demonstrated that this is not case for all colour change plants i.e. *Weigela middendorffiana.* It was also noted that although the size of the landing platform and its effect on proclivity to land was not measured, it was noted that *D. iulia,* a medium sized butterfly, as is *H. melpomene,* preferred both large mixed and small yellow to large red, whereas *H. melpomene* preferred large red to large mixed. This suggests, although not conclusively, that butterflies will feed on inflorescences of both sizes.

Overall we found differential preferences by our two focal species, with *D. iulia* visiting inflorescences many yellow flowers, viz. small yellow, yellow and orange (50:50) or control (un-manipulated mixed), more frequently while *H. melpomene* tended to frequent inflorescences with many red flowers; large red, large mixed, manipulated mixed (25:25:50-yellow: orange: red) and all red treatments. We also noted that when presented with small yellow, large mixed and large red inflorescences butterflies were more likely to visit the flowers of large red inflorescences only once. Similarly, when presented with control plants (un-manipulated mixed), manipulated mixed (25:25:50 - yellow:orange:red), yellow and orange (50:50) and all red, butterflies visited single flowers on all red inflorescences. Therefore, although unrewarding red flowers draw in pollinators from a long distance, only plants with rewarding flowers facilitate short distance feeding behaviours, while plants with both such as, large mixed and control (un-manipulated mixed), attracted the most visits overall.

In summary our results suggested that *L. camara* incorporates two main strategies to visually attract pollinators at long and short distances. First, they signal honestly as the rewards offered reliably correlated with colour stage. Secondly, by offering multiple coloured inflorescences with centrally located scarlet flower buds surrounded by pre-change yellow flowers and older post-change orange and older red flowers, plants behave like billboards communicating their attractiveness to pollinators at greater distances; a strategy that resulted in visitations by a diversity of pollinators at both long and short distances (Weiss 1991; Nuttman et al. 2005), the overall effect being that individual *L. camara* plants have increased fitness. Our study also highlighted species specific visitation preferences based on flower colour morphs presented, although both study species exhibit generalized learned preferences when it came to feeding, i.e., choosing flowers with greatest rewards. These visitation preferences may be due to inherent colour preferences of each butterfly species and linked to their abilities and genetic mechanisms to decipher colour (Hsu et al. 2001; Briscoe 2008). This study further identified areas of future work as we try to tease apart the specific visual signals that are used by each butterfly species and its impacts on pollination efficacy.

Acknowledgements

We are grateful to O. O'Dean, H. Stowe, H. Chaney, and S. France for field assistance and support during field work and to A. Dunlap, N. Muchhala and Y. Wu for their guidance and comments during the planning and execution of this research. This work was supported by a Small Grant from the Rufford Foundation, UK, and three CEIBA Biological Center Grants awarded to G. Maharaj, and three anonymous foundation grants to G.R. Bourne. This paper was constructed from a chapter in G. Maharaj's unpublished dissertation presented to the University of Missouri-St. Louis.

References

Ackerman JD (1986) Mechanisms and evolution of food-deceptive pollination systems in orchids. Lindleyana 1:108-113.

Barcant M (1970) Butterflies of Trinidad and Tobago. Collins, London.

Barrows EM (1976) Nectar robbing and pollination of *Lantana camara* (Verbenaceae). Biotropica 8:132-135.

Bolin JF, Maass E, Musselman LJ (2009) Pollination biology of *Hydnora africana* Thumb. (Hydnoraceae) in Namibia: brood-site mimicry with insect imprisonment. Journal of plant Science 170:157-163.

Borror DJ, White RE (1970) A Field Guide to the Insects. Houghton Mifflin Company, New York.

Bourne GR, Bourne CM (2010). The Biological Station and Its Programs. In: The CEIBA Reader. Yerfdog Publishing for CEIBA, Missouri, pp.44.

Briscoe AD (2008) Reconstructing the ancestral butterfly eye: focus on the opsins. Journal of Experimental Biology 211: 1805-1813.

Briscoe AD, Chittka L (2001) The evolution of color vision in insects. Annual review of entomology 46:471-510.

Campbell DR, Bischoff M, Lord JM, Robertson, AW (2012) Where have all the blue flowers gone: pollinator responses and selection on flower colour in New Zealand *Wahlenbergia albomarginata*. Journal of Evolutionary Biology 25:352-364.

Campbell JW, Hanula JL (2007). Efficiency of Malaise traps and colored pan traps for collecting flower visiting insects from three forested ecosystems. Journal of Insect Conservation 11:399-408.

Carrión-Tacuri J, Berjano R, Guerrero G, Figueroa ME, Tye A, Castillo JM (2012) Nectar Production by Invasive *Lantana camara* and Endemic *L. peduncularis* in the Galápagos Islands I. Pacific Science 66:435-445.

Darwin C (1877) Fritz Müller on flowers and insects. Nature 17:78.

DeVries PJ (1987) The butterflies of Costa Rica and their natural history: Papilionidae, Pieridae, Nymphalidae. Princeton University Press, Princeton New Jersey.

Ghisalberti EL (2000) Review *Lantana camara L.* (Verbenaceae). Fitoterapia 71: 467-486.

Gibernau M, Macquart D, Przetak G (2004) Pollination in the genus *Arum*: a review. Aroideana 27:147-166.

Goodrich KR (2012) Floral scent in Annonaceae. Botanical Journal of the Linnean Society 169: 262-279.

Gori DF (1989). Floral color change in *Lupinus argenteus* (Fabaceae): why should plants advertise the location of unrewarding flowers to pollinators? Evolution 42:870-881.

Graham LE, Graham JM, Wilcox LW (2003) Plant Biology. Prentice Hall, Pearson Education, Inc., Upper Saddle River, New Jersey.

Graham VE (1963) Tropical Wild Flowers. Hulton Educational Publications, London.

Hamilton WD, Brown SP (2001) Autumn tree colours as a handicap signal. Proceedings of the Royal Society of London. Series B: Biological Sciences 268:1489-1493.

Holland JS (2011) Gold dusters. National Geographic 219:114-131.

Hughes JH (1947) Forest Resources of British Guiana: Handbook of Natural Resources of British Guiana. Daily Chronicle Ltd., Georgetown.

Hsu R, Briscoe AD, Chang BS, Pierce NE (2001) Molecular evolution of a long wavelength-sensitive opsin in mimetic *Heliconius* butterflies (Lepidoptera: Nymphalidae). Biological Journal of the Linnean Society 72:435-449.

Ida TY, Kudo G (2003) Floral color change in *Weigela middendorffiana* (Caprifoliaceae): reduction of geitonogamous pollination by bumble bees. American Journal of Botany. 90:1751-1757.

Ida TY, Kudo G (2010) Modification of bumblebee behavior by floral color change and implications for pollen transfer in *Weigela middendorffiana*. Evolutionary Ecology 24:671-684.

IBM Corp. (2015) IBM SPSS Statistics for Windows, Version 23.0. Armonk, IBM Corp., New York

Johnson SD, Steiner KE (2000) Generalization versus specialization in plant pollination systems. Trends in Ecology and Evolution 15:140-143.

Kampny CM (1995) Pollination and flower diversity in Scrophulariaceae. The Botanical Review 61:350-366.

Keasar T, Pollak G, Arnon R, Cohen D, Shmida A (2006) Honesty of signaling and pollinator attraction: the case of flag-like bracts. Israel Journal of Plant Sciences 54: 119-128.

Kudo G, Hiroshi IS, Hirabayashi Y, Ida TY (2007) A test of the effect of floral color change on pollination effectiveness using artificial inflorescences visited by bumblebees. Oecologia 154: 119-128.

Lack AJ, Diaz A (1991) The pollination of *Arum maculatum L*: a historical review and new observations. Watsonia 18:333-342.

Maharaj G (2010) An Introduction to Butterflies of the Iwokrama Forest and Communities of the North Rupununi District. Darwin Initiative, University of Warwick, Warwick.

Marshall SA (2008) Five Hundred Insects: A Visual Reference. Firefly Books, Buffalo, New York.

Menzel R (1967) Untersuchungen zum Erlernen von Spektralfarben durch die Honigbiene (*Apis mellifica*). Zeit- schrift fur vergleichende Physiologie 56:22-62.

Menzel R (1985) Learning in honey bees in an ecological and behavioral context in B. Holldobler and M.

Lindauer, editors. Experimental behavioral ecology and sociobiology. Sinauer, Sunderland, Massachusetts, USA, pp. 55-74.

Lunau K, Maier EJ. (1995) Innate colour preferences of flower visitors. Journal of Comparative Physiology 177:1-9.

Milne L, Milne M (1980) National Audubon Society Field Guide to North American Insects and Spiders. Alfred A. Knopf, New York.

Munir AA (1996) A taxonomic review of *Lantana camara* L. and *L. montevidensis* (Spreng.) Briq. (Verbenaceae) in Australia. Journal of the Adelaide Botanic Garden 17:1-27.

Myiant A (1994) Common Weeds of Guyana. National Agricultural Research Institute, Guyana.

Nilsson LA (1992) Orchid pollination biology. Trends in Ecology and Evolution 7:255-259.

Nuttman CV, Semida FM, Zalat S, Willmer PG (2005) Visual cues and foraging choices: bee visits to floral colour phases in *Alkanna orientalis* (Boraginaceae). Biological Journal of the Linnean Society 87:427-435.

Oberrath R, Böhning-Gaese K (1999). Floral color change and the attraction of insect pollinators in lungwort (*Pulmonaria collina*). Oecologia 121:383-391.

Opler P (1992) A Field Guide to Eastern Butterflies. Houghton Mifflin Company, New York.

Ostler WK, Harper KT (1978) Floral ecology in relation to plant species diversity in the Wasatch Mountains of Utah and Idaho. Ecology 59:848-861.

Pereira AC, da Silva JB, Goldenberg R, Melo GA, Varassin IG (2011) Flower color change accelerated by bee pollination in *Tibouchina* (Melastomataceae). Flora-Morphology, Distribution, Functional Ecology of Plants. 206:491-497.

Proctor M, Yeo P, Lack A (1996) The Natural History of Pollination. Harper Collins, London.

Pyle MR (1981) National Audubon Society Field Guide to North American Butterflies. Alfred Knopf, New York.

R Development Core Team (2015) R: A language and environment for statistical computing. R Foundation for Statistical Computing, Vienna, Austria.

Raguso RA (2004) Flowers as sensory billboards: progress towards an integrated understanding of floral advertisement. Current opinion in plant biology 7:434-440.

Restall RL, Rodner C, Lentino R (2007a) Birds of northern South America: An Identification Guide, Volume 1: Species Accounts. Yale University Press, New Haven and London.

Restall RL, Rodner C, Lentino R (2007b) Birds of northern South America: An Identification Guide, Volume 2: Plates and Maps. Yale University Press, New Haven and London.

Schaefer HM, Schaefer V, Levey DJ (2004) How plant–animal interactions signal new insights in communication. Trends in Ecology & Evolution 19:577-584.

Seymour RS, Matthews PG (2006) The role of thermogenesis in the pollination biology of the Amazon waterlily *Victoria amazonica*. Annals of Botany 98: 1129-1135.

Sharma GP, Raghubanshi AS, Singh JS (2005) *Lantana* invasion: an overview. Weed Biology and Management 5:157-165.

Simpson BB, Neff JL (1981) Floral rewards: alternatives to pollen and nectar. Annals of the Missouri Botanical Garden 68:301-322.

Sison-Mangus MP, Bernard GD, Lampel J, Briscoe AD (2006) Beauty in the eye of the beholder: the two blue opsins of lycaenid butterflies and the opsin gene-driven evolution of sexually dimorphic eyes. Journal of Experimental Biology 209:3079-3090.

Smithe FB (1975) Naturalist's Color Guide. The American Museum of Natural History, New York.

von Arx M, Goyret J, Davidowitz G, Raguso RA (2012) Floral humidity as a reliable sensory cue for profitability assessment by nectar-foraging hawkmoths. Proceedings of the National Academy of Sciences 109:9471-9476.

Stavenga DG, Arikawa K (2006) Evolution of color and vision of butterflies. Arthropod structure & development 35(4):307-18.

Suzuki MF, Ohashi K (2014) How does a floral colour-changing species differ from its non-colour-changing congener? – a comparison of trait combinations and their effects on pollination. Functional ecology 28:549-560.

Waser NM, Chittka L, Price M, Williams N, & Ollerton J (1996). Generalization in Pollination Systems, and Why it Matters. Ecology, 77:1043-1060. doi:1. Retrieved from http://www.jstor.org/stable/2265575

Waser NM, Price MV (1981) Pollinator choice and stabilizing selection for flower color in *Delphinium nelsonii*. Evolution 35:376-390.

Waser NM, Price MV (1985) The effect of nectar guides on pollinator preference. Experimental studies with a montane herb. Oecologia 67:121-126.

Weiss MR (1991) Floral colour changes as cues for pollinators. Nature 354: 227-229.

Weiss MR (1995) Floral color change: a widespread functional convergence. American Journal of Botany 82:167-185.

Weiss MR (1997) Innate color preferences and flexible colour learning in the pipevine swallowtail. Animal Behaviour 53:1043-1052.

Weiss MR, Lamont BB (1997) Floral color change and insect pollination: a dynamic relationship. Israel Journal of Plant Sciences 45:185-199.

Wickler W (1968) Mimicry in Plants and Animals (translated from the German by R.D. Martin). World University Library. McGraw Hill, New York.

Willmer P, Stanley DA, Steijven K, Matthews IM, Nuttman CV (2009) Bidirectional flower color and shape changes allow a second opportunity for pollination. Current Biology 19: 919-923.

Wolff D. (2006) Nectar sugar composition and volumes of 47 species of Gentianales from a southern Ecuadorian montane forest. Annals of Botany 97:767-777.

Wright GA, Schiestl FP (2009) The evolution of floral scent: the influence of olfactory learning by insect pollinators on the honest signalling of floral rewards. Functional Ecology 23:841-851.

Yoshida K, Miki N, Momonoi K, Kawachi M, Katou K, Okazaki Y, ... Kondo T (2009) Synchrony between flower opening and petal-color change from red to blue in morning glory, *Ipomoea tricolor* cv. Heavenly Blue. Proceedings of the Japan Academy. Series B, Physical and biological sciences 85:187-197.

Yuan F, Bernard GD, Le J, Briscoe AD. (2010) Contrasting modes of evolution of the visual pigments in Heliconius butterflies. Molecular biology and evolution 27:2392-2405.

Zaccardi G, Kelber A, Sison-Mangus MP, Briscoe AD. (2006) Color discrimination in the red range with only one long-wavelength sensitive opsin. Journal of Experimental Biology 209:1944-1955.

19

TEMPORARY INCIDENCE OF POLYGAMOMONOECY IN THE BIODIESEL PLANT *JATROPHA CURCAS* L. IS ASSOCIATED WITH STYLE POLYMORPHISM PROMOTING AUTOMATIC SELF-POLLINATION

Y. Vaknin*[1], Y. Samocha[1,2], D. Eisikowitch[2]

[1]*Department of Agronomy and Natural Resources, Institute of Plant Sciences, Agricultural Research Organization, the Volcani Center, P.O. Box 6, Bet Dagan 50250, Israel*

[2]*Department of Plant Sciences, George S. Wise Faculty of Life Sciences, Tel-Aviv University, Tel-Aviv, Israel*

Abstract—Polygamomonoecy – the occurrence of perfect flowers alongside staminate and pistillate ones – in the monoecious biodiesel plant *Jatropha curcas* L. was reported as rare and often dismissed as having no specific functionality. Here we report that the incidence of perfect flowers is almost exclusively limited to a short period at the beginning of the blooming season and is directly associated with stigmas bending over fertile anthers, as early as in the mature bud stage. All perfect flowers have bent stigmas, while this trait is completely absent from female flowers produced later in the blooming season. In light of this species' self-compatibility, we suggest that the occurrence of perfect flowers with bent stigmas promotes automatic self-pollination, thus enhances reproductive success when conditions are unfavourable for pollination.

Keywords: Automatic self-pollination, hermaphrodite flower, Jatropha curcas, perfect flower, polygamomonoecy, sexual expression

INTRODUCTION

Jatropha curcas L., a native to tropical America, is a perennial shrub in the Euphorbiaceae. It was once used as a live fence to contain or exclude farm animals and as an oil source for soap production and lamp fuel (Openshaw 2000). Now it is cultivated as an oil crop for biodiesel production (de Jongh 2006; Divakara et al. 2010).

Jatropha curcas blooms in two major periods: the first in late spring to mid summer and the second in autumn (Wu et al. 2011; Wang & Ding 2012; Vaknin, personal information). It is monoecious with unisexual flowers, at various staminate (male) to pistillate (female) ratios, ranging from 4:1 to 29:1 (Raju & Ezradanam 2002; Bhattacharya et al. 2005; Chang-wei et al. 2007; Tewari 2007). The ratio decreases as the age of the plant increases (Prakash et al. 2007). Staminate flowers are green, plate shaped with ten diadelphous stamens, arranged in two tiers of five each (Fig. 1a). Pistillate flowers are also green, but are relatively larger with three bifid stigmas and a three-carpel ovary, which matures into a three-seeded capsule (Raju & Ezradanam; 2002; Fig. 1b). Both staminate and pistillate flowers have nectar glands in their base and are insect pollinated (Raju & Ezradanam 2002; Bhattacharya et al. 2005). The plant is reported to be self-compatible, possibly with a small rate (12%) of apomixis (Chang-wei et al. 2007). Previously, we described the pollination biology and breeding system of *J. curcas* (Samocha et al. 2014; Samra et al. 2014). Perfect (hermaphrodite) flowers (Fig. 1e) were reported by Dehgan & Webster (1979) as a rare occurrence, and since then no detailed analysis of these flowers has been provided other than by Abdelgadir et al. (2012), reporting on pollen from perfect flowers having lower viability, lower germination rate and shorter pollen tubes, compared to staminate flowers. However, our preliminary observations, during 2007 to 2009, revealed that perfect flowers were almost exclusively limited to a short period of 2-4 weeks, in late spring, at the beginning of the first blooming season. In 2010 we undertook more thorough observations of phenology and sexuality.

Our aim was to elucidate the significance of temporal polygamomonoecy in *J. curcas* by characterizing the occurrence of perfect flowers in association with seasonal changes in stigmatic traits.

MATERIALS AND METHODS

The studied plants were a mix of 4-year-old plants of wild germplasm originating from Brazil, Suriname, Niger, South Africa, India and China, none of which were ever bred or selected for a desirable trait. The plants were grown at the ARO research institute in Bet-Dagan, Israel (31°59' N, 34°49'E) at a density of 2 × 2 m. Local climate conditions are classified as Mediterranean, with hot dry summers and mild wet winters with an average of 524 mm rainfall per year (Israel Meteorological Service). Irrigation and fertilization

*Corresponding author; email: iftachv@volcani.agri.gov.il

FIGURE 1. *Jatropha curcas* flower types: Regular staminate (a) and pistillate (b) flowers and the same flowers with the sepals and petals removed (c and d, respectively). Early-blooming perfect flower, when fully open (e), at the mature bud stage (f) and with the sepals and petals removed (g). Early blooming pistillate flower with the sepals and petals removed (h). Arrows point at vestigial anthers.

were applied via drip irrigation throughout the summer and early autumn (30 cubic meters per day per hectare, with NPK fertilizer (20-10-20 0.1 kg per 10 litres).

Bloom phenology and floral traits

In early April 2010, the plants in the study site were surveyed daily. Plants with early developing inflorescences ($N = 50$) were marked and counted and one inflorescence was removed from each plant, for further analyses. Summer blooming, staminate and pistillate flowers, were similarly sampled. All inflorescences were analyzed for number of open staminate, pistillate and perfect flowers, number of flower buds, number of normal and/or aborted stamens for each flower type ($N = 10$), and number of bent stigmas for pistillate and perfect flowers (Fig. 2) ($N = 20$). Corolla width of all flower types was measured using digital callipers having an accuracy of $\pm$ 0.1 mm ($N = 10$). Pollen viability in all early-blooming-flower types as well as summer-blooming staminate flowers was estimated using the DAB (3.3' Diaminobenzidine) staining technique (Dafni et al. 2005) ($N = 10$). Stigmatic pollen germination was estimated using Aniline Blue Epifluorescence (Martin 1959) for early blooming pistillate and perfect flowers and for summer-blooming pistillate ones ($N = 10$). Crosses were made by hand, and total number of germinating pollen grains was scored. Stigmatic pollen germination was recorded with an Olympus IX 81 fully automated laser scanning confocal microscope (Olympus, Tokyo, Japan).

Fruit and seed traits

In early April 2010, a sample of 20-30 early-blooming pistillate and perfect flowers was marked. In late June 2010, all mature fruits, developed from these flowers, were harvested. Percentage fruit set and seed set were calculated. Fruit set is the proportion of carpels with at least one seed. Seed set is reported based on those fruits that set 1-3 seeds, i.e., given that the fruit as a whole set. Twenty seeds from each flower type were weighed and following the removal of seed coats, the kernels were weighted. Pistillate flowers ($N = 20$) were also marked during the summer bloom (June) and the developing fruits and seeds were similarly analyzed.

Statistical analyses

Statistical analyses were preformed with StatView® 5.0 (SAS Institute Inc. Cary, NC, USA) according to Sokal and Rohlf (1995). One-way analysis of variance (ANOVA) was used in order to compare how flower types differed in flower size, number of normal and aborted stamens, percentage of viable pollen, pollen germination on the stigma, seeds per fruit, and seed and kernel weight. Tukey-Kramer HSD was performed in order to compare the means two samples at a time. In order to normalize the data, arcsine square-root and square-root transformations were applied on percentages and counts, respectively. Differences are reported as significant when $P < 0.05$.

FIGURE 2. The stigma of a perfect flower bending over a fertile anther. Scale bar is 1 mm.

TABLE 1. Comparisons of various flower types during the early-bloom period (perfect, pistillate, and staminate flowers) and during the summer bloom period (pistillate and staminate flowers).

Bloom period	Early-bloom			Summer	
Flower type	Perfect	Pistillate	Staminate	Pistillate	Staminate
Corolla width (mm)	[a] 10.3 ± 0.5 a	8.9 ± 0.4 b	8.2 ± 0.4 bc	8.6 ± 0.1 bc	7.5 ± 0.2 c
Normal stamens (no.)	6.4 ± 1.1 b	0.0 c	9.3 ± 0.3 a	0.0 c	9.8 ± 0.1 a
Aborted stamens (no.)	2.4 ± 1.1 b	9.0 ± 0.3 a	0.1 ± 0.1 c	0.0 c	0.0 c
Bent stigmas (%)	100.0	36.0	-	0.0	-
% of seed-bearing flowers	38.7	61.3	0.0	100.0	-
% of open flowers	16.7	14.9	68.4	25.5	74.5
Staminate/pistillate ratio	5988.0	3685.0	-	26.0	-
Pollen viability (%)	94.4 ± 2.2 a	-	81.8 ± 4.4 b	-	95.9 ± 1.8 a
Stigmatic pollen germination (no.)	8.4 ± 3.6 a	4.2 ± 1.8 a	-	4.6 ± 1.0 a	-
Fruit set (%)	46.4	64.3	-	90.0	-
Seed set (%) in set fruits	33.3	57.1	-	90.2	-
Seeds per fruit in set fruits	2.20 ± 0.30 a	2.70 ± 0.10 a	-	2.70 ± 0.10 a	-
Seed weight (gr)	0.66 ± 0.01 a	0.71 ± 0.02 a	-	0.59 ± 0.02 b	-
Kernel weight (gr)	0.41 ± 0.01 a	0.43 ± 0.01 a	-	0.36 ± 0.02 b	-

[a] Data are shown as means ± SE. Post hoc comparisons between flower types were carried out for the entire year, early bloom and summer bloom. Means marked with the same letter were not significantly different.

RESULTS AND DISCUSSION

Bloom phenology and floral traits

During winter and early spring, *J. curcas* did not bloom. A relatively small portion of the plants (~10%) started blooming as early as April. From then through May was denoted as the early bloom period. During this period, the open flowers (N = 516) were of three types: perfect, pistillate and staminate (Fig. 1e, h and a, respectively) accounting for 17, 15 and 68% of the open flowers, respectively (Table 1). An additional 6023 flower buds were counted and sorted as 98.2% staminate, 1.5% pistillate, and 0.3% perfect flower buds. All open flower types developed stamens: normal fertile stamens, aborted non-fertile vestigial stamens or a combination of both (Table 1). The vestigial anthers were smaller than fertile ones and did not produce viable pollen. As the summer bloom period got well underway in June, with data gathered until July, the majority of the new inflorescences were composed only of pistillate and staminate flowers (Fig. 1b and a, respectively; N = 330 open flowers), and no vestigial stamens were observed. An additional 3382 flower buds were counted and sorted as 96% staminate and 4% pistillate flower buds.

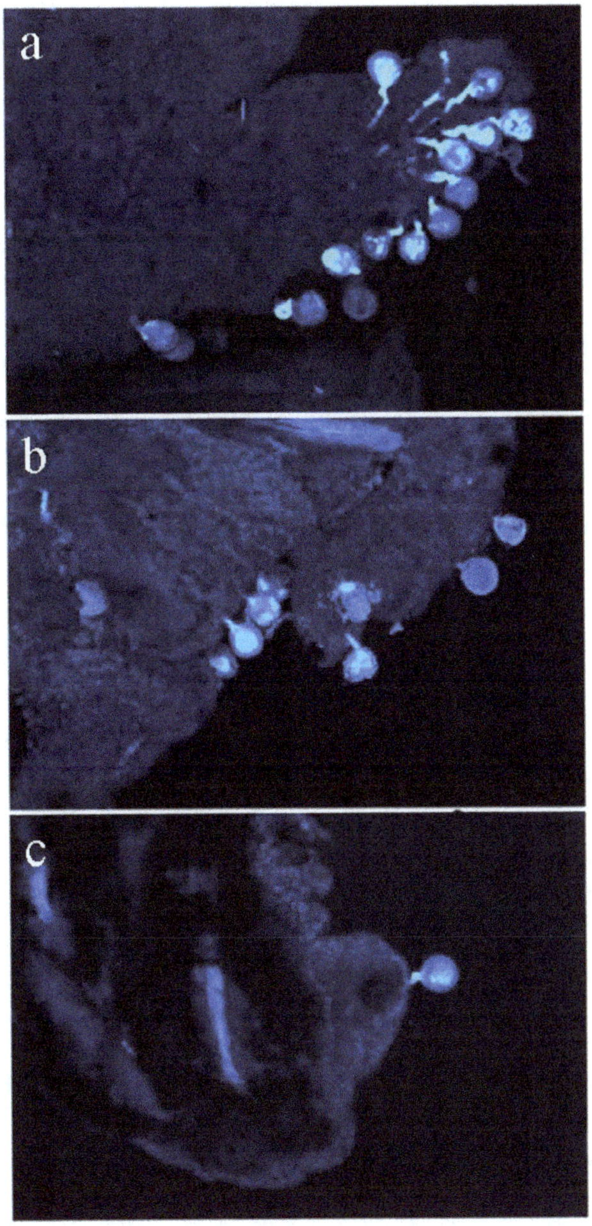

FIGURE 3. Pollen germination on the stigma of early blooming perfect flower (a), pistillate flower (b), and perfect flower at the bud stage (c). Scale bar is 100 µm.

Recent studies on *J. curcas* flowers revealed that pistillate flowers originate from bisexual primordia, with pistils and stamens growing simultaneously and the stamens stopping their development, either very early, leaving no trace in the mature flowers, or much later, thus forming antherodes with aborted pollen (Wu et al. 2011; Wang & Ding 2012).

Perfect flowers were significantly larger than other flower types (Table 1). According to Horovitz & Galil (1972), very often flowers of hermaphroditic plants have much larger petals than those of male-sterile plants. Early-blooming and summer-blooming pistillate flowers, were relatively similar in size, and were significantly larger than summer-blooming staminate flowers (Table 1). Similarly, Raju & Ezradanam (2002) described pistillate flowers as relatively larger than staminate ones.

Pollen viability of all pollen-producing flowers was relatively high (> 80%) with slightly lower values recorded for early-blooming staminate flowers (Table 1). Stigmatic pollen germination for perfect flowers (Fig. 3a) was higher than for early-blooming (Fig. 3b) and summer-blooming staminate flowers. However, the differences were not significant. Stigmatic pollen germination was also recorded on perfect flower buds (Fig. 3c); in other words, the flowers were cleistogamous. Our results were in contrast to those of Abdelgadir et al. (2012) who reported that pollen from perfect flowers has lower viability, lower germination rate and shorter pollen tubes, compared to staminate flowers. Possibly, the rare perfect flowers reported by Abdelgadir et al. (2012) were distinctly different from the ones we have encountered due to differences in environmental conditions.

Perfect flowers in *J. curcas* have been considered rare and abnormal with no specific function (Luo et al. 2007). Here we report on a unique phenomenon of early-blooming perfect flowers with stigmas bent over fertile stamens (Fig. 2) and pollen grains were deposited on the receptive surface (Fig 1e, f, and g) as early as in the mature bud stage (Fig 1f). Practically 100% of all early-blooming perfect flowers and about a third of the early-blooming pistillate flowers presented bent stigmas, whereas it was completely absent from summer-blooming pistillate flowers (Fig 1b and d). Automatic self-pollination is a known strategy adapted by plants to provide reproductive assurances when pollinators are scarce as in some orchids (Catling 1990) and the cleistogamous flowers of *Lamium amplexicaule* L. (Galil & Eisikowitch 1964; Lord, 1979).

An exogenous application of the plant growth regulator 6-benzyladenine (BA), a synthetic compound with cytokinin activity, on *J. curcas* (Pan & Zu 2011) and male grapevines (Negi & Olmo 1966) resulted in the induction of perfect flowers. Cytokinins are important factors in the regulation of flower bud initiation and development, and floral sex expression in certain fruit trees (Pan & Zu 2011). Therefore, the induction of perfect flowers in *J. curcas* could be associated with elevated cytokinins as the plant "breaks" its cold-induced dormancy in late spring.

Fruit and seed traits

Fruit set and seed set for summer-blooming pistillate flowers were ~50% higher than for early-blooming pistillate flowers and ~100% higher than for perfect flowers (Table 1). Pistillate flowers often produce more seeds than perfect ones, most likely due to the reallocation of energy away from pollen production to seed production (Shykoff et al. 2003). The reduction in fruit set and seed set in perfect flowers could not be explained by reduced pollen viability or stigmatic pollen germination (Table 1), but rather by factors that remain to be elucidated such as reduced embryo fertilization or increased level of aborted embryos or other forms of inbreeding depression.

Perfect flowers produced significantly lighter seeds than early blooming pistillate flowers. These results are in accordance with other plant species characterized by lighter seeds originating from perfect flowers (Shykoff et al. 2003). In *J. curcas*, smaller seeds will eventually develop into lower quality seedlings (Zaidman et al. 2010), a form of inbreeding depression (Charlesworth & Charlesworth 1979). This is not surprising as we have already shown that selfing in *J. curcas* will eventually lead to the production of smaller seeds (Samoch et al. 2014). Early-blooming perfect and pistillate flowers produced significantly heavier seeds and seed kernels in comparison with summer-blooming pistillate flowers.

As in *J. curcas*, most gynodioecious species are self-compatible. The perfect flowers more often self-fertilize (Meagher 2007; Ehlers & Schierup 2008), while the female flowers usually produce higher quality seeds resulting from being forced to outcross thereby limiting inbreeding depression in their offspring (e.g. Thompson & Tarayre 2000; Chang 2007). Our results, however, did not support that assumption, presumably since geitonogamous selfing levels were relatively high during the summer bloom period.

Summary

Perfect flowers with varying levels of male sterility and bent stigmas over fertile stamens occurred during a relatively short period in late spring when pollination services by the local fauna was scarce and opportunities for cross-fertilization were rare. The placement of a stigma and anthers in the same flower greatly enhances the chances of successful pollination and fertilization in one single floral visit, and the bending of the stigma over the anthers seems to act as a fail-safe mechanism to ensure reproductive success in the absence of pollinators. We suggest that the ability to change sex expression may be used in future breeding programs to enhance yield in this important biodiesel feedstock.

ACKNOWLEDGEMENTS

We thank Mrs. Yoaad Shbeita and Mr. Leonid Murakhovsky for their assistance in the field, Mr. Eduard Belausov for his assistance with the confocal microscopy, and Dr. Ilana Gelranter from Tel Aviv University for her statistical advice.

REFERENCES

Abdelgadir HA, Johnson SD, Van Staden J (2012) Pollen viability, pollen germination and pollen tube growth in the biofuel seed crop *Jatropha curcas* (Euphorbiaceae). South African Journal of Botany 79:132-139.

Ainsworth C (2000) Boys and girls come out to play: the molecular biology of dioecious plants. Annals of Botany 86:211-221.

Bhattacharya A, Datta K, Datta SK (2005) Floral biology, floral resource constraints and pollination limitation in *Jatropha curcas* L. Pakistan Journal of Biological Sciences 8:456-60.

Catling PM (1990) Auto-pollination in the Orchidaceae. In: Arditti J (ed) Orchid Biology: Reviews and Perspectives. Timber Press, Portland, pp 121-158.

Chang SM (2007) Gender-specific inbreeding depression in a gynodioecious plant, *Geranium maculatum* (Geraniaceae). American Journal of Botany 94:1193-1204.

Chang-wei L, Kun L, You C, Yong-yu S (2007) Floral display and breeding system of *Jatropha curcas* L. Forestry Studies in China 2007:114-9.

Charlesworth D, Charlesworth B (1979) The evolutionary genetics of sexual systems in flowering plants. Proceedings of the Royal Society of London. Series B. Biological Sciences 205:513-530.

Dafni A, Neppi M, Pacini E (2005) Pollen and stigma biology. In: Dafni A, Kevan PG, Husband B (eds) Practical pollination biology. Enviroquest, Cambridge, Canada, pp 83-147.

de Jongh J (2006) General data on *Jatropha*. In: Rijssenbeek W (ed) Handbook on *Jatropha curcas*, FACT foundation. Eindhoven, pp 4-9. URL: http://www.jatropha-alliance.org/fileadmin/documents/knowledgepool/Rijssenbeek_Jatropha_Handbook.pdf (Accessed March 2014).

Dehgan B, Webster GL (1979) Morphology and Intrageneric relationships of the genus Jatropha (Euphorbiaceae). vol. 74. University of California Publications in Botany.

Divakara BN, Upadhyaya HD, Wani SP, Laxmipathi Gowda CL (2010) Biology and genetic improvement of *Jatropha curcas* L.: A review. Applied Energy 87:732-742.

Ehlers BK, Schierup MH (2008) When gametophytic self-incompatibility meets gynodioecy. Genetics Research (Cambridge) 90:27-35.

Galil J, Eisikowitch D (1964) Cleistogamy in flowers. *Teva vaAretz* 6:2-12. (in Hebrew).

Horovitz A, Galil J (1972) Gynodioecism in East Mediterranean *Hirschfeldia incana* Cruciferae. Botanical Gazette 133:127-131.

Lord EM (1979) Physiological controls on the production of cleistogamous and chasmogamous flowers in *Lamium amplexicaule* L. Annals of Botany 44:757-766.

Luo C-W, Li K, Chen Y, Sun YY (2007) Floral display and breeding system of *Jatropha curcas* L. Forestry Studies in China 9:114-119.

Martin FW (1959) Staining and observing pollen tubes in the style by means of fluorescence. Stain Technology 34:125-128.

Meagher TR (2007) Linking the evolution of gender variation to floral development. Annals of Botany 100:165-176.

Negi S, Olmo H (1966) Sex conversion in a male *Vitis vinifera* L. by a kinin. Science 152:1624-1625.

Openshaw K (2000) A review of *Jatropha curcas*: an oil plant of unfulfilled promise. Biomass & Bioenergy 19:1-15.

Pan BZ, Xu ZF (2011) Benzyladenine treatment significantly increases the seed yield of the biofuel plant *Jatropha curcas*. Journal of Plant Growth Regulation 30:166-174.

Prakash, A.R., Patolia, J.S., Chikara, J. & Boricha, G.N. (2007) Floral biology and flowering behaviour of Jatropha curcas *Expert seminar on Jatropha curcas L. Agronomy and genetics*, pp. 26-28. URL: http://www.jatropha-alliance.org/fileadmin/documents/knowledgepool/PrakashPatolia_Floral_Biology_Jatropha_curcas.pdf (Accessed March 2014).

Raju AJS, Ezradanam V (2002) Pollination ecology and fruiting behaviour in a monoecious species, *Jatropha curcas* L. (Euphorbiaceae). Current Science India 83:1395-1998.

Samra S, Samocha Y, Eisikowitch D, Vaknin Y (in press) Can ants equal honeybees as effective pollinators of the energy crop *Jatropha curcas* L. under Mediterranean conditions? GCB Bioenergy 6:756-767.

Samocha Y, Eisikowitch D, Vaknin Y (in press) Pollen source effects on fruit and seed traits of *Jatropha curcas* L. - a renewable biodiesel feedstock. Bioenergy Research 7:1270-1279.

Shykoff JA, Kolokotronis S-O, Collin CL, Lopez-Villavicencio M (2003) Effects of male sterility on reproductive traits in gynodioecious plants: a meta-analysis. Oecologia 135:1-9.

Sokal RR, Rohlf FJ (1995) Biometry: The principles and practice of statistics in biological research. W.H. Freeman & Co, New York.

Tiwari DN (2007) *Jatropha and biodiesel*. 1st edn. Ocean Books Ltd, New Delhi.

Thompson JD, Tarayre M (2000) Exploring the genetic basis and proximate causes of female fertility advantage in gynodioecious *Thymus vulgaris*. Evolution 54:1510-1520.

Wang XR, Ding GJ (2012) Reproductive biology characteristic of *Jatropha curcas* (Euphorbiaceae). Revista de Biologia Tropical 60:1525-1533.

Wu J, Liu YA, Tang L, Zhang FL, Chen F (2011) A study on structural features in early flower development of *Jatropha curcas* L. and the classification of its inflorescences. African Journal of Agricultural Research 6:275-284.

Zaidman B, Ghanim M, Vaknin Y (2010) Effect of seed weight on seed vigor and early seedling growth of *Jatropha curcas*, a biodiesel plant. Seed Science and Technology 38:758-767.

20

THE FLORAL BAT LURE DIMETHYL DISULPHIDE DOES NOT ATTRACT THE PALAEOTROPICAL DAWN BAT

Gerald G. Carter* and Alyssa B. Stewart

Department of Biology, University of Maryland, College Park, MD 20742, USA

Abstract—In the neotropics, dimethyl disulphide (DMDS) is innately attractive to flower-visiting bats, and acts as a powerful bat lure in the scent bouquets of many bat-pollinated flowers. At first, DMDS appeared to be part of a general bat pollination syndrome. However, DMDS is absent in many bat-pollinated flowers of West Africa, and it is unclear whether palaeotropical flower-visiting bats are also attracted to it. Furthermore, DMDS was previously observed in neotropical, but not palaeotropical, populations of *Ceiba pentandra* (Malvaceae, Bombacoideae). We tested for an attraction to DMDS in the most common flower-visiting bat in Thailand, the dawn bat *Eonycteris spelaea*. We gave bats choices of *Ceiba pentandra* flowers, where one random flower was scented with DMDS. Rather than preferring the DMDS-treated flower, 21 of 22 bats chose an untreated flower, showing no attraction to DMDS. Alongside past evidence, this result suggests that the role of DMDS in bat pollination syndromes may result from an adaptive convergence that is limited to the neotropics. This hypothesis could be tested through comparative studies of (1) attraction across bats, (2) floral DMDS presence across bat-pollinated plants in Asia, and (3) floral DMDS measures across New and Old World populations of *Ceiba pentandra*.

Keywords: bat pollination, Ceiba pentandra, dimethyl disulphide, Eonycteris spelaea, floral scent, olfaction

INTRODUCTION

The odourous sulphur compound, dimethyl disulphide (DMDS), is a powerful attractant of neotropical nectarivorous bats (*Glossaphaga* sp., von Helversen et al. 2000; *Anoura* sp., N. Muchhala, pers. comm.). Flower-visiting bats use DMDS as a beacon to find flower locations and to locate the nectar reward within the flower. Captive-born long-tongued bats *Glossaphaga soricina* with no previous experience of the scent will hover and stick their heads into tubes of DMDS, which demonstrates that their attraction to the scent is innate (von Helversen et al. 2000; Carter et al. 2009).

DMDS appears to have evolved independently in the scent bouquets of many neotropical bat-pollinated flowers (Knudsen et al. 1995; Bestmann et al. 1997; von Helversen et al. 2000), but its evolutionary origin as a bat lure is unclear. Given that DMDS is attractive to the males of some mammalian species (e.g. Singer et al. 1976), one hypothesis is that its presence in flowers represents sensory exploitation of olfactory communication in bats or other mammals (von Helversen et al. 2000). DMDS may also have originally attracted flies (Jürgens et al. 2006; Shuttleworth & Johnson 2010; Stensmyr et al. 2002), and then later became innately attractive to bats. Based on early reports of floral odours in both the palaeotropics and neotropics, von Helversen *et al.* (2000) suggested that the peculiar and unpleasant odour of DMDS and other sulphur compounds might be common to bat-pollinated flowers in general. However, later observations suggest that DMDS may only attract bats in the Neotropics. In West Africa, Petterson et al. (2004) sampled the scent composition of seven bat-pollinated flowers and found that only the baobab *Adansonia digitata* (Malvaceae, Bombacoideae), possessed more than a trace amount of DMDS.

Perhaps most remarkable is the case of the silk-cotton tree *Ceiba pentandra* (Malvaceae, Bombacoideae), which is pollinated by bats on both hemispheres. In Central America, *C. pentandra* flowers possess a substantial amount of DMDS (von Helversen et al. 2000), but in West Africa, they contain no detectable DMDS (Petterson et al. 2004). In India, the short-nosed fruit bat (*Cynopterus sphinx*) sometimes feeds on nectar but it does not prefer the scent of DMDS over other odours or scentless controls (Elangovan et al. 2006). These two pieces of evidence suggest that DMDS may only be a component of the pollination syndrome for bats in the neotropics but not the palaeotropics.

Nectarivory in bats arose independently in the neotropical Phyllostomidae (subfamily Glossophaginae) and the palaeotropical Pteropodidae, lineages that diverged about 56 mya (Fleming et al. 2009; Jones et al. 2005; Teeling et al. 2005). The evolutionary origins of DMDS as a bat lure could therefore be clarified further by testing if an innate attraction to DMDS occurs in palaeotropical bats. Here, we tested whether the dawn bat (*Eonycteris spelaea*), a common palaeotropical nectarivorous bat, is attracted to DMDS.

*Corresponding author: gcarter@umd.edu

MATERIALS AND METHODS

We conducted 22 preference tests where single *E. spelaea* bats chose from a row of three ($N = 2$) or four ($N = 20$) *Ceiba pentandra* flowers. One flower was randomly scented with DMDS (Sigma-Aldrich, Inc, USA), and the others acted as untreated controls. We caught the bats ($N = 22$) in mist nets near Hat Yai, Thailand from 15 to 21 December 2014, and fed them with sugar water. The next night, we placed each bat in a 30 × 36 × 50 cm plastic-grid test cage. We placed all bats at the same starting point in the test cage. On the opposite side of the test cage we created a linear array of flowers spaced apart 5 cm. Each of the six arrays we made contained flowers from the same *C. pentandra* tree ($N = 4$ trees). To treat flowers, we either soaked 200 uL DMDS directly into one flower (3 trials) or pipetted it in a 0.5 mL tube positioned just below the flower (19 trials). The 0.5 mL tube's opening was directly below the flower's opening, but the odour tube remained outside the cage. After each trial, we used a fan to clear odours from the array for at least 3 min. To score choices, we used an infrared spotlight and a Sony Nightshot Camera to see which flower the bat first probed with its snout. Trials lasted 1-30 min. Using two-way binomial tests, we compared the selection rate of *E. spelaea* to the expected rate under random chance (25.73 %).

We also used a two-way binomial test to compare the *E. spelaea* selection rate to expected rates based on choices by the neotropical bat *Glossophaga soricina*. To calculate this expected rate, we used unpublished data from a pilot test that was most similar in design to this test. At the Montreal Biodome (see Carter et al. 2009) 19 naïve captive-born *G. soricina* chose from arrays of four tubes, one randomly marked with the same amount of DMDS as used in the tests above. Under these similar conditions, the mean DMDS choice rate across *G. soricina* bats weighted by number of visits per bat ($N = 1$ to 33) was 50.46% (chance = 25 %, 95% C.I. = 38-63 %, Carter et al. 2009, unpublished data).

RESULTS AND DISCUSSION

Eonycteris spelaea chose the flower marked with dimethyl disulphide in only one of 22 trials (selection rate = 4.5 %), fewer times than expected based on either random chance ($P = 0.025$, Fig. 1) or the *G. soricina* preference ($P < 0.0001$). This result suggests that *E. spelaea*, unlike *G. soricina*, is not innately attracted to DMDS. Instead, we found evidence that the bats avoided the DMDS-treated flower, presumably because the DMDS masked the flower's natural odour. This adds to a growing body of knowledge on how bat pollination syndromes differ between the neotropics and palaeotropics (Pettersson et al. 2004; Fleming et al. 2009).

Given that *C. pentandra* flowers possess DMDS in Central America (von Helversen et al. 2000), but not West Africa (Petterson et al. 2004), one possibility is that floral DMDS is present only in neotropical populations. Evidence suggests that mutations in floral traits can lead to rapid evolutionary shifts in pollination syndromes (Rosas-Guerrero et al. 2014). Shuttleworth and Johnson (2010) showed that mutations leading to the presence and absence of DMDS alone have likely led to evolutionary transitions among insect pollinators. They studied four closely related species of *Eucomis* (Hyacinthaceae) that are pollinated by either wasps or carrion flies (which are attracted to DMDS). All four *Eucomis* species have flowers of similar colour, nectar, and morphology, but differ in scent compounds. In a field test, they experimentally manipulated floral scents and demonstrated that the presence of dimethyl disulphide and dimethyl trisulphide led to shifts from wasp to fly pollinators. Hence, evolutionary shifts between pollinators can occur based on DMDS alone, without changes in floral shape, colour or nectar. Similar co-evolutionary transitions may also have occurred for palaeotropical and neotropical flower-visiting bats.

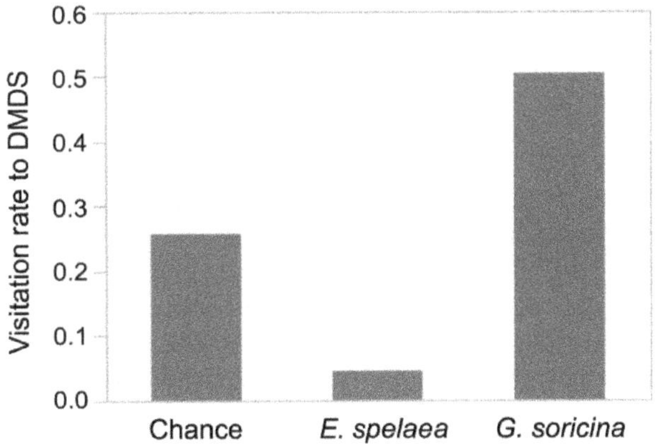

FIGURE 1. Observed and expected dimethyl disulphide selection rates. Rates of choices (probing with snout) to dimethyl disulphide odour cues are shown for the palaeotropical bat *Eonycteris spelaea*, alongside the expected levels based on chance and choices by the neotropical bat *Glossophaga soricina*.

Comparative studies would provide great insight into the coevolution of bat nectarivory and floral DMDS. First, to determine the extent of innate DMDS attraction in bats, it would be informative to test other captive-born or wild palaeotropical pteropodids as well as the neotropical Lonchophyllinae—a lineage which may have evolved their nectarivory independent of the Glossophaginae (Griffiths 1982; Datzmann et al. 2010; see also critical review by Dávalos et al. 2012). Second, to confirm a dichotomy in floral DMDS between neotropical and palaeotropical plant species, one could test for DMDS in other bat-pollinated flowers in Asia. Finally, the hypothesis of a neotropical versus palaeotropical dichotomy in floral DMDS within *C. pentandra* could be tested by mapping the occurrence of floral DMDS across the species range.

ACKNOWLEDGEMENTS

A. Stewart is supported by a National Science Foundation Graduate Research Fellowship. We thank D. Keawngam, W. Muangmanee, S. Srilopan, and M. Nowak for assisting with fieldwork, S. Bumrungsri for facilities and space. The University of Maryland Institutional Animal Care and Use Committee approved our work (Protocol # 615373). Two anonymous reviewers helped to improve the manuscript.

References

Bestmann HJ, Winkler L, von Helversen O (1997) Headspace analysis of volatile flower scent constituents of bat-pollinated plants. Phytochemistry 29:1359-1387.

Carter GG, Ratcliffe JM, Galef BG (2010) Flower bats (*Glossophaga soricina*) and fruit bats (*Carollia perspicillata*) rely on spatial cues over shapes and scents when relocating food. PLOS One 5:e10808.

Datzmann T, von Helversen O, Mayer F (2010) Evolution of nectarivory in phyllostomid bats (Phyllostomidae Gray, 1825, Chiroptera: Mammalia). BMC Evolutionary Biology 10:165.

Dávalos LM, Cirranello AL, Geisler JH, Simmons NB (2012) Understanding phylogenetic incongruence: lessons from phyllostomid bats. Biological Reviews 87, 991-1024.

Elangovan V, Priya EYS, Marimuthu G (2006) Olfactory discrimination ability in the short-nosed fruit bat *Cynopterus sphinx*. Acta Chiropterologica 8:247-253.

Fleming TH, Geiselman C, Kress WJ (2009) The evolution of bat pollination: a phylogenetic perspective. Annals of Botany 104:1017-1043.

Griffiths TA (1982) Systematics of the New World nectar-feeding bats (Mammalia, Phyllostomidae), based on the morphology of the hyoid and lingual regions. American Museum Novitates 2742:1-45.

Jones KE, Bininda-Emonds ORP, Gittleman JL (2005) Bats, clocks, and rocks: diversification patterns in Chiroptera. Evolution 59:2243-2255.

Jürgens A, Dötterl S, Meve U (2006) The chemical nature of fetid floral odours in stapeliads (Apocynaceae-Asclepiadoideae-Ceropegieae). New Phytologist 172:452-468.

Knudsen JT, Tollsten L (1995) Floral scent in bat-pollinated plants: a case of convergent evolution. Botanical Journal of the Linnean Society 119:45-57.

Pettersson S, Ervik F, Knudsen JT (2004) Floral scent of bat-pollinated species: West Africa vs. The New World. Botanical Journal of the Linnean Society 82:161-168.

Rosas-Guerrero V, Aguilar R, Martén-Rodríguez S, Ashworth L, Lopezaraiza-Mikel M, Bastida JM, Quesada M (2014). A quantitative review of pollination syndromes: do floral traits predict effective pollinators? Ecology Letters 17:388-400.

Singer AG, Agosta W, O'Connell R, Pfaffmann C, Bowen D, and Field F (1976). Dimethyl disulfide: an attractant pheromone in hamster vaginal secretion. Science 191:948-950.

Shuttleworth A, Johnson SD (2010) The missing stink: sulphur compounds can mediate a shift between fly and wasp pollination systems. Proceedings of The Royal Society B: Biological Sciences 277: 2811–2819.

Stensmyr MC, Urru I, Collu I, Celander M, Hansson BS, Angioy, AM (2002) Pollination: rotting smell of dead-horse arum florets. Nature 420:625-626.

Teeling EC, Springer MS, Madsen O, Bates P, O'Brien SJ, Murphy WJ (2005) A molecular phylogeny for bats illuminates biogeography and the fossil record. Science 307:580–584.

von Helversen O, Winkler L, Bestmann HJ (2000) Sulphur-containing 'perfumes' attract flower-visiting bats. Journal of Comparative Physiology A 186:143–153.

21

FLORAL LONGEVITY, NECTAR PRODUCTION, ANTHER DEHISCENCE, AND STIGMA RECEPTIVITY IN HASKAP (*LONICERA CAERULEA*)

S. Danae Frier[1,2], Christopher M. Somers[1] and Cory S. Sheffield[2,*]

[1]*Department of Biology, University of Regina, 3737 Wascana Parkway, Regina, SK S4S 0A2*
[2]*Royal Saskatchewan Museum, 2340 Albert Street, Regina, SK S4P 2V7*

Abstract—Haskap (*Lonicera caerulea* L.) is a temperate fruiting shrub grown commercially in northern regions of Europe, Asia, and North America. Haskap is self-incompatible and requires insect pollinators in order to set fruit; however, very little is currently known about its floral biology or pollinator specializations, particularly in North American cultivars. Here, we examine floral longevity, nectar dynamics, the timing of anther dehiscence, and stigma receptivity in flowers of greenhouse-grown 'Tundra', a Haskap cultivar developed and grown in Saskatchewan, Canada. Anthesis lasted 83.3 ± 25.9 hours (mean ± SD) in un-pollinated flowers; pollination caused early senescence within 34.3 ± 15.2 hours after pollination. Nectar was present from the onset of anthesis, and nectar volume peaked at 9-16 hours after opening. Nectar volume was maintained throughout anthesis and was not resorbed prior to abscission of the corolla from the ovary, and nectar removed during anthesis was replenished to the original volume. The stigma showed a reaction to hydrogen peroxide while still in the bud stage, suggesting it is receptive even before the flower opens. Early stigma receptivity, nectar production, and anther dehiscence maximize opportunities to be successfully pollinated, along with high floral longevity and pollination-triggered senescence. These results suggest that Haskap flowers utilize a generalist, rather than a specialized, pollination strategy. Observations that some flowers open in the evening or were already open in the morning suggest that nocturnal pollinators such as moths may be important, in addition to known diurnal pollinators.

Keywords: Haskap, Lonicera caerulea, floral traits, nectar, pollen, anthesis

INTRODUCTION

Haskap (*Lonicera caerulea* L.: Caprifoliaceae), also known commercially as blue honeysuckle or honeyberry, is an early flowering (April-May), temperate fruiting shrub native to northern parts of Europe, Asia and North America. Its tart blue berry-like fruit may have potential health benefits (Svarcova et al. 2007), and it is growing in popularity as a commercial crop, particularly in North America. Existing research on Haskap cultivation has focused largely on the shape, taste and harvestability of this fruit. However, despite Haskap being self-incompatible (Bors 2008) and requiring insect pollinators in order to set fruit (Bozek 2012), there has been little focus on its pollination biology. Studies of Haskap cultivars grown in Poland have found that it produces abundant pollen and nectar that is favoured by honey bees (*Apis mellifera*) and bumble bees (*Bombus* spp.) (Hymenoptera: Apidae), as well as a variety of solitary bees (Bożek & Wieniarska 2006; Bożek 2007). Similar pollinator guilds appear to visit North American cultivars, and field studies have suggested that bumble bees may be one of the most important pollinators for this cold-adapted crop (Frier et al. 2016). However, there is still little known about the full diversity of floral visitors of Haskap and their relative value to pollination, including the potential for nocturnal pollinators like moths, which are known to be important for other *Lonicera* species (Miyake & Yahara 1998). Additionally, although floral structure and fruit development is rather unique in Haskap (i.e. the fruit develops from the ovaries of both flowers in the inflorescence, which are enclosed by a cupule; Fig. 1), little is known about its specific floral traits or reproductive strategies.

Nectar and pollen are the most common pollinator rewards produced by flowers, and many pollinators, especially bees, rely entirely on these resources in one or all of their life stages (Proctor 1996; Armbruster 2012). Therefore, the nutritive composition, abundance, and availability of these rewards are often indicative of the pollinator guilds associated with the flowers (Fenster & Armbruster 2004). This may be especially true of nectar, which is produced specifically to attract pollinators (unlike pollen, where the reward function is secondary), and its production appears to adapt more quickly to different pollinator guilds than other floral traits (Ackermann & Weigend 2006). Many flowers are found to have distinct patterns of nectar production, nutritive composition, and resorption rates that may be correlated to the abundance, behaviour, and physiology of their associated pollinators (Galetto & Bernardello 1993, 1996; Galetto et al. 1997; Witt et al. 1999; Nepi et al. 2001; Fenster & Armbruster 2004; Wolff et al. 2006; Agostini et al. 2011; Kulloli et al. 2011; Amorim et al. 2013). This reflects the concept of pollination syndromes, whereby a suite of phenotypic floral traits are thought to correlate to the primary functional

*Corresponding author: cory.sheffield@sk.gc.ca

FIGURE 1. A typical Haskap inflorescence. The downward facing flowers are pale yellow and each has five petals, five stamens and one stigma; the ovaries of both flowers are enclosed by the bracteoles, which have fused into a cupule. The ovaries and bracteoles develop into the Haskap 'berry', actually a multiple fruit.

pollinator guild, and contrasts with the theory that most plants are in fact generalists that are visited by a wide variety of pollinator guilds (i.e., polyphily) (Waser et al. 1996; Ollerton & Watts 2000; Freitas & Sazima 2006; Johnson & Nicolson 2008; Petanidou et al. 2008; Ollerton et al. 2009). In either case, understanding the patterns of nectar production and other intra-floral aspects during anthesis may be especially important for agriculturally significant plants, where growers must decide whether to use managed pollinators, and what species will be most effective. Currently, the nectar dynamics of Haskap are entirely unknown; information on nectar production as well as other important aspects of anthesis will help refine and optimize the pollination strategy for this crop and ultimately help growers to maximize commercial yield.

The purpose of our study was to determine floral longevity, nectar dynamics, anther dehiscence, and stigma receptivity of Haskap flowers to learn more about the plant's reproductive strategy and its associated animal pollinators. Our specific objectives were to determine: (1) how long flowers are open, and whether floral longevity is affected by pollination; (2) when nectar is produced during anthesis, and whether total nectar production is affected by nectar removal; (3) when the anthers dehisce; and (4) how early in anthesis the stigma becomes receptive.

MATERIALS AND METHODS

Study site and plants

All experiments were performed during the month of February, 2015 at the University of Saskatchewan (Saskatoon, Saskatchewan), in a glass research greenhouse with daytime temperatures held at 15°C and nighttime temperatures at 10°C. The Haskap plants were grown under full spectrum lighting between 6 AM and 11 PM, with supplementation by natural light. We used a total of 8 Haskap bushes from the cultivar 'Tundra' for our experimental treatments, and pollen from unrelated plants originating from Japanese germplasm which is compatible with 'Tundra' was used as a pollen source (hereafter referred to as the pollinizer). Each potted plant was seven years old (maintained by the breeding program at the University of Saskatchewan), and kept in cold storage for a period of winter dormancy. Upon introduction to the greenhouse, the plants required approximately 10-14 days to begin flowering. Plants were watered every 2 days. As other researchers were using this space, a commercially purchased colony of *Bombus impatiens* Cresson was introduced to the greenhouse for pollination four days before the completion of our experiments. To prevent visitation in our experiments, flowers that remained to be sampled when the bees were present were covered with pollinator exclusion bags (Delnet® Pollination™ Bags) – this included 25/95 (26.3%) of the flowers used to assess anthesis length, 30/207 (14.4%) of the flowers used to assess nectar and pollen dynamics, and only 1/62 (1.6%) of the flowers used to assess the effect of nectar removal. Visual inspection of the data did not suggest that bagged flowers differed from un-bagged flowers and the data from both were pooled in our analysis.

Length of anthesis

Four experimental treatments were used to determine the duration of flowering in an inflorescence and the effects of pollination on anthesis length: (1) anthers removed, stigma not pollinated; (2) anthers left intact, stigma not pollinated; (3) anthers left intact, stigma hand-pollinated; and (4) anthers removed, stigma hand pollinated. In each case, the treatments were applied to both flowers in the inflorescence. Each potted plant received 3 replicates of each treatment, and the treatments were assigned in a randomized order as inflorescences opened. Anthers were removed immediately after the flowers opened, before they had dehisced, and hand pollination was performed in the morning, after the inflorescences had been open for at least 24 hours. Pollen was collected from a nearby pollinizer by pinching the anthers between fingertips and then dabbed directly on the stigmas of the experimental flowers. The inflorescences were observed once in the morning (9 AM), afternoon (2 PM) and evening (7 PM), and the beginning and end of anthesis were recorded, with the end marked by separation of the corolla of at least one of the flowers from the ovaries.

The duration (h) of anthesis for each treatment was compared using a Kruskal-Wallis test, followed by multiple comparisons using the Mann-Whitney U test, with p-values adjusted using the Bonferroni correction. In addition, we calculated the percentage of flowers (including flowers used in nectar and anther analysis) that were first observed to be open in the morning, versus the afternoon or evening. All statistical analysis was performed in R (R Core Team 2014).

Nectar, pollen, and stigma dynamics

In order to determine nectar and pollen dynamics, we measured nectar volume (pooled from both flowers) and recorded the number of dehisced anthers per inflorescence over the lifespan of the flowers. Sampling was done three times per day, at 9 AM, 2 PM, and 7 PM. Each inflorescence was only sampled once – once it had opened, it was randomly assigned a temporal period (9 AM, 2 PM, or 7 PM on the first, second, or third day of anthesis) in which to be sampled, and three flowers were assigned to each period per bush (9 periods, with 3 replicates, for a total of 27 flowers per bush). However, because flowers were observed to open throughout the day, flowers sampled at the same time of day may have been open for different lengths of time. To account for this, we analyzed the data by distributing the inflorescences into discrete 8 hour intervals (0-8hrs after the onset of anthesis, 9-16hrs, ..., 65-72hrs) based on their actual length of anthesis at the time of sampling.

Nectar sampling was performed by inserting a 5μl micro-capillary tube into the bottom of the corolla to draw up the nectar, and the nectar from both flowers in the inflorescence was collected into a single tube. The total nectar volume (V_n) was calculated by measuring how much of the tube was filled, and then applying the formula:

$$V_n = \frac{5\mu l}{L_c} * L_n$$

L_c is the length of the capillary tube until the 5 μl mark, and L_n is the length of the tube filled by the nectar sample.

To test for the effect of nectar removal on nectar production, three inflorescences were randomly selected from each bush. Starting at the beginning of anthesis, we removed the nectar from each of the inflorescences at 7 PM each day after opening, and recorded the volume of nectar as above. This was repeated daily until the end of anthesis.

Nectar production and anther dehiscence were analyzed using a Kruskal-Wallis test with the sampling interval as the predictor variable and nectar volume or number of dehisced anthers as the response, followed by multiple comparisons using the Mann-Whitney U test, with the *P*-values adjusted using the Bonferroni correction. The effect of nectar removal on total nectar production was analyzed using the same method as nectar production and anther dehiscence, with sampling day as the predictor and nectar volume as the response. To compare total nectar production from multiply-sampled flowers to single-sampled flowers, we added the total amount of nectar produced by each multiply-sampled inflorescence, using only inflorescences that were sampled 3 times total, to get the average nectar produced over the lifetime of the inflorescence. This was compared to the discrete 8-hour interval from single-sampled flowers that had the highest average maximum volume of nectar (i.e. the volume of nectar produced by a flower that is not visited), using a Mann-Whitney U test.

To determine when the stigmas become receptive with respect to stage of anthesis, we removed the stigmas from 3 inflorescences at different stages of opening, beginning with inflorescences that were still tightly closed. The stigmas were removed from each flower using forceps, and the styles were clipped towards the base as far down as possible. We ensured there were no pollen grains on the stigma using a hand lens, and then held the stigma in a 3% hydrogen peroxide solution and watched for the formation of bubbles on the stigma surface, which would indicate the presence of peroxidases (Dafni et al. 2005). This was repeated at progressively more advanced stages of anthesis until the stigmas showed receptivity. As old stigmas can react with hydrogen peroxide even though they are no longer receptive (Dafni & Maues 1998), only the beginning of receptivity was identified.

RESULTS

Length of anthesis

Anthesis began at various times throughout the day; however, the majority [i.e., 60.7% (162/267)] were first observed to be open in the morning (it is possible some of them opened earlier, during the night – see discussion), versus 18.3% in the afternoon, and 21.0% in the evening. Anthesis length differed between treatments ($H = 26.9$, df = 3, $P < 0.001$; Fig. 2), with pollinated inflorescences flowering for a shorter duration [59.5 ± 16.5 h (mean ± SD)] than un-pollinated inflorescences (83.4 ± 25.9 h). There was no effect of emasculation on length of anthesis. Following hand-pollination, anthesis lasted another 34.3 ± 15.2 hours.

Nectar, pollen, and stigma dynamics

Nectar was evident as soon as the individual flowers opened and may be produced even before the onset of anthesis. Nectar production peaked after 9-16 hours of anthesis, with maximum average nectar volume reached between 33-40 hours (Kruskal-Wallis test, $X^2 = 41.2$, df = 8, $P < 0.001$; Fig. 3). Nectar volume was maintained throughout anthesis and nectar was not resorbed. We observed that the corollas of the flowers would often abscise from the ovaries with nectar still present.

Anthers began to dehisce almost immediately after flowers opened, and proceeded rapidly over the first 24 hours, with the majority of anthers dehisced after 25-32 hrs (Kruskal-Wallis test, $X^2 = 130.2$, df = 8, $P < 0.001$; Fig. 3).

Due to the 14 hour gap between the last observation of the day and the first observation the next morning, it is possible that the 8 hour bins do not account for all possible error (i.e. flowers might actually belong in the following interval if they opened between 7 PM and 1 AM). Although we believe this to be a minority, we repeated the analysis on a subset of the data where we could be confident of the time

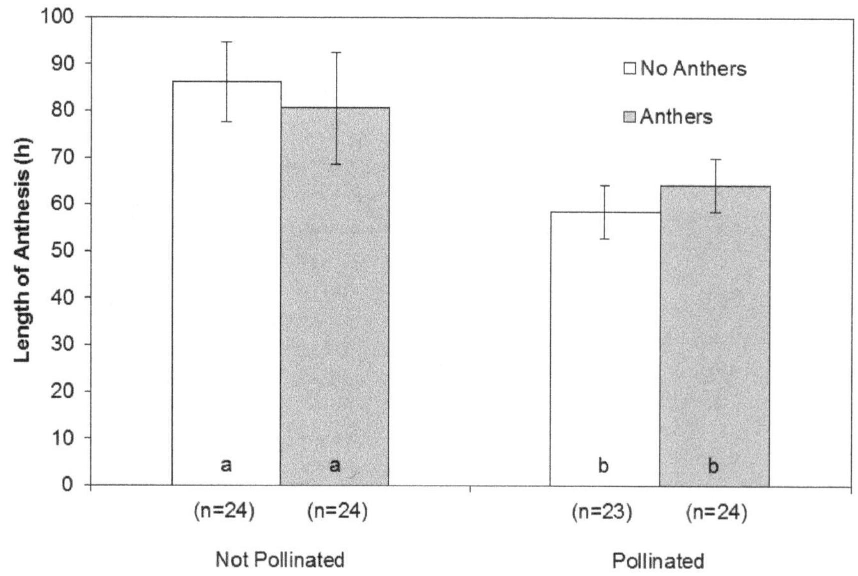

FIGURE 2. The mean length of anthesis (h), from opening of the bud to abscission of the corolla, for un-pollinated and pollinated Haskap inflorescences that had the anthers removed or left intact. Vertical lines represent the 95% confidence interval. Bars with different letters are significantly different according to pairwise comparisons using a Mann-Whitney U test with p-values adjusted for multiple comparisons using the Bonferroni correction.

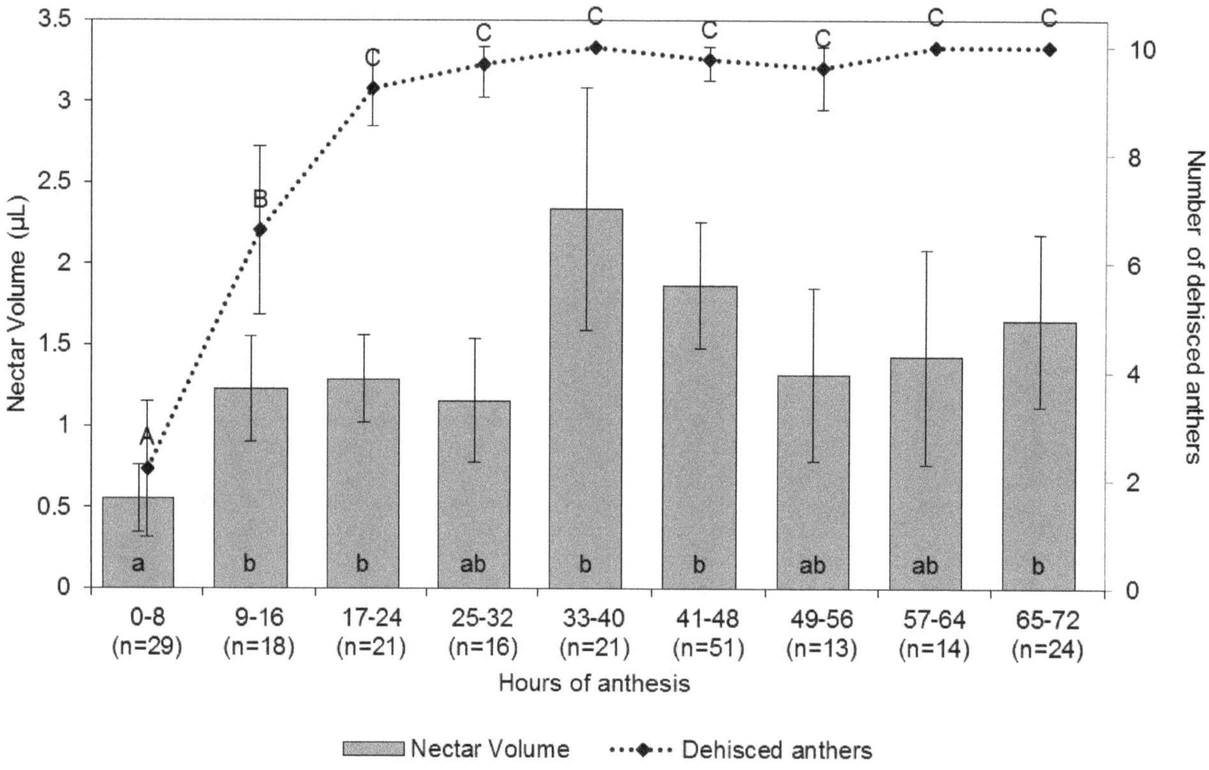

FIGURE 3. The mean amount of nectar (μL) (left Y-axis) and number of dehisced anthers (right Y-Axis) in Haskap (*Lonicera caerulea*) inflorescences per 8 hour interval after the onset of anthesis (X-axis). Vertical lines represent the 95% confidence interval. Plots with different letters are significantly different according to pairwise comparisons using a Mann-Whitney U test with p-values adjusted for multiple comparisons using the Bonferroni correction.

the flower opened – the results of this analysis show the same trend as the entire data set (results not shown).

After removal, nectar was wholly replenished throughout anthesis (Fig. 4). Maximum nectar volume recorded at the end of the second day was significantly higher than on day 1, supporting our previous finding that maximum nectar production occurs during the second day. Total nectar produced by inflorescences with their entire volume twice removed was significantly higher than single-sampled inflorescences (6.1 ± 3.8 μl compared to 2.3 ± 1.7 μl; Mann-Whitney U test, $W = 27$, $P < 0.001$).

Evidence of peroxidase activity on the stigmas was observed in flowers that were still tightly closed, suggesting that the stigma may be receptive even before the onset of anthesis.

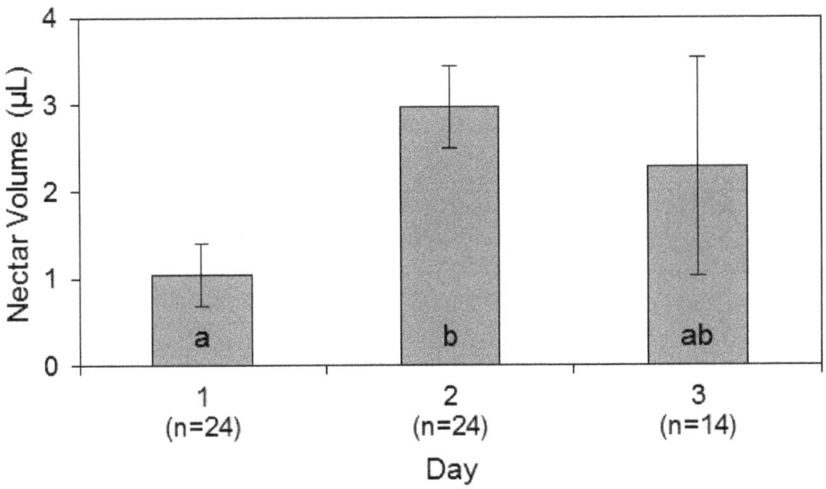

FIGURE 4. The mean amount of nectar (μL) in Haskap (*Lonicera caerulea*) inflorescences that had nectar removed once per day, up to 3 consecutive days. Vertical bars represent the 95% confidence interval. Plots with different letters are significantly different according to pairwise comparisons using a Mann-Whitney U test with p-values adjusted for multiple comparisons using the Bonferroni correction.

DISCUSSION

The timing of commencement of anthesis in Haskap is staggered; flowers open throughout the day, and likely during the night as well. The majority of flowers were first observed to be open by 9 AM; since observations were not made overnight, it is possible that many of the flowers observed at this time actually opened earlier, as anthesis appeared quite advanced in some of these flowers. Although this should be confirmed by future studies, it could indicate that nocturnal insects such as moths may be important contributors to Haskap pollination, in addition to diurnal bees and flies. This is further supported by our observation that roughly 20% of Haskap flowers also open in the evening.

The average duration of anthesis of the un-pollinated flowers is 83 hours (3.5 days; Fig. 2); however, pollination will significantly shorten this, triggering senescence of the flowers on an average of 34 hours following the pollination event. Our findings are slightly shorter than those of Bożek & Wieniarska (2006) who found that flowers of *Lonicera caerulea* var. *kamtschatica* lived 4-5 days, but they also noted that flowers excluded from pollinators were longer lived. This extended floral period may result in more opportunities for visitation by insects and increased pollination success, while senescence in response to pollination reduces the occurrence of repeat visits to flowers that are already pollinated, and may also reduce damage to the flowers, which can reduce seed set (Young 1988; Burquez & Corbet 1991). Our observation that emasculation has no effect on floral longevity in Haskap suggests it is completely self-incompatible, and self-pollination does not shorten the lifespan of the flower, nor does it result in significant stigma clogging that may prevent cross pollination.

However, although we did not quantify actual levels of self-pollination in this study, field studies have shown significant self-pollination of the stigma (Frier et al. 2016). It is possible that in greenhouse conditions very little self-pollination actually occurs, and in more realistic settings (with vigorous disruption by wind and handling by insect visitors) these factors may be more significant. This is made more likely by the fact that the anthers dehisce almost immediately following the onset of anthesis (Fig. 3), leaving little or no prior opportunity for cross-pollination. This strategy could maximize the chance that pollen is picked up by a floral visitor, but it also could increase the chance of self-pollen interfering with cross-pollination of the stigma and reducing reproductive success (Bertin & Sullivan 1988; Galen et al. 1989; Waser & Price 1991; Broyles & Wyatt 1993; Barrett 2002), especially because the anthers and stigma exist in very close proximity to one another. If this type of stigma clogging is common, this could suggest some competition or trade-off between male and female reproductive success in Haskap. It may be worthwhile to explore differences in style length among Haskap cultivars, as well as compared to wild varieties, as longer styles may be less likely to experience self-pollination.

Nectar production begins as soon as the flowers open, perhaps even slightly before, during the bud stage. We have observed bumble bees visiting Haskap flowers before they are entirely open, and our results suggest that the stigma is receptive at this stage as well. Early nectar production and stigma receptivity may have evolved to take advantage of these early visitors and increase the chance of successful pollination. Nectar production peaks between 8-16 hours of anthesis and is maintained throughout the lifetime of the inflorescence. Considerable variation in the amount of nectar is consistent with findings that variability in nectar production is correlated with large floral displays (Biernaskie & Cartar 2004), as risk-averse pollinators pay shorter visits to a single bush when nectar is variable (Biernaskie et al. 2002). As Haskap produces many flowers simultaneously and is self-incompatible, this strategy would help decrease the instance of geitonogamy and promote cross-pollination.

We found no evidence of nectar resorption, and the corolla abscised from the ovaries with the nectar load intact. However, we only analyzed nectar dynamics in un-pollinated flowers; in some flowers, reabsorption is triggered by pollination, presumably to reuse the energy resources in berry development (Luyt & Johnson 2002). However, Burquez and Corbet (1991) suggest that if the nectary is lost when the corolla dehisces, as in *Lonicera*, reabsorption is unlikely. Additionally, the entire nectar volume can be replaced several times throughout anthesis. This may be an adaptation to nectar robbers or ineffective pollinators, increasing the possibility of repeat visits to a single flower. We have commonly observed both honey bees and bumble bees nectar robbing the flowers, and there is evidence that many legitimate pollinator visits do not deposit sufficient pollen grains for full fertilization of the ovaries (Frier et al. 2016).

The results of this study suggest that Haskap flowers are likely generalized in their pollinator attraction strategy and may be pollinated by a wide variety of insect species and functional groups. Haskap flowering occurs very early in the year when few pollinators are active, and the characteristics described here may reflect adaptations to capitalize on every opportunity to receive a successful pollination visit. The flowers, which open throughout the day (and perhaps the night as well), remain open for up to four days, but successful pollination triggers early senescence. Nectar is produced immediately upon anthesis, potentially beginning in the large bud stage. After initial nectar production the volume is held relatively constant until senescence and any nectar removed during this period is replaced. The anthers begin to dehisce immediately after the flower opens and continue over the first day. The stigma appears receptive in the bud stage, and could potentially be pollinated before the flower opens.

As a plant with polyphilic flowers, Haskap is less likely to be pollinator limited than more specialized species, and it may be effectively pollinated by a wide variety of managed and wild insects. This means that Haskap could be successfully cultivated in many habitats and geographic locations, and not be limited to distributions or commercial use of specific pollinators. As the flowers are observed to open in the evening, it is likely that nocturnal insects such as moths are important pollinators of this crop, not just diurnal bees and flies. Haskap growers should take advantage of this generalist system by developing and maintaining healthy pollinator habitat in and around Haskap orchards, as species rich pollinator populations may be essential to realizing optimum fruit yields from this crop (Frier et al. 2016).

Acknowledgements

We gratefully acknowledge the Saskatchewan Ministry of Agriculture and the Canada-Saskatchewan Growing Forward 2 bi-lateral agreement, delivered by the Agriculture Council of Saskatchewan, for funding and support in this project. We also thank Bob Bors and Bayartulga Lkhagvasuren for providing access to the facilities and plants used in this project.

References

Ackermann M, Weigend M (2006) Nectar, floral morphology and pollination syndrome in *Loasaceae* subfam. *Loasoideae* (Cornales). Annals of Botany 98:503–514.

Agostini K, Sazima M, Galetto L (2011) Nectar production dynamics and sugar composition in two Mucuna species (Leguminosae, Faboideae) with different specialized pollinators. Die Naturwissenschaften 98:933–42.

Amorim FW, Galetto L, Sazima M (2013) Beyond the pollination syndrome : nectar ecology and the role of diurnal and nocturnal pollinators in the reproductive success of *Inga sessilis* (Fabaceae). Plant Biology 15:317–327.

Armbruster WS (2012) Evolution and ecological implications of "specialized" pollinator rewards. In: Patiny S (ed) Evolution of Plant-Pollinator Relationships. Cambridge University Press, pp 44–67.

Barrett SCH (2002) Sexual interference of the floral kind. Heredity 88:154–159.

Bertin RI, Sullivan M (1988) Pollen interference and cryptic self-fertility in *Campsis radicans*. American Journal of Botany 75:1140–1147.

Biernaskie JM, Cartar RV (2004) Variation in rate of nectar production depends on floral display size: a pollinator manipulation hypothesis. Functional Ecology 18:125–129.

Biernaskie JM, Cartar R V, Hurly TA (2002) Risk-averse inflorescence departure in hummingbirds and bumble bees: could plants benefit from variable nectar volumes? Oikos 98:98–104.

Bors B (2008) Growing Haskap in Canada. University of Saskatchewan Fruit Program

Bozek M (2012) The effect of pollinating insects on fruiting of two cultivars of *Lonicera caerulea* L . Journal of Apicultural Science 56:5–11.

Bożek M (2007) Pollen productivity and morphology of pollen grains in two cultivars of honeyberry (*Lonicera kamtschatica* (Sevast.) Pojark.). Acta Agrobotanica 60:73–77.

Bożek M, Wieniarska J (2006) Biologia kwitnienia i wydajność cukrowa kwiatów dwóch odmian *Lonicera kamtschatica* (Sevast.) Pojark / Blooming biology and sugar efficiency of two cultivars of *Lonicera kamtschatica* (Sevast.) Pojark. Acta Agrobotanica 59:177–182.

Broyles S, Wyatt R (1993) The consequences of self-pollination in *Asclepias exaltata*, a self-incompatible milkweed. American Journal of Botany 80:41–44.

Burquez A, Corbet SA (1991) Do flowers reabsorb nectar? Functional Ecology 5:369–379.

Dafni A, Kevan PG, Husband BC (2005) Practical Pollination Biology. Enviroquest Ltd., Cambridge, Ontario.

Dafni A, Maues MM (1998) A rapid and simple procedure to determine stigma receptivity. Sexual Plant Reproduction 11:177–180.

Fenster C, Armbruster W (2004) Pollination syndromes and floral specialization. Annual Review of Ecology, Evolution, and Systematics 35:375–403.

Freitas L, Sazima M (2006) Pollination biology in a tropical high-altitude grassland in Brazil: interactions at the community level. Annals of the Missouri Botanical Garden 93:465–516.

Frier SD, Somers CM, Sheffield CS (2016) Comparing the performance of native and managed pollinators of Haskap (*Lonicera caerulea*: Caprifoliaceae), an emerging fruit crop. Agriculture, Ecosystems and Environment 219:42–48.

Galen C, Gregory T, Galloway L (1989) Costs of self-pollination in a self-incompatible plant, *Polemonium viscosum*. American Journal of Botany 76:1675–1680.

Galetto L, Bernardello L (1993) Nectar secretion pattern and removal effects in three species of Solanaceae. Canadian Journal of Botany 71:1394–1398.

Galetto L, Bernardello G (1996) Characteristics of nectar secretion by *Lycium cestroides, L. ciliatum* (Solanaceae), and their hybrid. Plant Species Biology 11:157–163.

Galetto L, Bernardello G, Rivera GL (1997) Nectar, nectaries, flower visitors, and breeding system in five terrestrial Orchidaceae from central Argentina. Journal of Plant Research 110:393–403.

Johnson SD, Nicolson SW (2008) Evolutionary associations between nectar properties and specificity in bird pollination systems. Biology letters 4:49–52.

Kulloli SK, Chandore AN, Aitawade MM (2011) Nectar dynamics and pollination studies in three species of Lamiaceae. Current Science 100:509–516.

Luyt R, Johnson S (2002) Postpollination nectar reabsorption and its implications for fruit quality in an epiphytic orchid. Biotropica 34:442–446.

Miyake T, Yahara T (1998) Why does the flower of *Lonicera japonica* open at dusk? Canadian Journal of Botany 76:1806–1811.

Nepi M, Guarnieri M, Pacini E (2001) Nectar secretion , reabsorption , and sugar composition in male and female flowers of *Cucurbita pepo.* International Journal of Plant Sciences 162:353–358.

Ollerton J, Alarcón R, Waser NM, Price M V., Watts S, Cranmer L, Hingston A, Peter CI, Rotenberry J (2009) A global test of the pollination syndrome hypothesis. Annals of Botany 103:1471–1480.

Ollerton J, Watts S (2000) Phenotype space and floral typology: towards an objective assessment of pollination syndromes. Det Norske Videnskaps-Akademi. I. Matematisk Naturvidenskaplige Klasse, Skrifter, Ny Serie 39:149–159.

Petanidou T, Kallimanis AS, Tzanopoulos J, Sgardelis SP, Pantis JD (2008) Long-term observation of a pollination network: Fluctuation in species and interactions, relative invariance of network structure and implications for estimates of specialization. Ecology Letters 11:564–575.

Proctor M (1996) The Natural History of Pollination. Whitecap Books.

Svarcova I, Heinrich J, Valentova K (2007) Berry fruits as a source of biologically active compounds: the case of *Lonicera caerulea.* Biomedical Papers 151:163–174.

R Core Team (2014) R: A language and environment for statistical computing.

Waser NM, Chittka L, Price M V., Williams NM, Ollerton J (1996) Generalization in pollination systems, and why it matters. Ecology 77:1043–1060.

Waser N, Price M (1991) Reproductive costs of self-pollination in *Ipomopsis aggregata* (Polemoniaeae): are ovules usurped? American Journal of Botany 78:1036–1043.

Witt T, Geyer R, Gottsberçjer G (1999) Nectar dynamics and sugar composition in flowers of *Silene* and *Saponaria* species (Caryophyllaceae). Plant Biology 1:334–345.

Wolff D, Witt T, Ju A, Gottsberger G (2006) Nectar dynamics and reproductive success in *Saponaria officinalis* (Caryophyllaceae) in southern Germany. Flora 201:353–364.

Young HJ (1988) Differential importance of beetle species pollinating *Dieffenbachia longispatha* (Araceae). Ecology 69:832–844.

22

GENERALIZED VERSUS SPECIALIZED POLLINATION SYSTEMS IN *OENOTHERA* (ONAGRACEAE)

Krakos, Kyra N.[1]* & Fabricant, Scott A.[2]

[1]*College of Arts and Sciences, Maryville University, 650 Maryville University Drive, St. Louis, Missouri*
[2]*Department of Biological Science, Macquarie University, North Ryde, New South Wales, Australia*

Abstract— Although generalized and specialized plants are often discussed as alternative states, the biological reality may better be viewed as a continuum. However, estimations of pollinator specificity have been confounded in some studies by the assumption that all floral visitors are pollinators. Failure to account for pollen load can lead to inaccurate conclusions regarding the number of pollinators with which a species actually interacts. The aim of this study was to clarify the distribution of pollination-system specialization within one clade, using a more rigorous assessment of pollen flow. The genus *Oenothera* has long been used as a model system for studying reproductive biology, and it provides a diversity of pollination systems and a wealth of historical data. Both floral visitation rate and pollen-load analysis of sampled pollinators, combined into a metric of pollen flow, were used to quantify the pollination systems of 26 *Oenothera* taxa. Metric of pollinator specialization were calculated as functions of both total pollinator taxa, and as pollinator functional groups. We found that for *Oenothera*, the number of floral visitors highly overestimates the number of pollinators, and is inadequate for determining or predicting pollination system specialization. We found that that pollination systems were distributed on a gradient from generalized to specialized, with more pollinator-specialized plant taxa, especially when estimated using pollinator functional groups. These results are in conflict with previous studies that depict most plant species as generalists, and this finding may be related to how prior studies have estimated specialization.

Keywords: Pollination systems; pollinator specialization; pollinator functional groups; Oenothera

INTRODUCTION

The rapid diversification and apparent specialization of the angiosperms in the early Cretaceous is traditionally explained by the co-evolution of plants with their insect pollinators (Crane et al. 1995; Grimaldi 1999; De Bodt et al. 2005; Solds et al. 2008; Soltis et al. 2008). Studies in the latter half of the twentieth century tended focused on the tightly coupled relationships of a plant and its pollinators (Grant & Grant 1965; Faegri & Pijl 1966; Stebbins 1970) and depicted these interactions as highly specialized, meaning that a given plant species relied on a small number of pollinator species. However, beginning in the last decade of the twentieth century, pollination biology research started challenging these traditional ideas and debating the specialization of pollination systems (Ollerton 1996; Waser et al. 1996; Johnson & Steiner 2000; Bascompte et al. 2003; Fenster et al. 2004; Sahli & Conner 2006; Tripp & Manos 2008; Mitchell et al. 2009; Ollerton et al. 2009). This debate continues unabated, and often revolves on how one measures the efficacy of flower visitors (Padyšáková et al. 2013). Although pollination generalists and specialists are often discussed as alternative states, the biological reality may be better viewed as a continuum of generalization to specialization (Johnson & Steiner 2000).

*Corresponding author; email: kkrakos@maryville.edu

A major impediment to understanding the apparent paradox of specialized plants with generalized pollination systems is the lack of a standardized method for measuring pollination system specialization (Ne'eman et al. 2010). Traditionally, one counted the number of pollinator taxa visiting a plant species (Waser et al. 1996). This method may be misleading in the case of a "generalist" plant species that is visited by multiple pollinator species, if all of the pollinators belong to a functional group defined by a single morphology or foraging behaviour. The use of pollinator functional groups, which are defined as multiple taxa that share features (such as body size or tongue length) that determine their functionality as pollinators, provides a more accurate characterization of a plant's pollination biology (Fenster et al. 2004) and can drastically alter the perceived degree of specialization. For instance, Waser et al. (1996) analyzed Robertson's (Robertson 1928) pollinator survey and reported that 91% of 375 native plants in Illinois were visited by more than one insect species and therefore were generalists. However, Waser et al. (1996) did not differentiate between insects that were visitors and insects that were pollinators. Reanalysis of the same data indicated that when the insects were grouped into functional groups and non-pollinating visitors were excluded, 75% of the flowering plants only used one pollinator type and could therefore be considered specialized by that criterion (Fenster et al. 2004).

Calculating the degree of pollination specialization based solely on visitation, meaning the animals that land on the flower, can also be misleading because not all flower visitors are pollinators. A plant may be visited by dozens of potential pollinators, but critical pollen transfer may be accomplished by few pollinators (King et al. 2013). In addition, a frequent visitor may carry a small pollen load, while a less frequent visitor may carry a large pollen load (Mayfield et al. 2001). In *Oenothera cinerea* (Onagraceae), when both visitation and pollen load were examined, Clinebell et al. (2004) found a high degree of specialization to a few major pollinators: of 45 species of floral visitor, only 5 carried major pollen loads, and 32 carried little or no pollen. However, relatively few recent studies evaluate pollination based on both visitation and pollen load (Armbruster 1985; Armbruster et al. 1989; Clinebell et al. 2004; Forup et al. 2008; Reynolds & Fenster 2008; Bosch et al. 2009; Reynolds et al. 2009; Jacobs et al. 2010; Popic et al. 2013), and failure to account for pollen load can lead to inaccurate assumptions regarding the number of pollinators with which a plant species actually interacts.

Furthermore, many angiosperm traits, including pollination system, are shared due to common ancestry, and results from comparative studies can be biased by phylogenetic constraint and niche conservatism (Sanderson & Donoghue 1996; Sakai et al. 1997; Freckleton 2000; Vamosi et al. 2003; Machado & Lopes 2004). A well-resolved phylogeny can provide a framework for comparing pollination systems while controlling for shared evolutionary history (Armbruster 1992; Nosil & Mooers 2005). Furthermore, performing an analysis of pollination system across one sub-generic clade will help shed light on the degree of evolutionary lability of pollinator specialization.

Onagraceae, specifically the genus *Oenothera*, has long served as a model system for the evolution of flowering plant reproductive biology (Raven 1979; Raven 1988; Hoch et al. 1993; Clinebell et al. 2004). The diversity of pollination systems within *Oenothera* make it ideal for testing hypotheses of pollination system specialization. Recent molecular phylogenetic studies have clarified phylogenetic relationships within *Oenothera* (Levin et al. 2003; Hoggard et al. 2004; Levin et al. 2004; Wagner et al. 2007; Wagner et al. 2013), notably, the once segregate genera *Gaura*, *Calylophus* and *Stenisiphon* now appear within a monophyletic *Oenothera* (Raven & Gregory 1972; Raven 1988). We focused on 26 species in sections *Kneiffia*, *Megapterium*, *Peniophyllum*, *Paradoxus* and *Gaura*, hereafter referred to as the "*Gaura* clade." The 26 species of the *Gaura* clade are widely-distributed in North America and Mexico (Raven & Gregory 1972; Straley 1977; Raven 1979), and they exhibit a broad array of floral form, both diurnal and nocturnal flowering, and diverse pollinators, including noctuid moths, antlions, bees, flies, wasps, butterflies, and hawkmoths (Raven & Gregory 1972; Straley 1977; Raven 1979; Nonnenmacher 1999; Moody-Weis & Heywood 2001; Clinebell et al. 2004). The *Gaura* clade provides a system in which we can make a more rigorous assessment of *Oenothera* pollination systems to clarify the degree of specialization while controlling for similarity due to shared ancestry.

The aim of this study is to examine the pollination systems of 26 species of taxa in the *Gaura* clade of *Oenothera*, and identify pollination specialization within the clade. We use these 26 focal species to test the hypothesis that insect visitation rate is sufficient to characterize pollination system specialization. We also examine the distribution of the pollination system of these *Oenothera* to explore the hypothesis that most flower species have generalized pollination systems. We predict that defining pollination systems using pollinator functional groups will result in a distribution that shows that most pollination systems for these *Oenothera* are specialized, and that functional groups will be informative about which pollinator group interacts with the plant most often. To place our findings and metric in the broader literature, we also review recent pollination literature in order gain a broad view of how pollination specialization is measured. We determine if they use visitation rate alone as a measurement of pollination, or visitation and a secondary assessment such as stigma contact, pollen load transferred, etc.

MATERIALS AND METHODS

Characterization of pollination system in contemporary published studies

Using the Boolean search terms "pollination AND ecology", we searched in Web of Science (Thompson Reuters 2010) for all publications from 2004 to 2012. We examined the research objectives of 978 records and found that 326 of these records measured pollination systems as part of their study (Tab. S1). For these 326 records, we read the methods and determined whether the number of animal visitors alone was used to characterize the pollination systems, or whether both visitation and pollen load were used to determine the pollination system.

Study system and field sites

We studied 26 species of *Oenothera*, all within the *Gaura* clade, in sites throughout the Northeast, Midwest, and Southwest of the United States. Fieldwork was conducted from April 2007 to August 2010. Our survey data were combined with archived data and captured pollinators stored at the Missouri Botanical Garden, especially data collected between 1999 and 2004 by R. Clinebell from sites in the Midwest United States and Mexico, and data from P. Raven and D. Gregory between 1964 and 1966 from Texas and Mexico. Each of the 26 species of Oenothera had between one and three study sites, for a total of 44 sites spread over 11 U.S. states and 3 Mexican states (Tab. S2). For each population, we conducted pollination observations and collected insects for later pollen load analyses. Vouchers of the *Oenothera* species were collected from each site and deposited with the Missouri Botanical Garden herbarium.

Measuring Pollination

Pollination system was determined based on both visitation rates and pollen-load analysis. For each population of *Oenothera*, we conducted a series of 20 min observation periods at multiple randomly-chosen inflorescences, recording the total number of visits, type of visitor, and behaviour of visitors, including physical contact between an insect and stigma. Observations were conducted on four occasions during each species' flowering season, and took place at peak pollinator activity times of the day or night. For species whose flowers remained viable across both day and night periods, a series of observations were carried out for both night and day. We recorded between 63 and 1110 (median: 231) observations per species (Tab. 1). A sample of insect visitors was collected after observation periods to determine average pollen load. Insects were collected only after observed stigma contact to maximize the probability of collecting true pollinators. The insect visitors to the flower were collected using a net and a killing jar charged with ethyl acetate vapor, and pinned for later pollen analysis. The sampling of the insects included at least five collections per species.

To assess the identity and number of pollen grains carried by each visitor to an *Oenothera* species, we made a library of pollen grains from flowering plants at each study site. Dehiscent stamens were placed on labeled glass slides and the pollen was teased out with probes. Pollen was then stained with 1-2 drops of Calbera's fluid to make a semi-permanent mount (Goldblatt et al. 1998; Bernhardt et al. 2003). In the lab, each insect collected on the *Oenothera* species was placed on a separate glass slide and washed with a few drops of 70% ethanol. Any remaining pollen was scraped off the insect, including the scopal loads in the bees. The insect specimen was removed from the slide and the slide was allowed to air dry. Washed insect specimens were then dried, pinned, and saved for identification by regional entomologists. The pollen on the slide was stained with 1-2 drops of Calbera's fluid (Goldblatt et al. 1998) and mounted with a cover slip. Slides were then viewed with a light microscope to count pollen grains and compare to the pollen library for identification. Foreign pollen was identified and counted as well; however, most loads were homogenous. Data on the location, amount, and type of pollen were all recorded; however, only the load from the contact area of the insect with the stigma was used for these analyses.

Earlier collections of insect visitors collected on various *Oenothera* species are stored at the Missouri Botanical Garden. These include collections by R. Clinebell, P. Raven, and D. Gregory. We conducted the pollen-load analysis on these insect collections. Visitation rates for these visitors are found in records kept at the Missouri Botanical Garden (unpublished data).

Quantifying specialization: "S-score" and "F-score"

In quantifying specialization, we focused on the movement of pollen, and therefore the male reproductive success of the plant. Pollinators were characterized by their rate of visitation and pollen carrying efficiency. The degree of pollinator specialization of a plant species, which we have termed the S-score, is defined as the minimum number of pollinator taxa that account for 95% of the *Oenothera* pollen flow. Pollen flow is a measure to correct for the disparity between frequency and efficacy of pollinators. To calculate pollen flow, we multiplied the visitation rate (visits/inflorescence/20min) by the pollen load (number of *Oenothera* pollen grains carried by an animal visitor) for each visitor species, and then summed this across visitor species:

$$\text{Pollen Flow} = \sum (\text{Visitation Ratespx} * \text{PollenLoadspx})$$

We then determined the number of visiting species that accounted for 95% of the total pollen flow, and designated that as the "S-score" for those specific *Oenothera* taxa. We also measured generalization by placing the visitors into functional groups based on family and size (Tab. 1). For example, all noctuid moths of a similar size that visited during the same time period were considered as one functional group. We then determined the number of functional groups that accounted for 95% of the total pollen flow and designated that as the "F-Score". We designated 95% of the total pollen flow as our metric in order to capture the broadest set of pollinators that are moving pollen. It is also the 95% probability that pollen will be carried by a member of the designated pollinators/functional group members.

Statistical Analyses

To test whether visitation alone was sufficient to characterize pollination systems for a plant taxon, we compared the total number of all floral visitors with the number of pollinators, defined as those visitors that carried the plant species pollen and made stigma contact. To assess if visitors were significantly more numerous than pollinators, we used a Wilcoxon signed rank test, as well as a paired t-test after log transformation for normality. A Wilcoxon signed rank test was also used to test if F-scores were significantly lower than S-scores. A linear regression between visitors and pollinators was calculated using log-transformed data. This linear model, after diagnostic checks for patterns and normality in fitted values and residuals, was used to create a prediction interval for S-score based on number of visitors.

RESULTS

Characterization of pollination systems in contemporary published studies

Of the 326 records examined, 66.6% used only observed visitation rates of insects or birds to plants as a method to characterize pollination (Tab. S1). The remaining 33.4% used various techniques of measuring pollen load, flow, or deposition. There was no trend of increasing reliance on secondary pollen measurement over time.

Pollination System

Of the 26 *Oenothera* species examined, *O. curtiflora*, *O. sessilis*, and *O. simulans* were completely autogamous. *Oenothera simulans* and *O. sessilis* had visitors, but none that carried any pollen and contacted a stigma. *Oenothera macrocarpa*, *O. nealleyi*, *O. filiformis*, *O. dodgeniana*, and *O. gaura* all used both night and day pollinators. *Oenothera linifolia*, *O. pilosella*, *O. perennis*, *O. riparia*, *O. glaucifolia*, *O. demareei* and *O. lindheimeri* were all day pollinated. *Oenothera patriciae*, *O. triangulata*, *O. xenogaura*, *O.*

TABLE 1. Species measured, number of flowers observed for 20-minute periods, number of insects with pollen loads counted, and functional groups identified per species.

Species name	Species code	Flowers observed	Pollen loads counted	Functional groups
O. anomala	A	176	71	Hawk moth; Antlion
O. arida	B	63	25	Noctuid moth
O. cinerea ssp. *cinerea*	C	904	330	Noctuid moth; Halictid bee; Bumble bee
O. dodgeniana	D	308	66	Halictid bee; Noctuid moth
O. curtiflora	E	400	0	Autogamous
O. demareei	F	311	132	Honeybee; Bumble bee
O. filiformis	G	1110	212	Halictid bee; Noctuid moth
O. gaura	H	241	126	Noctuid moth; Fly; Wasp; Halictid bee
O. glaucifolia	I	228	182	Halictid bee; Wasp; Fly; Syrphid
O. harvardii	J	125	30	Hawk moth
O. hexandra	K	236	108	Noctuid moth; Honeybee; Fly
O. lindheimeri	L	180	72	Honeybee; Wasp
O. linifolia	M	152	52	Fly; Halictid bee
O. macrocarpa ssp. *macrocarpa*	N	331	155	Noctuid moth; Hawk moth; Halictid bee; Wasp; Honeybee
O. patriciae	O	228	29	Noctuid moth
O. perennis	P	236	56	Halictid bee; Honeybee; Bumble bee
O. pilosella ssp. *pilosella*	Q	185	69	Honeybee; Halictid bee
O. riparia	R	285	41	Halictid bee; Honeybee; Bumble bee
O. sessilis	S	137	0	Autogamous
O. simulans	T	290	19	Autogamous
O. sinuosa	U	330	50	Noctuid moth
O. suffrutescens	V	373	100	Noctuid moth; Honeybee; Halictid bee; Bumble bee
O. nealleyi	W	296	46	Noctuid moth; Halictid bee
O. suffulta	X	133	109	Noctuid moth
O. triangulata	Y	155	19	Noctuid moth
O. xenogaura	Z	349	41	Noctuid moth

suffulta, O. sinuosa, O. cinerea ssp. cinerea, O. hexandra, O. havardii, O. arida, O. anomala, and *O. suffrutescens* were all exclusively night pollinated. The main functional groups for all 26 species are listed in Tab. 1.

Visitation versus Pollination

For 3 of the *Oenothera* species, the number of visitors equalled the number of pollinators. In all other species, visitation alone was not sufficient to characterize accurately the pollination system (Fig. 1). There was a significant difference between visitation and S-score using both non-parametric (Wilcoxon V = 253, $P < 0.0001$, two-tailed), and transformed parametric ($t_{25} = 7.382$, $P < 0.0001$) analyses. There was also a significant difference in how specialized the pollination systems are when calculated using taxa (S-score) versus when using functional groups (F-score) (Wilcoxon V = 120, $P = 0.0006$). There was a significant relationship between number of visitors and S-score for *Oenothera* plants (S-score = Visitors0.55, $F_{1,24} = 48.13$, $P < 0.0001$). However, number of total visitors could only explain a moderate amount of the variation in S-score ($R^2 = 0.67$), and this scatter results in a very wide prediction interval (Fig. 1). The majority of non-autogamous species (56.5%) were found to have between 1 and 3 pollinators, suggesting more specialization than generalization in this system (Fig. 2). This number increases to 82.6% when pollinators are considered as functional groups, while the majority (65.2%) only required the services of 1 or 2 functional groups to achieve pollen flow (Fig. 3).

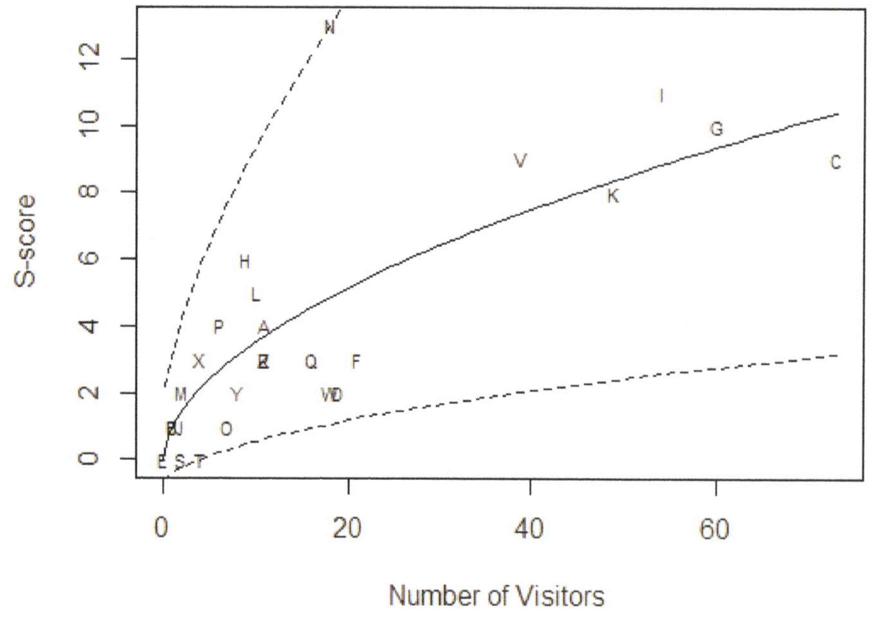

FIGURE 1. Scatterplot of S-score versus number of visitors for each of the 26 *Oenothera* taxa measured. The solid line is a linear regression of log-transformed visitor count versus S-score. The dotted lines demarcate the 95% prediction interval for S-score given a novel visitor count. For species names, see the species codes in Tab. 1. For an extended definition and calculation of S-score (pollinator specialization score), see text.

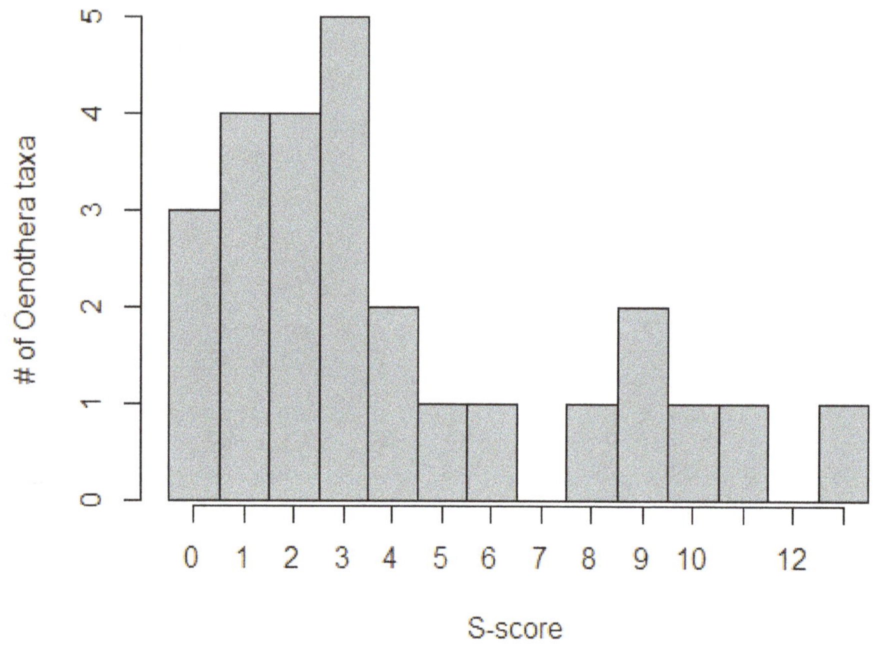

FIGURE 2. Histogram of the number of 26 *Oenothera* taxa in this study with a particular S-score (the number of pollinating species responsible for 95% of the pollen flow).

DISCUSSION

The pollination systems of *Oenothera* have been studied for several decades and serve as a model system for studying plant reproduction (Raven & Gregory 1972; Raven 1979; Raven 1988; Wagner et al. 2007). This study has furthered our understanding of *Oenothera* as a model system for studying pollination biology and provided insight into how specialization of pollination systems is measured. We examined the pollination systems of 26 taxa of *Oenothera* and focused on how to most accurately define the degree of specialization of these pollination systems. We found that these species attract a wide range of major pollinator groups including fly, bee, moth, hawk moth, wasp, and antlion. These species cover a broad range of pollination system types, both in temporal and spatial variation, and are a good representation of North American pollination. However, most taxa studied had large disparities between number of visiting insect species and number of important contributors to pollen flow, and this is likely to have implication for selection on floral traits. In contrast to Waser et al. (1996), we find that *Oenothera* pollination systems, as representative of North American pollination systems, are more specialized than generalized, similar to findings by Fenster et al. (2004).

It has been suggested that visitation is still an accurate way to measure specialization of pollination systems because the number of visitors is proportional to the number of actual pollinators (Cayenne Engel & Irwin 2003), and so one

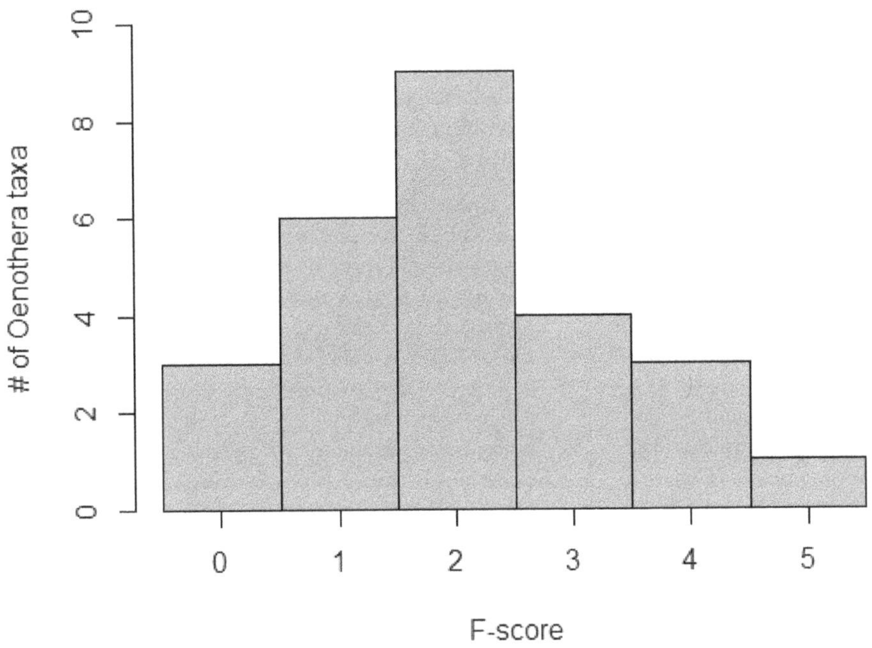

FIGURE 3. Histogram of the number of 26 *Oenothera* taxa in this study with a particular F-score (number of pollinator functional groups).

could make relative comparisons between plant species based on just visitation. Of the 23 non-autogamous *Oenothera* species we studied, for only 3 of those species did the numbers of visitors equal the number of pollinators, while in the other 20 species, the number of actual pollinators was lower than the number of visitors. We found that while visitors were proportional to pollinators on a log scale, the high variation in visitors for a given S-score results in a very wide prediction interval (Fig. 1). Therefore, visitation rate is a very blunt instrument, and is not reliable for predicting degree of pollinator specialization (as inferred from pollen flow). For example, species with an S-score of 1 can have between 1 and 7 visitor species, while an S-score of 2 is associated with between 2 and 19 visiting insect species, and this variability increases with S-score. Visiting species richness explained approximately 67% of the variance in S-score; these results are similar to the correlations found by Salhi & Connor (2006). We suggest that not only does visitation highly over-estimate the number of pollinators, but it is also not a sufficiently accurate measurement of specialization of pollination systems in a proportional or comparative way either.

Pollination systems, in particular their degree of generalization/specialization, are most often measured as the number of taxa involved in the plant-pollinator interaction. However, the use of functional groups, in which the visiting taxa are grouped by some morphological characteristic that defines how pollen load is delivered to a plant species, is perhaps a more informative way to examine pollination systems. This may be especially important for studies that examine pollination syndromes, given that pollination syndromes are predicting the selective agent on a floral trait. One definition is that a pollination system is considered specialized when a single functional group is responsible for greater than 75% of the pollination visits (Fenster et al. 2004). In this study, when we grouped the insect visitors by major taxon groups and size, we found that the *Oenothera* pollination systems were more specialist than generalist. By the definition of Fenster et al. (2004), 17 of the *Oenothera* species have specialized pollination systems. Of the remaining species, 8 used only 2 functional groups of pollinators to reach 75% of the pollination visits, and only one species, *O. gaura*, uses 3 functional groups.

We decided to measure pollination specialization in a way that would show the continuous nature of pollination systems. The F-score calculated was the number of functional groups responsible for 95% of pollen flow, and thus is somewhat more stringent than the metric of Fenster et al. (2004). The majority of the *Oenothera* species were toward the specialist end of the continuum, i.e. having a functional group score (F-score) of 1 or 2 (Fig. 3). The highest F-score was a 5, and only two species, *O. macrocarpa* and *O. glaucifolia*, had this score. While F-scores were statistically lower than S-scores, the number of functional groups was not just a proportionally smaller set of the pollinators; of the 26 *Oenothera* pollination systems we studied, 8 species had the same number of pollinating taxa as they did functional groups. Furthermore, some taxa do not contribute sufficiently to the pollen flow to be included in the S-score; however, when the taxa are grouped by functional groups, they can become a dominant contributor to the pollen flow. For example, *O. cinerea ssp. cinerea* is pollinated by several species of small halictid bees, bumble bees and noctuid moths. As individual species, the noctuid moths contribute little to total pollen flow, but when all 14 species of noctuid are considered one functional group, they collectively become the second most important group contributing to pollen flow. Overall, we found that the use of functional groups gives the most accurate representation of how specialized these *Oenothera* pollination systems are, with respect to morphological specialization to a specific type of pollinator.

One reason that pollination systems are often seen as generalized is because they are defined only using visitation rates of all potential pollinators to a plant. In our literature review, we found that the majority of pollination studies in a recent nine-year period used only visitation rate to characterize a pollination system. However, we found that visitation rate highly over-estimates the number of taxa pollinating a plant species. Insects visit flowers for a variety of reasons (Buchmann & Nabhan 1996). The flower may be a mating site or a source of food or shelter, and such visits may not involve anther or stigma contact.

This current study has several important strengths and limitations. The greatest strengths are in its number of species, their respective diversity of pollination systems, and broad geographic range, all phylogenetically controlled by studying members of one clade within one genus. This is compared to other recent studies which focus on few taxa (Adler & Irwin 2006; Fumero-Cabán & Meléndez-Ackerman 2007) or unrelated taxa (Bosch et al. 2009; King et al. 2013). However, this study focuses on pollen flow, and does not address pollen deposition as those studies do (Adler & Irwin 2006; Fumero-Cabán & Meléndez-Ackerman 2007; King et al. 2013). This is because we have utilized archived data on visitation rate and pollen flow to maximize our taxonomic coverage. By only considering visitors that make stigma contact, we partially account for this discrepancy. Furthermore, *Oenothera* species have low ovule count per flower, rendering voluminous pollen transfer unnecessary (Raven & Gregory 1972). However, as effective pollinators will be a subset of total pollinators, our measure is intrinsically conservative. Therefore, *Oenothera* plants may be even more specialized in pollination than we have ascertained.

For several of the species, we have multiple years of pollination data, but with such a broad study, this was not available with all the taxa. This is a second potential limitation of the study. We have sufficient data to gauge the stability of pollination systems between years for two species, *O. filiformis* and *O. macrocarpa*. We found that, although the taxa or functional group of pollinator differed, the total number of pollinator species or functional groups active in a single year did not change. This is in agreement with recent pollination network studies that show that, while the type of species interacting may change from year to year, the overall number of interactions tends to remain constant (Memmott et al. 2004; Petanidou et al. 2008). When looking at functional groups, a snapshot approach can be sufficient to characterize a pollination system in terms of how specialized it may be (Alarcon et al. 2008). Therefore, we made an informed assumption that a 'snapshot' approach, involving a single season of detailed pollination data, is sufficient for the broad scale comparison of this project.

One further difficulty in applying these results to other floral systems is that *Oenothera* pollen grains are large compared to other flowering species. Pollen size could be a trait that limits the number of pollinators that can manipulate and carry pollen. This could possibly filter out smaller visitors that would be pollinators if the pollen were smaller. Perhaps smaller pollen may encourage more generalized pollination systems. In addition, the viscine threads that hold together *Oenothera* pollen could potentially affect the size of pollen load carried by a pollinator. Future studies should look at a comparison of specialization of pollination systems between different floral systems.

Understanding the degree of specialization of pollination systems is important when making inferences about a plant's evolutionary history. If pollinators are a selective pressure that has led to such a diversity of floral form, then plant-pollinator interactions are expected to be highly specialized, and specialized pollination systems should be seen for a majority of flowering species (Ollerton 1996). Of course, pollinators are not the only factor in the adaptation of floral forms. For example, life history, breeding system, successional status, and abundance all play roles, but pollinators are considered a dominant influence in the evolution of floral specialization (Endress 1994; Crane et al. 1995; De Bodt et al. 2005; Soltis et al. 2008). How specialized a pollination system is also plays a critical role when making conservation decisions for plant species (Johnson & Steiner 2000; Biesmeijer et al. 2006; Winfree 2008; Ashworth et al. 2009; Bascompte 2009). Concluding that a plant has a generalized pollination system, when it is actually highly specialized, could lead to poor management decisions and result in a loss of plant diversity in the face of a changing climate or habitat loss. Therefore, this study may serve as an example of a more informative way to determine pollination system in future research.

ACKNOWLEDGEMENTS

Logistical support was provided by Missouri Dept of Conservation, Illinois Dept of Conservation, The Nature Conservancy, Shaw Nature Preserve, Selman Living Labs, Center for Plant Conservation, Great Meadows National Wildlife Refuge, and Harvard University Herbarium. We thank Mike Arduser, Phil Kroenig, and Richard Heitzmann for insect identification. A network of organizations, land owners, and botanists all contributed to the extensive fieldwork necessary for this work. We thank Nels Holmberg, Tim Smith, Martha Younkin, Tom Nagel, James Trager, Gloria and Ron Hoggard, George Yatskievych, Marty Kemper, Marc Phipps, Linda Wallace, Bill Caire, Mike Powell, Martin Terry, Patrick Conner, Dave and Judy Dodgen, Matthew Albrecht, Theo Witsell, Cindy Osborne, Carol Ann McCormick, Richard LeBlond, Alan Weakley, Katherine Kennedy, and Anna Strong. Thank you to Dr. Peter Hoch, Dr. Peter Raven, and Dr. Nicole Miller-Struttman for comments and discussion throughout the writing of the manuscript. We thank the following agencies for funding this work: National Science Foundation (DDIG 0910202), the Howard Hughes Foundation, American Women in Science, Botanical Society of America, Washington University in St. Louis, Sigma Xi, Prairie Biotic Research Inc., Missouri Native Plant Society, Webster Groves Nature Society, and the Missouri Botanical Garden.

APPENDICES

Additional supporting information may be found in the online version of this article:

APPENDIX I: References used for literature review on pollination study methodologies.

APPENDIX II: Table of field sites used in pollinator surveys.

REFERENCES

Adler LS, Irwin RE (2006) Comparison of pollen transfer dynamics by multiple floral visitors: Experiments with Pollen and Fluorescent Dye. Annals of Botany 97:141-150.

Alarcon R, Waser NM, Ollerton J (2008) Year-to-year variation in the topology of a plant-pollinator interaction network. Oikos 117:1796-1807.

Armbruster WS (1985) Patterns of character divergence and the evolution of reproductive ecotypes of *Dalechampia scandens* (Euphorbiaceae). Evolution 9(4): 733-752.

Armbruster, WS (1992) Phylogeny and the evolution of plant-animal interactions. BioScience 42(1): 12-20.

Armbruster WS, Keller S, Matsuki M, Clausen TP (1989). Pollination of *Dalechampia magnoliifolia* (Euphorbiaceae) by male Euglossine bees. American Journal of Botany 76(9): 1279-1285.

Ashworth L, Quesada M, Casas A, Aguilar R, Oyama K (2009) Pollinator-dependent food production in Mexico. Biological Conservation 142:1050-1057.

Bascompte J (2009) Mutualistic networks. Frontiers in Ecology and the Environment 7:429-436.

Bascompte J, Jordano P, Melian CJ, Olesen JM (2003) The nested assembly of plant-animal mutualistic networks. Proceedings of the National Academy of Sciences of the United States of America 100:9383-9387.

Bernhardt P, Sage T, Weston P, Azuma H, Lam M, Thien LB, Bruhl J (2003) The pollination of Trimenia moorei (Trimeniaceae): Floral volatiles, insect/wind pollen vectors and stigmatic self-incompatibility in a basal angiosperm. Annals of Botany 92:445-458.

Biesmeijer JC, Roberts SPM, Reemer M, Ohlemuller R, Edwards M, Peeters T, Schaffers AP, Potts SG, Kleukers R, Thomas CD, Settele J, Kunin WE (2006) Parallel declines in pollinators and insect-pollinated plants in Britain and the Netherlands. Science 313:351-354.

Bosch J, Martín González AM, Rodrigo A, Navarro D (2009) Plant–pollinator networks: adding the pollinator's perspective. Ecology Letters 12:409-419.

Buchmann SL, Nabhan FP (1996) The Forgotten Pollinators. Shearwater Books, Washington D.C.

Cayenne Engel E, Irwin RE (2003) Linking pollinator visitation rate and pollen receipt. American Journal of Botany 90:1612-1618.

Clinebell RR, Crowe A, Gregory DP, Hoch PC (2004) Pollination ecology of *Gaura* and *Calylophus* (Onagraceae, Tribe Onagreae) in western Texas, USA. Annals of the Missouri Botanical Garden 91:369-400.

Crane PR, Friis EM, Pedersen KR (1995) The origin and early diversification of angiosperms. Nature 374:27-33.

De Bodt S, Maere S, Van de Peer Y (2005) Genome duplication and the origin of angiosperms. Trends in Ecology & Evolution 20:591-597.

Endress PK (1994) Floral structure and evolution of primitive angiosperms - recent advances. Plant Systematics and Evolution 192:79-97.

Faegri K, Pijl Lvd (1966) The principles of pollination ecology. Pergamon, Oxford, UK

Fenster CB, Armbruster WS, Wilson P, Dudash MR, Thomson JD (2004) Pollination syndromes and floral specialization. Annual Review of Ecology Evolution and Systematics 35:375-403.

Forup ML, Henson KSE, Craze PG, Memmott J (2008) The restoration of ecological interactions: plant–pollinator networks on ancient and restored heathlands. Journal of Applied Ecology 45:742-752.

Freckleton R (2000) Phylogenetic tests of ecological and evolutionary hypotheses: checking for phylogenetic independence. Functional Ecology 14:129-134.

Fumero-Cabán JJ, Meléndez-Ackerman EJ (2007) Relative pollination effectiveness of floral visitors of *Pitcairnia angustifolia* (Bromeliaceae). American Journal of Botany 94:419-424.

Goldblatt P, Manning JC, Bernhardt P (1998) Adaptive radiation of bee-pollinated *Gladiolus* species (Iridaceae) in southern Africa. Annals of the Missouri Botanical Garden 85:492-517.

Grant V, Grant KA (1965) Flower pollination in the *Phlox* family. Columbia University Press, New York, NY

Grimaldi D (1999) The co-radiations of pollinating insects and angiosperms in the Cretaceous. Annals of the Missouri Botanical Garden 86:373-406.

Hoch PC, Crisci JV, Tobe H, Berry PE (1993) A cladistic analysis of the plant family Onagraceae. Systematic Botany 18:31-47.

Hoggard GD, Kores PJ, Molvray M, Hoggard RK (2004) The phylogeny of *Gaura* (Onagraceae) based on ITS, ETS, and trnl-F sequence data. American Journal of Botany 91:139-148.

Jacobs J, Clark S, Denholm I, Goulson D, Stoate C, Osborne J (2010) Pollinator effectiveness and fruit set in common ivy, *Hedera helix* (Araliaceae). Arthropod-Plant Interactions 4:19-28.

Johnson SD, Steiner KE (2000) Generalization versus specialization in plant pollination systems. Trends in Ecology & Evolution 15:140-143.

King C, Ballantyne G, Willmer PG (2013) Why flower visitation is a poor proxy for pollination: measuring single-visit pollen deposition, with implications for pollination networks and conservation. Methods in Ecology and Evolution 4:811-818.

Levin RA, Wagner WL, Hoch PC, Hahn WJ, Rodriguez A, Baum DA, Katinas L, Zimmer EA, Sytsma KJ (2004) Paraphyly in tribe Onagreae: Insights into phylogenetic relationships of Onagraceae based on nuclear and chloroplast sequence data. Systematic Botany 29:147-164.

Levin RA, Wagner WL, Hoch PC, Nepokroeff M, Pires JC, Zimmer EA, Sytsma KJ (2003) Family-level relationships of Onagraceae based on chloroplast rbcL and ndhF data. American Journal of Botany 90:107-115.

Machado IC, Lopes AV (2004) Floral traits and pollination systems in the Caatinga, a Brazilian tropical dry forest. Annals of Botany 94:365-376.

Mayfield MM, Waser NM, Price MV (2001) Exploring the 'most effective pollinator principle'with complex flowers: bumblebees and *Ipomopsis aggregata*. Annals of Botany 88:591-596.

Memmott J, Waser NM, Price MV (2004) Tolerance of pollination networks to species extinctions. Proceedings of the Royal Society of London Series B-Biological Sciences 271:2605-2611.

Mitchell RJ, Irwin RE, Flanagan RJ, Karron JD (2009) Ecology and evolution of plant-pollinator interactions. Annals of Botany 103:1355-1363.

Moody-Weis JM, Heywood JS (2001) Pollination limitation to reproductive success in the Missouri evening primrose, *Oenothera macrocarpa* (Onagraceae). American Journal of Botany 88:1615-1622.

Ne'eman G, Jurgens A, Newstrom-Lloyd L, Potts SG, Dafni A (2010) A framework for comparing pollinator performance: effectiveness and efficiency. Biological Reviews 85:435-451.

Nonnenmacher HF (1999) The comparative floral ecology of vernal and autumnal Onagraceae in and near Konza Prairie Research Natural Area, Kansas. PhD, St. Louis University, St. Louis

Nosil P, Mooers AO (2005) Testing hypotheses about ecological specialization using phylogenetic trees. Evolution 59:2256-2263.

Ollerton J (1996) Reconciling ecological processes with phylogenetic patterns: The apparent paradox of plant-pollinator systems. Journal of Ecology 84:767-769.

Ollerton J, Alarcón R, Waser NM, Price MV, Watts S, Cranmer L, Hingston A, Peter CI, Rotenberry J (2009) A global test of the pollination syndrome hypothesis. Annals of Botany 103:1471-1480.

Padyšáková E, Bartoš M, Tropek R, Janeček Š (2013) Generalization versus specialization in pollination systems: visitors, thieves, and pollinators of *Hypoestes aristata* (Acanthaceae). PLoS One 8:e59299.

Petanidou T, Kallimanis AS, Tzanopoulos J, Sgardelis SP, Pantis JD (2008) Long-term observation of a pollination network: fluctuation in species and interactions, relative invariance of network structure and implications for estimates of specialization. Ecology Letters 11:564-575.

Popic TJ, Wardle GM, Davila YC (2013) Flower-visitor networks only partially predict the function of pollen transport by bees. Austral Ecology 38:76-86.

Raven PH (1979) A survey of reproductive biology in Onagraceae. New Zealand Journal of Botany 17:575-593.

Raven PH (ed) (1988) Onagraceae as a model of plant evolution. Chapman and Hall, New York

Raven PH, Gregory DP (1972) A revision of the genus *Gaura* (Onagraceae). Memoirs of the Torrey Botany Club 23:1-96.

Reynolds RJ, Fenster CB (2008) Point and interval estimation of pollinator importance: a study using pollination data of *Silene caroliniana*. Oecologia 156(2): 325-332.

Reynolds RJ, Westbrook MJ, Rohde AS, Cridland JM, Fenster CB, Dudash, MR (2009). Pollinator specialization and pollination syndromes of three related North American *Silene*. Ecology 90(8): 2077-2087.

Robertson C (1928) Flowers and insects: lists of visitors of four hundred and fifty-three flowers. Charles Robertson, Carlinville, IL

Sahli HF, Conner JK (2006) Characterizing ecological generalization in plant-pollination systems. Oecologia 148:365-372.

Sakai AK, Weller SG, Chen M-L, Chou S-Y, Tasanont C (1997) Evolution of gynodioecy and maintenance of females: the role of inbreeding depression, outcrossing rates, and resource allocation in *Schiedea adamantis* (Caryophyllaceae). Evolution:724-736.

Sanderson MJ, Donoghue MJ (1996) Reconstructing shifts in diversification rates on phylogenetic trees. Trends in Ecology & Evolution 11:15-20.

Solds DE, Bell CD, Kim S, Soltis PS (2008) Origin and early evolution of angiosperms. Year in Evolutionary Biology 1133: 3-25

Soltis DE, Bell CD, Kim S, Soltis PS (2008) Origin and early evolution of angiosperms. Annals of the New York Academy of Sciences 1133:3-25.

Stebbins GL (1970) Adaptive radiation of reproductive characteristics in angiosperms, I: pollination mechanisms. Annual Review of Ecology and Systematics 1:307-326.

Straley GB (1977) Systematics of *Oenothera* sect. *Kneffia* (Onagraceae). Annals of the Missouri Botanical Garden 64:381-424.

Tripp EA, Manos PS (2008) Is floral specialization an evolutionary dead-end? Pollination system transitions in *Ruellia* (Acanthaceae). Evolution 62:1712-1736.

Vamosi J, Otto S, Barrett S (2003) Phylogenetic analysis of the ecological correlates of dioecy in angiosperms. Journal of Evolutionary Biology 16:1006-1018.

Wagner WL, Hoch PC, Raven PH (2007) Revised classification of the Onagraceae. Systematic Botany Monographs 83:1-240.

Wagner W, Krakos K, Hoch PC. (2013) Taxonomic changes in *Oenothera* sections *Gaura* and *Calylophus* (Onagraceae) Phytokeys 28: 61-72.

Waser NM, Chittka L, Price MV, Williams NM, Ollerton J (1996) Generalization in pollination systems, and why it matters. Ecology 77:1043-1060.

Winfree R (2008) Pollinator-dependent crops: an increasingly risky business. Current Biology 18:R968-R969.

23

SPECIALIST NECTARIVOROUS BIRDS (*CINNYRIS OSEA*) STEAL NECTAR WHEREAS OMNIVOROUS BIRDS ARE POLLEN TRANSFER VECTORS OF *ANAGYRIS FOETIDA*

Ron Haran*[1], Ido Izhaki[2], Amots Dafni[3]

[1]*16 Ibn Shaprut Street, Jerusalem, Israel*

[2]*Department of Evolutionary and Environmental Biology, University of Haifa, Haifa, Israel 3498838*

[3]*Institute of Evolution, Department of Evolutionary and Environmental Biology, University of Haifa, Haifa, Israel 3498838*

Abstract—The ornithophilous *Anagyris foetida* is almost the only shrub in the eastern Mediterranean region that flowers in winter. We assessed the pollination potential of flower visitors by observing pollen loads on trapped animals and contacts with floral parts. Resident and wintering omnivorous bird species, together with honey bees (*Apis mellifera*), were found to carry the shrub's pollen throughout its long flowering period in Israel and thus might be legitimate pollinators. Spring-migrating omnivorous birds passing through Israel carried pollen. The Palestine Sunbird (*Cinnyris osea*) is the only specialist nectarivorous bird in the Middle East. It carries little pollen and is therefore primarily a nectar thief of *A. foetida*.

Keywords: Anagyris foetida, Palestine Sunbird, legitimate pollination, nectar theft, Mediterranean forest, Honeybees

INTRODUCTION

Mutualism between plant and pollinators occurs when the plant benefits from the pollinator's transport of pollen grains and the pollinator benefits from a food source in the form of nectar and/or pollen. Previous studies showed that specialized nectar-feeding birds have several morphological characteristics that enable them to feed efficiently and to act as legitimate pollinators of flowers with compatible morphologies. Characteristics of bills and tongues influence the speed and efficiency with which nectar is extracted from flowers, whereas body mass and the length and shape of the wings and legs influence the speed, efficiency and method of movement between flowers, and the techniques of perching or hovering (Collins & Paton 1989; Paton & Collins 1989; Geerts & Pauw 2009; Myers et al. 2010). On the other hand, morphological mismatch between the flower and visitor can lead to nectar theft, which is the consumption of nectar from flowers without contributing to its pollination (Inouye 1983; Irwin et al. 2010). Also, a growing array of research provides evidence of non-specialist nectar-feeding birds, typically passerines, that serve as pollen transport vectors, both in Europe and the Middle East (e.g., Valido et al. 2004; da Silva et al. 2014; Cecere et al. 2011) and more globally (e.g., Ragusa-Netto 2002; Kunitake et al. 2004; Johnson & Nicolson 2008; Fang et al. 2012).

Anagyris foetida is a leguminous shrub or small tree, deciduous in summer, with a Mediterranean and Irano-Turanian distribution (Valtueña et al. 2008). In Spain it is pollinated by three species of Passeriformes (Ortega et al. 2005; Valtueña et al. 2007; Valtueña et al. 2008). It is not uncommon to find ornithophily in Fabaceae, although the family is mainly bee-pollinated (Stone et al 2003; Tandon et al. 2003; Ragusa-Netto 2002; Arroyo 1981). Its flowers show typical ornithophilous attributes, such as diurnal anthesis, no odour, showy corollas, pseudo-tubular flowers with copious nectar located at the base of the floral tube, hard protective tissues surrounding the ovary and other flower organs and pendulous flowers (Faegri & Pijl 1979; Proctor et al. 1996; Ortega et al. 2005; Willmer 2011). The flowers copiously produce pollen and dilute nectar. Plants are self-compatible, and selfing may occur via geitonogamy or vector-facilitated self-pollination may occur, although the level of fruit production is generally higher after outcrossing (Ortega et al. 2005). *Anagyris foetida* has adaptations that can be found in autumnal and winter-flowering species, such as long exposure of flowers and ecologically generalized pollination system, which allow for the simultaneous use of floral resources by unrelated pollinators (Dafni 1996; Olesen & Jordano 2002; Galloni et al. 2008; Padyšáková et al. 2013).

Anagyris foetida pollination has not previously been studied in the eastern Mediterranean, where climate, diversity of bird species, and subspecies of other bird species differ from those in Spain. The shrub occurs in the central and northern parts of Israel and flowers from mid-November to mid-March (Shmida 2005).

Sunbirds are a group of obligate nectar-feeding birds that can act as pollinators or thieves when consuming nectar from different plant species (Coombs et al. 2007; Coombs & Peter 2009; Padyšáková et al. 2013). Sunbirds generally perch while extracting nectar from flowers (Collins & Paton 1989). The Nectariniidae family is one of the major groups of pollinating

*Corresponding author: ronharan@zahav.net.il

birds in Africa and Asia (Cronk & Ojeda 2008; Da Silva et al. 2014).

The Palestine Sunbird (*Cinnyris osea*) is the only member of the Nectariniidae that occurs in the Mediterranean. Its world distribution includes the west Arabian Peninsula as far as Yemen, Israel, Jordan, Lebanon, and Syria. In Israel, it had a very limited distribution within the Syrian-African Rift Valley until the 1940's. Following the urbanization and development of agriculture and domestic gardens in the last century, Palestine Sunbirds spread out all over Israel (Yom-Tov 2014). In late autumn and winter, the sunbirds tend to wander away from their breeding habitats to look for food sources (Hadoram 1976; Cheke & Mann 2010). The maximum distance record of a recaptured sunbird is 109 km from Jerusalem to Megadim (Israeli Bird Ringing Center Database, IBRC Code: IL007/16). The sunbirds' tendency to wander explains the relatively recent encounter with *A. foetida* flowers.

Cinnyris osea is a small bird weighing 6-7 grams, with a long, curved bill measuring 16-18 mm and a typically long tongue split at the end to enable nectar consumption (Yom-Tov 2014). The Palestine Sunbird is known to be either a pollinator or a thief (Eisikowitch & Nahari 1982; Vaknin et al. 1996; Tadmor et al. 2004). It usually feeds from a perching position and only occasionally while hovering. Apart from nectar, it eats insects and spiders, especially during the breeding season. This bird is a potential pollinator of *Loranthus acaciae* and might be its only pollinator in winter (Vaknin et. al 1996).

In winter and early spring, omnivorous birds can also be found in the Mediterranean forest. These include resident, wintering, and transient species of various families, including the Sylviidae, Turdidae, Pycnonotidae, Fringillidae and Passeridae.

This study evaluates the likely relative effectiveness of specialist sunbirds and omnivorous birds as pollinators of *A. foetida*.

MATERIALS AND METHODS

Study Areas

We studied two populations of native *A. foetida,* 17 km apart. Both were located in the Judean Plain: Ella Valley (31°40'58" N, 14°00'34" E), 290 m above sea level, and Hulda (31°49'14" N, 21°33'34" E), 120 m above sea level. In Ella Valley, ca 90 individuals of *A. foetida* are part of the Mediterranean shrub vegetation, spread across an area of ca 3 ha. In Hulda, ca 1000 individuals dominate an area of 5 ha. The average annual rainfall in Ella Valley and Hulda is 402 mm and 492 mm, respectively. In both study areas, the range of temperatures in winter is -4°C to 30°C but on only 10% of the days during winter did high temperatures remain below 14 °C. The average high temperature was 19.18 °C ± 4.54 SD and 19.49 °C ± 4.38 SD in Ella Valley and Hulda, respectively. Both study areas support honey bee hives. The Ella Valley study area is also subjected to grazing livestock in spring and summer.

Video recordings

Sunbirds start foraging on *A. foetida* at dawn, concentrating on the upper parts of the plants. Video recordings were done by an observer from a distance of ca 4 m with a high-definition video camera (HC-W850 Panasonic) mounted on a tripod. The video recordings were analysed according the following criteria: time spent on individual flowers, the perching position above or bellow the flower, and whether the bird opened the keel in the foraging process.

Pollen loads on the birds

Mist nets were erected in very close proximity to *A. f-oetida* bushes. Birds of all species were mist netted in compliance with the Israeli Bird Ringing Center (IBRC) regulations in order to check pollen loads on their bodies. Twenty-four ringing sessions (9 in Ella Valley and 15 in Hulda), were conducted between 3.12.2016 and 10.03.2017. Each session used on average 120 m of net and ran for four hours from dawn. All the birds were identified and ringed, and weight, wing length, age and sex were recorded. An adhesive transparent strip (cellotape) of 4 cm^2 was applied to the parts of the body on which the pollen could be transported: throat, forehead and bill, and then stuck onto a microscope slide. All the pollen grains that were identified as *A. foetida* were counted under the microscope. Levels of pollen load were evaluated from 0 to 3 whereby: 0 = no pollen grains; 1 = < 10 pollen grains; 2 = < 40 pollen grains; and 3 = > 200 pollen grains.

Bees

Honey bees' (*Apis mellifera*) foraged on *A. foetida* flowers only when temperatures were higher than ca 14°C (Joshi & Joshi 2010; Abou-Shaara et al. 2012). Three parameters were recorded while visiting an individual flower: duration of flower visits, pollen or nectar collection, and success or failure in opening the keels of the flowers. The honeybee visits were observed by eye from 20–50 cm. Some video recordings were taken as well.

RESULTS

Phenology

Anagyris foetida flowers bloomed from the last week of November 2016 until mid-March 2017 (Fig. 1). Three resident bird species (Palestine Sunbird (*Cinnyris osea*), Sardinian Warbler (*Sylvia melanocephala*), White-spectacled Bulbul (*Pycnonotus xanthopygos*)) and three wintering species Blackcap (*Sylvia atricapilla*), Chiffchaff (*Phylloscopus collybita*), Spanish Sparrow (*Passer hispaniolensis*)) were present throughout the flowering period of *A. foetida* (Fig. 1). Four migrant species (Common Whitethroat (*Sylvia communis*), Lesser Whitethroat (*Sylvia curruca*), Ruppell's Warbler (*Sylvia rueppelli*), Orphean Warbler (*Sylvia crassirostris*) were present only from mid-February onward. All ten of these species were also observed foraging on the plant.

The arrival of sunbirds to the research sites preceded anthesis by ca two weeks (e.g., the first catch of sunbirds in

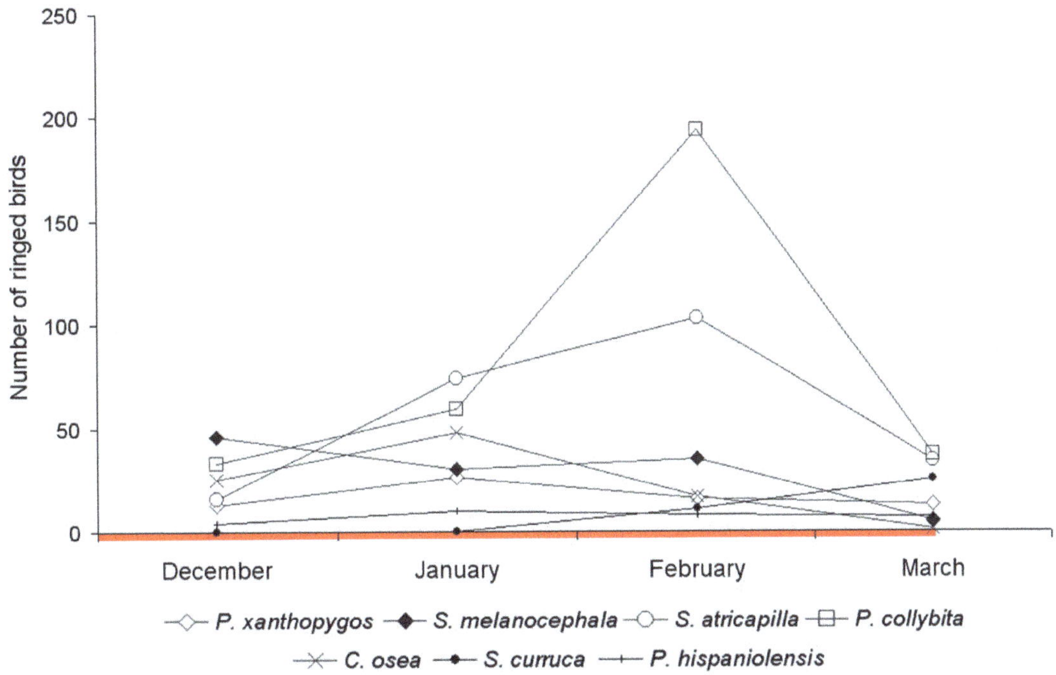

FIGURE 1: Birds presence during the flowering season of *A. foetida*, 2016/2017. (In red the flowering period of *A. foetida*)

FIGURE 2: Downward (A) and upward (B) approaches of Palestine Sunbird to *A. foetida* flower

Ella Valley was 18 November 2014, 23 November 2015, 1 December 2016). A sharp decline in the presence of sunbirds at the research sites occurred in mid-February.

Foraging behaviour of sunbirds on A. foetida

297 different visits of Palestine Sunbirds to *A. foetida* flowers in 18 different sessions were recorded. The birds did not hover but perched on the branches to collect the nectar. In 235 visits (79.12%), the birds approached the flower from above, inserting their bills between the short standard and the wings, elevating the angle of the flower to ease nectar consumption. After the visit, the flower parts returned to their original position (Fig. 2A). In 62 visits (20.88%), the sunbirds approached the flowers from below following the same procedure (Fig. 2B). 26 visits (8.75%) were recorded to flowers with apparently high level of nectar. The time spent consuming nectar from these flowers was 3.86 sec ± 1.04 sec SD. All other visits were much shorter (ca 0.5 sec), probably to flowers having less nectar or none at all.

When young, *A. foetida* flowers are closed, green-yellowish and their keel has two blackish spots. After anthesis the flowers open and become yellow with reddish-black spots (Valtueña et al. 2008, Ortega et al. 2005). The sunbirds bypassed all yellow flowers with reddish spots on the standard and with open keels. Instead, they nectar in the greenish flowers, of which most had closed keels. The sunbirds did not make any visible contact with the anthers or stigmas

Pollen load

Pollen loads were sampled from 643 birds from 10 species: 3 resident, 3 wintering and 4 "pure" migrating species (Table 1A and Fig. 3). The Palestine Sunbird and Chiffchaff only rarely carried substantial amounts of pollen. The Blackcap, Bulbul and Sardinian Warbler were quite often found to carry large amounts of pollen grains (level 3 in Table 1A). The migrating birds, which appeared at the research sites only at the end of the flowering season, were often found with large amounts of pollen grains (level 3 in Table 1A). The

TABLE I: (A) Level of pollen load found on different bird species. (B) Level of pollen load found on recaptured birds. (Pollen load: 0 = 0 pollen grains, 1 = < 10 pollen grains, 2 < 40 pollen grains, 3 > 200 pollen grains)

(A)

Species	N	Status	0 load	1 load	2 load	3 load	% 2+3 load	% 3 load
Blackcap (*Sylvia atricapilla*)	209	wintering	29	4	9	167	84.2	79.9
Chiffchaff (*Phylloscopus collybita*)	144	wintering	123	17	2	2	2.7	1.4
Common Whitethroat (*Sylvia communis*)	11	migrating	4	1	0	6	54.5	54.5
Lesser Whitethroat (*Sylvia curruca*)	35	migrating	15	0	0	20	57	57
Orphean Warbler (*Sylvia crassirostris*)	4	migrating	1	0	1	2	75	50
Palestine Sunbird (*Cinnyris osea*)	90	resident	60	23	5	2	7.7	2.2
Ruppell's Warbler (*Sylvia rueppelli*)	2	migrating	0	0	0	2	100	100
Sardinian Warbler (*Sylvia melanocephala*)	82	resident	20	3	2	57	71.9	69.5
Spanish Sparrow (*Passer hispaniolensis*)	14	wintering	10	0	1	3	28.5	21.5
White-spectacled Bulbul (*Pycnonotus xanthopygos*)	52	resident	5	1	4	42	88.5	80.7

(B)

Species	N	0 load	1 load	2 load	3 load	% 2+3 load	% 3 load
Blackcap (*Sylvia atricapilla*)	30	1	0	2	27	96.6	90
Chiffchaff (*Phylloscopus collybita*)	34	30	4	0	0	0	0
Lesser Whitehroat (*Sylvia curruca*)	3	0	0	0	3	100	100
Sardinian Warbler (*Sylvia melanocephala*)	33	3	2	1	27	84.8	81.8
White-spectacled Bulbul (*Pycnonotus xanthopygos*)	5	0	0	0	5	100	100

Spanish Sparrow can be regarded as an occasional carrier of pollen grains.

Recaptures

Some birds did not carry pollen grains the first time they were captured. We assumed that they were new arrivals to the research area. After getting familiar with the *A. foetida* flower, these birds began consuming its nectar and were found to carry pollen grains the second time caught (Table 1B). Only recaptures between ringing sessions are included in Table 1B. Apart from the Chiffchaff, birds from each species were found to have more pollen on average on the second capture. No Palestine Sunbird was recaptured.

Honeybees

190 visits of Honeybees to *A. foetida* flowers were recorded in both study sites when the ambient temperature between 10:30 and noon was 15–20°C. In 167 visits (87.9%) they tried to open the keels of the flowers, indicating that they were seeking pollen (Fig. 4). The average time they spent on a flower was 8.91 ± 6.81 sec (range of 2-29, N= 78) if they failed to open the keels and 19.51 ± 14.33 sec (3-70, N= 78) if they succeeded. Alternatively, 12.1% of the bees appeared to be seeking nectar only; these inserted their heads under the standard.

DISCUSSION

If an animal picks up pollen from one flower and its body is adapted for pollen transport to the stigma of another flower of the same species, it can be regarded as a true pollinator (Ortega et al. 2005). In order to extract nectar in a legitimate way from the fabaceous flower of *A. foetida,* a bird has to insert its bill into the flower under the short standard and into the pseudo-tubular perianth all the way into the hypanthium.

FIGURE 3. Pollen of *A. foetida* on feathers of (A) *Sylvia melanocephala*, (B) *Sylvia communis*, (C) *Sylvia crassirostris*, (D) *Sylvia curruca*, (E) *Sylvia rueppelli*, (F) *Pycnonotus xathophygos*, (G) *Passer hispaniolensis*, (H) *Sylvia atricapilla*

When legitimate visitors enact such a visit (Fig. 5), their head or throat touches the closed keel and opens it (Ortega et al.2005). In a successful pollination visit to *A. foetida* (Valtueña et al. 2008), two major events occur: 1. the stigma cuticle is ruptured, which is necessary to allow pollen germination 2. pollen is deposited on the bird's feathers.

The presence of pollen grains was easily observed with the naked eye on the bird's forehead, throat or bill (Fig. 3.) Captured birds frequently carried thousands of grains (level 3 in Table 1A). We consider a species to be a major potential pollinator if it fulfils the following criteria: 1. the species was present in the field in large numbers (can reach over 10 birds per ringing session) during major periods of the flowering season; 2. Over 70% of the species' individuals were found to carry large amounts (level 2 and level 3) of pollen on their feathers. Of the ten bird species sampled (Table 1A), only the Blackcap and the Sardinian Warbler meet these criteria as major potential pollinators of the *A. foetida*.

Palestine Sunbird

The bill length of *C. osea* (16-18 mm) is longer than the tubular structure of the *A. foetida* flower (ca 10 mm). Warblers' bills are shorter than those of *C. osea*: 13 mm for

FIGURE 4. Honey bees foraging pollen of *A. foetida.* (A) Trying to open the keel. (B) Inside the keel.

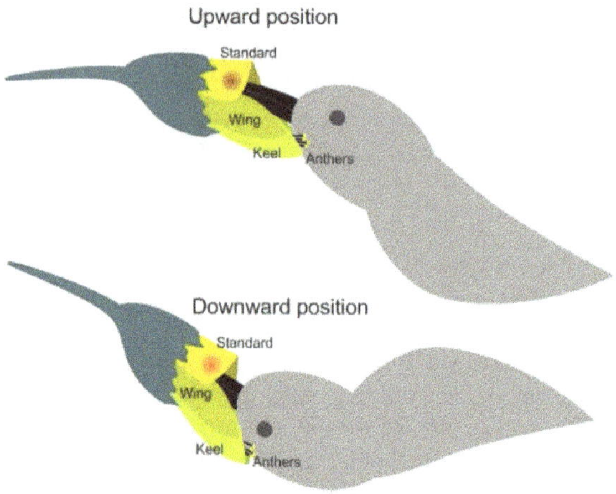

FIGURE 5. Diagram of downward and upward legitimate approaches of pollinating birds to *A. foetida* flower.

S. curruca, 14.7 mm for *S. atricapilla*, 14.3 mm for *S. melanocephala* (Hadoram et al. 2001). Video recordings of sunbirds foraging for nectar in *A. foetida* flowers show that the bird usually inserts its bill in the proper position to collect nectar without damaging the flower tissues. But, due to its long and curved bill, it does not touch the keel from either perching positions (Fig. 2). Only two netted Palestine Sunbirds (2.2% of trapped sunbirds) showed high amounts of pollen grains (level 3 in Table 1A). Thus, only rarely does it act as a pollinator in carrying pollen grains on its feathers or bill. Therefore, our study showed that, despite being the most specialized nectar feeder, Palestine Sunbirds are primarily nectar thieves of *A. foetida.*

The immense quantity of flowers—which can reach over 100,000 flowers for the season on an individual plant of *A. foetida* (Valutena et al. 2008)—and the large quantity of daily nectar production— that can reach over 40 μl per flower (Ortega et al. 2005)—are an ideal combination of features for foraging sunbirds (Wolf et al. 1975).

The role of omnivorous birds as pollen transport vectors

The omnivorous, all-year resident and wintering, bird species were present in the research area throughout the flowering season of *A. foetida* (Fig. 1). In autumn and winter (October–March) these birds rely almost entirely on fleshy fat-rich fruits and play an important role as seed dispersers of western and eastern Mediterranean plants (Herrera 1995; Izhaki 2002). Nectarivory by non-nectarivorous species is observed when resources such as insects, fruits or seeds are in short supply and nectar is abundant, e.g., during winter or early, cold springs (Valido et al. 2004). We found heavy pollen loads on foreheads, throats, or bills of these birds, indicating a high probability of successful pollination of *A. foetida.*

The White-spectacled Bulbul is a resident generalist species seen in groups of up to five birds in both research areas. *Anagyris foetida* pollen grains were found mainly on its bill and forehead (Fig. 3). As in the case of the Chinese Bulbul (*Pycnonotus sinensis*) (Fang et al. 2012) and the Red-whiskered Bulbul (*Pycnonotus jocosus*) (Olesen et al. 1998), the White-spectacled Bulbul was seen both consuming flower parts and inserting its bill in the proper position. More than 80% of trapped birds carried a large amount of pollen grains (Table 1A) and thus can be considered a potential pollinator.

Two subspecies of Sardinian Warblers, *S. m. momus* and *S. m. melanocephala* (local breeding and wintering/migrating, respectfully), were present in the research areas throughout the flowering season and bore pollen on both their foreheads and throats. We suggest that there are high chances that this species is a pollinator of *A. foetida* as approximately 70% of the birds were caught with high levels of its pollen grains (Table 1A) and more than 80% of recaptured birds had high levels of its pollen grains (Table 1B). *S. melanocephala* was also found to be nectar feeding from, and an effective pollinator of, *Isoplexis canariensis* in the Canary Islands (Rodrigues & Valido 2008).

Among the wintering species, only the Blackcap (*Sylvia atricapilla*) could be regarded as a major potential pollinator. 80% of the trapped birds carried a considerable amount of pollen grains on both their foreheads and throats (Table 1A) and had a high likelihood of touching the stigma (Fig. 3). Blackcaps were abundant throughout the flowering season. Recaptured birds carried a high amount of pollen grains in 90% of the cases (Table 1B). The Blackcap was found to visit *Isoplexis canariensis* and *Musschia wollastonii* Lowe flowers for nectar in Macronesia (Valido et al. 2004).

The Chiffchaff (*Phylloscopus collybita*) was caught in large quantities throughout the flowering season, but fewer than 2% carried substantial amounts of pollen grains (Table 1A, Table 1B). Chiffchaffs might best be regarded as non-pollinating visitors of *A. foetida* in Israel, where the weather might enable it to rely on its regular insect diet without consuming nectar.

The Spanish Sparrow (*P. hispaniolensis*) was seen and caught rarely, and only ca 20% (Table 1A) carried large amounts of pollen grains of *A. foetida* on their throats and bills (Fig. 3). It can be concluded that this species might be an occasional pollinator of this plant. This species was found pollinating introduced bird flowers such as *Strelitzia, Hibiscus, Aloe, Nicotiana* and *Agave* in Macaronesia (Valido et al. 2004).

Migrating birds

Various studies have already shown that many migrating birds are looking for nectar as an easy, energetic resource. They may act as pollinators in the process (Schwilch et al. 2001; Salewski et al. 2006; Cecere et al. 2011; Da Silva et al. 2014).

The beginning of spring migration coincided with the end of the flowering season of *A. foetida*. In the overlapping period, four species of migrating warblers were trapped with *A. foetida* pollen grains on their feathers (Fig. 3): Lesser Whitethroat, Common Whitethroat, Orphean Warbler and Ruppell's Warbler (Table 1A). The Lesser Whitethroat was more common than all the other migrating species. Pollen was found on both its forehead and throat (Fig. 3). 54.5% of the individuals caught carried level 3 of *A. foetida* pollen (Table 1A). Therefore, the Lesser Whitethroat is the only migrant that can be regarded as a major pollinator.

Honey bee (Apis mellifera) as a pollen vector

The distance between the beehives and the *A. foetida* bushes was 0–350 m in Hulda and 40-250 m in Ella Valley. Every day when temperatures were high enough (above ca 14°C), the bees foraged for food. As the *A. foetida* was almost the only flowering plant in the area, the bees were all over it, despite the fact that it is considered as only moderately attractive for bees looking for pollen (Keasar & Shmida 2009).

88% of the observed honey bees collected and carried pollen grains after successfully opening of the keel. It is likely they touch and rupture the stigma while entering into the keel and collecting pollen grains.

The high presence of honey bees on *A. foetida* flowers throughout the flowering season makes them important pollinators of this plant. A mixed pollination system between Hymenoptera and passerine birds was found in the Canary Islands in a variety of plants (Ortega et al. 2011).

Conclusions

Anagyris foetida, having an ecologically generalized pollination system, attracts a large array of potential pollinators including resident, wintering and migrating birds, and honey bees.

The variety of visitors to *A. foetida* and their role as potential pollinators differes from what Ortega et. al. (2005) found in Spain. This is probably due to differences in biogeography in terms of latitude, climate, diversity of bird species and subspecies, and the routes of bird migration.

Of the three bird species found in Spain as major pollinators, the Blackcap and Sardinian Warbler have higher chances of being important pollinators of *A. foetida* in Israel as well. However, the Chiffchaff appears to be a poor pollinator of *A. foetida* in Israel. Climate differences are probably also responsible for the different roles honey bees have in the pollination of *A. foetida*. The mild Israeli winters enable honey bees to be active outside their hives from late morning and to forage pollen and nectar.

To conclude, all omnivorous bird species (excluding the Chiffchaff) were highly suspected as legitimate pollinators, whereas only the nectarivorous Palestine sunbird acted as a nectar thief.

ACKNOWLEDGEMENTS

We would like to thank Tomas Rodriguez- Riano and Dan Eisikowitch for their helpful comments on a draft of this paper. We would also like to thank Carmel Zitronblat and many other bird ringers who joined the efforts of the winter ringing sessions.

REFERENCES

Abou-Shaara HF, Al-Ghamdi AA & Mohamed AA (2012) Tolerance of two honey bee races to various temperature and relative humidity gradients. Environmental and Experimental Biology 10: 133-138.

Arroyo, K (1981) Breeding systems and pollination biology. Advances in Legume Systematics. Part 2: 723-769.

Cecere JG, Spina F, Jenni-Eiermann S & Boitani L (2011) Nectar: an energy drink used by European songbirds during spring migration. Journal of Ornithology 152:923-931.

Cheke RA & Mann CF (2010) Sunbirds: A Guide to the Sunbirds, Flowerpeckers, Spiderhunters and Flowerpeckers of the World. Bloomsbury Publishing.

Collins BG & Paton DC (1989) Consequences of differences in body size, wing length and leg morphology for nectar-feeding birds. Australian Journal of Ecology 14: 269-289.

Coombs G, Mitchell S & Peter C (2007). Pollen as a reward for birds. The unique case of weaver bird pollination in *Strelitzia reginae*. South African Journal of Botany 73: 283.

Coombs V & Peter CI (2009) Do floral traits of *Strelitzia reginae* limit nectar theft by sunbirds? South African Journal of Botany 75: 751-756.

Cronk Q & Ojeda I (2008) Bird-pollinated flowers in an evolutionary and molecular context. Journal of Experimental Botany 59: 715-727.

Da Silva LP, Ramos JA, Olesen JM, Traveset A & Heleno RH (2014) Flower visitation by birds in Europe. Oikos 123: 1377-1383.

Dafni A (1996) Autumnal and winter pollination adaptations under Mediterranean conditions. Bocconea 5: 171-181.

Eisikowitch D & Nahari N (1982) The discrimination of *Malvaviscus Arboreus* flowers by the Sunbird, *Nectarinia osea*. Plant Biology 31: 55-58.

Faegri K & Pijl LV (1979) The Principles of Pollination Ecology, 3rd edition. Pergamon, Oxford.

Fang Q, Chen YZ & Huang SQ (2012) Generalist passerine pollination of a winter-flowering fruit tree in central China. Annals of Botany 109: 379-384.

Galloni M, Podda L, Vivarelli D, Quaranta M. & Cristofolini G (2008) Visitor diversity and pollinator specialization in Mediterranean legumes. Flora-Morphology, Distribution, Functional Ecology of Plants 203, 94-102.

Geerts S, & Pauw A (2009) African sunbirds hover to pollinate an invasive hummingbird-pollinated plant. Oikos 118:573-579.

Hadoram S (1996) The Birds of Israel. Academic Press.

Hadoram S, Gargallo G & Helbig AJ (2001) *Sylvia* warblers: Identification, Taxonomy and Phylogeny of the Genus *Sylvia*. Princeton University Press.

Herrera CM (1995) Dispersal systems in the Mediterranean: ecological, evolutionary, and historical determinants. Annual Review of Ecology and Systematics 26: 705-727.

Inouye DW (1983) The ecology of nectar robbing. The Biology of Nectaries. Columbia University Press 153-173.

Irwin RE, Bronstein JL, Manson JS & Richardson L (2010) Nectar Robbing: Ecological and Evolutionary Perspectives. Annual Review of Ecology, Evolution, and Systematics 41: 271-292.

Izhaki I (2002) The role of fruit traits in determining fruit removal in East Mediterranean ecosystems. In: Levey DJ, Silva WR, Galetti M (eds) Dispersal and Frugivory: Ecology, Evolution and Conservation. CAB International Publishing, Wallingford, Oxfordshire, UK, pp 161-175.

Joshi NC & Joshi PC (2010) Foraging behaviour of *Apis* spp. On apple flowers in a subtropical environment. New York Science Journal 3: 71-76.

Johnson SD & Nicolson SW (2008) Evolutionary associations between nectar properties and specificity in bird pollination systems. Biology Letters 4: 49-52.

Keasar T & Shmida A (2009) An evaluation of Israeli forestry trees and shrubs as potential forage plants for bees. Israel Journal of Plant Sciences 57: 49-64.

Kunitake YK, Hasegawa M, Miyashita T & Higuchi H. (2004) Role of a seasonally specialist bird *Zosterops japonica* on pollen transfer and reproductive success of *Camellia japonica* in a temperate area. Plant Species Biology 19: 197-201.

Myers S, Brown G & Kleindorfer S (2010) Divergence in New Holland Honeyeaters (*Phylidonyris novaehollandiae*): evidence from morphology and feeding behavior. Journal of Ornithology 151: 287-296.

Olesen JM, Rønsted N, Tolderlund U, Cornett C, Mølgaard P, Madsen J & Olsen CE (1998) Mauritian red nectar remains a mystery. Nature 393: 529-529.

Olesen, JM & Jordano, P (2002) Geographic patterns in plant–pollinator mutualistic networks. Ecology 83: 2416-2424.

Ortega-Olivencia A, Rodríguez-Riaño T, Valtueña FJ, López J & Devesa JA (2005) First confirmation of a native bird-pollinated plant in Europe. Oikos 110: 578-590.

Ortega-Olivencia A., Rodríguez-Riaño T, Pérez-Bote JL, López J, Mayo C, Valtueña FJ, & Navarro-Pérez M (2011) Insects, birds and lizards as pollinators of the largest-flowered *Scrophularia* of Europe and Macaronesia. Annals of Botany 109: 153-167.

Padyšáková E, Bartoš M, Tropek R & Janeček Š (2013) Generalization versus specialization in pollination systems: visitors, thieves, and pollinators of *Hypoestes aristata* (Acanthaceae). PLOS One 8.4 e59299.

Paton DC & Collins BG (1989) Bills and tongues of nectar-feeding birds: A review of morphology, function and performance, with intercontinental comparisons. Austral Ecology 14: 473-506.

Proctor M, Yeo P & Lack A (1996) The Natural History of Pollination. Harper Collins Publishers.

Ragusa-Netto J (2002). Exploitation of *Erythrina dominguezii* Hassl. (Fabaceae) nectar by perching birds in a dry forest in western Brazil. Brazilian Journal of Biology 62: 877-883.

Rodríguez-Rodríguez MC & Valido A (2008) Opportunistic nectar-feeding birds are effective pollinators of bird-flowers from Canary Islands: experimental evidence from *Isoplexis canariensis* (Scrophulariaceae). American Journal of Botany 95: 1408–1415.

Salewski V, Almasi B & Schlageter A (2006) Nectarivory of Palearctic migrants at a stopover site in the Sahara. British Birds 99: 299–305.

Schwilch R, Mantovani R, Spina F & Jenni L (2001) Nectar consumption of warblers after long distance flights during spring migration. Ibis 143: 24-32.

Stone GN, Raine NE, Prescott M & Willmer PG (2003) Pollination ecology of acacias (Fabaceae, Mimosoideae). Australian Systematic Botany 16: 103-118.

Tadmor-Melamed H, Markman S, Arieli A, Distl M, Wink M & Izhaki I (2004) Limited ability of Palestine Sunbirds *Nectarinia osea* to cope with pyridine alkaloids in nectar of Tree Tobacco *Nicotiana glauca*. Functional Ecology 18: 844–850.

Tandon R, Shivanna KR & Mohan Ram HY (2003). Reproductive biology of *Butea monosperma (Fabaceae)*. Annals of Botany 92: 715-723.

Vaknin Y, Tov YY & Eisikowitch D (1996) Flowering seasonality and flower characteristics of *Loranthus acacia* Zucc. (Loranthaceae): implications for advertisement and bird pollination. Sexual Plant Reproduction 9: 279–285.

Valido A, Yoko LD, & Olesen JM (2004) Bird–flower interactions in the Macaronesian islands. Journal of Biogeography 31: 1945-1953.

Valtueña FJ, Ortega-Olivencia A & Rodríguez-Riaño T (2007) Nectar production in *Anagyris foetida* (Fabaceae): two types of concentration in flowers with hanging droplet. International Journal of Plant Science 168: 627-638.

Valtueña FJ, Ortega-Olivencia A, Rodríguez-Riaño T & Lopez J (2008) Reproductive biology in *Anagyris foetida* L.(Leguminosae), an autumn–winter flowering and ornithophilous Mediterranean shrub. Botanical Journal of the Linnaean Society 157: 519-532.

Willmer P (2011) Pollination and Floral Ecology. Princeton University Press.

Wolf LL, Hainsworth FR & Gill FB (1975) Foraging Efficiencies and Time Budgets in Nectar – Feeding Birds. Ecology 56: 117-128.

Yom-Tov Y (2014) The Orange-tufted (Palestine) sunbird (in Hebrew). Carta-Jerusalem, Israel.

24

CONTROLLED POLLINATIONS REVEAL SELF-INCOMPATIBILITY AND INBREEDING DEPRESSION IN THE NUTRITIONALLY IMPORTANT PARKLAND TREE, *PARKIA BIGLOBOSA*, IN BURKINA FASO

Kristin Marie Lassen[a], Erik Dahl Kjær[a], Moussa Ouédraogo[b], Yoko Luise Dupont[c], Lene Rostgaard Nielsen[a]

[a]*Department of Geosciences and Natural Resource Management, Faculty of Science, University of Copenhagen, Rolighedsvej 23, 1958 Frederiksberg C, Denmark*

[b]*Centre National de Semences Forestières, Route de Kaya, 01 BP 2682 Ouagadougou, Burkina Faso*

[c]*Department of Bioscience, Aarhus University, Vejlsøvej 25, 8600 Silkeborg, Denmark*

Abstract—The socioeconomically important fruit tree *Parkia biglobosa* is becoming less abundant in the West African savannah, possibly due to poor regeneration. This decline can be self-enforcing if lower densities of fertile trees result in increasing self-pollination followed by increased abortion rates or poor regeneration due to inbreeding depression. Hence, we have studied the reproductive success and seedling viability of *P. biglobosa* after controlled self- and cross-pollination based on a full diallel crossing design with eight trees. Controlled cross-pollination tripled the pod set compared to open-pollinated capitula, suggesting that fruiting of *P. biglobosa* trees in the study area is already seriously pollen limited. Self-pollination and specific pairs of trees resulted in very few pods, suggesting a high level of self-incompatibility. Cross-pollination resulted in larger pods with more and heavier seeds than self-pollinated pods. The total amount of sugar in the fruit pulp was correlated with both the number of healthy and total seeds per pod. Growth rate of self-pollinated seedlings was lower than the cross-pollinated ones, suggesting significant inbreeding depression. Because the wild fruit trees play an important role in human nutrition, these results give rise to serious concerns. We recommend that future studies investigate how the level of cross-pollination can be increased and how the regeneration of *P. biglobosa*, whether natural or planted, can be improved.

Keywords: Controlled pollination; Fruit quality; Inbreeding depression; Pollen limitation; Self-incompatibility

INTRODUCTION

The majority of tropical tree species rely on animals for pollination (Ollerton et al. 2011), and many possess a system of self-incompatibility (Bawa et al. 1985; Ward et al. 2005). The high dependency on animal pollinators in the tropics (Ollerton et al. 2011), combined with pollination deficit due to historical decline of wild pollinators (e.g. Goulson et al. 2015) and limited alternative options for cross-pollination, may decrease fruit production and quality of important crop trees. In addition, anthropogenic impacts, such as farming (monoculture, pesticides), landscape fragmentation, and climate change are likely to reduce abundance and species diversity of pollinators (Kennedy et al. 2013). Understanding mechanisms of pollination and optimising fruit production in tropical fruit trees are crucial, given that many livelihoods depend on crop yield from animal-pollinated tropical trees.

Local decline of pollinators and/or increased distances between conspecific trees can reduce seed set due to pollen limitation, if fruit and seed set are limited by the supply of compatible pollen, rather than resource availability (Ashman et al. 2004; Knight et al. 2005; Aizen & Harder 2007). Another potential consequence is increased selfing, which may decrease fruit production, and reduce the fitness of seedlings due to inbreeding depression (Husband & Schemske 1996). Hence, we expect a decrease in reproductive success, combined with reduction of seedling fitness when conspecific trees are more widely spaced in a disturbed landscape.

Parkia biglobosa (Jacq.) R. Br. ex G. Don (Fabaceae: Mimosoideae), is a West African parkland tree with high nutritional importance for the rural people due to its sweet fruit pulp and seeds with high protein content (Uwaegbute 1996; Hall et al. 1997). In Burkina Faso, pods from *P. biglobosa* are an important food source (Lykke et al. 2002), especially during periods of food scarcity (Nyadanu et al. 2017). Due to over-exploitation (Gaisberger et al. 2017), low regeneration (Ræbild et al. 2012), and reduced annual rainfalls (Maranz 2009; Funk et al. 2012), there is a high risk of seriously decreasing tree densities (Gaisberger et al. 2017) as already witnessed by the local population (Lykke et al. 2002). This may increase selfing (Lassen et al. 2017), but the potential effects of increased self-pollination on seedling growth and nutritional content of the fruits of *P. biglobosa* are unknown. Related species possess a self-incompatibility system (Hinata et al. 1993), and hence increased selfing is expected to reduce seed set. Studies of other plant species have documented positive correlations between seed number and fruit traits including fruit weight, size, and oil content (Hopping 1976; Roldán Serrano & Guerra-Sanz 2006; Abrol

*Corresponding author: kristin_lassen@yahoo.dk

2012). Thus, increased self-pollination, by causing reduced seed number, may also reduce the nutritional value of *P. biglobosa* fruits.

To understand productivity of this important crop tree species, it is important to determine if *P. biglobosa* is pollen limited, and how increased levels of self-pollination will affect fruit and seed set, nutrition contents and seedling fitness of this species. We addressed these questions by performing controlled self- and cross-pollinations of *P. biglobosa* capitula in order to compare 1) the reproductive success of selfed, crossed, and open-pollinated capitula, 2) the carbon and nitrogen contents of seeds and sugar content of fruit pulp from selfed and outcrossed pods, and 3) the germination percentage and seedling vigour of selfed and outcrossed seeds. We discuss the probability of *P. biglobosa* experiencing pollen limitation, self-incompatibility and inbreeding depression.

MATERIALS AND METHODS

Plant species

Parkia biglobosa is pollinated mainly by bees and bats (Baker & Harris 1957; Hopkins 1983; Ouédraogo 1995; Lassen et al. 2012; Lassen et al. 2017). It is predominantly outcrossing (Ouédraogo 1995; Sina 2006), although Lassen et al. (2017) found that selfing can occur in areas with low tree density. *Parkia biglobosa* flowers for around four weeks in the dry season in Burkina Faso from January to April, depending on latitude, and with year to year variation (pers. obs.).

Flowers are grouped in ball-shaped capitula with around 2,200 tiny bright red flowers packed closely on a bulbous receptacle hanging on a long peduncle (Hopkins 1983). Nectar is produced by sterile flowers close to the peduncle and accumulates in a nectar ring (Hopkins 1983). At the study site, buds opened during the afternoon. The capitula started producing nectar around 18:45 h (local time, UTC + 0 h) and shedding pollen around 19:30 h (pers. obs.). Each tiny flower has ten anthers and one style. *Parkia biglobosa* is andromonoecious, and the functionally male capitula have styles, which fail to elongate (Hopkins 1981). Pollen is shed in polyads with 32 pollen grains (pers. obs.) clumped together. The cup-shaped stigma can hold only one polyad; therefore all seeds per pod are full siblings (Lassen et al. 2014). According to our observations, based on $N = 15$ ovaries distributed on three trees in Burkina Faso, the ovary of a hermaphroditic capitulum contained a mean of 23 ovules (SE = 0.63, min-max: 16-29). The hermaphroditic capitula are protandrous (Ouédraogo 1995) and the female phase begins around 23:00-24:00 h when the stigmas have extended to reach the same level or above that of the anthers (pers. obs.). Furthermore, flowering within a capitulum is highly synchronized. Each capitulum blooms one night and during the morning both hermaphroditic and functionally male capitula start to wilt (pers. obs.). In the present study, each capitulum is treated as one unit.

One week after pollination and fertilisation, tiny green pods are visible, and after around two months the indehiscent, brown pods are mature (pers. obs.). Even though each hermaphroditic flower has the potential of producing a pod, only a few pods per capitulum develop (Hopkins 1984).

Study site

The present study took place in the village Pinyiri (syn. Kacheli) (11°14'34.89"N, 1° 8'1.73"W), eight km north of Pô, Nahouri province. The site is within the Sudanian climatic zone with a unimodal rainy season and an average precipitation (1981-2010) of 900-1,000 mm (Sanfo 2012). In 2011, preceding the fruiting season of *P. biglobosa* in 2012, the annual precipitation in Pô was 927 mm (Météo 2015).

Controlled pollination experiment

Prior to the experiment, swollen buds (expected to open the following night) were covered with cheesecloth with an inner band of chicken wire to keep the nettings from touching the flowers. We used inflorescences from the lowest part of the tree crown, which could be reached from a ladder. In the experiment, only the most apical bud within the compound inflorescence was used. During 8-18 March 2012, the crossing experiment was carried out as a diallelic cross (i.e. all trees were crossed with each other) of eight trees (mean DBH = 1.4 m, SD = 0.81), although one of the trees was not used as a pollen donor. Each treatment was replicated nine times per tree.

Treatments:

- Open = open-pollinated capitula (control, $N = 72$)
- Self = self-pollinated capitula ($N = 72$)
- Cross = cross-pollinated capitula ($N = 423$)
- 'Both' = half of a capitulum was pollinated with self-pollen and the other half with cross-pollen (only on six trees due to lack of capitula, $N = 54$)

Prior to the pollination treatments, we checked the sex (hermaphroditic or functionally male) of the flowers on each receiving capitulum, since only hermaphroditic capitula can develop pods. When in doubt, we measured the distance between stigmas and anthers, as this difference was suspected to influence the functional sex of the capitulum, due to the probability of the many densely packed anthers acting as a carpet keeping pollen away from the shorter stigmas. The dividing line between hermaphroditic and functionally male capitula was unknown at the time we carried out the controlled pollinations. Hence, these were carried out regardless of the assigned sex, from midnight until early morning (03-04 h) using capitula directly as pollen brushes by dabbing the donor capitula on the receiving capitula, which were previously protected by bags. No flowers were emasculated due to the high number of anthers per capitulum. To standardise the dose of pollen, one donor capitulum was used on three receiving capitula (1:3) for the treatments of self- and cross-pollinations (approximately one third of the donor on each of the receiving capitulum). To study the effect of self- and cross-pollination under identical conditions, we applied self and cross-pollen to flowers on the same capitulum (but not the same flowers). We refer to this treatment as 'both', and for this treatment we used two donor capitula (self and cross) for two receiving capitula (half self and half cross on each capitulum) (1:1). Half of the peduncle was marked with a black marker (for the cross-pollination) and we performed the two kinds of pollination by eye. After pollination, the capitula were re-bagged. The trees were visited daily for five weeks; dropped capitula were collected and examined for sex (hermaphroditic or functionally male) and signs of developing pods and/or predation.

After harvest, pods were counted and weighed. Their lengths were measured as the mean of each side of the pod, excluding the pedicel. We did not count the seeds of these pods (see below).

DNA extraction and genotyping

To assess the level of self-pollination (on purely self-pollinated capitula and on 'both' capitula), we extracted DNA, genotyped seeds ($N = 142$) and compared their genotypes with the genotype of the mother trees. Likewise, genotypes of the eight experimental trees were compared to genotypes of a sub-sample of pods, to test if the cross had been performed correctly. DNA was extracted directly from dehulled seed; endosperm is absent because the cotyledons provide resources to the embryo (Hopkins 1983). DNA extraction used the DNeasy 96 Plant Kit (QIAGEN, Hombrechtikon, Switzerland), following the manufacturer's protocol. The genotyping was based on ten microsatellite primer pairs developed for *P. biglobosa* (Lassen et al. 2014) with PCR reactions and fragment analysis following Lassen et al. (2017).

Comparison of self- and cross-pollinated pods

The self-pollinated capitula in the controlled pollination experiment yielded extremely few pods. Thus, in order to compare the quality of seeds in *P. biglobosa* after self- and cross-pollination we assessed the quality from 48 pods obtained from a previous experiment from the same study site (Lassen et al. 2017). These pods were collected from eleven trees and were randomly selected, although the selection was balanced with an equal number of pods per tree being self- and cross-pollinated. Seeds were categorised as 'healthy', 'eaten' (hole in the husk and/or seed remains due to predation by parrots and/or worms), 'aborted' (weight < 0.05 g and/or with a flat shape), 'missing' (empty cavity in the pulp), and from these four numbers a total number of seeds was established.

Moisture, carbon and nitrogen content of seeds

One seed from each of 48 pods (24 selfed and 24 outcrossed) was analysed for moisture, carbon and nitrogen content. The seeds were weighed with and without testa, placed in open 2 ml Eppendorf tubes and dried at 65°C for three days until stable weight. Moisture content of seeds without testa was calculated as fresh weight minus dry weight, divided by fresh weight and multiplied by 100 (%). Then the dehulled seeds were ground in a mortar and weighed into tin capsules with 5 mg in each sample. Total carbon (C) and nitrogen (N) content were measured using the Dumas principle: samples were combusted at 1,850°C on a FLASH 2000 NC Analyzer (Thermo Scientific) according to the manufacturer's manual. To calibrate the measurements, we used a standard ('spruce needles' Forest Foliar Coordinating Centre, FFCC), and two reference samples ('maple leaves' FFCC, run twice, and 'corn gluten organic' Sercon, run for every nine samples of *P. biglobosa*). The FLASH 2000 Analyzer is comparable with the Kjeldahl method (Krotz & Giazzi 2014).

We have converted the amount of nitrogen (N) to crude protein by multiplying N with the commonly used factor 6.25 (AOAC 1990), although Ezeagu et al. (2002) found a lower nitrogen-to-protein conversion factor on 4.97 for Fabaceae seeds (mean for ten species) and Yeoh & Wee (1994) found an even lower conversion factor on 4.23 for leaves of *Parkia timoriana* (DC.) Merr. (syn. *P. javanica*).

Pod pulp analysis

We analysed the sugar content in the pulp of the same pods as above, except for two self-pollinated pods, which had no pulp (i.e. 22 self- and 24 cross-pollinated pods were used in this analysis). We kept the pulp from each pod separately. The pulp was dried (103°C for 3 h), ground and sifted, and two sub-samples of 100 mg per pod were used. The soluble sugars were extracted and analysed by HPLC according to the method described by Liu et al. (2004).

Germination rate and seedling vigour of selfed and crossed seeds

In order to compare seed viability under optimal conditions, we tested the germination by using healthy-looking seeds from the 24 self-pollinated pods (90 seeds) and 24 cross-pollinated pods (163 seeds) and assessed the seedling vigour by growth and dry weight. Each seed was weighed and scarified because of the hard seed coat (testa), which must be broken before the seeds can germinate (Etejere et al. 1982). The seeds were germinated in plastic boxes in a growth chamber in a 25°C day and night, and 12 h light/darkness regime. We defined the seeds as germinated when the radicle protruded for 3 mm, and we monitored the seed germination daily.

On the 19th day after sowing, seedlings were weighed and planted individually in pots (Ø = 13 cm) with planting peat soil. The pots were placed randomly in a greenhouse at 28°C day and 20°C night and with a 12 h light/darkness regime from 08:00 h (local time, UTC + 1 h). The seedlings were watered daily with demineralised water without fertiliser for the first 2 months, and thereafter with fertiliser. Plant height (from soil level to top of main stem, or to the highest stem in case of more stems), stem diameter (measured with an electronic caliper) and number of pinnae (i.e. primary division of a bipinnate compound leaf) were measured on five occasions (43, 76, 104, 144, and 222 days after sowing). At the last measuring, we included fresh weight of the seedlings before drying them at 80°C (Osonubi & Fasehun 1987). After 24 hours (until stable weight), we recorded dry weight of the entire plant, shoots (stem plus leaves) only, and roots only, in order to calculate the shoot:root ratio.

Data analysis

In the controlled pollination experiment, reproductive success was evaluated as 1) numbers of immature pods (reflecting the success of pollination) and mature pods (additionally reflecting available resources by the mother-tree) per hermaphroditic capitulum and 2) weight and length of these pods. Analysis of variance was performed based on average values per treatment and tree applying the statistical SAS software v.9.4 (SAS Institute 2011). The mean number of pods (immature and mature) per hermaphroditic capitulum included capitula without any pods. When testing differences between treatments for the controlled pollination experiment, we used the general linear model as implemented in the GLM procedure:

1. $Y_{gh} = \text{Treatment}_g + \text{Mother-tree}_h + \varepsilon_{gh}$

where Y_{gh} is the response variable, treatment g = (open, self, cross, 'both'), and mother-tree h = 1...8. Treatment_g was considered a fixed effect, whereas Mother-tree_h was considered a random effect with residual ε_{gh} assumed independent and $N(0,\sigma_e^2)$. We assessed and accepted the model assumptions by visual inspection of the residuals.

For testing differences among pairs of pollen donors (i.e. male parent) and mother-trees (i.e. female parent), we analysed number of pods per single capitulum (not averaged per tree), using the general linear model as implemented in the GLM procedure:

2. $Y_{ij} = \text{Pollen donor}_i + \text{Mother-tree}_j + \text{Pollen donor}_i{*}\text{Mother-tree}_j + \varepsilon_{ij}$

where Y_{ij} is the response variable (log transformed), pollen donor i = (T10, T14, T22, T76, T90, T92, and T93), mother-tree j = 1...8, and pollen $\text{donor}_i{*}\text{mother-tree}_j$ = the interaction between the pollen donor and the mother-tree. All effects were considered fixed with residual ε_{ij} assumed independent and $N(0,\sigma_e^2)$. We assessed and accepted the model assumptions by visual inspection of the residuals.

When testing differences between self- and cross-pollination for several parameters related to plant fitness, the results were averaged per type of pollination and tree, and we used a similar general linear model as above:

3. $Y_{kl} = \text{Pollination type}_k + \text{Mother-tree}_l + \varepsilon_{kl}$

where Y_{kl} is the response variable, pollination type k = (self, cross), and mother-tree l = 1...11. $\text{Pollination type}_k$ was considered a fixed effect, whereas Mother-tree_l was considered a random effect with residual ε_{kl} assumed independent and $N(0,\sigma_e^2)$. Again, we assessed and accepted the model assumptions by visual inspection of the residuals.

When testing the difference between carbon and nitrogen contents of selfed versus outcrossed seeds, we used the following general linear model:

4. $Y_{klm} = \text{Pollination type}_k + \text{Mother-tree}_l + \text{Seed weight}_m + \varepsilon_{klm}$

where Y_{klm} is the response variable, pollination type k = (self, cross), mother-tree l = 1...11, and seed weight included as covariate. Seed weight varied from 0.0821 – 0.2733 g per seed. $\text{Pollination type}_k$ was considered a fixed effect, whereas Mother-tree_l was considered a random effect with residual ε_{klm} assumed independent and $N(0,\sigma_e^2)$. Again, we assessed and accepted the model assumptions by visual inspection of the residuals.

We used Fisher's exact test (as implemented in SAS procedure FREQ) to test differences between self- and cross-pollination for number of germination seeds and surviving seedlings.

The relationship between number of seeds per pod and sugar content in fruit pulp was analysed by calculation of Pearson correlation coefficients (using SAS procedure CORR).

RESULTS

Assessment of the diallel crossing experiment

In the controlled pollination experiment, 451 capitula were hermaphroditic, 139 capitula were functionally male, and 45 capitula were 'mixed' (i.e. containing both hermaphroditic and functionally male flowers in different ratios). Of capitula with hermaphroditic flowers, 83% developed pods.

The measurement of distances between stigmas and anthers of the capitula (N = 245 capitula) coupled with the reproductive success (setting fruit or not) revealed that capitula having anthers protruding > 5 mm longer than stigmas were typically functionally male (91%). Hence, this measure could be used as a rule of thumb.

Of the four pollination treatments (open, self, cross, and 'both'), self-pollination led to a significantly lower proportion of capitula with at least one immature pod (100%, 19%, 96%, 97%, respectively). In addition, the mature pod set also differed, and the experiment yielded 2,643 pods for the purely cross-pollinated capitula (423 pollinated capitula) and only 2 pods for the purely self-pollinated capitula (72 pollinated capitula). Genotyping of a subset of pods confirmed that cross-pollinated pods indeed were results of cross-pollinations while a few matured 'self-pollinated' pods, turned out to be cross-pollinated, probably due to small amounts of 'carryover' pollen. For treatment 'both', in which 502 pods were matured, no pods developed in the self-pollinated halves of the capitula and the subset of genotyped pods showed no self-pollination (N = 133).

Effects of pairs of trees and pollen doses on fruit set

In the controlled cross-pollination treatments, we found highly significant effects of both mother-trees (female parent, $F_{(7,231)} = 9.1$, $P < 0.001$) and pollen donors (male parent, $F_{(6,231)} = 4.7$, $P < 0.001$) on the number of immature pods per hermaphroditic capitulum. However, the number of mature pods was only significantly affected by the mother-trees ($F_{(7,199)} = 13.5$, $P < 0.001$). Two trees (P14 and P22) were equally good as mother-trees and pollen donors, whereas three trees had highest reproductive success as mother-trees (i.e. producing many pods, P76, P92, and P93) and two trees were best as pollen donors (i.e. fathering many pods, T10 and T90) (Tab. 1 and 2).

Interactions between mother-trees and pollen donors were highly significant for both immature ($F_{(34,231)} = 3.4$, $P < 0.001$) and mature pods ($F_{(34,199)} = 3.4$, $P < 0.001$), i.e. fruit set depended on the combination of mother-trees and pollen

TABLE 1. Mean number (± SE) of immature pods per hermaphroditic capitulum (i.e. small pods before maturation) of *Parkia biglobosa* (incl. capitula without pods). Self-pollination (grey colour) is shown in the last row. Tree P33 did not give pollen to the other trees.

		Female parent (i.e. mother-tree)								
Treatment		P10	P14	P22	P33	P76	P90	P92	P93	Across trees
Open[a]		4.1 (4.80)	5.6 (4.52)	1.3 (4.52)	2.8 (4.52)	7.1 (4.52)	2.9 (4.52)	2.8 (4.52)	4.3 (4.52)	3.9 (0.41)
Male parent (i.e. pollen donor)	T10	-	42.5 (5.54)	41.7 (7.83)	5.0 (7.83)	50.0 (4.52)	10.3 (7.83)	23.9 (5.13)	61.6 (4.80)	39.4 (4.74)
	T14	15.5 (4.80)	-	15.8 (4.52)	9.6 (4.80)	31.1 (4.52)	1.0 (6.07)	23.5 (5.54)	22.6 (5.13)	17.8 (2.08)
	T22	26.8 (5.54)	17.3 (5.13)	-	18.8 (6.07)	26.1 (4.52)	7.4 (5.13)	9.5 (5.54)	26.5 (4.80)	19.4 (2.23)
	T33	na	na	na	-	na	Na	na	na	na
	T76	10.7 (5.13)	15.3 (5.54)	15.4 (4.80)	na[d]	-	9.5 (5.54)	1.0 (7.83)	27.5 (5.54)	14.3 (1.70)
	T90	12.4 (6.07)	1.1 (5.13)	22.0 (6.07)	12.0 (4.80)	39.0 (4.52)	-	42.5 (6.78)	20.3 (7.83)	20.9 (3.11)
	T92	18.2 (5.54)	23.7 (5.54)	14.2 (6.07)	18.0 (13.57)	11.6 (5.13)	8.5 (5.54)	-	25.8 (6.78)	16.4 (1.97)
	T93	26.5 (5.54)	25.8 (4.52)	9.0 (6.07)	7.7 (7.83)	23.1 (4.52)	14.6 (5.13)	45.5 (9.59)	-	21.0 (2.48)
Cross[b]		18.2 (2.27)	20.7 (2.46)	17.6 (2.01)	11.5 (1.64)	30.9 (3.56)	8.8 (0.93)	22.5 (3.78)	33.1 (3.77)	21.2 (1.12)
Self[c]		0.0 (0.00)	0.0 (0.00)	0.4 (0.30)	0.3 (0.25)	0.4 (0.18)	0.1 (0.12)	0.5 (0.50)	0.0 (0.00)	0.2 (0.07)

[a]Open is open-pollination (control). [b]Cross denotes the mean of cross-pollination (i.e. across T10-T93 except self-pollination). [c]Self signifies self-pollination. [d]All nine P33-capitula pollinated with T76 were functionally male.

donors (Tab. 1 and 2). For instance, using P10 as a pollen donor resulted in more than twice as many immature pods per hermaphroditic capitulum on P93 compared to P92 (61.6 versus 23.9), while for P90 the result was vice versa (20.3 versus 42.5, Tab. 1).

Effects of pollen doses on fruit set (excluding self-pollination) and results of the open-treatment are shown in Tab. 3. The number of capitula with at least one immature pod was not influenced by the different pollen doses while the number of pods per hermaphroditic capitulum was significantly higher in hand-pollinated capitula (cross and 'both') compared to the open-pollinated capitula. The hand-pollinated capitula differ in that the 'both' treatment had around 3 times more cross-pollen than the cross treatment, but only on half of the capitulum. Hence the figures can be made comparable by multiplying those for the 'both' treatment with ⅔: (30.8 x ⅔ =) 20.5 for immature pods and (13.5 x ⅔ =) 9.0 for mature pods, which are close to the actual figures for the cross treatment on 20.4 and 9.8 for immature and mature pods, respectively (Tab. 3). However, with an increasing number of pods per capitulum the abortion rate (immature minus mature pods) also increased. For pod weight and length, the differences between pollen doses were non-significant (Tab. 3).

In one pair of trees (P14 and P90), we found that cross-pollination resulted in a similar low pod set as for self-pollination, independently of which tree was mother-tree and which was pollen donor, suggesting that these two trees were not compatible. Finally, we found another pair of trees (P76 and P92) with very few pods when P92 was the mother-tree, but only a reduced pod set when P76 was the mother-tree (Fig. 1, Tab. 1 and 2), which suggest an incompatibility system.

Comparison of self- and cross-pollinated pods

Because the diallelic crossing experiment resulted in only two matured self-pollinated pods, we used 24 self-pollinated and 24 cross-pollinated pods originating from another experiment at the same study site, as stated above. Self-pollinated pods were significantly shorter and weighed less than cross-pollinated pods (Tab. 4). Furthermore, although not significant, the self-pollinated pods had half as many healthy seeds, significantly more aborted seeds and significantly fewer total seeds compared to the cross-pollinated pods (Tab. 4). The amounts of eaten seeds and missing seeds were low, and showed no difference between selfed and outcrossed pods (Tab. 4).

Table 2. Mean number (± SE) of mature pods per hermaphroditic capitulum (i.e. harvested pods) of *Parkia biglobosa* (incl. capitula without pods). Self-pollination (grey colour) is shown in the last row. Tree P33 was not used as a pollen donor.

Treatment	Female parent (i.e. mother-tree) P10	P14	P22	P33	P76	P90	P92	P93	Across trees
Open[a]	4.1 (1.93)	4.7 (1.93)	1.3 (1.93)	2.7 (1.93)	6.8 (1.93)	2.9 (1.93)	2.4 (1.93)	3.7 (1.93)	3.6 (0.39)
Male parent (i.e. pollen donor): T10	-	7.3 (2.36)	16.0 (5.79)	2.0 (4.09)	13.4 (1.93)	2.3 (3.34)	15.7 (2.19)	23.8 (2.05)	13.7 (1.84)
T14	8.3 (2.05)	-	12.0 (4.09)	4.8 (2.36)	15.6 (2.05)	0.8 (2.59)	17.5 (2.36)	10.2 (2.36)	10.1 (1.33)
T22	9.2 (2.36)	5.2 (2.59)	-	7.0 (2.59)	7.3 (2.19)	4.3 (2.19)	6.8 (2.59)	17.1 (2.19)	8.4 (1.04)
T33	na	na	na	-	na	na	na	na	na
T76	7.9 (2.05)	9.0 (2.36)	8.0 (2.36)	na[d]	-	4.8 (2.36)	1.0 (3.34)	18.8 (2.36)	8.9 (1.09)
T90	8.0 (2.36)	0.9 (2.19)	7.0 (2.59)	5.3 (2.36)	20.0 (2.05)	-	25.3 (3.34)	6.7 (3.34)	9.9 (1.54)
T92	13.5 (2.36)	11.8 (2.36)	11.0 (2.59)	1.0 (5.79)	8.0 (2.19)	3.4 (2.59)	-	15.5 (2.89)	10.1 (1.20)
T93	13.3 (2.36)	11.0 (1.93)	4.2 (2.59)	1.7 (3.34)	9.3 (1.93)	6.9 (2.19)	18.5 (4.09)	-	9.1 (1.00)
Cross[b]	9.8 (0.85)	7.7 (0.86)	8.3 (1.27)	4.6 (0.76)	12.4 (1.26)	4.1 (0.60)	14.0 (2.22)	16.6 (1.65)	10.0 (0.50)
Self[c]	0.0 (0.00)	0.0 (0.00)	0.3 (0.19)	0.0 (0.00)	0.3 (0.16)	0.0 (0.00)	0.0 (0.00)	0.0 (0.00)	0.1 (0.05)

[a]Open is open-pollination (control). [b]Cross denotes the mean of cross-pollination (i.e. across T10-T93 except self-pollination). [c]Self signifies self-pollination. [d]All nine P33-capitula pollinated with T76 were functionally male.

TABLE 3. Effect of open-pollination (unknown pollen doses) and two known pollen doses on the percentage of hermaphroditic capitula with at least one immature pod, number of immature and mature pods per hermaphroditic capitulum (incl. capitula without pods), and the pod weight and length of *Parkia biglobosa*, including *F*-tests and significance levels. Except in the open-pollinated treatment, capitula were bagged until hand-pollination and re-bagged following pollination treatment.

Type of pollination	Pollen doses	No. of capitula[a], *N*	Capitula with ≥ 1 immature pod[b], %	No. of immature pods/capitulum[b]	No. of mature pods/capitulum[b]	No. of pods[c], *N*	Pod weight[c], g	Pod length[c], cm
Open[d]	unknown	71	100.0 (1.15)	3.9 (2.87)	3.6 (1.02)	258	14.0 (0.62)	22.6 (0.74)
Cross[e]	⅓	292	95.7 (1.15)	20.4 (2.87)	9.8 (1.02)	2,643	11.9 (0.62)	20.7 (0.74)
'Both'[f]	1	40	98.0 (1.41)	30.8 (3.52)	13.5 (1.25)	502	12.8 (0.77)	21.2 (0.90)
$F(P)_{Pollination}$			$F_{(2,12)}=3.5$ (ns)	$F_{(2,12)}=18.8$ (***)	$F_{(2,12)}=20.7$ (***)		$F_{(2,12)}=2.9$ (ns)	$F_{(2,12)}=1.8$ (ns)
$F(P)_{Tree}$			$F_{(7,12)}=0.7$ (ns)	$F_{(7,12)}=2.3$ (ns)	$F_{(7,12)}=3.0$ (*)		$F_{(7,12)}=5.6$ (**)	$F_{(7,12)}=5.2$ (**)

[a]Number of capitula used per type of pollination. [b]Values are least squares (LS) means with the standard error (SE) in brackets of the LS estimate. Significance level: ***= $P<0.001$, **= $P<0.01$, *= $P<0.05$, and ns= $P>0.05$. [c]Number of pods harvested. [d]Open was open-pollinated capitula, which were neither bagged nor hand-pollinated (control). [e]Cross was cross-pollinated capitula with one capitulum giving pollen to three capitula (1:3 ratio). [f]'Both' was capitula, which were self-pollinated on one half and cross-pollinated on the other half, with one capitulum giving pollen to two half capitula (1:1 ratio), but only cross-pollinated pods developed.

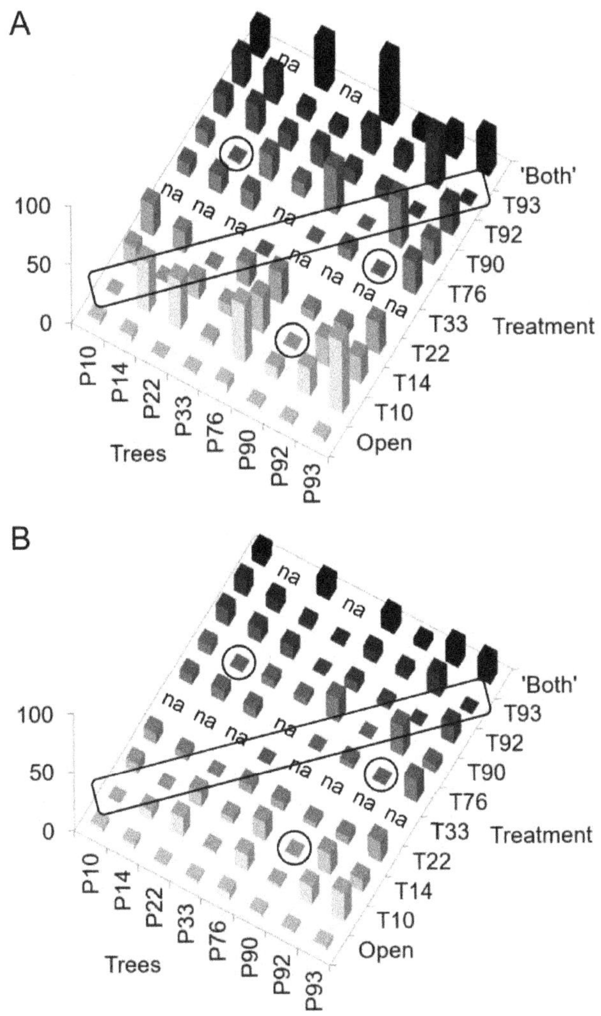

FIGURE 1. Overview of the number of immature (A) and mature (B) pods per hermaphroditic capitula for *Parkia biglobosa*, distributed on treatments (i.e. pollen donors) and trees. The box encompasses the events of self-pollination, which all resulted in very few pods, and the circles indicate low numbers of pods resulting from cross-pollinations. 'Open' indicate open-pollination (control) and 'Both' is a treatment in which capitula were pollinated with self-pollen on one half and cross-pollen on the other half. However, 'Both' yielded exclusively outcrossed pods. Pollen donor T33 was only applied to tree P33, pollen donor T76 was applied to only functionally male capitula on tree P33 and neither tree P14 or tree P33 had not enough capitula for the 'Both'-treatment.

Carbon & protein contents in seeds, and sugar contents in fruit pulp

Raw, dehulled seeds from cross-pollinated seeds had slightly more carbon compared to self-pollinated seeds (54.3% versus 53.4%), while percentage of protein (dry weight) was slightly lower (43.3% versus 45.2%, Tab. 4). The carbon content was highly dependent on seed weight ($F_{1,35}$ = 12.4, $P < 0.001$) with larger seeds containing more carbon. This was not the case for amount of protein ($F_{1,35}$ = 1.0, P = 0.3), hence larger seeds contained more carbon whereas protein content was constant, regardless of seed size.

Sugar content was higher in fruit pulp for the cross-pollinated pods, but only significantly so for glucose and fructose (Tab. 4). Numbers of healthy seeds and amounts of glucose, fructose and total sugars, respectively, were positively and significantly correlated (Tab. 5). No correlation was found between the amount of sucrose and number of seeds per pod (Tab. 5).

Germination and seedling growth of self- and cross-pollinated seeds

Seed weight and seedling growth was highly variable in both types of pollination (self and cross). The mean weight of cross-pollinated seeds was significantly higher than self-pollinated seeds, while germination percentage and germination speed did not differ significantly between pollination types (Tab. 4). Nine selfed and nine cross-pollinated seeds germinated but died before the first measurement of seedlings, and most seedlings, which died during the trial, perished before the second measurement. From the initial 90 self-pollinated and 163 cross-pollinated seeds, significantly fewer self-pollinated seedlings (69) than cross-pollinated (147) seedlings survived until the trial was terminated (Tab. 4). Initial fresh weights (19 days after sowing) and fresh and dry weights at harvest were significantly higher for the cross-pollinated seedlings compared to the self-pollinated ones. The shoot:root ratio was independent of the type of pollination (Tab. 4).

The growth of the seedlings is shown in Fig. 2 (A, B, and C) and the final height, stem diameter and number of pinnae 222 days after sowing in Tab. 4. Means of height, diameter, and numbers of pinnae were always lower for self-pollinated seedlings compared to cross-pollinated ones, and these differences in growth increased with time (Fig. 2). Seven months after sowing (222 days), the effect of type of pollination was significant for plant height and stem diameter, but not for number of pinnae ($F_{1,7}$ = 4.3, P = 0.08) (Tab. 4).

DISCUSSION

Parkia biglobosa is known to be mainly outcrossing (Ouédraogo 1995; Sina 2006; Lassen et al. 2017), and this study showed that self-pollination reduced the number of pods produced, pod size, number of seeds, sugar content in pulp, seed weight, and weight of seedlings (Tab. 1, 2, and 5). In addition, the diallel cross revealed that some combinations of mother-trees and pollen donors were more productive than others. The findings of variation in the success of male and female reproductive organs, i.e., an individual plant being good at either setting pods or at fathering pods on other conspecifics, has also been reported in other plant species such as the self-incompatible Trumpet creeper *Campsis radicans* (L.) Seem. (Bignoniaceae) (Bertin 1982) and the self-incompatible Crested dogstail grass *Cynosurus cristatus* L. (Poaceae) (Ennos & Dodson 1987).

Pollen limitation in Parkia biglobosa

Controlled pollinations with different doses of cross-pollen conducted in the present study yielded significantly more immature and mature fruits than in open-pollinated

TABLE 4. Influence of self- versus cross-pollination on the fitness of various parameters of pods and seeds of *Parkia biglobosa* including *F*-tests, Fisher's exact tests, and their significance levels.

	Type of pollination					
	Self, *N*	Cross, *N*	Self[a]	Cross[a]	$F(P)_{Pollination}$	Inbreeding depression[b], %
Pod length, cm	24	24	15.8 (1.16)	20.0 (1.16)	$F_{(1,10)} = 6.4$ (*)	21.0
Pod weight, g	24	24	7.9 (1.03)	12.5 (1.03)	$F_{(1,10)} = 10.1$ (**)	36.8
Husk weight, g	24	24	4.0 (0.41)	5.5 (0.41)	$F_{(1,10)} = 7.5$ (*)	27.3
Pulp[c] weight, g	22	24	2.5 (0.40)	4.2 (0.37)	$F_{(1,9)} = 8.7$ (*)	40.5
Seed weight, g	24	24	1.6 (0.29)	2.6 (0.26)	$F_{(1,9)} = 6.9$ (*)	38.5
Seeds per pod:						
Healthy seeds, *n*	24	24	6.6 (1.87).	12.4 (1.87).	$F_{(1,10)} = 4.8$ (ns)	46.8
Eaten seeds, *n*	24	24	1.4 (0.70)	0.7 (0.70)	$F_{(1,10)} = 0.6$ (ns)	-100.0
Aborted seeds[d], *n*	24	24	2.8 (0.46)	1.3 (0.46)	$F_{(1,10)} = 5.2$ (*)	-115.4
Missing seeds[e], *n*	24	24	1.1 (0.29)	1.5 (0.29)	$F_{(1,10)} = 0.9$ (ns)	26.7
Total seeds, *n*	24	24	11.9 (0.99)	15.9 (0.99)	$F_{(1,10)} = 8.1$ (*)	25.2
Moisture in seeds, %	24	24	3.5 (0.14)	3.1 (0.14)	$F_{(1,10)} = 3.3$ (ns)	-12.9
C and N in seeds, dry weight:						
Carbon in raw seeds, %	24	24	53.4 (0.29)	54.3 (0.29)	$F_{(1,10)} = 4.8$ (ns)[f]	1.6
Nitrogen in raw seeds, %	24	24	7.2 (0.10)	6.9 (0.10)	$F_{(1,10)} = 5.0$ (*)	-4.4
Protein in raw seeds[g], %	24	24	45.2 (0.60)	43.3 (0.60)	$F_{(1,10)} = 5.0$ (*)	-4.4
Sugars in fruit pulp, dry weight:						
Sucrose[c], %	22	24	28.5 (0.81)	29.5 (0.74)	$F_{(1,9)} = 0.9$ (ns)	3.4
Glucose[c], %	22	24	3.5 (0.68)	5.6 (0.62)	$F_{(1,9)} = 5.6$ (*)	37.5
Fructose[c], %	22	24	4.2 (0.51)	5.9 (0.47)	$F_{(1,9)} = 5.6$ (*)	23.8
Total sugar[c], %	22	24	36.2 (1.81)	41.3 (1.66)	$F_{(1,9)} = 4.3$ (ns)	12.3
Seed germination:						
Seed weight, g	117	196	0.16 (0.006)	0.20 (0.005)	$F_{(1,8)} = 32.9$(***)	20.0
Seed germination[h], %	117	196	76.9	83.2	$P < 0.2$ (ns)	7.6
Days to germination	90	163	4.4 (0.28)	4.3 (0.23)	$F_{(1,8)} = 0.1$ (ns)	-2.3
Seedlings:						
Fresh weight 19 days[i], g	88	161	0.85 (0.04)	1.02 (0.03)	$F_{(1,8)} = 10.2$ (*)	16.7
Survival 222 days[hi], %	90	163	76.7	90.2	$P < 0.005$ (**)	15.0
Height 222 days[i], cm	69	147	20.3 (0.97)	23.2 (0.74)	$F_{(1,7)} = 5.8$ (*)	12.5
Stem diameter 222 days[i], mm	69	147	3.9 (0.20)	4.6 (0.15)	$F_{(1,7)} = 6.2$ (*)	15.2
Number of pinnae 222 days[i]	69	147	35.0 (2.13)	40.5 (1.61)	$F_{(1,7)} = 4.3$ (ns)	13.5
Fresh weight 222 days[i], g	69	147	17.0 (2.03)	23.4 (1.53)	$F_{(1,7)} = 6.3$ (*)	27.4
Dry weight 222 days[i], g	69	147	6.1 (0.72)	8.4 (0.54)	$F_{(1,7)} = 6.5$ (*)	27.4
Shoot, dry weight g	69	147	2.8 (0.41)	4.0 (0.31)	$F_{(1,7)} = 4.9$ (ns)	30.0
Root, dry weight g	69	147	3.3 (0.48)	4.5 (0.37)	$F_{(1,7)} = 3.6$ (ns)	26.7
Shoot:root ratio (dry weight)	69	147	1.0 (0.15)	0.9 (0.11)	$F_{(1,7)} = 0.1$ (ns)	-11.1

[a]Values are least squares (LS) means with the standard error (SE) in brackets of the LS estimate. Significance level: ***= $P<0.001$, **= $P<0.01$, *= $P<0.05$, and ns= $P>0.05$. [b]Inbreeding depression is calculated as: (Cross-self) /cross*100. [c]Missing data for two self-pollinated pods (no pulp). [d]Aborted seeds: weight<0.05 g and/or with a flat shape. [e]Missing seeds: empty cavity in the pulp. [f]$P = 0.054$. [g]Nitrogen-to-protein conversion factor = 6.25 (AOAC, 1990). [h]Fisher's exact test. [i]Days after sowing.

capitula (Tab. 3). These results document pollen limitation of fruit set in *P. biglobosa* during the study year, indicating that fruit set could be increased by increasing pollen load above the natural level of pollination (but see discussion below). Freely exposed capitula attracted up to 50 honey bees foraging simultaneously per capitulum. As honey bees have been found to be good pollinators of *P. biglobosa* (Lassen et al. 2017), pollen limitation may be due to insufficient deposition of compatible cross-pollen compared to deposition of self- and incompatible cross-pollen. The density of *P. biglobosa* was relatively high (1.2 trees/ha), and seven of the eight mother-trees were separated by less than 60 m to the nearest *P. biglobosa* tree. However, the mother-trees were large (mean crown area = 472 m², SD = 170) with many capitula, and this profuse blooming may have increased geitonogamy of the open-pollinated capitula. Higher fruit production of controlled cross-pollination compared to controlled self-pollination and open-pollination has also been reported for other tropical tree species e.g. the Coligallo palm, *Calyptrogyne ghiesbreghtiana* H. Wendl. (Cunningham 1996) (Arecaceae) and five species of neotropical *Inga* trees (Koptur 1984) (Fabaceae), where it has been linked to limitation of compatible pollen of self- and open-pollinations.

TABLE 5. Correlation between number of seeds per pod and amount of sugar (sucrose, glucose, and fructose) in pulp for 46 pods (22 self-pollinated and 24 cross-pollinated) of *Parkia biglobosa* illustrated by Pearson correlation coefficients and significance levels.

	Sucrose	Glucose	Fructose	Total sugars
No. of healthy seeds	0.19 (ns)	0.48 (***)	0.38 (**)	0.48 (***)
No. of aborted seeds	0.08 (ns)	-0.48 (***)	-0.42 (**)	-0.28 (ns)
No. of total seeds	0.15 (ns)	0.26 (ns)	0.19 (ns)	0.31 (*)

Significance level: ***=$P<0.001$, **= $P<0.01$, *= $P<0.05$, and ns= $P>0.05$.

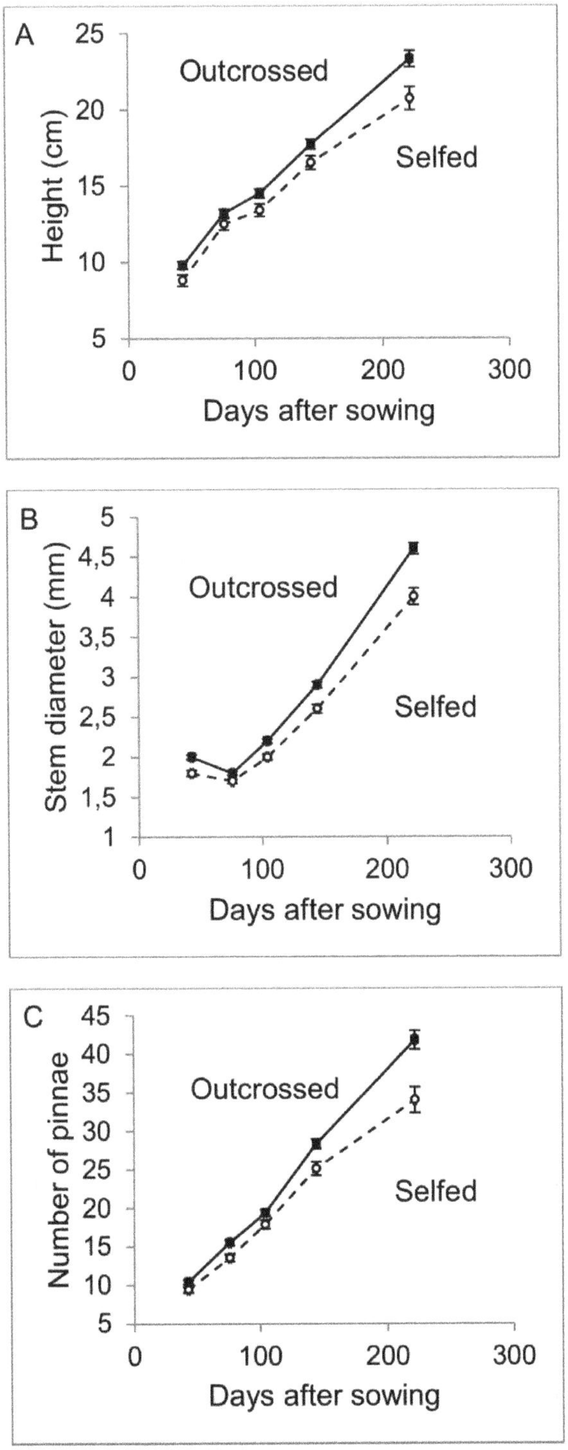

FIGURE 2. Growth of seedlings of *Parkia biglobosa* originating from 69 self-pollinated and 147 cross-pollinated seeds, showing the temporal development (days after sowing) of A) mean height, B) mean stem diameter, and C) mean number of pinnae. Error bars indicate ± 1 standard error of the mean. Only seedlings, which survived until harvest at 222 days after sowing are included.

Although half of the immature pods were aborted in controlled cross-pollinations, pollen dose was highly important for number of mature pods in *Parkia biglobosa*. The converting calculations between treatments with different pollen doses ('cross' and 'both') proposed a linear relationship between pollen dose and fruit set, suggesting that even more pods could have been initiated and matured, if the high dose of the cross-pollen in the 'both'-treatment (1:1) had been applied to the whole capitulum, and not only to half of it, or if the pollen dose had been even higher (e.g., 3:1). Most likely, the increased abortion of pods (i.e., immature pods minus mature pods) with higher pollen doses were due to lack of resources, suggesting that there is an upper limit for the number of produced pods per capitulum. However, we found no differences in pod weight or pod length between pods from open-, cross- or 'both'-treatments. The experiment had some limitations, and hence we cannot conclude whether the trees were truly pollen limited. First, the experiment lasted only one night at each tree; second, only a part of the blooming capitula per tree was included in the experiment; third, the experiment was only performed in one season. It is possible that the increased fruit set of hand-pollinated capitula came at the cost of open-pollinated capitula within the same tree (Obeso 2002). Nevertheless, some plant species show pollen limitation both on a whole plant level and in subsequent years (Ashman et al. 2004). More detailed hand-pollination experiments are needed to test whether a higher fruit production can be obtained in *P. biglobosa* by supplying cross-pollen night after night and year after year.

In the current study, a cross-pollinated capitulum was pollinated with only one cross-pollen donor, and hence the pod abortion was not due to selection between pollen donors, as has sometimes been suggested when explaining high rates of abortion (Bookman 1984). These patterns suggest a general lack of maternal resources to mature all or most of the initiated cross-pollinated pods.

Self-incompatibility in Parkia biglobosa

Self-incompatibility of *P. biglobosa,* as suggested by our study, is supported by other studies: Using controlled self- and cross-pollination (N = 15 trees), Ouédraogo (1995) concluded that *P. biglobosa* is largely a self-incompatible species: self-pollination was possible, but outcrossing was more successful. Likewise, Sina (2006) found high values of multi-locus outcrossing (N= 238 trees), also consistent with partial self-incompatibility. *Parkia* is a pantropical genus of around 35 species, of which most are believed to be bat-pollinated and the rest insect-pollinated (Hopkins 1998). The breeding system of most *Parkia* species has not been investigated (Bumrungsri et al. 2008), but their high pollen:ovule ratios are in the range characteristic of outcrossing (Cruden 1977; Hopkins 1984). Piechowski (2007) has tested *Parkia pendula* in Brazil for selfing, and as no pods were produced in spontaneous or controlled self-pollinations, he concluded that this species was self-incompatible. Furthermore, a study of two Asian species of *Parkia*, *P. speciosa* and *P. timoriana,* involving spontaneous (i.e. bagged capitula) and controlled self-pollination treatments, suggested that both species were self-incompatible (Bumrungsri et al. 2008). Finally, our finding of two cross-incompatible tree pairs (P14 & P90, and P76 & P92) fit with the self-incompatibility being controlled by a few specific loci (de Nettancourt 1977; Seavey and Bawa 1986).

In spite of the high pollen dose in the 'both'-treatment, no self-pollinated flowers developed into pods in this treatment. Since only one polyad pollinates one flower in *P. biglobosa* (Lassen et al. 2014), competition and selection between pollen donors is likely to take place between flowers rather than within flowers (Bawa & Buckley 1989). Pod set after self-pollination of the entire capitulum was rare. The 24 selfed pods contained significantly more aborted seeds and fewer total seeds, perhaps due to late acting self-incompatibility and/or early acting inbreeding depression. The 'missing' seeds were thought to be seeds that aborted very early, leaving only the empty cavity in the pulp, but we found few cavities per pod, and no differences between types of pollination.

Inbreeding depression in Parkia biglobosa

Initial seedling growth has been shown to depend on seed size, possibly due to the size of the cotyledons (Blackman 1919; Howe & Richter 1982; Boot 1996). Considering the correlation between carbon content and seed weight, and the higher seed weight of cross-pollinated seeds, we hypothesise that the nutritional differences between selfed and outcrossed seeds were due to relatively larger cotyledons in the cross-pollinated seeds. Because the differences in growth between self- and cross-pollinated seedlings increased with time in this study, we expect that the self-pollinated seedlings suffered from inbreeding depression. Few cases of inbreeding depression in early germination stages have been documented in other plant species while inbreeding depression at later life stages ('seed production of parent' and 'growth and reproduction') has more often been reported (Husband & Schemske 1996; Hardner & Potts 1995). A study by Mašková and Herben (2018) showed that larger-seeded species consistently had lower root:shoot ratios, explained by an advantage of faster development of shoots in asymmetric above-ground competition. We found no difference between shoot:root ratios of selfed and outcrossed seedlings, perhaps due to lack of competition in the greenhouse. In the present study, the test of germination and growth took place under presumably optimal growth conditions (available water, light, nutrition and no competition), but the survival of seedlings was significantly lower for self-pollinated seeds. Walters & Reich (2000) observed that for ten tree species survival of seedlings in low light and/or low levels of N increased with seed weight. Therefore, it is likely that under natural conditions, inbreeding depression may have been more evident.

We expect that increased selfing in natural populations of *P. biglobosa* will negatively affect propagation by seeds, resulting in decreasing densities of adult trees in the future.

Quality of seeds and pod pulp from self- and cross-pollinated capitula

Pollination has been shown to impact the quality of fruits in different species (IPBES 2016). Because seeds and pulp from *P. biglobosa* are important food resources consumed by people and animals, it is highly relevant to understand the impact of self- versus cross-pollination on fruit and seed quality.

The content of protein in raw *P. biglobosa* seeds without testa (dry weight) of 43%- 45% (for outcrossed and selfed seeds, respectively) is similar to 43% found by one study (Ekpenyong et al. 1977), but much higher than reported by other studies: 27% (Esenwah & Ikenebomeh 2008), 30% (syn. *Parkia filicoidea* Welw.) (Fetuga et al. 1974) and 34% (Ijarotimi & Keshinro 2012). We found that self-pollinated seeds were more protein-rich but weighed less than cross-pollinated seeds. Hence, the increase in protein content can probably be explained by a simple concentration effect as found for various species grown under stress (Wang & Frei 2011). Inverse relationships of protein content and starch content of grains and yield, respectively, have been documented inmaize hybrids (*Zea mays* subsp. *mays* L.) (Poaceae) (Idikut et al. 2009). Likewise, other studies of maize have found significantly higher protein content in self-pollinated kernels and significantly higher starch content in cross-pollinated kernels (Letchworth & Lambert 1998; Sulewska et al. 2014). Total seed protein produced by self-pollinated pods was much lower than for crossed pods, as self-pollinated pods contained much fewer healthy seeds (Tab. 4).

The positive correlation between seed number per pod and sugar content in pulp observed in the current study was also found in a similar study of *P. biglobosa* in The Gambia (Lassen et al. 2012). However, the percentages of sugars (dry weight) were much higher in the fruit pulp from The Gambia compared with the pulp from Burkina Faso (total sugar: 60% versus 36-41%) (Lassen et al. 2012). In the literature, carbohydrate content (dry weight) in fruit pulp of *P. biglobosa* is reported to be from around 40% in Nigeria (Nadro & Umaru 2004) to around 85% in Mali (Nordeide et al. 1996). The positive correlation between number of healthy seeds and total amount of sugar in the pulp in the current study may be due to seeds acting as sinks during pod development, attracting nutrients to their own growth and to that of the surrounding pod (Stephenson 1981; Lee 1988; Marcelis & Hofman-Eijer, 1997). Similarly, Valantin-

Morison et al. (2006) found a less-pronounced sweetness of the flesh in fruits with few filled seeds compared to fruits with a normal number of filled seeds in cantaloupe melon (*Cucumis melo* L.). In self-compatible, sweet orange (*Citrus sinensis* var. Red Junar) (Rutaceae), Partap (2000) found a higher number of seeds and more juice with a higher sugar content after honey bee pollination compared to wind-pollination.

Overall, our results suggest that fruit production, nutritional value of *P. biglobosa* pulp and seeds, and the fitness of seedlings, will decrease with increased levels of self-pollinations and affect the rural human populations negatively. We propose more research into how to increase the regeneration (natural or planted) of *P. biglobosa.* Our results could also be expanded by testing combinations of some of the acknowledged plus-trees (i.e. the superior trees) of *P. biglobosa* and by grafting the best combinations together, making it easier for the pollinators to bring about more cross-pollination with highly compatible pollen.

Acknowledgements

We thank the staff of the Centre National de Semences Forestières (CNSF) for their cooperative spirit. Thanks to Alassane Ouédraogo, Madi Tiemtoré, and Philbert Zoungrana for field assistance and to the farmers in Pinyiri for allowing us to use their trees. We also thank Sofie Fiona Hansen, Ruth Bruus Jakobsen, and Annalise Metz for measuring the pods; head laboratory technician Lene Korsholm Jørgensen from the Department of Plant and Environmental Sciences (UCPH) for analysing the content of carbohydrates in the fruit pulp of *P. biglobosa*; laboratory coordinator Preben Frederiksen from the Department of Geosciences and Natural Resources (UCPH) for analysing the content of carbon and nitrogen in the seeds of *P. biglobosa*; gardener Kurt Dahl and greenhouse supervisor Theodor Emil Bolsterli for taking good care of the *P. biglobosa* seedlings in the greenhouse (UCPH). The present paper is part of a PhD study financed by the Danish International Development Agency (Danida (FFU), research project no. 10-106-LIFE).

References

Abrol DP (2012) Pollination biology. Biodiversity conservation and agricultural production. Springer, New York, USA.

Aizen MA, Harder LD (2007) Expanding the limits of the pollen-limitation concept: Effects of pollen quantity and quality. Ecology 88:271-281.

AOAC (Association of Official Analytical Chemists) (1990) Official methods of Analysis of the AOAC. (No. 954.01). Volume 1. Association of Official Analytical Chemists Inc., Arlington, USA.

Ashman T-L, Knight TM, Steets JA, Amarasekare P, Burd M, Campbell DR, Dudash MR, Johnston MO, Mazer SJ, Mitchell RJ, Morgan MT, Wilson WG (2004) Pollen limitation of plant reproduction: Ecological and evolutionary causes and consequences. Ecology 85:2408-2421.

Baker HG, Harris BJ (1957) The pollination of *Parkia* by bats and its attendant evolutionary problems. Evolution 11:449-460.

Bawa KS, Buckley DP (1989) Seed: ovule ratios, selective seed abortion and mating systems in Leguminosae. In: Stirton CH, Zarucchi JL (eds.) Advances in legume biology. Monographs of systematic botany of the Missouri Botanical Garden No. 29. Missouri Botanical Garden, St. Louis, USA, pp 243-262.

Bawa KS, Perry DR, Beach JH (1985) Reproductive biology of tropical lowland rain forest trees. 1. Sexual systems and incompatibility mechanisms. American Journal of Botany 72:331-345.

Bertin RI (1982) Paternity and fruit production in trumpet creeper (Campsis radicans). American Naturalist 119:694-709.

Blackman VH (1919) The compound interest law and plant growth. Annals of Botany 33:353-360.

Bookman SS (1984) Evidence for selective fruit production in *Asclepias*. Evolution 38:72-86.

Boot RGA (1996) The significance of seedling size and growth rate of tropical rain forest tree seedlings for regeneration in canopy openings. In: Swaine MD (ed.) The ecology of tropical forest tree seedlings. UNESCO, Paris, France, pp 267-283.

Bumrungsri S, Harbit A, Benzie C, Carmouche K, Sridith K, Racey P (2008) The pollination ecology of two species of *Parkia* (Mimosaceae) in southern Thailand. Journal of Tropical Ecology 24:467-475.

Cruden RW (1977) Pollen-Ovule ratios: A conservative indicator of breeding systems in flowering plants. Evolution 31:32-46.

Cunningham SA (1996) Pollen supply limits fruit initiation by a rain forest understorey palm. Journal of Ecology 84:185-194.

de Nettancourt D (1977) Incompatibility in angiosperms. Springer-Verlag, New York, USA.

Ekpenyong TE, Fetuga BL, Oyenuga VA (1977) Fortification of maize flour-based diets with blends of cashewnut meal, African locust bean meal and sesame oil meal. Journal of the Science of Food and Agriculture 28:710-716.

Ennos RA, Dodson RK (1987) Pollen success, functional gender and assortative mating in an experimental plant population. Heredity 58:119-126.

Esenwah CN, Ikenebomeh MJ (2008) Processing effects on the nutritional and anti-nutritional contents of African locust bean (*Parkia biglobosa* Benth.) seed. Pakistan Journal of Nutrition 7:214-217.

Etejere EO, Fawole MO, Sani A (1982) Studies on the seed germination of *Parkia clapertoniana*. Turrialba 32:181-185.

Ezeagu IE, Petzke JK, Metges CC, Akinsoyinu AO, Ologhobo AD (2002) Seed protein contents and nitrogen-to-protein conversion factors for some uncultivated tropical plant seeds. Food Chemistry 78:105-109.

Fetuga BL, Babatunde GM, Oyenuga VA (1974) Protein quality of some unusual protein foodstuffs. Studies on the African locust-bean seed (*Parkia filicoidea* Welw.). British Journal of Nutrition 32:27-36.

Funk C, Rowland J, Adoum A, Eilerts G, White L (2012). A climate trend analysis of Burkina Faso, U.S. Geological Survey, South Dakota, USA.

Gaisberger H, Kindt R, Loo J, Schmidt M, Bognounou F, Da SS, Diallo OB, Ganaba S, Gnoumou A, Lompo D, Lykke AM, Mbayngone E, Nacoulma BMI, Ouedraogo M, Ouédraogo O, Parkouda C, Porembski S, Savadogo P, Thiombiano A, Zerbo G, Vinceti B (2017) Spatially explicit multi-threat assessment of food tree species in Burkina Faso: A fine-scale approach. PLoS ONE 12:e0184457.

Goulson D, Nicholls B, Botias C, Rotheray E (2015) Bee declines driven by combined stress from parasites, pesticides and lack of flowers. Science 347:1255957.

Hall JB, Tomlinson HF, Oni PI, Buchy M, Aebischer DP (1997) *Parkia biglobosa*: a monograph. School of Agricultural and Forest Sciences Publications Number 9, University of Wales, Bangor, UK.

Hardner CM, Potts BM (1995). Inbreeding depression and changes in variation after selfing in *Eucalyptus globulus* ssp. *globulus*. Silvae Genetica 44:46-54.

Hinata K, Watanabe M, Toriyama K, Isogai A (1993) A review of recent studies on homomorphic self-incompatibility. International Review of Cytology 143:257-296.

Hopkins HC (1981) Taxonomy and reproductive biology of, and evolution in the bat-pollinated genus *Parkia*. PhD thesis, Oxford University, St. Hilda's College, UK.

Hopkins HC (1983) The taxonomy, reproductive biology and economic potential of *Parkia* (Leguminosae: Mimosoideae) in Africa and Madagascar. Botanical Journal of the Linnean Society 87:135-167.

Hopkins HC (1984) Floral biology and pollination ecology of the neotropical species of *Parkia*. Journal of Ecology 72:1-23.

Hopkins HCF (1998) Bat pollination and taxonomy in *Parkia* (Leguminosae: Mimosoideae). In: Hopkins HCF, Huxley CR, Pannell CM, White F (eds.) The biological monograph. The importance of field studies and functional syndromes for taxonomy and evolution of tropical plants. The Royal Botanical Gardens, Kew, UK, pp 31-55.

Hopping ME (1976) Effect of exogenous auxins, gibberellins, and cytokinins on fruit development in Chinese gooseberry (*Actinidia chinensis* Planch.). New Zealand Journal of Botany 14:69-75.

Howe HF, Richter WM (1982) Effects of seed size on seedling size in *Virola surinamensis*; a within and between tree analysis. Oecologia 53:347-351.

Husband BC, Schemske DW (1996) Evolution of the magnitude and timing of inbreeding depression in plants. Evolution 50:54-70.

Idikut L, Atalay AI, Kara SN, Kamalak A (2009) Effect of hybrid on starch, protein and yields of maize grain. Journal of Animal and Veterinary Advances 8:1945-1947.

Ijarotimi OS, Keshinro OO (2012) Comparison between the amino acid, fatty acid, mineral and nutritional quality of raw, germinated and fermented African locust bean (*Parkia biglobosa*) flour. Acta Scientiarum Polonorum Technologia Alimentaria 11:151-165.

IPBES (2016) The assessment report on pollinators, pollination and food production of the Intergovernmental Science-Policy Platform on Biodivversity and Ecosystem Services. Potts SG, Imperatriz-Fonseca VL, Ngo HT (eds). Secretary of the Intergovernmental Science-Policy Platform on Biodivversity and Ecosystem Services, Bonn, Germany.

Kennedy CM, Lonsdorf E, Neel MC, Williams NM, Ricketts TH, Winfree R, Bommarco R, Brittain C, Burley AL, Cariveau D, Carvalheiro LG, Chacoff NP, Cunningham SA, Danforth BN, Dudenhöffer J-H, Elle E, Gaines HR, Garibaldi LA, Gratton C, Holzschuh A, Isaacs R, Javorek SK, Jha S, Klein AM, Krewenka K, Mandelik Y, Mayfield MM, Morandin L, Neame LA, Otieno M, Park M, Potts SG, Rundlöf M, Saez A, Steffan-Dewenter I, Taki H, Felipe Viana B, Westphal C, Wilson JK, Greenleaf SS, Kremen C (2013) A global quantitative synthesis of local and landscape effects on wild bee pollinators in agroecosystems. Ecology Letters 16:584-599.

Knight TM, Steets JA, Vamosi JC, Mazer SJ, Burd M, Campbell DR, Dudash MR, Johnston MO, Mitchell RJ, Ashman T-L (2005) Pollen limitation of plant reproduction: Pattern and process. Annual Review of Ecology, Evolution and Systematics 36:467-497.

Koptur S (1984) Outcrossing and pollinator limitation of fruit-set - breeding systems of neotropical *Inga* trees (Fabaceae, Mimosoideae). Evolution 38:1130-1143.

Krotz L, Giazzi G (2014) Technical comparison of the Thermo Scientific FLASH 2000 Nitrogen/Protein Analyzer with the traditional Kjeldahl method Technical Note 42215. Thermo Fisher Scientific, Milan, Italy.

Lassen KM, Kjær ED, Ouédraogo M, Nielsen LR (2014) Microsatellite primers for *Parkia biglobosa* (Fabaceae: Mimosoideae) reveal that a single plant sires all seeds per pod. Applications in Plant Sciences 2:1400024.

Lassen KM, Ouédraogo M, Dupont YL, Kjær ED, Nielsen LR (2017) Honey bees ensure the pollination of *Parkia biglobosa* in absence of bats. Journal of Pollination Ecology 20:22-34.

Lassen KM, Ræbild A, Hansen H, Brødsgaard CJ, Eriksen EN (2012) Bats and bees are pollinating *Parkia biglobosa* in The Gambia. Agroforestry Systems 85:465-475.

Lee TD (1988) Patterns of fruit and seed production. In: Lovett Doust J, Lovett Doust L (eds.) Plant reproductive ecology. Patterns and strategies. Oxford University Press, New York, USA, pp 179-202.

Letchworth MB, Lambert RJ (1998) Pollen parent effects on oil, protein, and starch concentration in maize kernels. Crop Science 38:363-367.

Liu FL, Jensen CR, Andersen MN (2004) Drought stress effect on carbohydrate concentration in soybean leaves and pods during early reproductive development: its implication in altering pod set. Field Crops Research 86:1-13.

Lykke AM, Mertz O, Ganaba S (2002) Food consumption in rural Burkina Faso. Ecology of Food and Nutrition 41:119-153

Maranz S (2009) Tree mortality in the African Sahel indicates an anthropogenic ecosystem displaced by climate change. Journal of Biogeography 36:1181-1193.

Marcelis LFM, Hofman-Eijer LRB (1997) Effects of seed number on competition and dominance among fruits in *Capsicum annuum* L. Annals of Botany 79:687-693.

Mašková T, Herben T (2018) Root:shoot ratio in developing seedlings: How seedlings change their allocation in response to seed mass and ambient nutrient supply. Ecology and Evolution 2018;00:1-8. https//doi.org/10.1002/ece3.4238.

Météo (2015) Pluviometrie mensuelle (mm) dans Guilongou (Ziniare) et Pô 2010-2012 et temperature mensuelle (°C) dans Ouagadougou Aéroport et Pô 2011-2012, Direction Générale de la Météorologie du Burkina, Ouagadougou, Burkina Faso.

Nadro M, Umaru H (2004) Comparative chemical evaluation of locust bean (*Parkia biglobosa*) fruit pulp harvested during the dry and wet season. Nigerian Journal of Biotechnology 15:42-47.

Nordeide MB, Hatloy A, Folling M, Lied E, Oshaug A (1996) Nutrient composition and nutritional importance of green leaves and wild food resources in an agricultural district, Koutiala, in Southern Mali. International Journal of Food Sciences and Nutrition 47:455-468.

Nyadanu D, Adu Amoah R, Obeng B, Kwarteng AO, Akromah R, Aboagye LM, Adu-Dapaah H (2017) Ethnobotany and analysis of food components of African locust bean (*Parkia biglobosa* (Jacq.) Benth.) in the transitional zone of Ghana: implications for domestication, conservation and breeding of improved varieties. Genetic Resources and Crop Evolution 64:1231-1240.

Obeso JR (2002) The costs of reproduction in plants. New Phytologist 155:321-348.

Ollerton J, Winfree R, Tarrant S (2011) How many flowering plants are pollinated by animals? Oikos 120:321-326.

Osonubi O, Fasehun FE (1987) Adaptations to soil drying in woody seedlings of African locust bean, (*Parkia biglobosa* (Jacq.) Benth.). Tree Physiology 3:321-329.

Ouédraogo AS (1995) *Parkia biglobosa* (Leguminosae) in West Africa; biosystematics and improvement, Landbouwuniversiteit

Wageningen (Wageningen Agricultural University), Wageningen, The Netherlands.

Partap U (2000) Foraging behaviour of *Apis cerana* on sweet orange (*Citrus sinensis* var Red Junar) and its impact on fruit production. In: Asian bees and beekeeping. Progress of research and development. Proceedings of the 4th Asian Apicultural Association, Kathmandu, Nepal, pp 174-177.

Piechowski D (2007) Reproductive ecology, seedling performance, and population structure of *Parkia pendula* in an Atlantic forest fragment in Northeastern Brazil, Universität Ulm, Köln, Germany.

Ræbild A, Hansen UB, Kambou S (2012) Regeneration of *Vitellaria paradoxa* and *Parkia biglobosa* in a parkland in Southern Burkina Faso. Agroforestry Systems 85:443-453.

Roldán Serrano A, Guerra-Sanz JM (2006) Quality fruit improvement in sweet pepper culture by bumblebee pollination. Scientia Horticulturae 110:160-166.

Sanfo JB (2012) Apport de la Direction Générale de la Météorologie dans le processus d'information sur l'eau 13ème sommet de l'information sur l'eau, 11-13 April 2012, Ouagadougou, Burkina Faso.

SAS Institute (2011) The SAS system for Windows. Release 9.4, Cary, North Carolina, USA.

Seavey SR, Bawa KS (1986) Late-acting self-incompatibility in angiosperms. Botanical Review 52:195-219.

Sina S (2006) Reproduction et diversité génétique chez *Parkia biglobosa* (Jacq.) G.Don, Wageningen University, Wageningen, the Netherlands.

Stephenson AG (1981) Flower and fruit abortion: Proximate causes and ultimate functions. Annual Review of Ecology and Systematics 12:253-279.

Sulewska H, Adamczyk J, Cygert H, Rogacki J, Szymanska G, Smiatacz K, Panasiewicz K, Tomaszyk K (2014) A comparison of controlled self-pollination and open pollination results based on maize grain quality. Spanish Journal of Agricultural Research 12:492-500.

Uwaegbute AC (1996) African locust bean (*Parkia filicoidea* Welw.). In: Nwokolo E, Smartt J (eds.) Food and feed from legumes and oilseeds. Chapman & Hall, London, United Kingdom, pp 124-129.

Valantin-Morison M, Vaissiere BE, Gary C, Robin P (2006) Source-sink balance affects reproductive development and fruit quality in cantaloupe melon (*Cucumis melo* L.). Journal of Horticultural Science and Biotechnology 81:105-117.

Walters MB, Reich PB (2000) Seed size, nitrogen supply, and growth rate affect tree seedling survival in deep shade. Ecology 81:1887-1901.

Wang Y, Frei M (2011) Stressed food – The impact of abiotic environmental stresses on crop quality. Agriculture, Ecosystems & Environment 141:271-286.

Ward M, Dick CW, Gribel R, Lowe AJ (2005) To self, or not to self... a review of outcrossing and pollen-mediated gene flow in neotropical trees. Heredity 95:246-254.

Yeoh HH, Wee YC (1994) Leaf protein contents and nitrogen-to-protein conversion factors for 90 plant species. Food Chemistry 49:245-250.

25

IDENTIFICATION OF PLANT SPECIES FOR POLLINATOR RESTORATION IN THE NORTHERN PRAIRIES

Diana B. Robson[1,*], Cary Hamel[2] and Rebekah Neufeld[3]

[1]Manitoba Museum, 190 Rupert Avenue, Winnipeg, MB Canada R3B 0N2
[2]Nature Conservancy of Manitoba, 611 Corydon Ave., Winnipeg, MB, Canada, R3L 0P3
[3]Nature Conservancy of Manitoba, 207-1570 18th Street, Brandon, MB, Canada R7A 5C5

Abstract—Research on diurnal plant–pollinator interactions indicates that a small number of generalist plants provide a disproportionately high amount of floral resources to pollinating insects. Identifying these generalist plants would help prairie restoration specialists select species that will provide forage for the majority of pollinator taxa. Field research in three Canadian fescue (*Festuca hallii*) prairie preserves that were at most 3.3 km away from each other was conducted in 2014 and 2015 to create pooled, weighted, plant–insect visitor matrices for each site. Using these matrices, generalization (G) scores were calculated for each plant species to help assess their importance to wild insect visitors as this method controls for differences in insect abundances over the year. The three species with the highest average generalization scores were *Solidago rigida, Erigeron glabellus* and *Symphyotrichum laeve.* Species accumulation curves were created to determine how many plant species would need to be present before most pollinator taxa would have at least one acceptable forage species. This research indicates that the 16 plant species (33% of the total) with the highest average generalization scores were visited by 90% of the observed pollinator taxa. To detect exceptionally attractive plant species while accounting for natural differences in abundance, we calculated the insect, bee and fly visitation rates per inflorescence. There was several specialized plant species that were visited frequently by bees. Most of these specialized plants had purple or yellow, tubular flowers, and bloomed in mid to late summer when bee populations were most numerous.

Keywords: Fescue prairie, generalists, insect visitors, pollination, restoration, seed mix

INTRODUCTION

Fescue prairie is a unique North American grassland formation dominated by *Festuca hallii* (Coupland & Brayshaw 1953; Coupland 1961). Interest in conserving and restoring Canada's fescue prairies is increasing due to the relative rarity of this ecosystem; agricultural conversion, industrial activity and exotic species encroachment have drastically reduced the amount of native prairie remaining. Complete restoration of all species that were native to an ecosystem is the ideal end goal as literature suggests that the more species present, the greater the resiliency of the community (LaBar et al. 2014). Due to the short growing season and potential for both late spring and early fall frosts in the northern prairies, using locally-adapted seeds is recommended to avoid phenological mismatching (Chuine 2010). Currently, the seeds of native grasses are the ones most commonly available from commercial seed growers in western Canada. Unfortunately, grasses are less important to pollinators than plants that offer large amounts of nectar. The addition of insect-pollinated herbs or shrubs to seed mixes is needed to ensure that pollinators will be attracted to restored areas (Menz et al. 2011). Some plants may be exceptionally attractive to pollinators; research is needed to identify them.

Research on plant–pollinator interactions has found that some plants are super-generalists that act as the skeleton of an ecosystem (Jordano et al. 2003; Bascompte & Jordano 2007). Essentially, super-generalists enable less common species to persist (Memmott et al. 2004; Saavedra et al. 2011). Usually these super-generalist plants are visited by a wide diversity of insect taxa due to their large floral displays, radially symmetrical flowers and easily accessible nectar (Elle et al. 2012; Koski et al. 2015). Ecological theory suggests that including super-generalist plants in a seed mix will result in a more diverse and stable plant–pollinator community than adding species at random (Montoya et al. 2012; LaBar et al. 2014; Harmon-Threatt & Hendrix 2015). As a result, plant–pollinator interaction matrices are increasingly being used to help identify the most important plants for restoration purposes (Forup et al. 2008; Devoto et al. 2012; Montoya et al. 2012; Russo et al. 2013). Indeed, including the most generalized plants to restoration seed mixes does increase pollinating insect diversity and abundance over randomly chosen mixes (LaBar et al. 2014; Kremen & M'Gonigle 2015; M'Gonigle et al. 2015).

Following Isaacs et al. (2009) note that regional research efforts are needed to identify the best native plant species to support the pollinator communities in each region, the goal of this study was to identify plant species most attractive to pollinators to use for restoration of Canada's fescue prairies. To conduct our evaluation we collected plant–pollinator interaction data from three fescue prairie preserves in south

*Corresponding author: drobson@manitobamuseum.ca

western Manitoba during the summers of 2014 and 2015. To identify the super-generalist plant species, and make it easier to compare species with a different number of available insect pollinators, we calculated a generalization score (G) for each species developed by Medan et al. (2006). This generalization score takes into account not just the number of insect-visiting taxa (S), but their resource usage (RU) (i.e. the proportion of all available insect taxa that visit the plant), and evenness (E). Plants with high generalization scores are visited equally by a large number of the available insect taxa compared to plants with low generalization scores, which tend to be visited by only a few insect taxa and thus are more specialized. We identified the species with the highest generalization scores in these fescue prairies and considered them to be top candidates for inclusion in fescue prairie seed mixes. However, Harmon-Threatt and Hendrix (2015) and Russo et al. (2013) noted that there are certain specialist plants (often legumes) that are visited very frequently by bees and may be important to include in a seed mix to support these insects. Further, it is possible that plant density will affect the G score simply by increasing the likeliness that an encounter will be observed or that insects will visit that plant species. Therefore, we also determined the visitation rate per inflorescence (instead of the rate per plot) by all the pollinators, by just bee taxa, and by just fly taxa. We created species accumulation curves (Ebeling et al. 2008) to help determine how many of the most generalized plants would have to be added to a seed mix before most insect taxa (90%) had at least one plant species to forage on. Lastly, we graphed the phenology of all plant species observed to determine the flowering sequence and detect any gaps.

MATERIALS AND METHODS

Study Sites

Three prairie conservation sites in the Aspen Parkland Ecoregion south of Riding Mountain National Park, Manitoba were surveyed: Elk Glen (50.849444°N, 100.819417°W), Cleland (50.830693°N, 100.788083°W) and Crown (50.834333°N, 100.787750°W). The plots at Elk Glen were 2.8 to 3.3 km away from the plots at the Cleland and Crown sites; plots at the latter two sites were all within 600 m of each other. The soils in this area are moderately calcareous consisting of mainly sand and silt derived from glacial till. The sites contain mainly mixed wood forests with some fescue prairie patches; past grazing practices are thought to have reduced the fescue grass component in the Riding Mountain region (Trottier 1986). As the prairies nearest the roads were heavily invaded by non-native species such as *Bromus inermis* and *Poa pratensis*, our choices regarding plot locations were limited to areas that were accessible only by foot, which constrained the number of plots that could be surveyed in a day. There are at least 155 vascular plant species that occur in the area, but many of these species are found in the wetlands or forested parts of the sites. The close proximity of the sites means that the climate and landscapes were similar. The Elk Glen and Cleland sites are owned and managed as nature preserves by the Nature Conservancy of Canada (NCC) but both were previously grazed by cattle, the latter as recently as 2013. The Crown land is managed by the Manitoba government for wildlife and agricultural values and may have been historically grazed by cattle several decades ago. As there were some differences in the plant communities due to their grazing history, they were all considered separate sampling sites.

Vegetation Surveys

In 2014, we established six 4 m^2 permanent plots in each of the three sites on flat to gently sloping land. Small plots rather than transects were chosen to enable: (1) future comparison of these data with other Manitoba timed pollinator surveys in tall-grass prairie (Robson 2008, 2013), (2) observations of interactions between less common plants and insects (Gibson et al. 2011), and (3) calculation of visits/inflorescence in a reliable way. The plots were randomly selected and at least 10 m apart. In 2014, sampling was conducted for four consecutive days per site, which was repeated four times in mid-June, -July, -August and -September (16 days total). In 2015, sampling was conducted on the previously established plots for four consecutive days per site, repeated four times at the beginning and at the end of June, in early July and in late August (16 days total). In total 15 more plant species were observed in 2015 than in 2014 as a result of conducting the surveys on different calendar days. Plant richness and number of inflorescences of each species in the plots was recorded each sampling day.

Floral Visitor Surveys

Flower-visiting insect sampling occurred on the same days and in the same plots as the vegetation surveys: four consecutive days for four months in each of 2014 and 2015 for a total of 32 days. Each plot was surveyed for 10 minutes each sampling day thus the total time spent surveying was 96 hrs. Due to natural variability in abundance, some plant species were present in more plots than others and thus were observed for a longer period of time. Surveys were conducted between 09:30 and 17:00 when insect foraging activity is at a maximum (Kevan & Baker 1983) thus nocturnal insects were not observed. During the survey periods the mean temperature recorded at the nearest Environment Canada weather station (i.e. Brandon, MB) was 17.7°C (± 0.9 SE), the mean relative humidity 53.0% (± 1.9 SE) and the mean wind speed 10.0 km/hr (± 0.62 SE). The order in which the plots were visited was randomized each day. Some of the flower-visiting insects (e.g. crabronid wasps in the genus *Ectemnius*) may have been predators of other flower-visiting insects. Regardless of whether the wasps were foraging for pollen, nectar or prey, they were considered potential pollinators and their visitations recorded. Ambush bugs (*Phymata* spp.) and crab spider (*Misumena* spp.) visits were not recorded as they remain stationary on one inflorescence for a long time and thus were unlikely to act as pollinators.

All insect visitations to any inflorescence in the plot were recorded but the quality of the visit was not assessed and therefore may not have resulted in pollination of the plant. However, even if an insect visit did not fulfill the reproductive goal of the plant, the plant was probably important for the insect's survival; this is an aspect of plant–

pollinator interaction that we wanted to capture. The first time an insect taxon was observed a specimen was obtained and given a unique collection number. When the same (or what appeared to be a very similar) taxon was observed later on, the collection number was used to link that insect visit to the plant. Although this technique does not allow for complete identification "on the wing" (resulting in an underestimate of insect taxa) it does enable evaluation of insect visitation frequency, which was then used to determine the visitation rate for each plant species by all insect taxa, by bees only and by flies only (Parachnowitsch & Elle 2005). All insect voucher specimens were identified by zoologists using reference specimens at the Manitoba Museum (MM) and the Wallis Roughley Entomology Museum in Winnipeg, Manitoba; the specimens were deposited in MM's zoology collection.

Data Analysis

The daily inflorescence density at each site was calculated by counting the number of inflorescences of each species in each plot, dividing by the plot area (4 m^2) and averaging the values of all plots. The average inflorescence density at all three sites over the length of the study (32 days) for each species was then calculated. Using the density data we created a phenology graph. To do this, we determined the calendar day (from pooled 2014 and 2015 data) when each plant species began and ceased blooming in our plots. Differences in flowering times from year to year are known to vary in this region due to differences in snow cover, spring temperatures and late frosts, thus bloom times should only be considered approximate. Further, as we did not conduct the surveys every single day, the length of the flowering period is likely slightly longer than indicated.

For each of the three sites the data were used to create two year cumulative, weighted (i.e. by the number of visits recorded) plant–insect visitor matrices. These matrices were used to calculate a generalization (G) score (Medan et al. 2006) for each plant species using pooled data over two years (Appendix I) to facilitate comparison between them as:

$$G = RU \times E$$

Where

RU (Resource use) is:

$$RU = \frac{\#\ of\ visiting\ insect\ taxa}{\#\ of\ available\ insect\ taxa}$$

E (evenness) is:

$$E = \frac{-\sum_i p_i \cdot \ln p_i}{\ln S}$$

where p_i is the proportion of all interactions corresponding to the ith individual visiting insect of a given plant, and S is the total number of individual insects visiting it. When evenness is maximal, E is equal to 1 and when it approaches 0, interaction frequencies are very unequal. When a plant had only one insect visitor, the index could not be computed because it required dividing by zero; these plants were excluded from the analysis. The available insect taxa were all the taxa observed at that site at any time over the two year study. The G scores from each site were then averaged to get a single G score for each plant species. Although pooling the data over two years obscures some of the variation, we believe the results reflect the core of the most common insect visitors and interactions in this area. When a plant flowered at more than one site, the overall G score was the average of all the scores. The G, RU and E scores varied between 0 and 1. Low G scores indicated specialists while higher ones indicated generalists. All plants were ranked according to their G scores. If two species had the same G score, the plant that was visited at a higher rate by all insects was listed first. We then graphed species accumulation curves using Microsoft Excel where the plants were arranged in decreasing order of their G scores. The cumulative number of insect taxa that visited the accumulative number of plant species was determined using the matrix data. These data were used to determine the minimum number of plant species that would need to be added to an ecosystem to ensure that most (~90%) insect taxa had at least one preferred forage plant species.

The density of each plant species at the sites was different, which obscured evaluation of their importance to the insect visitors. To address this, we calculated the insect visitation rate per inflorescence to reduce the chance that density was affecting the results. The visitation rates are the averages over the number of days the plant was in flower. We also calculated the bee and fly visitation rates per inflorescence to determine which plant species were particularly attractive to them. We used linear regression analysis to determine the relationship between plant density and G, and plant density and insect visitation rate per inflorescence by all insects, by bees only and by flies only. Logarithmic transformations to homogenize variances were applied to the data. These statistical tests were performed using Analyze-it software (Analyze-it Software Ltd. 2009). The percentage of all visits each insect taxon made at each site was determined using pooled data from 2014 and 2015, and then averaged when the taxon occurred at more than one site. Lastly, we obtained lists of species available from local Canadian seed suppliers to determine which plants are currently available for use in restorations.

RESULTS

During the study, a total of 110 insect taxa were observed visiting the flowers of 48 forbs and/or shrubs found within the study plots (Appendix II). The average generalization score of all 48 plant species was 0.07 with the highest value 0.19 and the lowest 0.01 (Tab. 1). *Solidago rigida* had the highest G score and received the greatest number of insect visitor taxa (38). *Erigeron glabellus* has the second highest G score even though it was visited by fewer insect taxa than *Symphyotrichum laeve* and *Solidago nemoralis* due to a greater evenness. Three of the species with the highest G scores also had high insect visitation rates per inflorescence: *Solidago rigida, S. canadensis* and *Symphyotrichum laeve*. *Solidago rigida* and *Symphyotrichum laeve* were also visited frequently by bees. However, many of the plant species with the highest visitation rates had relatively low G scores (≤ 0.06) including: *Cirsium*

TABLE I. Average generalization scores (G), insect visitation rates, number of insect visitor taxa (S), and inflorescence (infl.) density and flowering period data for vascular plant species in three fescue prairie sites in southern Manitoba (sorted by G score then insect visitation rate). Plant species with only one insect visitor were excluded. Numbers in bold are the ten highest values for that variable.

Scientific name	G (mean ±SE)[1]	Insect[2] visitation rate (mean visits infl.$^{-1}$day^{-1}±SE)	Bee[3] visitation rate (mean visits infl.$^{-1}$day^{-1}±SE)	Diptera visitation rate (mean visits infl.$^{-1}$day^{-1}±SE)	S (total # of insect visitor taxa)	Infl. density (mean #/m^2±SE)	Flowering period[4]
Solidago rigida	**0.19**±0.03	**0.50**±0.11	**0.35**±0.08	**0.13**±0.04	**38**	**0.34**±0.11	M-L
Erigeron glabellus	**0.15**±0.06	0.19±0.07	0.07±0.02	**0.09**±0.05	**22**	0.16±0.04	E-M-L
Symphyotrichum laeve	**0.14**±0.01	**0.41**±0.09	**0.30**±0.08	**0.10**±0.03	**31**	**0.41**±0.11	L
Solidago nemoralis	**0.14**±0.08	0.23±0.09	0.14±0.06	**0.08**±0.04	**32**	**0.51**±0.14	M-L
Linum lewisii	**0.14**±0.0	0.19±0.08	0.08±0.04	0.03±0.02	11	0.07±0.02	E-M
Solidago canadensis	**0.12**±0.01	**0.31**±0.09	0.14±0.05	**0.09**±0.04	**19**	0.05±0.02	L
Campanula rotundifolia	**0.12**±0.04	0.07±0.01	0.06±0.01	0.0±0.0	**20**	**0.36**±0.10	M-L
Achillea millefolium	**0.12**±0.02	0.06±0.01	0.02±0.01	0.04±0.01	**20**	**0.32**±0.08	E-M-L
Symphoricarpos occidentalis	**0.12**±0.03	0.03±0.02	0.01±0.01	0.02±0.02	**26**	**0.29**±0.09	M-L
Apocynum androsaemifolium	0.10±0.0	0.20±0.10	0.09±0.04	0.07±0.05	7	0.03±0.01	M
Astragalus laxmannii	**0.10**±0.04	0.09±0.06	0.07±0.06	0.0±0.0	14	**0.36**±0.14	M
Prunus virginiana	0.09±0.0	**0.50**±0.29	**0.40**±0.31	0.04±0.02	9	0.07±0.03	E
Drymocallis arguta	0.09±0.03	0.23±0.05	**0.18**±0.05	0.02±0.01	**18**	0.12±0.03	M
Rudbeckia hirta	0.09±0.03	0.22±0.06	0.06±0.04	**0.16**±0.06	**18**	0.12±0.03	M-L
Lathyrus venosus	0.09±0.0	0.12±0.07	0.11±0.06	0.0±0.0	6	0.03±0.01	E-M
Cerastium arvense	0.09±0.01	0.06±0.02	0.02±0.01	0.03±0.02	11	0.19±0.06	E
Symphyotrichum ericoides	0.08±0.0	0.01±0.00	0.0±0.0	0.01±0.01	6	**0.71**±0.31	L
Liatris ligulistylis	0.07±0.04	**0.24**±0.12	**0.18**±0.09	0.02±0.02	8	0.03±0.01	L
Erigeron strigosus	0.07±0.0	0.15±0.09	0.0±0.0	**0.15**±0.08	4	0.04±0.01	M
Vicia americana	0.07±0.0	0.11±0.04	0.10±0.04	0.02±0.02	6	0.10±0.03	E-M
Cirsium drummondii	0.06±0.05	**1.18**±0.41	**1.17**±0.41	0.0±0.0	10	0.02±0.01	E-M
Agastache foeniculum	0.06±0.03	**0.46**±0.16	**0.45**±0.16	0.0±0.0	10	0.08±0.02	M-L
Monarda fistulosa	0.06±0.02	**0.37**±0.14	**0.32**±0.12	0.0±0.0	12	0.16±0.05	M-L
Liatris punctata	0.05±0.0	0.22±0.08	0.11±0.06	0.04±0.04	4	0.04±0.02	L
Helianthus pauciflorus	0.05±0.0	0.19±0.08	0.10±0.07	**0.16**±0.06	4	0.04±0.02	L
Hedysarum boreale	0.05±0.01	0.18±0.05	**0.18**±0.05	0.0±0.0	5	0.23±0.07	M
Astragalus crassicarpus	0.05±0.04	0.15±0.04	0.04±0.03	0.0±0.0	7	0.17±0.06	E-M
Geum triflorum	0.05±0.02	0.05±0.02	0.05±0.02	0.0±0.0	7	**0.59**±0.15	E
Zizia aptera	0.04±0.03	0.19±0.06	0.06±0.02	**0.11**±0.04	13	0.12±0.05	E
Agoseris glauca	0.04±0.01	0.22±0.08	0.13±0.07	0.06±0.05	5	0.02±0.01	M-L
Allium stellatum	0.04±0.02	0.04±0.03	0.04±0.02	0.01±0.01	4	0.04±0.02	L
Lithospermum canescens	0.04±0.02	0.02±0.01	0.01±0.0	0.0±0.0	10	**0.68**±0.22	E
Dalea purpurea	0.03±0.01	**0.79**±0.35	**0.78**±0.35	0.01±0.0	8	0.13±0.05	L

Table I continued.

Scientific name	G (mean ±SE)[1]	Insect[2] visitation rate (mean visits infl.$^{-1}$day^{-1}±SE)	Bee[3] visitation rate (mean visits infl.$^{-1}$day^{-1}±SE)	Diptera visitation rate (mean visits infl.$^{-1}$day^{-1}±SE)	S (total # of insect visitor taxa)	Infl. density (mean #/m^2±SE)	Flowering period[4]
Tragopogon dubius	0.03±0.0	0.13±0.12	0.08±0.08	0.04±0.04	2	0.01±0.01	E
Gaillardia aristata	0.03±0.02	0.12±0.06	0.08±0.05	0.03±0.03	4	0.02±0.01	M-L
Oxytropis campestris	0.03±0.0	0.11±0.02	0.09±0.02	0.0±0.0	4	0.19±0.06	M
Galium boreale	0.03±0.02	0.04±0.03	0.01±0.0	0.01±0.01	5	0.27±0.09	M
Rosa acicularis	0.02±0.01	**0.55**±0.26	0.02±0.01	**0.53**±0.26	6	0.03±0.01	E-M
Astragalus agrestis	0.02±0.01	0.19±0.05	0.16±0.02	0.0±0.0	3	0.04±0.01	E
Penstemon gracilis	0.02±0.01	0.14±0.11	0.14±0.10	0.0±0.0	3	0.02±0.01	M
Heuchera richardsonii	0.02±0.01	0.11±0.04	0.11±0.04	0.0±0.0	4	0.02±0.01	E-M
Pediomelum esculentum	0.02±0.01	0.11±0.05	0.10±0.05	0.0±0.0	3	0.04±0.01	E-M
Polygala senega	0.02±0.02	0.01±0.01	0.0±0.0	0.0±0.0	4	0.21±0.08	E
Fragaria virginiana	0.01±0.01	0.01±0.01	0.01±0.0	0.0±0.0	2	0.06±0.02	E

[1]G = generalization score. A G score (see Methods section for the formula) was calculated for each of the three sites using pooled data over two years (Appendix I) and averaged if the plant occurred at more than one site.
[2]Includes the visits by all insect taxa pooled over two years and averaged from all three sites.
[3]Includes only visits by bees (Hymenoptera) in the Andrenidae, Apidae, Colletidae, Halictidae and Megachilidae pooled over two years and averaged from all three sites.
[4]E= Early (June), M=Mid (July), L= Late (August-September). Determined using pooled data over two years from all three sites.

drummondii, Dalea purpurea, Agastache foeniculum and *Monarda fistulosa.* This is because these plants were visited frequently by two of the most frequently observed insect taxa - *Bombus ternarius* and *B. sandersoni* (Tab. 2) - which greatly decreased the evenness and consequently the G score. *Rosa acicularis* was among the most frequently visited plants but its visitors were mainly root maggot flies (*Drymeia* sp.) not bumblebees (*Bombus* spp.).

The average density of each plant varied from a high of 0.7 inflorescences/m^2 for *Symphyotrichum ericoides* to < 0.1 inflorescences/m^2 for *Tragopogon dubius.* Differences in inflorescence density may have influenced the G scores as these two variables were strongly positively correlated ($y = 0.3393x + 2.202$, $R^2 = 0.2186$, $P < 0.001$). This may be in part because species with low densities were observed for less time than more abundant plants, potentially decreasing the number of insect visitor taxa observed and influencing the G score. However, inflorescence density was not correlated with the insect visitation rate per inflorescence ($y = -0.1839x + 2.6452$, $R^2 = 0.0287$, $P = 0.25$), bee visitation rate per inflorescence ($y = -0.0224x + 2.4753$, $R^2 = 0.0004$, $P = 0.25$) or fly visitation rate per inflorescence ($y = -0.0017x + 2.3504$, $R^2 = 6E^{-06}$, $P = 0.99$) suggesting that factors other than abundance are being used by some insects to select plants. Indeed some species with low densities had some of the highest visitation rates per inflorescence (e.g. *Prunus virginiana* and *Solidago canadensis*).

The insect taxa observed in this study belonged to five orders: 7 beetle species (Coleoptera), 39 fly species (Diptera), 3 bug species (Hemiptera), 39 bee, ant and wasp species (Hymenoptera) and 22 butterfly and moth species (Lepidoptera) (Tab. 2). The Hymenoptera consisted of mostly bees (29 species), seven species of wasps, two species of ants and one sawfly. On average, *Bombus* species together made 54.7% of the visits at each site, with *Bombus ternarius* alone responsible for over 32.0%. *Bombus* species also visited the greatest number of plant species: 12.4% over the whole course of the study on average compared to 3.7% for all other taxa. The root maggot fly *Drymeia* sp. was the most frequently observed fly taxon making on average 3.8% of all insect visits, mainly to *Rosa acicularis.* Of the fly species observed, the tachinid *Peleteria* sp. visited the highest number of plant species (12). Most insect taxa (56.0%) were observed in only one time period (e.g. early, mid or late summer) but 28.0% were observed in two periods and 15.0% in all three. The taxa most frequently observed were also the ones observed over the longest period of time.

The species accumulation curve for all insect visitor taxa indicates that most (90%) would have at least one appropriate forage plant once 16 of the most generalized plants were present (Fig. 1). Fly taxa reached the 90% point after nine of the most generalized plant species were added, and bees after 13 species. Figure 2 shows the flowering range of all the plants observed being visited by insects during the survey period. Of the 16 most generalized plant species (see Table I) seven began flowering in June, six in July and three in August.

TABLE 2. The time period when 20 insect visitor taxa with the highest percentage of visits and number of plant species visited (S) in three fescue prairie sites in southern Manitoba were observed.

Order	Family	Common Name	Scientific Name	Time period[1]	Visits (mean %/site±SE)[2]	S (total # of plant species visited)
Hymenoptera	Apidae	Tri-colored bumblebee	*Bombus ternarius*	E-M-L	32.4±5.3	18
Hymenoptera	Apidae	Sanderson bumblebee	*Bombus sandersoni*	E-M-L	9.5±1.7	21
Hymenoptera	Apidae	Nevada bumblebee	*Bombus nevadensis*	E-M-L	4.8±1.5	18
Diptera	Muscidae	Root maggot fly	*Drymeia* sp.	E-M	3.8±1.9	3
Diptera	Tachinidae	Tachinid fly	*Tachina* sp.	M-L	3.2±0.2	7
Hymenoptera	Halictidae	Sweat bee	*Augochlorella aurata*	E-M-L	2.5±0.99	14
Hymenoptera	Apidae	Northern amber bumblebee	*Bombus borealis*	E-M-L	2.5±0.86	13
Diptera	Tachinidae	Tachinid fly	*Chaetogaedia* sp.	L	2.5±0.95	7
Hymenoptera	Apidae	Yellow-banded bumblebee	*Bombus terricola*	E-M-L	2.2±0.49	7
Diptera	Bombyliidae	Bee fly	*Poecilanthrax alycon*	L	1.9±0.60	8
Hymenoptera	Apidae	Confusing bumblebee	*Bombus perplexus*	L	1.8±0.90	6
Diptera	Tachinidae	Tachinid fly	*Peleteria* sp.	L	1.6±0.20	12
Hymenoptera	Megachilidae	Unarmed leaf-cutter bee	*Megachile inermis*	E-M-L	1.6±0.51	10
Hymenoptera	Apidae	Red-belted bumblebee	*Bombus rufocinctus*	L	1.5±0.73	4
Diptera	Syrphidae	Hover fly	*Toxomerus geminatus*	E-M-L	1.3±0.08	11
Hymenoptera	Halictidae	Sweat bee	*Lasioglossum succinipenne*	E-M-L	1.3±0.58	10
Diptera	Bombyliidae	Bee fly	*Villa nigra*	L	1.3±0.43	3
Hymenoptera	Megachilidae	Small-handed leaf-cutter bee	*Megachile gemula*	E-M-L	1.2±0.33	10
Lepidoptera	Lycaenidae	Silvery blue	*Glaucopsyche lygdamus*	E-M	1.2±1.10	8
Hymenoptera	Andrenidae	Miserable andrena	*Andrena miserabilis*	E-M-L	1.2±0.39	6
Other Diptera	n/a	Flies	32 spp.	E-M-L	11.6	27
Other Hymenoptera (bees)	n/a	Bees	17 spp.	E-M-L	8.0	29
Other Lepidoptera	n/a	Butterflies/moths	21 spp.	E-M-L	7.6	25
Other Hymenoptera (ants & wasps)	n/a	Ants/wasps	10 spp.	E-M-L	2.6	12
Coleoptera/ Hemiptera	n/a	Beetles/bugs	10 spp.	E-M-L	2.6	9

[1]E= Early (June), M=Mid (July), L= Late (August-September). Determined using pooled data over two years from all three sites.
[2]The % of visits was calculated for each taxa at each of three sites using pooled data over two years and averaged if the insect was seen at more than one site.

DISCUSSION

The results of this study and many others (Forup et al. 2008; Devoto et al. 2012; Montoya et al. 2012; Russo et al. 2013) suggest that just a small number of the most generalized entomophilous plant species supply most of the insect pollinator taxa in a community with appropriate forage. This likely has to do with the functional redundancy of some plant species for the pollinator community (Goldstein & Zych 2016). For example, many of the legumes (e.g. *Astragalus* spp., *Vicia americana, Hedysarum*

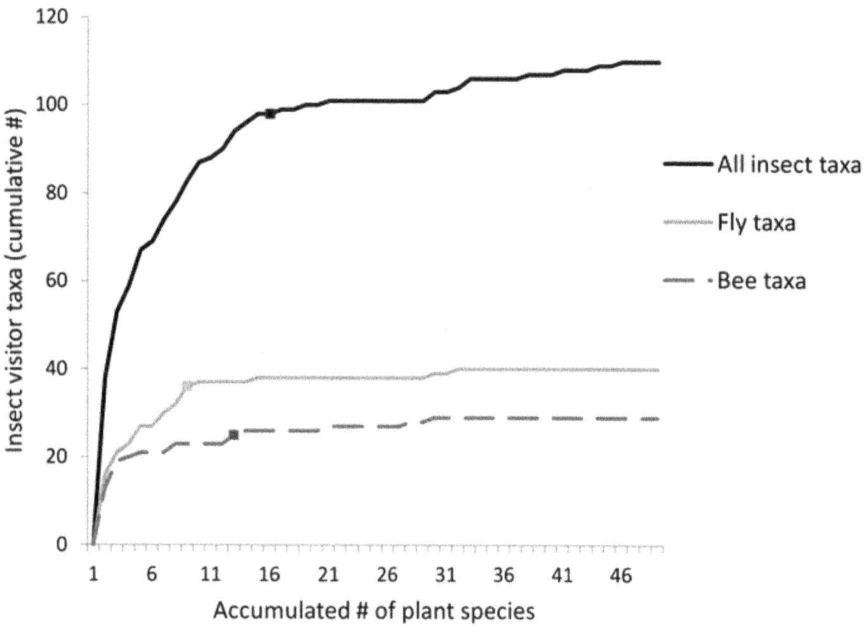

FIGURE 1. Species accumulation curves for all insect visitor taxa recorded (110), also broken down for the taxa of flies (39) and bees (29) as the most to the least generalized plants are added. Squares represent the points at which the plant species present were visited by 90% of all insect taxa in that group.

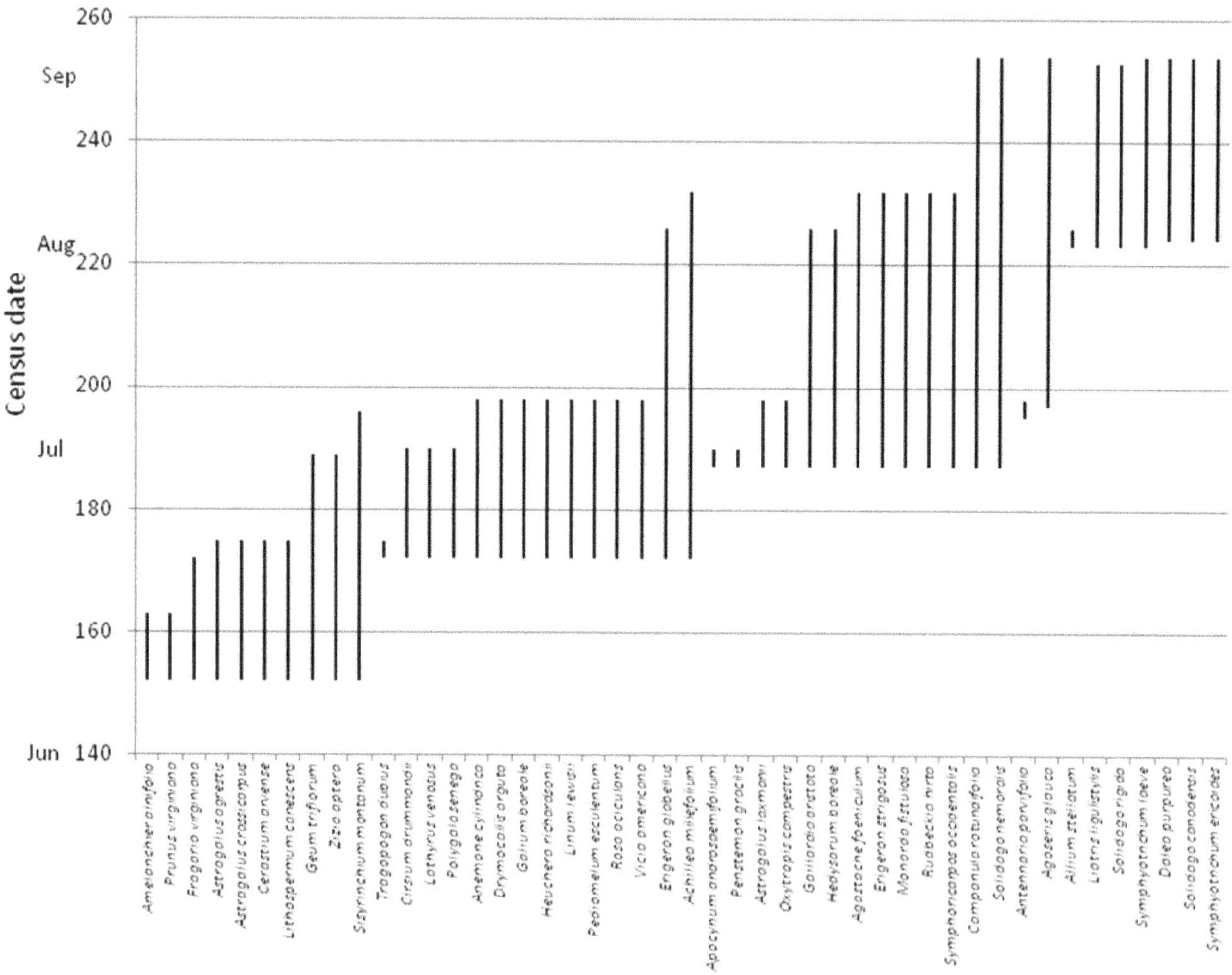

FIGURE 2. Flowering phenology of 48 plant species in south western Manitoba observed over 102 calendar days ordered by earliest to latest flowering period. The line indicates the flowering duration in the research plots using merged 2014 and 2015 survey data.

boreale and *Dalea purpurea*) were visited by the exact same suite of insect taxa. However, the aforementioned legumes flower at different times so at least three species would be needed to provide the insect community with consistent leguminous forage over the year. This is important to remember as restoration ecologists are often under pressure to keep costs down. Wilkerson et al. (2014) found that although using a seed mix with nine species had the highest germination, cover and floral resources, it was not as cost effective as a mix with only five species due to the higher price. However, the actual diversity of pollinators was not assessed in the Wilkerson study. Although adding fewer species may be more cost effective, doing so may result in inadequate floral resources: in our study less than 1% of all early summer insect visits would have occurred if only the top five plant species (in terms of their G scores) were present. A lack of early blooming flowers inhibits the development of healthy bumble bee colonies, which start out in spring with a single queen (Goulson 2003; Goulson et al. 2015; Isaacs et al. 2009). By determining G scores and preparing species accumulation curves with our data, we were able to identify the minimum number of species that would be needed to provide forage for most insect taxa over the entire growing season.

In addition to species with high generalization scores, it may be beneficial to add more specialized plant species to a seed mix if providing consistent forage for bees is desirable. Several studies have shown that *Bombus* spp. preferred plants with relatively deep nectar tubes (Stang et al. 2006; Forup et al. 2008; Elle et al. 2012), and researchers have found that adding such plants (particularly legumes) to restored areas results in increased richness and abundance of this genus (M'Gonigle et al. 2015; Carvell et al. 2011). Although four of the top ten plant species most frequently visited by bees were among the 16 most generalized plants, six species were not. These six plants had purple flowers and short to long nectar tubes that could only be accessed by relatively longer-tongued insects: *Cirsium drummondii, Dalea purpurea, Agastache foeniculum, Monarda fistulosa, Hedysarum boreale* and *Liatris ligulistylis*. The differences in the species with the highest G scores and insect visitation rates per inflorescence are due to bee preferences for the aforementioned flowers. In general, plants with the highest visitation rates were visited mainly by bees but one species was frequently visited by flies: *Rosa acicularis*. This rose had a low G score but a high visitation rate mainly due a species of *Drymeia* that swarmed the flowers once they bloomed, quickly removing all pollen. Therefore, specialized plant species should not be discounted for inclusion in wildflower plantings, if they are frequently visited. The visitation rate per inflorescence was thus a valuable metric to identify popular species.

Unfortunately, two of the plant species with the highest generalization scores appear to be currently unavailable from Canadian native seed supply companies (NPSS 2013): *Astragalus laxmannii* and *Cerastium arvense*. Seeds of these species are likely unavailable because they are difficult to harvest or germinate (Baskin & Baskin 1998). Further, seeds from the plant with the highest visitation rate per inflorescence (*Cirsium drummondii*) appear to be unavailable as the plant is uncommon and, due to its prickly leaves, not generally considered a desirable plant (although its popularity with bees may encourage restoration specialists to reconsider it). In the event these plants cannot be acquired, potential substitutes that flower in the same season with slightly lower G scores or visitation rates, and are visited by a similar subset of pollinators, could be selected instead. For example, *Geum triflorum* or *Astragalus crassicarpus* could be substituted for *Cerastium arvense* as they all flower at the same time. Possible substitutes for the mid-summer species -*Astragalus laxmannii* and *Cirsium drummondii*- are two leguminous species: *Vicia americana* and/or *Hedysarum boreale*. Spreading native hay is another possible strategy to introduce propagules for a restoration (Donath et al. 2007), but this may be difficult due the high number of invasive plants present in most fescue prairies.

There are several limitations of this study that need to be considered. Although moths are important pollinators in northern ecosystems (MacGregor et al. 2015), nocturnal insect activity was not monitored and, as a result, the importance of some plant species was likely underestimated. In addition, the 48 plant species were not observed for an equal length of time due to their natural variability in abundance. The possibility for modification to the G score to adjust for differences plant abundance was not addressed in the original reference (Medan et al. 2006), so additional observations are needed for the rarer species. An assumption was made that if these plants are grown as part of a prairie restoration, the insect taxa that were observed under natural conditions would eventually begin to inhabit the site. However, some studies comparing plant–pollinator communities in restored habitats have found that although they eventually function in a similar way, species composition between sites was somewhat different, possibly due to habitat differences (Forup and Memmott 2005; Forup et al. 2008; Menz et al. 2011; Tarrant et al. 2012). Further, there are strong temporal variations in insect pollinator composition from year to year (Dupont et al. 2009; MacLeod et al. 2016; CaraDonna et al. 2017). Thus, a restored prairie may not have the exact same composition as a native reference community. However, Forup et al. (2008) argues that restoration success should be based less on pollinator composition and more on functional similarity. This is because increased species richness may not be strongly correlated with functional diversity (Mayfield et al. 2010; Cadotte et al. 2011). Assessing the functional diversity of an ecosystem is increasingly being used to help set restoration goals (Thorpe & Stanley 2011; Giannini et al. 2016) and assess community health (Cadotte et al. 2011).

Considering all our data and the aforementioned studies, we propose one possible species list for fescue prairie restoration (Tab. 3). It includes 16 of the most generalized plants, plus four that are more specialized but frequently visited: *Agastache foeniculum, Dalea purpurea, Monarda fistulosa* and *Rosa acicularis*. All of the species in Tab. 3 were also among the top ten most generalized plants at the individual sites except one (*Liatris ligulistylis*). The plants on this list are in eight families: Asteraceae (7 species), Rosaceae (4), Fabaceae (3), Lamiaceae (2), and one species each in the Apocynaceae, Campanulaceae, Caprifoliaceae and Linaceae.

TABLE 3. Suggested minimum fescue prairie seed mix, and data regarding the expected insect community that would be supported by it.

Flowering period	Species	Insect taxa (cumulative #)[1]	Insect visits (cumulative %)[1]	Most attracted insect taxa[1]
Early	*Geum triflorum*[2]	7	2.0	Eusocial & solitary bees
	Prunus virginiana	13	3.0	Beetles & bees
	Rosa acicularis	18	4.2	Flies & solitary bees
	Lathyrus venosus	21	4.5	Butterflies & eusocial bees
	Linum lewisii	30	5.5	Flies & butterflies
Early-mid	*Apocynum androsaemifolium*	35	5.9	Solitary bees & flies
	Drymocallis arguta	41	7.6	Solitary bees & beetles
	Vicia americana[3]	42	9.1	Butterflies & eusocial bees
Mid	*Rudbeckia hirta*	53	10.9	Flies & butterflies
	Achillea millefolium	61	12.4	Beetles & flies
	Campanula rotundifolia	67	14.5	Solitary & eusocial bees
	Erigeron glabellus	68	16.5	Flies & butterflies
	Monarda fistulosa	69	20.1	Eusocial bees & butterflies
	Symphoricarpos occidentalis	74	25.8	Flies & eusocial bees
Late	*Solidago rigida*	86	43.3	Flies & beetles
	Agastache foeniculum	86	45.6	Eusocial & solitary bees
	Solidago canadensis	90	47.2	Beetles & solitary bees
	Solidago nemoralis	96	55.5	Beetles & flies
	Symphyotrichum laeve	98	69.7	Eusocial bees & flies
	Dalea purpurea	98	77.6	Eusocial bees & butterflies

[1]Based on 2014-2015 insect visitation observations in fescue prairie.
[2]This species is a suggested substitute if seeds of *Cerastium arvense* are unavailable from commercial Canadian seed suppliers
[3]This species is a suggested substitute if seeds of *Astragalus laxmannii* are unavailable from commercial Canadian seed suppliers

Three of the species were shrubs (i.e. *Prunus virginiana, Symphoricarpos occidentalis* and *Rosa acicularis*) and the remainder are herbaceous. Almost 90% of all insect taxa observed visited at least one of the 20 plants on this list. All of the major insect flower visitor groups (i.e. eusocial & solitary bees, beetles, butterflies and flies) would have at least one plant they favoured in each flowering period (i.e. early, early-mid, mid and late summer). Many of the species we recommended were previously noted as being important pollen and nectar sources for pollinators in North America including: *Agastache foeniculum, Campanula rotundifolia, Dalea purpurea, Monarda fistulosa, Prunus virginiana, Solidago* spp., *Symphoricarpos occidentalis* and *Symphyotrichum* spp. (Isaacs et al. 2009; Mader et al. 2011; Evans 2013). Substitutions could, of course, be made to accommodate different environmental conditions but care should be taken to select species that flower at the same time and support a similar insect community. For example, *Liatris ligulistylis* could be added instead of *Dalea purpurea* where the soils are relatively moist as the latter species appears to be less tolerant of such conditions (Anderson & Schelfhout 1980). Differences in seed availability may also affect the precise mix. This seed mix should be considered the minimum number of forb/shrub species for any fescue prairie restoration. The addition of other plant species may be important to meet other restoration goals, such as increasing the total plant richness or providing appropriate floral resources for rare pollinators such as oligolectic bees (Sheffield et al. 2014) or butterflies such as *Danaus plexippus* (Landis 2014). Additional research on pollination in fescue prairie is needed to: (1) identify pollinators of plant species we missed, (2) determine the plant species required by rarer pollinators, (3) determine if there is regional variability in the pollinator community, and (4) determine what role landscape variability may have on pollinator nesting.

ACKNOWLEDGEMENTS

The authors gratefully acknowledge the financial support of the Nature Conservancy of Canada and The Manitoba Museum Foundation Inc. Thanks to the Government of Manitoba for permitting research on their lands. We are grateful for Manitoba Museum staff and volunteer assistance with specimen processing. Special thanks to Sarah Semmler for preparation and identification of insect specimens. We are also grateful to the reviewers of this manuscript for their thoughtful comments, which greatly improved our paper.

APPENDICES

Additional supporting information may be found in the online version of this article:

APPENDIX I. Generalization scores (G) for vascular plant species in three fescue prairie sites in southern Manitoba.

APPENDIX II. Two-year pooled plant–insect visitor matrix in three fescue prairie sites, Manitoba.

REFERENCES

Analyze-it Software, Ltd. (2009) http://www.analyse-it.com/.

Anderson RC, Schelfhout S (1980) Phenological patterns among tallgrass prairie plants and their implication for pollinator competition. The American Midland Naturalist 104:253-263.

Bascompte J, Jordano P (2007) Plant-animal mutualistic networks: the architecture of biodiversity. Annual Review of Ecology, Evolution and Systematics 38:567-593.

Baskin CC, Baskin JM (1998) Seeds: ecology, biogeography, and, evolution of dormancy and germination. Elsevier

Cadotte MW, Carscadden K, Mirotchnick N (2011). Beyond species: functional diversity and the maintenance of ecological processes and services. Journal of Applied Ecology 48:1079-1087.

CaraDonna PJ, Petry WK, Brennan RM, Cunningham JL, Bronstein JL, Waser NM, Sanders NJ (2017) Interaction rewiring and the rapid turnover of plant–pollinator networks. Ecology Letters 20:385-94.

Carvell C, Osborne JL, Bourke AFG, Freeman SN, Pywell RF, Heard MS (2011) Bumble bee species' responses to a targeted conservation measure depend on landscape context and habitat quality. Ecological Applications 21:1760-1771.

Chuine I (2010) Why does phenology drive species distribution? Philosophical Transactions B Royal Society of London 365:3149-3160.

Coupland RT (1961) A reconsideration of grassland classification in the northern great plains of North America. Journal of Ecology 49:135-167.

Coupland RT, Brayshaw TC (1953) The fescue grassland in Saskatchewan. Ecology 34:386-405.

Devoto M, Bailey S, Craze P, Memmott J (2012) Understanding and planning ecological restoration of plant–pollinator networks. Ecology Letters 15:319-328.

Donath TW, Bissels S, Hölzel N, Otte A (2007) Large scale application of diaspore transfer with plant material in restoration practice–Impact of seed and microsite limitation. Biological Conservation 138:224-234.

Dupont YL, Padrón B, Olesen JM, Petanidou T (2009) Spatio-temporal variation in the structure of pollination networks. Oikos 118:1261-9.]

Ebeling A, Klein AM, Schumacher J, Weisser WW, Tscharntke T (2008) How does plant richness affect pollinator richness and temporal stability of flower visits? Oikos 117:1808-15.

Elle E, Elwell SL, Gielens GA (2012) The use of pollination networks in conservation. Botany 90:525-534.

Evans MM (2013) Influences of grazing and landscape on bee pollinators and their floral resources in rough fescue grassland. M.Sc. Dissertation, University of Calgary, Calgary, Alberta.

Forup ML, Memmott J (2005) The restoration of plant–pollinator interactions in hay meadows. Restoration Ecology 13:265-274.

Forup ML, Henson KS, Craze PG, Memmott J (2008) The restoration of ecological interactions: plant–pollinator networks on ancient and restored heathlands. Journal of Applied Ecology 45:742-752.

Giannini TC, Giulietti AM, Harley RM, Viana PL, Jaffe R, Alves R, Pinto CE, Mota NF, Caldeira CF, Imperatriz-Fonseca VL, Furtini AE (2016) Selecting plant species for practical restoration of degraded lands using a multiple-trait approach. Austral Ecology.

Gibson RH, Knott B, Eberlein T, Memmott J (2011) Sampling method influences the structure of plant–pollinator networks. Oikos 120:822-831.

Goldstein J, Zych M (2016). What if we lose a hub? Experimental testing of pollination network resilience to removal of keystone floral resources. Arthropod-Plant Interactions 10:263-271.

Goulson D (2003) Bumblebees: behaviour and ecology. Oxford Univ. Press.

Goulson D, Nicholls E, Botías C, Rotheray EL (2015) Bee declines driven by combined stress from parasites, pesticides, and lack of flowers. Science 347:1255957.

Harmon-Threatt AN, Hendrix SD (2015) Prairie restorations and bees: the potential ability of seed mixes to foster native bee communities. Basic and Applied Ecology 6:64-72.

Isaacs R, Tuell J, Fiedler A, Gardiner M, Landis D (2009) Maximizing arthropod-mediated ecosystem services in agricultural landscapes: the role of native plants. Frontiers in Ecology and the Environment 7:196-203.

Jordano P, Bascompte J, Olesen JM (2003) Invariant properties in coevolutionary networks of plant-animal interactions. Ecology Letters 6:69-81.

Kevan PG, Baker HG (1983) Insects as flower visitors and pollinators. Annual Review of Entomology 28:407-453.

Kremen C, M'Gonigle LK (2015). Small-scale restoration in intensive agricultural landscapes supports more specialized and less mobile pollinator species. Journal of Applied Ecology 52:602-610.

Koski MH, Meindl GA, Arceo-Gómez G, Wolowski M, LeCroy KA, Ashman TL (2015) Plant–flower visitor networks in a serpentine metacommunity: assessing traits associated with keystone plant species. Arthropod-Plant Interactions 9:9-21.

LaBar T, Campbell C, Yang S, Albert R, Shea K (2014) Restoration of plant–pollinator interaction networks via species translocation. Theoretical Ecology 7:209-220.

Landis TD (2014) Monarch waystations: propagating native plants to create travel corridors for migrating monarch butterflies. Native Plants Journal 15:5-16.

MacGregor CJ, Pocock MJ, Fox R, Evans DM (2015) Pollination by nocturnal Lepidoptera, and the effects of light pollution: a review. Ecological Entomology 40:187-198.

MacLeod M, Genung MA, Ascher JS and Winfree R (2016) Measuring partner choice in plant–pollinator networks: using null models to separate rewiring and fidelity from chance. Ecology 97:2925-2931.

Mader E, Shepard M, Vaughan M, Hoffman Black S, LeBuhn G (2011) Attracting native pollinators: Protecting North America's bees and butterflies. The Xerces Society, North Adams, Massachusetts.

Mayfield MM, Bonser SP, Morgan JW, Aubin I, McNamara S, Vesk PA (2010) What does species richness tell us about functional trait diversity? Predictions and evidence for responses of species and functional trait diversity to land-use change. Global Ecology and Biogeography 19:423-431.

Medan D, Basilio AM, Devoto M, Bartoloni NJ, Torretta JP, Petanidou T (2006) Measuring generalization and connectance in temperate, year-long active systems. In: Waser N, Ollerton J (eds) Plant–pollinator interactions: from specialization to generalization. University of Chicago Press, Chicago, Illinois, pp 245-259.

Memmott J, Waser NM, Price MV (2004) Tolerance of pollination networks to species extinctions. Proceedings of the Royal Society of London Series B: Biological Sciences 271:2605-2611.

Menz MH, Phillips RD, Winfree R, Kreme C, Aizen MA, Johnson SD, Dixon KW (2011) Reconnecting plants and pollinators: challenges in the restoration of pollination mutualisms. Trends in Plant Science 16:4-12.

M'Gonigle LK, Ponisio LC, Cutler K, Kremen C (2015) Habitat restoration promotes pollinator persistence and colonization in intensively managed agriculture. Ecological Applications 25:1557-1565.

Montoya D, Rogers L, Memmott J (2012) Emerging perspectives in the restoration of biodiversity-based ecosystem services. Trends in Ecology and Evolution 27:666-672.

NPSS (Native Plant Society of Saskatchewan) (2013) Native plant material and services supplier list [online] URL: http://www.npss.sk.ca/docs/2_pdf/Native_Plant_Source_List_2013_-_revised.pdf (accessed 28 September 2016).

Parachnowitsch AL, Elle E (2005) Insect visitation to wildflowers in the endangered Garry Oak, *Quercus garryana*, ecosystem of British Columbia. The Canadian Field Naturalist 119:245-253.

Robson DB (2008) The structure of the flower-insect visitor system in tall-grass prairie. Botany 86:1226-1278.

Robson DB (2013) An assessment of the potential for pollination facilitation of a rare plants by common plants: *Symphyotrichum sericeum* (Asteraceae) as a case study. Botany 91:34-42.

Russo L, DeBarros N, Yang S, Shea K, Mortensen D (2013) Supporting crop pollinators with floral resources: network-based phenological matching. Ecology and Evolution 3:3125-3140.

Saavedra S, Stouffer DB, Uzzi B, Bascompte J (2011) Strong contributors to network persistence are the most vulnerable to extinction. Nature 478:233-235.

Sheffield CS, Frier SD, Dumesh S (2014) The bees (Hymenoptera: Apoidea, Apiformes) of the Prairie Ecozone, with comparisons to other grasslands of Canada. In: Giberson J, Carcamo HA (eds) Arthropods of Canadian grasslands volume 4: Biodiversity and systematics part 2, Biological Survey of Canada, pp 427-467.

Stang M, Klinkhamer PG, Van Der Meijden E (2006) Size constraints and flower abundance determine the number of interactions in a plant–flower visitor web. Oikos 112:111-121.

Tarrant S, Ollerton J, Rahman ML, Tarrant J McCollin D (2012) Grassland restoration on landfill sites in the east midlands, United Kingdom: an evaluation of floral resources and pollinating insects. Restoration Ecology 21:560-568.

Thorpe AS, Stanley AG (2011) Determining appropriate goals for restoration of imperilled communities and species. Journal of Applied Ecology 48:275-279.

Trottier GC (1986) Disruption of rough fescue, *Festuca hallii*, grassland by livestock grazing in Riding Mountain National Park, Manitoba. Canadian Field-Naturalist 100:488-495.

Wilkerson ML, Ward KL, Williams NM, Ullmann KS, Young TP (2014) Diminishing returns from higher density restoration seedings suggest trade-offs in pollinator seed mixes. Restoration Ecology 22:782-789.

Permissions

All chapters in this book were first published in JPE, by Enviroquest Ltd; hereby published with permission under the Creative Commons Attribution License or equivalent. Every chapter published in this book has been scrutinized by our experts. Their significance has been extensively debated. The topics covered herein carry significant findings which will fuel the growth of the discipline. They may even be implemented as practical applications or may be referred to as a beginning point for another development.

The contributors of this book come from diverse backgrounds, making this book a truly international effort. This book will bring forth new frontiers with its revolutionizing research information and detailed analysis of the nascent developments around the world.

We would like to thank all the contributing authors for lending their expertise to make the book truly unique. They have played a crucial role in the development of this book. Without their invaluable contributions this book wouldn't have been possible. They have made vital efforts to compile up to date information on the varied aspects of this subject to make this book a valuable addition to the collection of many professionals and students.

This book was conceptualized with the vision of imparting up-to-date information and advanced data in this field. To ensure the same, a matchless editorial board was set up. Every individual on the board went through rigorous rounds of assessment to prove their worth. After which they invested a large part of their time researching and compiling the most relevant data for our readers.

The editorial board has been involved in producing this book since its inception. They have spent rigorous hours researching and exploring the diverse topics which have resulted in the successful publishing of this book. They have passed on their knowledge of decades through this book. To expedite this challenging task, the publisher supported the team at every step. A small team of assistant editors was also appointed to further simplify the editing procedure and attain best results for the readers.

Apart from the editorial board, the designing team has also invested a significant amount of their time in understanding the subject and creating the most relevant covers. They scrutinized every image to scout for the most suitable representation of the subject and create an appropriate cover for the book.

The publishing team has been an ardent support to the editorial, designing and production team. Their endless efforts to recruit the best for this project, has resulted in the accomplishment of this book. They are a veteran in the field of academics and their pool of knowledge is as vast as their experience in printing. Their expertise and guidance has proved useful at every step. Their uncompromising quality standards have made this book an exceptional effort. Their encouragement from time to time has been an inspiration for everyone.

The publisher and the editorial board hope that this book will prove to be a valuable piece of knowledge for researchers, students, practitioners and scholars across the globe.

List of Contributors

Carolin Mayer
Earth and Life Institute, Université Catholique de Louvain, Croix du Sud 2 box 14, 1348 Louvain-la-Neuve, Belgium

Lynn Adler
Department of Plant, Soil and Insect Science, 270 Stockbridge Rd., University of Massachusetts, Amherst, MA 01002, USA

W. Scott Armbruster
School of Biological Sciences, University of Portsmouth, Portsmouth PO1 2DY, UK / Institute of Arctic Biology, University of Alaska, Fairbanks, AK 99775, USA

Amots Dafni
Department of Evolutionary and Environmental Biology, Institute of Evolution, Faculty of Sciences, Haifa University, Haifa 31905, Israel

Connal Eardley
Agricultural Research Council, Queenswood 0121, South Africa/School of Biological and Conservation Sciences, University of KwaZulu-Natal, Scottsville, Pietermaritzburg, 3209, South Africa

Jeff Ollerton
School of Science and Technology, University of Northampton, Newton Building, Avenue Campus, Northampton, NN2 6JD, UK

Laurence Packer
Department of Biology, York University, 4700 Keele St, Toronto, Ontario, M3J 1P3 Canada

Axel Ssymank
Federal Agency for Nature Conservation (BfN), Konstantinstrasse 110, 53179 Bonn, Germany

Jocelyn R. Holt and Paul Wilson
Department of Biology, California State University, 18111 Nordhoff Street, Northridge, CA 91330-8303

Christy A. Brigham
Santa Monica Mountains National Recreation Area, 401 West Hillcrest Drive, Thousand Oaks, CA 91360

Paola Ugoletti, Michael B. Jones and Jane C. Stout
Trinity Centre for Biodiversity Research, Trinity College Dublin, Dublin 2, Ireland
School of Natural Sciences, Botany Building, Trinity College Dublin, Dublin 2, Ireland

Darren Reidy
School of Natural Sciences, Botany Building, Trinity College Dublin, Dublin 2, Ireland
Killarney National Park Education Centre, Knockreer House, Killarney, Ireland

Peter Bernhardt
Department of Biology, Saint Louis University, St. Louis, Missouri 63103, USA

Retha Meier
Department of Educational Studies, Saint Louis University, St. Louis, Missouri 63103

Nan Vance
USDA Forest Service, Pacific Northwest Research Station, 3200 SW Jefferson Way, Corvallis, Oregon 97331, USA

Valérie Cawoy, Carolin Mayer and Anne-Laure Jacquemart
Research group " Genetics, reproduction, populations " - Earth and Life Institute - Université catholique de Louvain, 2 Croix du Sud, 1348 Louvain-la-Neuve, Belgium

Mathieu Jonard
Research group " Forest sciences " - Earth and Life Institute - Université catholique de Louvain, 2 Croix du Sud, 1348 Louvain-la-Neuve, Belgium

Victoria A. Wojcik
Pollinator Partnership, 423 Washington Street, 5th Floor San Francisco, CA 94111

Stephen Buchmann
Pollinator Partnership, 423 Washington Street, 5th Floor San Francisco, CA 94111
Departments of Entomology, and Ecology and Evolutionary Biology, The University of Arizona, Tucson, AZ 85743

Elena S. Tartaglia and Steven N. Handel
Department of Ecology, Evolution and Natural Resources, Rutgers, The State University of New Jersey, 1 College Farm Road, New Brunswick, New Jersey, 08901-1582

Rebekka Lundgren and Ørjan Totland
Norwegian University of Life Sciences, Department of Ecology and Natural Resource Management, Ås, Norway

Amparo Lázaro
Norwegian University of Life Sciences, Department of Ecology and Natural Resource Management, Ås, Norway
Mediterranean Institute for Advanced Studies, Miquel Marques 21, 07190 Esporles, Balearic Islands, Spain

Amy M. Iler
Department of Evolution, Ecology, and Organismal Biology, The Ohio State University, 318 W 12th Ave, Columbus, OH 43210 USA
Department of Biology, University of Maryland, College Park, Maryland 20742 USA

Karen Goodell
Department of Evolution, Ecology, and Organismal Biology, The Ohio State University, 1179 University Dr, Newark, OH 43055 USA

Asif Sajjad, Shafqat Saeed and Muhammad Amjad Bashir
Department of Entomology, University College of Agriculture, Bahauddin Zakariya University Multan, Pakistan

Emily S. Darling
Department of Biological Sciences, Simon Fraser University, 8888 University Drive, Burnaby, British Columbia, Canada V5A 1S6

Spencer C. H. Barrett
Department of Ecology and Evolutionary Biology, University of Toronto, 25 Willcocks Street, Toronto, Ontario, Canada M5S 3B2

Elaine Y. Luo
Department of Ecology and Evolutionary Biology, University of Toronto, 25 Harbord Street, Toronto, ON, M5S 3G5, Canada

Jane E. Ogilvie and James D. Thomson
Department of Ecology and Evolutionary Biology, University of Toronto, 25 Harbord Street, Toronto, ON, M5S 3G5, Canada
Rocky Mountain Biological Laboratory, Crested Butte, CO 81224-0519, USA

Leif L. Richardson
Department of Biology, Dartmouth College, Hanover, New Hampshire 03755, USA

Judith L. Bronstein
Department of Ecology and Evolutionary Biology, University of Arizona, Tucson 85721, USA

Blandina F. Viana, Eduardo Mariano Neto, Patrícia A. Ferreira and Luis M. Primo
Instituto de Biologia, Universidade Federal da Bahia, Salvador, Bahia, 40170-210, Brazil

Danilo Boscolo
Universidade Federal de São Paulo - UNIFESP, Campus Diadema, Zip 09972-270, Diadema, São Paulo, Brazil

Luciano E. Lopes
Centro de Ciências Biológicas e da Saúde, Universidade Federal de São Carlos, São Carlos, 13565-905, Brazil

Ariadna V. Lopes
Centro de Ciências Biológicas, Universidade Federal de Pernambuco, Recife, Pernambuco, 50372-970, Brazil

Camila M. Pigozzo
Centro Universitário Jorge Amado, Salvador, Bahia, 41745-130, Brazil

Tania Jogesh, Arthur Zangerl and May R. Berenbaum
Department of Entomology, University of Illinois at Urbana-Champaign, Urbana, IL

Margaret C. Stanley
Centre for Biodiversity and Biosecurity, School of Biological Sciences, Tamaki Campus, University of Auckland, Auckland, New Zealand

Wyndee A. Guzman, Paul Wilson
Department of Biology, California State University, Northridge, CA 91330-8303

Deirdre Loughnan and Benjamin Gilbert
Department of Ecology and Evolutionary Biology, University of Toronto, 25 Harbord Street, Toronto, ON, M5S 3G5, Canada

James D. Thomson and Jane E. Ogilvie
Department of Ecology and Evolutionary Biology, University of Toronto, 25 Harbord Street, Toronto, ON, M5S 3G5, Canada
Rocky Mountain Biological Laboratory, Crested Butte, CO 81224-0519, USA

Godfrey R. Bourne
Department of Biology, University of Missouri–St. Louis, One University Boulevard, St. Louis, MO 63121–4400, USA
CEIBA Biological Center, Linden Highway, Madewini, Guyana

Gyanpriya Maharaj
Department of Biology, University of Missouri–St. Louis, One University Boulevard, St. Louis, MO 63121–4400, USA
Centre for the Study of Biological Diversity and Department of Biology, University of Guyana, Turkeyen Campus, Greater Georgetown, Guyana
CEIBA Biological Center, Linden Highway, Madewini, Guyana

Y. Vaknin
Department of Agronomy and Natural Resources, Institute of Plant Sciences, Agricultural Research Organization, the Volcani Center, Bet Dagan 50250, Israel

Y. Samocha
Department of Agronomy and Natural Resources, Institute of Plant Sciences, Agricultural Research Organization, the Volcani Center, Bet Dagan 50250, Israel
Department of Plant Sciences, George S. Wise Faculty of Life Sciences, Tel-Aviv University, Tel-Aviv, Israel

D. Eisikowitch
Department of Plant Sciences, George S. Wise Faculty of Life Sciences, Tel-Aviv University, Tel-Aviv, Israel

Gerald G. Carter and Alyssa B. Stewart
Department of Biology, University of Maryland, College Park, MD 20742, USA

Christopher M. Somers
Department of Biology, University of Regina, 3737 Wascana Parkway, Regina, SK S4S 0A2

S. Danae Frier
Department of Biology, University of Regina, 3737 Wascana Parkway, Regina, SK S4S 0A2
Royal Saskatchewan Museum, 2340 Albert Street, Regina, SK S4P 2V7

Cory S. Sheffield
Royal Saskatchewan Museum, 2340 Albert Street, Regina, SK S4P 2V7

Krakos, Kyra N.
College of Arts and Sciences, Maryville University, 650 Maryville University Drive, St. Louis, Missouri

Fabricant, Scott A.
Department of Biological Science, Macquarie University, North Ryde, New South Wales, Australia

Ron Haran
16 Ibn Shaprut Street, Jerusalem, Israel

Ido Izhaki
Department of Evolutionary and Environmental Biology, University of Haifa, Haifa, Israel 3498838

Amots Dafni
Institute of Evolution, Department of Evolutionary and Environmental Biology, University of Haifa, Haifa, Israel 3498838

Index

www.ingramcontent.com/pod-product-compliance
Lightning Source LLC
LaVergne TN
LVHW060523210926
848485LV00005B/1134
* 9 7 8 1 6 3 9 8 7 4 3 9 2 *